W9-BQZ-525

ANNUAL REVIEW OF
FLUID MECHANICS

# ANNUAL REVIEW OF FLUID MECHANICS

MILTON VAN DYKE, *Co-Editor*
Stanford University

J. V. WEHAUSEN, *Co-Editor*
University of California, Berkeley

JOHN L. LUMLEY, *Associate Editor*
Cornell University

*VOLUME 13*

*1981*

ANNUAL REVIEWS INC. 4139 EL CAMINO WAY PALO ALTO, CALIFORNIA 94306 USA

ANNUAL REVIEWS INC.
Palo Alto, California, USA



REPRINTS The conspicuous number aligned in the margin with the title of each article in this volume is a key for use in ordering reprints. Available reprints are priced at the uniform rate of $2.00 each postpaid. The minimum acceptable reprint order is 5 reprints and/or $10.00 prepaid. A quantity discount is available.

*International Standard Serial Number: 0066-4189*
*International Standard Book Number: 0-8243-0713-5*
*Library of Congress Catalog Card Number: 74-80866*



PRINTED AND BOUND IN THE UNITED STATES OF AMERICA

# PREFACE

A special pleasure in editing this series comes from dealing each year with the author of the prefatory chapter, which has been a feature our readers often comment upon favorably. At the outset we adopted from our parent *Annual Review of Biochemistry* this tradition of starting each volume not with a review but with an article of a quite different kind: a vignette of scientific history, personal reminiscence, or philosophy written by a distinguished figure in fluid mechanics. Twice we have altered the format to let former associates do the writing: Irmgaard Flügge-Lotz and Wilhelm Flügge wrote about Prandtl, and Mabel and William Sears wrote about von Kármán. The roster of such authors, extending from Sydney Goldstein in the first volume to Max Munk in the present one, includes some of the most interesting fluid dynamicists of this century, and there are still others with fascinating stories to tell.

We remember carrying on a delightful correspondence with Henri Villat in his ninety-second year, which led to his "As Luck Would Have It" in Volume 4. Later we visited Jan Burgers to persuade him to reminisce about his early days in Delft for Volume 7. We remarked that he must be gratified at the surge of interest in the Burgers equation, to which he replied with typical modesty that it is "really a very simple equation" (and we had the wit to reply, "So is the Laplace equation"). But pleasure is inevitably mixed with sadness in this search. Ten years ago Sydney Chapman promised to tell us what it was like to be the last research student of Osborne Reynolds, but his death robbed us of that article.

When we meet each year (currently in November) to plan the next volume, we review our birthday list to decide which Grand Old Man of Fluid Mechanics should be invited to tell his story. (Grand Old Women are regrettably rare in our field.) We welcome suggestions from our readers for this special article, as well as for the regular reviews.

Our thanks go to Stanley Corrsin, whose term of service on the Editorial Committee ended with the planning of this volume.

THE EDITORS

*Annual Review of Fluid Mechanics*
*Volume 13, 1981*

# CONTENTS

ANNUAL REVIEWS INC. is a nonprofit corporation established to promote the advancement of the sciences. Beginning in 1932 with the *Annual Review of Biochemistry*, the Company has pursued as its principal function the publication of high quality, reasonably priced Annual Review volumes. The volumes are organized by Editors and Editorial Committees who invite qualified authors to contribute critical articles reviewing significant developments within each major discipline. Annual Reviews Inc. is administered by a Board of Directors whose members serve without compensation.



Publications

*Annual Reviews of Anthropology, Astronomy and Astrophysics, Biochemistry, Biophysics and Bioengineering, Earth and Planetary Sciences, Ecology and Systematics, Energy, Entomology, Fluid Mechanics, Genetics, Materials Science, Medicine, Microbiology, Neuroscience, Nuclear and Particle Science, Nutrition, Pharmacology and Toxicology, Physical Chemistry, Physiology, Phytopathology, Plant Physiology, Psychology, Public Health,* and *Sociology.*

Special Publications: *History of Entomology* (1973), *The Excitement and Fascination of Science* (1965), *The Excitement and Fascination of Science, Volume Two* (1978), *Annual Reviews Reprints: Cell Membranes, 1975–1977* (published 1978), and *Annual Reviews Reprints: Immunology, 1977–1979* (published 1980).

For the convenience of readers, a detachable order form/envelope is bound into the back of this volume.

March, 1980, Rehoboth Beach, Maryland
Photograph by Mark Bernard

Max M. Munk.

*Ann. Rev. Fluid Mech. 1981. 13:1–7*


# MY EARLY AERODYNAMIC RESEARCH—THOUGHTS AND MEMORIES

*8166

*Max M. Munk*
221 Munsen Street, Rehoboth Beach, Delaware 19971

## *I Days with Prandtl*

It makes me happy to remember that long ago, at the beginning of aviation, Heaven granted me the opportunity to make significant contributions to aerodynamics. My papers were published and the results were used in all countries. But the variable-density wind tunnel, which I invented and gave to the US, was used for many measurements that were kept secret and strengthened the US militarily in the last great war.

Directly after Engineering graduation, I came to work with Ludwig Prandtl. I found a most primitive, nonstreamlined, and underpowered wind tunnel, the only one in the country. And still, we learned so much from it, because we started with ignorance. The necessary features of wing sections were already recognized. Two researches had already been completed. Airfoil models with just straight or circular-arc wing sections were exposed to the airstream and the air forces measured and plotted against the angle of attack. It was hoped that the tests would suggest some new thought. No specific question was asked and so no answer was given by the wind tunnel. No aimful thinking was involved, and, hence, nothing new was learned.

The other research was of the opposite kind. Spindle-shaped models were carefully computed to have meridians suitable for producing the flow of a perfect fluid. The wind tunnel was asked whether the actual pressure distribution was sufficiently equal to the one computed for a perfect fluid. It was asked whether the study of the motion of a perfect fluid was helpful for practical aerodynamics. The wind tunnel answered with a loud Yes. This was a very great achievement of Prandtl, one for which he did not get enough credit.

The research was carried out with infinite diligence and skill. Alas, after the work was completed, the young man in charge of it was



0066-4189/81/0115-0001$01.00

drafted into the Army. The First World War was in full swing, and the promising young man was killed in battle. His name was Fuhrmann.

Prandtl was not vastly outstanding in any one field, but he was eminent in so many fields. He understood mathematics better than many mathematicians do; he was a very good physicist and engineer. Above all else stood the remarkable nobility of his character. His unselfishness and friendliness to his young associates and toward everybody contributed to his well-deserved fame.

Later in life, I met occasionally another man famed for his contributions to early aviation. Orville Wright was in many respects the contrary of Prandtl. By his own admission, he was just a simple mechanic. But he had in common with Prandtl diligence and aimful thinking. Above all else he shared with Prandtl the great nobility of character. He loved his country and had the strength to forgive those who had trespassed against him.

Each day Prandtl came to have lunch with his young associates. He told us so much, and we learned so much from him. One day he told us that salmons can only eat in salt water. During their long trip upstream toward the mountains, they have to use up their fat as fuel. From that, their water resistance can be estimated, and comes out very small. Years later I recalled this, and wishing to check up, wrote to the US Agricultural Department, asking whether salmons can eat in sweet water. After many weeks, the answer came. I was informed that my inquiry had been referred to their aquatic zoologist. After many more weeks, his answer came. The aquatic zoologist informed me that he did not know. The aquatic zoologist was not a Ludwig Prandtl nor an Orville Wright.

## *II Aerodynamic Research*

All of my Goettingen work was published in secret reports, by the German army, for this was during the First World War. Nevertheless, they were translated in England a week after appearance and distributed there and in the US. The first tests referred to struts, the compression members connecting the upper and lower wings of a biplane. The object was to ascertain whether streamlining was really beneficial, for that was still controversial. I used oversize models to increase the Reynolds number. The first airplanes were all biplanes. It was believed that the greatest thinness of airfoil was necessary. This went so far that the propeller blades of the Parseval blimps were just fabric stretched by the centrifugal force. These thin airfoils could not be made stiff and strong enough without support, so a girder was formed by a pair of wings connected by struts and a large amount of cables. The drag of these struts and cables was considerable.

The small, underpowered, and primitive wind tunnel showed that thicker airfoils were by far less objectionable than thought. Thus, the tests led to a revolutionary step in airplane design. The monoplane with cantilever wings appeared. The thicker airfoils also had larger maximum lift coefficients and thus made up by smaller area requirements for a little larger drag. Wing-section research remained active for many years. After all, the wings are the most important part of the airplane, with the engines next. Airplane engines were at first water cooled.

While this testing went on, I heard much about boundary layers. Their discovery is considered by many as the greatest achievement of Prandtl. The motion of a perfect fluid, one without friction, is governed by a linear partial differential equation for which solutions can be computed. Viscosity makes the differential equation nonlinear, and thus makes it impossible to solve them. The viscosity of air is small. Prandtl thought that matter over by abstract, living thinking, not by shuffling mathematical symbols about. He conceived that as the viscosity became smaller and smaller, its influence must become more and more unnoticeable except in regions quite close to the surface of the immersed body. The pressure distribution at the surface can thus be computed as if the air was nonviscous, and the flow inside the thin boundary layer can be computed in view of its thinness and of the known pressure distribution. It was a beautiful result of sound, abstract, aimful thinking, an example of the clear thinking of Prandtl. But it did not work. The solutions for the steady flow of viscous air are of course steady, but the actual flows even in the thin boundary layers are not steady. They are turbulent; the laminar steady flow is not stable. The theory of the turbulent boundary layer relies on experimental data. Prandtl solved the differential equation but not the problem.

We have thus arrived at turbulence, the consequence of viscosity. This sounds like a contradiction and is one. Viscosity smooths out turbulence and tends to make the flow laminar. Kinetic energy is changed into heat. How can it make turbulence? It does not. It is the surface friction that makes turbulence. The thinner the boundary layer, the larger the velocity gradient at right angles to the surface. A disturbing air particle has not far to travel to pick up energy, not from viscosity but from the moving surface. It carries it into the fluid, to make things worse until the smoothing action of viscosity counteracts the increase of turbulence.

This picture points to a fundamental difference between laminar air friction and turbulent air friction. The laminar surface friction is evenly distributed over the surface, but the turbulent friction is not. It is decidedly spotty in a very small scale and rapidly changing at every point from maximum local friction to perhaps locally negative friction. That should make us think. The elastic property of the surface probably has little to do with the laminar surface friction. But it may be quite different with turbulent surface friction. There are reports that hummingbirds have a surprisingly small air resistance. An ornithologist may know more about that. But it is not impossible that the fluffy soft feathers of the bird have something to do with that. That should be looked into. Technical applications of possible results may not be likely. But true research is guided by curiosity, not by technical applications. Modern history shows that first comes curiosity and only later technical application, sometimes the most unexpected one.

Next to the boundary layer, there was also much talk about lighter-than-air and heavier-than-air, and which would win. The terms are today almost forgotten. The present generation could not understand the question. But it took many years to realize the possibilities of airplanes. They were then slow and small, with an open cockpit. It was mathematically proven that larger airplanes could never be built. The Zeppelin airships flew over England. There were no anti-aircraft guns. The airplanes are now large enough to hold 300 passengers and the dirigibles have disappeared. They could never fly fast, nor economically reach high altitudes. They could not economically cross the Rocky Mountains. I have not even seen a small blimp for a long time. But times change and we change with them. It is not absolutely certain that the dirigibles are gone for good. We have now better materials, and the frame may be made a little larger, of high grade steel. For dirigibles have one advantage over airplanes which fits present needs. They can transport heavy loads with very little energy required. They have only small engines. They need no power for keeping them aloft. And the larger they are, the more favorable is the power requirement. The last word about dirigibles is perhaps not yet spoken.

## *III Wing Theory*

My interest in words, their etymology, and original meaning is enormous. I have spent much more time studying words than studying mathematics. Mathematics comes from within, but words have to be looked up. My interest in words came in handy when new names had to be selected. I was the first to whom it occurred to compute the portion of the propeller thrust used for throwing air down so that the airplane may stay up. I gave long thought to what would be the best name. "Induced drag" was a happy choice. The term is now international and in the dictionaries. It will be used long after I shall be gone and my own name forgotten.

My principal paper on the induced drag was still under the spell of Prandtl's vortex theory. Everything that Prandtl said was correct, but it was not the right approach. My paper was full of clumsy mathematics because I had to integrate what never should have been obtained by differentiation. My simple result spelled the end of the vortex theory. I kept in mind, instead, the two-dimensional flow patterns in vertical planes at right angles to the direction of flight. The apparent additional mass of the airfoil in this case gives the answer immediately. My airship formula, now used for missiles, was obtained with hardly any mathematics. Technical success requires integration, not differentiation.

The only paper of mine requiring plenty of mathematics was my wing-section theory paper. Its history is quite interesting. The first mathematical wing-section paper was by Kutta. He computed the airfoil forces on straight lines and on circular arcs used as wing sections. His results were really important, but they were not understood at that time, and he did not get the credit which he deserved. His technique consisted in changing the theoretical flow across a circular obstacle in two-dimensional flow of a perfect fluid into the airfoil flow by conformal transformation in the complex plane. Joukowsky thought of improving the method by a slight change, so that his airfoil shapes resembled actual wing sections. But he was widely misunderstood, and his wing sections were not that because their tail ends were entirely too thin. His success was merely artistic. However, he inspired von Kármán to work out a conformal transformation of the circle into a wing section having a thicker end. Von Kármán's paper required an excessive amount of mathematics, and was technically useless. I do not believe that anybody except the author himself ever read the paper.

It made me think, though. It occurred to me that it would never get us anywhere if we fitted the wing section to the mathematical method. On the contrary, we should start with the wing section, with any technically valuable wing section, and fit the mathematics to the wing section. I abandoned the conformal transformation and relied on understanding the aerodynamic action of every portion of the wing section separately. This I laid down mathematically and integrated. The resulting integrals, for the zero-lift angle of attack and for the so-called aerodynamic center (also named by me) were found very useful and the result was found surprisingly accurate.

This is my only paper that found entry into the mathematical literature. My wing-section paper is discussed by Bateman in his excellent book on partial differential equations. He states, correctly, that I obtained the effect of the shape not by changing the shape of the boundary, but by changing the boundary condition.

In the same section, Bateman also demonstrated that mathematicians make as many mistakes as all of us do. Wishing to add something of his own, he computed the fluid resistance of the two-dimensional airfoil section. But that drag, in a perfect fluid and two-dimensional flow, is zero. Where would the energy go? Bateman forgot the force component parallel to the chord.

## *IV Fermat's Last Theorem*

Undertaking research for the advance of mathematics is more difficult than using established mathematics. It requires more curiosity, diligence, and aimful thinking. The researcher's character in general, I believe, has also much to do with it. The pure in heart shall see. The Latin mundo corde describes it even better. Pure is the cleanness of fire, but mundus is order, which is the world.

I proceed to illustrate this by discussing a little theorem, the proof of which has baffled the mathematical profession for the last 300 years. The theorem has no practical application, and that may explain its status.

Fermat left behind a note in which he stated that the equation

$$a^p + b^p = c^p, \qquad p > 2,$$

cannot be satisfied by integers. He also said that the proof is marvelous. Since then, a vigorous search for the proof has gone on, but in vain. Oystein Ore, in his excellent book, says that the proof will never be found. Michael S. Mahoney, only a few years ago, goes beyond that. He proclaims, to the astonishment of the mathematical world, that the proof does not exist, that Fermat's Last Theorem is fundamentally proofless.

The fox in the fable does not go that far. He only says that the grapes are sour, he does not deny their existence. Since Fermat's Last Theorem is excluding, not predicting, it must necessarily be the consequence of another excluding theorem, having to do with powers and roots. Nothing comes to mind respecting powers and roots for integers. But it does regarding fractions. An exact power of the form

$$a + \left(\frac{b}{q}\right)^p$$

must have a root of the form

$$c + \frac{b}{q}$$

provided $q - 1$ is not divisible by $p$. All root fractions other than $b/q$ are excluded. Here $p$ is a prime number and $p>2$. The fraction may be positive or negative.

It follows that this exclusion attaches also to the periods of a decimal fraction representing the fraction in question, regardless of whether the periodic sequence occurs behind or in front of the decimal point.

The root extraction of one such periodic sequence does not lead to the theorem. We must have two different sequences and they must be intimately interconnected. Consider the endings of the two numbers $z$ and $u$, with

$$x=kp^s+a, \qquad y=x^p, \qquad z=x/q, \qquad u=z^p$$

in which $s$ is a very large number and $p$ is an odd prime number serving as digital base as well as an exponent. The endings of $z$ and of $u$ are each a pair of digit periodic sequences separated by the decimal point. The single-root theorem is applied to the two decimal fractions and also to the two sequences left of the decimal point. The two fractions belonging to the ending sequences of $u$ may both be exact powers, but not both at the same time, in the same number $z$. Their difference is $(a/q)^p$, that of $z$ is $q/p$. This is Fermat's theorem applied to the ending of $u$. And why are not both fractions exact powers at the same time? Because then the single-root theorem would exclude $z$ from being the root of $u=z^p$.

A closer look at all details of this argument does not belong here. It confirms the soundness of the reasoning and the absence of a gap in the proof. The proof is really marvelous, so short, so direct, leaning chiefly on arithmetic. It must be the marvelous proof that Fermat mentions. The proof is clear, convincing, and revealing. It was that to him, it is that to me, and it will be that to many others. Of course to those who are not interested, or to whom it is not given to enjoy the beauty of abstract mathematical relations, the theorem must remain proofless.

It is now enough. Basic research requires rare thoughts and rare thinkers. Let us hope there will always be enough around and that they will find opportunity to use their thoughts for the good of our country.

*References*

Bateman, H. 1944. *Partial Differential Equations*. New York: Dover

Mahoney, M. S. 1972. Fermat's mathematics: Proofs and conjectures. *Science* 178:30

Munk, M. M. 1977. *Congruence Surds and Fermat's Last Theorem*. New York: Vantage

Ore, O. 1946. *Number Theory and Its History*. New York: McGraw-Hill

*Ann. Rev. Fluid Mech. 1981. 13:9-32*


# COASTAL SEDIMENT PROCESSES

*8167

*Kiyoshi Horikawa*
Department of Civil Engineering, University of Tokyo, Bunkyo-ku, Tokyo 113, Japan

## 1 INTRODUCTION

The study of coastal sediment processes has been systematized over the past half century by coastal geomorphologists, while the dynamics of nearshore sediment movement has been treated only during the past twenty years by workers in the fields of coastal engineering and nearshore oceanography. For our purposes here, the term "coastal sediment processes" will cover the time history of the numerous phenomena related to sediment movement in the coastal area. From the perspective of engineering science, coastal sediment phenomena are closely related to various important practical problems such as siltation of harbor basins and beach erosion.

It was not so long ago that most coastal projects were carried out by trial and error because of a lack of knowledge of the underlying mechanisms of coastal sediment processes. During the last few decades, large quantities of data have been taken in studies of the transport of coastal sediment through field and laboratory investigations. Although these data are helpful, the phenomena are very complex and as yet most are understood only in a qualitative sense. Therefore, more basic research effort is required before we can confidently deal with practical problems. The extensive research activities carried out in recent years by groups all over the world hold promise of a better understanding of coastal sediment processes. This article gives a review of selected past studies and describes the present state of knowledge in this field.

## 2 VARIOUS APPROACHES AND RELATED ELEMENTS

Coastal sediment processes are extremely complex and include phenomena having quite different scales in space and in time. For purposes of



0066-4189/81/0115-0009$01.00

**Table 1** Classification of coastal process scales

| | Macroscale | Mesoscale | Microscale |
|---|---|---|---|
| Time scale | year | day/hour | second |
| Space scale | kilometer | meter | millimeter |

clarification, Horikawa (1970) proposed classifying coastal phenomena into three categories as macroscale, mesoscale, and microscale. These scales can be roughly described as shown in Table 1. Specific examples for each scale will be given in the following sections.

Theoretically speaking, the complete superposition of microscale phenomena should compose the mesoscale phenomena and that of the mesoscale phenomena, the macroscale phenomena. At present, the above connections cannot be made. It is therefore suggested that coastal engineers devote their efforts to clarifying the mechanisms of these different scale phenomena in parallel, and to trying to fill in the great gaps still existing among them.

## *2.1 Macroscale Approach*

To treat macroscale phenomena, the approach of the geologist and geomorphologist is quite helpful for understanding the general tendencies of coastal processes. But the time scale of their interest is usually too long for engineering purposes; hence coastal engineers have developed their own measures and devices for obtaining data on the relatively short-term variation of coastal processes. This "short" term, roughly the span of human life, is still quite long when considering normal engineering practice. Nevertheless, the historical sea-level change, ground-level variation due to crustal movement, as well as changes due to artificial interference, must all be taken into consideration at the locations where they occur.

In addition to natural processes, coastal engineers should be concerned with effects produced mainly by the construction of coastal structures as well as by various other similar sources. These man-made influences are occasionally the main causes of the beach erosion occurring on coasts everywhere.

Based on the above considerations, it is frequently realized that various natural and artificial causes combine at a given location. But in the future, man-made effects may steadily increase their contribution to the beach-erosion problem. Therefore, long-term coastal changes should be predictable with sufficient accuracy before starting any coastal construction project.

## *2.2 Mesoscale Approach*

Changes in shoreline and sea-bottom topography, bar and cusp formation, and nearshore currents all fall in the category of mesoscale phenomena. Numerous investigations have been carried out on these subjects, and certain details are understood at least qualitatively. Because our understanding of nearshore dynamics has tremendously improved over the last decade, fluid motion may be predicted fairly precisely under appropriate assumptions. But actual coastal sediment processes have not been sufficiently investigated due to the complexity of coastal sediment movement and the formidable nature of the coastal environment. For example, no precise expression for the coastal-sediment transport rate is yet available. For this reason we cannot at present combine the fundamental equations of fluid motion and the conservation of sediment material to calculate the bottom change in the nearshore area.

In order to surmount this problem for practical applications, numerous longshore-transport formulas have been proposed based in large part on empirical results. Combining one of these formulas and the continuity equation derived under simplified conditions yields a prediction of shoreline change, some details of which will be given later. This kind of approach is quite useful for practical purposes, but the assumptions made in the above treatments are not always correct.

In order to simulate changes in the sea-bottom configuration, we start with the following equation:

$$(1-\lambda)(\partial h/\partial t)-\partial q_x/\partial x-\partial q_y/\partial y=0, \tag{2.1}$$

where $x$ and $y$ denote the horizontal axes, $\lambda$ is the porosity of the sediment, $h$ the water depth at a particular point $(x, y)$ at a particular time $t$, and $q_x, q_y$ the sediment transport rates through a unit length in either the $y$ or $x$ direction, respectively, per unit time. To close the above equation, $q_x$ and $q_y$ should be evaluated using the wave characteristics (wave height, wave period, and wave angle), water depth $h$, and grain characteristics (grain size $d$, density $\rho_s$, and porosity $\lambda$). As stated above, at present no reliable formulas for $q_x$ and $q_y$ are available.

Sediment particles will be transported by fluid motion. Fluid motion in the nearshore area is mainly caused by wave action. Therefore wave transformation in shallow water, that is, wave refraction, wave diffraction, wave reflection, and wave breaking should be determined beforehand. The computation of wave transformation can be successfully done, at least in a first approximation. The next step is the prediction of the nearshore current, which is presently possible with adequate accuracy.

## 2.3 Microscale Approach

In order to correlate the sediment-transport rate with wave action, the detailed mechanisms of fluid and sediment movement must be known. However, our knowledge on these subjects is still inadequate, and further work is needed.

The important problems to be solved are listed below, and most are classified in the microscale regime. These problems are

1. Characteristics of oscillatory boundary-layer flow due to the coexistence of waves and currents, such as the velocity distribution and bottom shear stress, under hydrodynamically smooth or rough conditions,
2. Vertical distribution of the momentum-exchange coefficient (or the eddy viscosity) and of the diffusion coefficient in the wave field,
3. Vertical distribution of the mass-transport velocity,
4. Wave deformation and velocity field on a sloping bottom,
5. Interaction between waves and wave-induced currents,
6. Ripple formation and ripple size, disappearance of ripples and formation of sheet flow,
7. Vertical distribution of the suspended-sediment concentration inside and outside the breaker zone,
8. Detailed mechanisms of sediment transport due to waves and currents.

## 2.4 Interactions among Waves, Currents, Topography, and Sediment Transport

Figure 1 schematizes the relations among the various elements such as waves, nearshore currents, sediment transport, and topography, etc. The

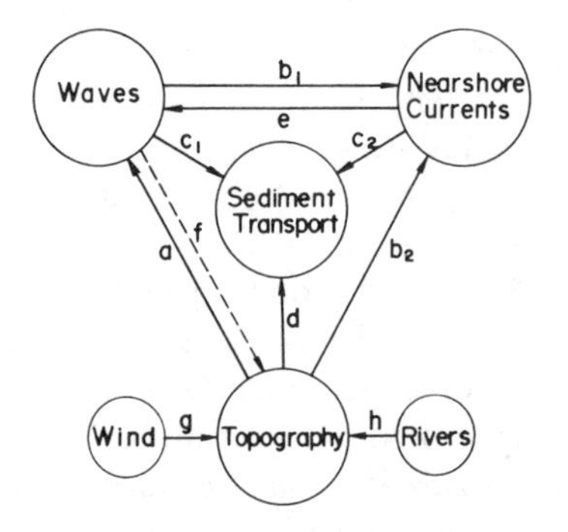


*Figure 1* Relationship among waves, nearshore currents, sediment transport, and topography (adapted from Horikawa, Harikai & Kraus 1979).

aim of our investigations is to clarify these complex relationships. In order to treat them, a suitable approach must be selected from among the three discussed above.

## 3 NEARSHORE DYNAMICS

Phenomena observed in the nearshore area are extremely complex, and seem to elude any kind of analytical treatment. Since the end of the 1940s, a number of workers have treated these problems through laboratory and field investigations, and numerous semi-empirical relations or correlations have been found involving nearshore phenomena. Our next interest is to understand their mechanisms, if only partially.

The radiation-stress concept proposed by Longuet-Higgins & Stewart (1960, 1962) opened a new era in the analysis of nearshore dynamics. Lundgren (1963) independently derived and named the same term as the wave thrust. In the present article, treatments relating directly to coastal sediment processes will be discussed briefly.

### *3.1 Dynamical Equations and Definition of Radiation Stresses*

Following Phillips (1977), the dynamical equations for the coexistent field of waves and a current with vertically uniform velocity will be given. These are the conservation equations of mass, momentum, and energy derived by integrating the corresponding basic equations with respect to the vertical axis from the bottom ($z=-h$) to the water surface ($z=\zeta$) and taking a time average over many wave periods,

$$\frac{\partial}{\partial t}\left[\rho\left(h+\bar{\zeta}\right)\right]+\frac{\partial \tilde{M}_\alpha}{\partial x_\alpha}=0, \tag{3.1}$$

$$\frac{\partial \tilde{M}_\alpha}{\partial t}+\frac{\partial}{\partial x_\beta}\left(\tilde{U}_\alpha \tilde{M}_\beta+S_{\alpha\beta}\right)=T_\alpha+R_\alpha, \tag{3.2}$$

$$\begin{aligned}\frac{\partial}{\partial t}&\left[\frac{1}{2}\tilde{U}_\alpha\tilde{M}_\beta+\frac{1}{2}\rho g(\bar{\zeta}^2-h^2)-\frac{1}{2}M_\alpha^2/\rho\left(h+\bar{\zeta}\right)+\bar{E}\right]\\&+\frac{\partial}{\partial x_\alpha}\left[\tilde{M}_\alpha\left(\frac{1}{2}\tilde{U}_\beta^2+g\bar{\zeta}\right)-\frac{1}{2}\tilde{U}_\alpha M_\beta^2/\rho\left(h+\bar{\zeta}\right)+U_\alpha\bar{E}\right.\\&\left.+F_\alpha+U_\beta S_{\alpha\beta}\right]+(\text{rate of energy loss})=0,\end{aligned} \tag{3.3}$$

where the horizontal axes $x_1$ and $x_2$ are taken on the still-water level and the vertical axis $z$ in the upward direction, hence $\alpha$, $\beta=1$ and 2. The notations $\rho$, $\bar{\zeta}$, and $\bar{E}$ represent the fluid density, mean sea-level rise due to the interference between waves and current, and wave energy per unit surface area. In the present treatment, the horizontal velocity component $u_\alpha$ was split into the uniform velocity component $U_\alpha$ and the horizontal component $u'_\alpha$ of the wave-particle velocity. Also, the definitions

$$\tilde{M}_\alpha = \overline{\int_{-h}^{\zeta} \rho u_\alpha dz}, \quad \hat{M}_\alpha = \overline{\int_{-h}^{\zeta} \rho U_\alpha dz} = \rho(h+\bar{\zeta})U_\alpha,$$

and $M_\alpha = \overline{\int_{-h}^{\zeta} \rho u'_\alpha dz}$

were introduced. Therefore $u_\alpha = U_\alpha + u'_\alpha$, and $\tilde{M}_\alpha = \hat{M}_\alpha + M_\alpha$. The other terms included in the above equations are

$$S_{\alpha\beta} = \overline{\int_{-h}^{\zeta} (\rho u'_\alpha u'_\beta + p\delta_{\alpha\beta})\, dz} - \tfrac{1}{2}\rho g(h+\bar{\zeta})^2 \delta_{\alpha\beta}$$

$$- M_\alpha M_\beta / \rho(h+\bar{\zeta}), \tag{3.4}$$

$$T_\alpha = -\rho g(h+\bar{\zeta})\partial\bar{\zeta}/\partial x_\alpha, \tag{3.5}$$

$$R_\alpha = \overline{\int_{-h}^{\zeta} \partial\tau_{\beta\alpha}/\partial x_\beta dz} + \bar{\tau}_{\zeta\alpha} - \bar{\tau}_{h\alpha}, \tag{3.6}$$

$$F_\alpha = \rho \overline{\int_{-h}^{\zeta} u'^2_\alpha \left[\tfrac{1}{2}u'^2_i + g(z-\bar{\zeta}) + p/\rho\right] dz}, \tag{3.7}$$

where $p$ is the pressure, $\delta_{\alpha\beta}$ the Kronecker delta (i.e., $\delta_{11} = \delta_{22} = 1$, and $\delta_{12} = \delta_{21} = 0$), $\tau_{\beta\alpha}$ the stress component inducing the interference between waves and current, and $\bar{\tau}_{\zeta\alpha}$, $\bar{\tau}_{h\alpha}$ the mean shear stresses at the free surface and at the bottom respectively.

The term $S_{\alpha\beta}$ is the radiation-stress tensor, which corresponds to the excess momentum flux tensor, $T_\alpha$ the horizontal force per unit surface area induced by the free-surface gradient, $R_\alpha$ the frictional term consisting of the lateral and boundary frictional terms, and $F_\alpha$ the mean energy flux by the fluctuating motion alone.

The radiation-stress tensor $S$ for the case of a train of waves $\zeta = \frac{1}{2} H\cos(x_1 k\cos\theta + x_2 k\sin\theta - \sigma t)$ is expressed by

$$S = \bar{E}\begin{bmatrix} (c_g/c)\cos^2\theta + \frac{1}{2}(2c_g/c-1) & \frac{1}{2}(c_g/c)\sin 2\theta \\ \frac{1}{2}(c_g/c)\sin 2\theta & (c_g/c)\sin^2\theta + \frac{1}{2}(2c_g/c-1) \end{bmatrix} \tag{3.8}$$

where $H$ is the wave height, $k$ the wave number, $\sigma$ the angular frequency,

$\bar{E}=\rho g H^2/8$, $\rho$ the fluid density, $g$ the gravitational acceleration, $\theta$ the wave direction angle, $c$ the wave celerity, and $c_g$ the group velocity.

## 3.2 *Wave Set-Down and Wave Set-Up*

As a typical application of the radiation-stress concept to nearshore phenomena, we will consider the simple case where waves arrive normal to the shoreline ($y$ axis) from offshore to a beach with a uniformly gentle slope. Taking the $x$ axis from offshore toward shore, and considering the steady state, Equation (3.1) becomes $d\tilde{M}_x/dx=0$, that is $\tilde{M}_x=\hat{M}_x+M_x=\rho(h+\bar{\zeta})U_x+M_x=\text{constant}$. Due to the existence of the beach, the constant should be zero. Therefore Equation (3.2) can be written in the following form under the assumption that the frictional term $R_x$ is negligible:

$$dS_{xx}/dx=-\rho g(h+\bar{\zeta})\,d\bar{\zeta}/dx. \tag{3.9}$$

Integration of the above equation outside and inside the surf zone separately under the appropriate assumptions yields the result that the mean sea level is reduced from the deep-water level to a certain value at the breaking point, then rises up toward the shore. The former and latter phenomena are called "wave set-down" and "wave set-up" respectively.

Laboratory investigations (Bowen, Inman & Simmons 1968) have confirmed that the above theoretical treatment predicts well the observed wave set-down and wave set-up, but that there are still some discrepancies remaining, especially in the vicinity of the breaking point.

## 3.3 *Longshore-Current Velocity Distribution*

It has been realized for many years that the longshore current is the most important agent for the longshore sediment transport. In order to establish a relationship between the longshore current and the longshore-sediment-transport rate, it is necessary to know the horizontal and vertical longshore-current velocity distributions in the nearshore area. Here we will take the simplest case, where the bottom slope is uniform, the depth contour is parallel to a straight shoreline, and the wave-induced current motion is steady. The $x$ and $y$ axes are taken perpendicular (towards offshore) and parallel to the shoreline, respectively. The basic equation in this case is simply written as follows:

$$dS_{xy}/dx=R_y. \tag{3.10}$$

We will consider the nearshore zone in separate regions, outside and inside the surf zone. Outside the surf zone, we can assume that the energy flux is conserved and that Snell's law is applicable. Applying this law, we can derive $S_{xy}$=constant. Therefore Equation (3.10) becomes

$$R_y = dS_{xy}/dx = 0. \tag{3.11}$$

Inside the surf zone, the wave height $H$ is well expressed by $H=\gamma(h+\bar{\zeta})$, where $\gamma$ is a constant. Bowen (1969a) and Longuet-Higgins (1970) evaluated $dS_{xy}/dx$ under approximations leading to the equations,

$$R_y = \begin{cases} (1/4)\rho g \gamma^2 (h+\bar{\zeta}) \sin\alpha_b \cos\alpha_b \, d(h+\bar{\zeta})/dx & \text{(Bowen)} \quad (3.12a) \\ (5/16)\rho g \gamma^2 (h+\bar{\zeta}) \sin\alpha \, d(h+\bar{\zeta})/dx & \text{(Longuet-Higgins)} \quad (3.12b) \end{cases}$$

where $\alpha$ is the angle between the wave crest and bottom contour, and the subscript b denotes the value at the breaking point.

The next task is to determine the expression for the frictional term $R_y$. Considering Equation (3.6), and neglecting the frictional stress at the surface, $\bar{\tau}_{\zeta\alpha}$, we can construct the next equation as a general one,

$$R_y = \partial(\varepsilon_\mu \partial v/\partial x)/\partial x - \bar{\tau}_{hy}, \tag{3.13}$$

where $v$ is the longshore current velocity and $\varepsilon_\mu$ is the momentum-exchange coefficient. The first and second terms on the right-hand side are the lateral and bottom friction terms, respectively.

For the lateral friction term, Bowen and Longuet-Higgins used the expressions,

$$\partial(\varepsilon_\mu \partial v/\partial x)/\partial x = \begin{cases} \rho(h+\bar{\zeta})\Lambda_h d^2 v/dx^2 & \text{(Bowen)} \quad (3.14a) \\ d\left[\mu_e (h+\bar{\zeta})\, dv/dx\right]/dx & \text{(Longuet-Higgins)} \quad (3.14b) \end{cases}$$

$$\mu_e = N\rho x\sqrt{g(h+\bar{\zeta})}\,, \qquad 0 < N < 0.016,$$

where $\Lambda_h$ and $N$ are constants, and $x$ is measured from shoreline. According to these expressions, the momentum-exchange coefficient is assumed to increase monotonically from the shoreline to offshore beyond the breaker line. This behavior is not physically realistic. That is to say, the momentum-exchange coefficient should increase from the shoreline to the breaker line, but decrease in a certain manner beyond the breaker line.

For the bottom frictional term, Bowen and Longuet-Higgins took the expressions,

$$\bar{\tau}_{hy} = \begin{cases} \rho C v & \text{(Bowen)} \qquad (3.15a) \\ \frac{1}{2}\rho f_w |\bar{u}_{wh}| v = \rho f_w u_{bm} v/\pi & \text{(Longuet-Higgins)} \qquad (3.15b) \end{cases}$$

where $C$ and $f_w$ are both friction coefficients, $|\bar{u}_{wh}|$ the absolute-time average, and $u_{bm}$ the amplitude of the orbital velocity at the bottom induced by wave action. The former expression is based on a linear friction law, while the latter is based on a quadratic law with the assumption that the longshore-current velocity $v$ is small compared with the wave orbital velocity.

By using the above expressions for the lateral and bottom friction terms, Bowen and Longuet-Higgins obtained realistic velocity-distribution curves of the longshore current. Their contribution to the coastal engineering field was extremely important because they demonstrated that the wave-induced current can be calculated analytically. However, the assumptions listed below were made and, as a result, the applicability of the solutions is restricted to a certain range. These assumptions are: 1. The bottom beach slope is uniform; 2. The incident wave angle is small; 3. The lateral and bottom frictional terms are valid under certain limited conditions; and 4. The wave height inside the surf zone is simply expressed as being proportional to the water depth.

The last assumption seems to be applicable to only a uniformly sloping bottom. According to recent investigations (Mizuguchi, Tsujioka & Horikawa 1978, Hotta & Mizuguchi 1978) the proportionality constant $\gamma$ is not constant in a real surf zone; it depends on the bottom configuration and also the breaker type.

Jonsson (1966, 1976), Kajiura (1964, 1968), and Kamphuis (1975) treated the frictional stress along a horizontal bottom due to waves. The diagram prepared by Jonsson (1976) is commonly used in longshore-current computations. Inside the surf zone, waves are superimposed obliquely on the longshore current. For a case where the wave orbital velocity is fairly large compared to the longshore-current velocity, the present treatment would be acceptable. However, in a case where the longshore-current velocity is comparable with the wave orbital velocity, it is expected that the bottom frictional term would be of a different form and that a new frictional coefficient would be required. Neither of these quantities has been given satisfactorily.

Concerning the effect of incident wave angle on the longshore-current velocity distribution, Liu & Dalrymple (1978), and Kraus & Sasaki (1979) independently treated the phenomenon and found a systematic influence of the angle of wave approach on the longshore-current velocity distribution.

### 3.4 Nearshore Currents

There are complex currents induced by numerous elements in the nearshore area. Among these currents, the wave-induced currents are most closely related to coastal sediment processes. Mass transport associated with waves, longshore currents, and rip currents are all wave-induced currents that form the nearshore current. When waves arrive almost perpendicular to the shoreline, the region bounded by two adjacent rip currents forms a unit cell. When the incident wave angle increases, the closed cell disappears and forms a meandering current or a longshore current.

If we use the horizontal velocity components $u$ and $v$ in place of $\tilde{U}_x$ and $\tilde{U}_y$, we can write $\tilde{M}_x=\rho(h+\bar{\zeta})u$ and $\tilde{M}_y=\rho(h+\bar{\zeta})v$. Introducing these expressions into Equations (3.1) and (3.2) and considering the steady state, we can derive the following equations:

$$\partial\left[(h+\bar{\zeta})u\right]/\partial x+\partial\left[(h+\bar{\zeta})v\right]/\partial y=0, \tag{3.16}$$

$$u\partial u/\partial x+v\partial u/\partial y=\left[T_x+R_x-(\partial S_{xx}/\partial x+\partial S_{xy}/\partial y)\right]/\rho(h+\bar{\zeta}), \tag{3.17a}$$

$$u\partial v/\partial x+v\partial v/\partial y=\left[T_y+R_y-(\partial S_{xy}/\partial x+\partial S_{yy}/\partial y)\right]/\rho(h+\bar{\zeta}), \tag{3.17b}$$

where $\rho=$constant. In order to satisfy Equation (3.16) a scalar function $\psi$ is sometimes used, defined by

$$u(h+\bar{\zeta})=\partial\psi/\partial y,$$

$$v(h+\bar{\zeta})=-\partial\psi/\partial x. \tag{3.18}$$

The scalar function $\psi$ was introduced by Arthur (1962) and called the transport stream function.

Remembering that $T_x/\rho(h+\bar{\zeta})=-g\partial\bar{\zeta}/\partial x$ and $T_y/\rho(h+\bar{\zeta})=-g\partial\bar{\zeta}/\partial y$, we cross-differentiate the first and second equations in Equations (3.17a, b) with respect to $y$ and $x$, respectively, subtract the second equation from the first, and write the result using the vorticity component $\omega=\partial v/\partial x-\partial u/\partial y$,

$$\begin{aligned}&-D\left[u\partial(\omega/D)/\partial x+v\partial(\omega/D)/\partial y\right]\\&\quad=\partial(R_x/\rho D)/\partial y-\partial(R_y/\rho D)/\partial x\\&\qquad-\partial\left[\{\partial S_{xx}/\partial x+\partial S_{xy}/\partial y\}/\rho D\right]/\partial y\\&\qquad+\partial\left[\{\partial S_{xy}/\partial x+\partial S_{yy}/\partial y\}/\rho D\right]/\partial x,\end{aligned} \tag{3.19}$$

where $D = h + \bar{\zeta}$. The term on the left-hand side, and the first and second terms on the right-hand side of Equation (3.19) are the nonlinear, frictional, and forcing terms respectively.

Bowen (1969b) treated the idealized nearshore-current pattern of a closed cell produced by waves arriving normal to the shoreline. He introduced linear expressions for the bottom friction terms and solved the resultant equation under the assumption that the wave height varies periodically along the shore. The important feature of Bowen's treatment was that it paved the way for numerical calculations of the nearshore-current system.

Noda (1974) was the first to carry out numerical calculations for nearshore currents using realistic bottom contours and arbitrary incident-wave conditions. Bottom friction was included in this model, but nonlinear terms were neglected. In order to evaluate the frictional terms, he applied a quadratic resistance law with a constant-friction coefficient. Sasaki (1975) used essentially the same model with a variable-friction coefficient based on Jonsson's diagram for oscillatory flow, and calculated values of the friction coefficient at each mesh point. The calculated results gave rather good agreement with field-observation data in a qualitative sense.

## 4 MECHANISM OF COASTAL SEDIMENT MOVEMENT

In the treatment of coastal sediment transport, for simplicity it is quite common to consider separately sediment movement perpendicular and parallel to the shoreline. Needless to say, the two kinds of sediment movement are closely related to each other. However, onshore-offshore sediment movement is considered to be more significant for the short-term variation of coastal processes, while longshore sediment movement is more significant for the long-term variation of the coastal topography.

In the following section, we will discuss several topics related to the mechanisms of coastal sediment movement from the perspective of fluid mechanics. However, the essential point and goal of the present subject, that is, the evaluation of the sediment-transport rate, has not yet been fully clarified. The main reason for such an unfortunate state is the almost insuperable difficulty in measuring the rate and the direction of sediment transport in the coastal zone, which must be performed in conjunction with measuring the waves, currents, and bathymetry.

### *4.1 Inception of Sediment Movement*

The water depth at which sediment particles are first influenced by water motion is about 150 to 200 m, depending upon the environmental conditions, especially the wave characteristics. For engineering application, knowledge of the water depth where sediment particles move appreciable distances under wave action is important for determining the initial point of beach-profile change in the offshore region.

Considerable research has been conducted during the last three decades concerning the critical water depth for the inception of sediment movement. A detailed discussion of past treatments of this subject can be found elsewhere (Horikawa 1978b).

Horikawa & Watanabe (1967) treated the problem in the following simple manner. The equilibrium condition for a particle on the sea bottom can be expressed by

$$R_{H}=(W-R_{V})\tan\phi, \tag{4.1}$$

where $R_{H}$ and $R_{V}$ are the horizontal and vertical components of the wave force acting on the particle, $W$ is the immersed weight of the particle, and $\phi$ is the critical angle of static friction of a grain on the bottom in water. The values of $W$ and $R_{H}$ can be evaluated by the following equations:

$$W=(4\pi/3)(\rho_{s}-\rho)g(d/2)^{3}, \tag{4.2}$$

$$R_{H}=(1/4)K\pi d^{2}\tau_{bm}, \tag{4.3}$$

where $d$ and $\rho_{s}$ are the grain diameter and density respectively, $\rho$ is the fluid density, $g$ the acceleration of gravity, $\tau_{bm}$ the amplitude of tangential stress acting on a grain due to the wave, and $K$ a coefficient taking on different values depending upon the shape of the sediment particle and the type of movement. The magnitude of $R_{V}$ was assumed negligibly small compared with $W$, and the following relationship was obtained:

$$\psi_{m}=\tau_{bm}/s\rho gd=2/(3K) \tag{4.4}$$

where $s=(\rho_{s}-\rho)/\rho$ and $\psi_{m}$ is a kind of Shields parameter. The theory of Kajiura (1968) for the bottom shear stress due to oscillatory flow was applied and comparisons made with experimental data on the inception of sediment-particle movement due to wave action. Based on these results, criteria were proposed for general movement in laminar or turbulent flow over a smooth surface, as well as for turbulent flow over a rough surface. Here, general movement is defined as the state where most of the sediment particles in the first layer are in motion. Based on the above, Horikawa & Sasaki (1970) compiled tables for the direct determination of the critical water depth for general sand motion under various wave conditions for sand grains having diameters between 100 $\mu$m and 1 mm.

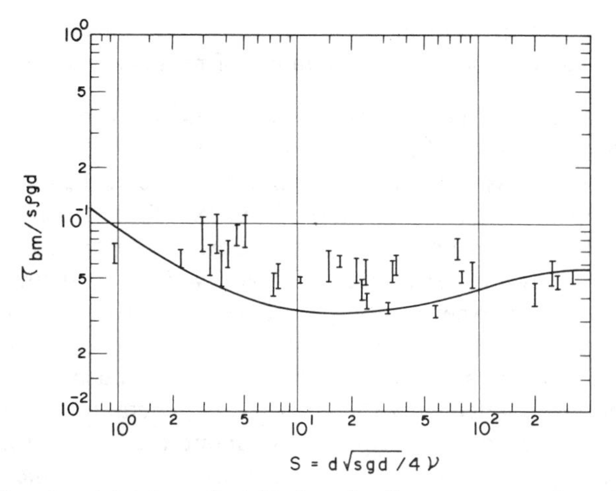


*Figure 2* Experimental data on the initiation of sediment movement in oscillatory flow (adapted from Madsen & Grant 1976).

Madsen & Grant (1976) reanalyzed past results leading to the diagram shown in Figure 2, in which the ordinate is the Shields parameter $\psi_m = \tau_{bm}/s\rho gd$ and the abscissa is $S = d\sqrt{(sgd)}/4\nu$, where $\nu$ is the kinematic viscosity of the fluid. Considering the scatter in the data, we may well be able to assume $\psi_m$ = constant. From this fact, Equation (4.4) still seems to be valid.

## *4.2 Onshore-Offshore Sediment Movement*

In this section we will consider onshore-offshore sediment movement in a region at a depth shallower than the critical water depth. Dingler & Inman (1976) examined the bed form and type of onshore-offshore sediment movement in the nearshore area. In the offshore zone, sediment transport is induced by wave action as a combination of bed load and suspended load, and sand ripples on the bed play an important role in the mode of sediment transport. On the other hand, the sediment-transport mode inside the surf zone is considerably different from that outside. Agitation due to breaking waves is so strong that bed materials are suspended, and the concentration of suspended sediment in the vicinity of the breaking point is enormous. The amount of suspended sediment is strongly dependent on the breaker type (Kana 1978). Further shoreward, especially in the swash zone, sheet flow is predominant and sand ripples disappear.

In order to predict the time history of bottom-profile change, we must know the direction and the amount of net sediment transport due to wave action. However, at present, our knowledge on these subjects is still inadequate, and only a small number of results are available.

4.2.1 SAND RIPPLE FORMATION It has been known for many years that sand ripples are generated on the sea bed under the influence of waves. Bagnold (1946) was probably the first to systematically study the process of ripple formation. He carried out experiments using an oscillating board swung in an arc. Sediment particles of various sizes and specific gravities were spread over the board which was swung in a still body of water. The experimental procedure originated by Bagnold was continued by workers at the University of California, Berkeley. Using this kind of apparatus, Manohar (1955) obtained much data on sand ripples, and investigated fully the processes of generation, development, and disappearance of sand ripples. At about the same time, Inman (1957) measured the sizes of wave-generated ripples on the southern California coast.

The shapes of wave-generated ripples are rather symmetrical and are completely different from the shapes of ripples generated by unidirectional flow. Therefore the shape of a wave-generated ripple can be simply expressed by its height $\eta$ and length $\lambda$. Hom-ma & Horikawa (1962) applied dimensional analysis to determine the functional relationships among ripple size, grain size, water depth, and local wave characteristics. These relationships are

$$F_1(\eta/\lambda, \lambda/d_0)=0, \tag{4.5a}$$

$$F_2(d_0/\lambda, u_{bm}d_0/\nu, w_0 d/\nu)=0, \tag{4.5b}$$

where $d_0 = H/\sinh(2\pi h/L)$, $u_{bm} = \pi H/[T\sinh(2\pi h/L)]$, $h$ is water depth, $H$, $T$, and $L$ are the wave height, wave period, and wave length, $\nu$ is the fluid kinematic viscosity, $w_0$ the fall velocity of a grain, and $d$ the grain size. A subsequent study by Horikawa & Watanabe (1967) confirmed that the relationships expressed in Equations (4.5a, b) were also applicable to materials other than natural sand. Furthermore, Carstens, Neilson & Altinbilek (1969), Mogridge & Kamphuis (1973), and Dingler & Inman (1976) have continued investigations on the characteristics of wave-generated ripples in nearshore sands. It has been realized that sand ripples have an important role in sediment movement by wave action.

4.2.2 SEDIMENT TRANSPORT PATTERNS ON A BEACH Looking at the bottom configuration of a gently sloping beach, we can find a certain region where wave-generated sand ripples can be seen. In such a region the sediment-transport pattern is strongly influenced by the existence of the sand ripples in the following manner: Incident waves arriving to shore possess nonlinear characteristics, that is to say, the wave profile is peaked at a crest and flattened at a trough. Corresponding to these wave characteristics, the onshore velocity displays a large maximum over a shorter interval than that of the offshore velocity. Because of the asymmetrical characteristics of the velocity field, the onshore-offshore sediment transport due to wave action is not in balance over a wave period.

Sediment particles are, generally speaking, transported in the form of either bed load or suspended load. Owing to the wave nonlinearity discussed above, the net bed load in the present case would be in the onshore direction. On the other hand, suspended sediment particles are picked up by vortices generated behind ripples and transported onshore during the passage of a wave crest and offshore during the passage of a wave trough. Inman & Bowen (1963) demonstrated that offshore sediment transport occurs due to ripple asymmetry and resulting differences in intensity of wave-induced vortices. According to laboratory measurements of suspended sediment transport conducted by Sunamura, Bando & Horikawa (1978), the net suspended-sediment transport from solely these processes is in the offshore direction.

Based on observations of transport in a wave flume, Horikawa, Sunamura & Shibayama (1977) classified sediment-transport patterns due to wave action into the following four types:

Type 1: Bed-load transport is dominant, but no suspended sand cloud exists. Therefore the net sediment transport is in the onshore direction.

Type 2: Suspended sand clouds are formed, hence both bed load and suspended load should be considered. The net sediment-transport direction is either onshore or offshore depending on the dominance of the sediment-transport mode.

Type 3: Suspended-sediment transport is dominant. Suspended sand clouds are formed only on the onshore sides of ripples. Hence the net sediment transport is in the offshore direction.

Type 4: Suspended-sediment transport is dominant. Suspended sand clouds are formed on both sides of ripples, but the net sediment transport is in the offshore direction.

Shibayama & Horikawa (1980) presented limits for each transport type which are well described by the two parameters $\psi_m$ and $u_{bm}/w_0$. Here $u_{bm}$ is the amplitude of the near-bed fluid velocity and $w_0$ the fall velocity of a sand grain.

Tunstall & Inman (1975) treated fluid-energy dissipation due to vortices formed by oscillatory flow over a wavy boundary using a standing-vortex theory. Sunamura, Bando & Horikawa (1978) recorded sand-particle movement over a rippled bed using a 16-mm high-speed motion-analysis camera, and verified quantitatively that the movement of the suspended sand clouds is one of the main factors in producing net sediment transport. Sawamoto & Yamaguchi (1979) studied the motion of vortices above a rippled bed using potential-flow theory and estimated the suspended-sediment concentration above such a rippled bed. Shibayama & Horikawa (1980) introduced a dissipation model of time-dependent vortex circulation formed by wave action on a wavy boundary and calculated the water-particle path. Visual observation verified that the calculated water-particle path well represents the motion of the suspended sediment particles. Owing to the above efforts, the oscillatory velocity field in the vicinity of a wavy boundary can be analyzed, hence the motion of an individual suspended sediment particle seems to have been clarified in more detail.

4.2.3 SEDIMENT-TRANSPORT RATE At present, the available data for the bed-load-transport rate are only those obtained by Manohar (1955), Kalkanis (1964), and Abou-Seida (1965). Using these data and assuming that a Brown-type formula for the bed-load transport rate under a unidirectional-flow condition is applicable to the instantaneous bed-load transport rate by oscillatory flow, Madsen & Grant (1976) presented an empirical relationship for the average rate of sediment transport in an oscillatory flow, as shown in Figure 3. The ordinate, $\bar{\phi}=q_b/w_0 d$ is the

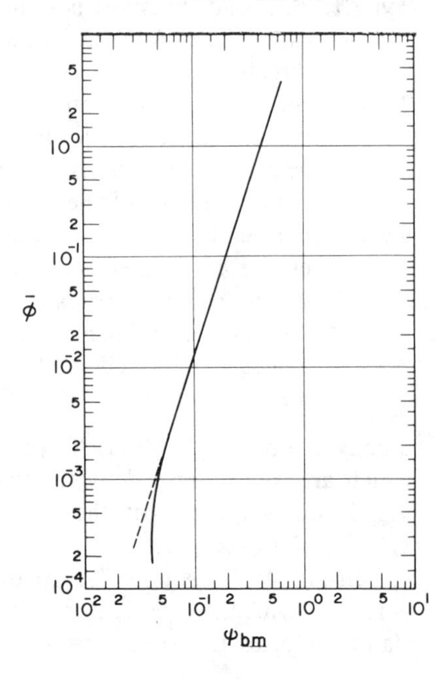


*Figure 3* Empirical relationship for the average rate of sediment transport in oscillatory flow (adapted from Madsen & Grant 1976).

average bed-load transport rate, $q_b$, nondimensionalized by the fall velocity $w_0$ and the grain diameter $d$, while the abscissa, $\psi_m$, is the Shields parameter. Their proposed curve is expressed by

$$\bar{\phi} = 12.5\psi_m^3. \tag{4.6}$$

Watanabe, Riho & Horikawa (1979) tested the above relationship between $\bar{\phi}$ and $\psi_m$ outside the surf zone on a sloping sand bed in a wave flume by analyzing experimental data of wave characteristics and bottom-configuration change. The curves obtained are well fit by Equation (4.6).

Shibayama & Horikawa (1980) used the same procedure as Madsen & Grant to evaluate the suspended-sediment transport rate and obtained a similar relationship. Their treatment made use of the fact that bed-load particles are picked up, confined within vortices formed behind sand ripples, moved by the wave oscillatory flow, and finally deposited on the bed under gravitational force. In addition to these, Nielsen, Svendsen & Staub (1978) conducted intensive studies on the suspended-sediment transport mechanism. Sleath (1978) measured the quantity of bed materials in motion as bed load at each instant of the oscillatory-wave cycle.

At any rate, these relationships can be applied only to the sediment movement outside the surf zone. In order to look at the bottom change inside the surf zone, we must have some appropriate formula for the sediment transport rate in the sheet-flow region.

## *4.3 Longshore Sediment Transport*

During the last two decades many approaches have been attempted to determine the longshore sediment-transport rate. Usually, one of the following general procedures is taken: 1. measuring the amount of deposition at the upstream side of a coastal structure such as a jetty or a breakwater, or of siltation inside a harbor basin; or 2. measuring the amount of the same kinds of sediment in a model basin. However, there are numerous problems to be solved, including the accuracy of the hydrographic survey, the limited extent of the surveyed area, the accuracy of the wave-energy evaluation, and the influence of grain size on the transport rate.

The longshore-transport rate should bear a close relationship with the longshore-current velocity, but the exact relation between them has not yet been established. On the other hand, many attempts have been made to correlate directly the wave-energy flux in the alongshore direction, $E_y$, with the rate of longshore transport, $Q_y$, according to the following form,

$$Q_y = \alpha E_y^m, \tag{4.7}$$

where $\alpha$ and $m$ are constants to be determined empirically. Caldwell (1956) initiated the above treatment using field data. Subsequent to his work, numerous formulas have been presented during the last two decades. A comparison of these formulas was given by Galvin & Vitale (1976) and Horikawa (1978a).

One of the defects of Equation (4.7) is that it is not dimensionally correct. In order to eliminate the above defect, Inman & Bagnold (1963) suggested the use of the immersed-weight-transport rate $I_\ell$ instead of the volume-transport rate. Komar & Inman (1970) established the following relationship between $I_\ell$ and the longshore power at the breaking point, $P_\ell$:

$$I_\ell = K P_\ell. \tag{4.8}$$

The values of $I_\ell$ and $P_\ell$ are defined respectively as,

$$I_\ell = a'(\rho_s - \rho) g S_\ell, \tag{4.9}$$

$$P_\ell = \left(\bar{E} c_g\right)_b \sin\alpha_b \cos\alpha_b, \tag{4.10}$$

where $S_\ell$ is the sand-volume-transport rate, $\rho_s$ the sand density, $\rho$ the fluid density, $a'$ the correction factor for the pore space of the beach sand (approximately 0.6), $g$ the acceleration of gravity, $\bar{E}$ the wave energy density, $c_g$ the group velocity, $\alpha$ the angle between a wave crest and bottom contour, and the subscript "b" indicates the condition at the breaking point. The proportionality constant $K$ is dimensionless and has been found equal to about 0.77 for a limited number of measurements. Komar & Inman (1970) correlated $I_\ell$ with the longshore-current velocity based on the Bagnold (1963) model, and presented the following relationship,

$$I_\ell = 0.28\left(\bar{E} c_g\right)_b (\cos\alpha_b)\bar{v}/u_{bm}, \tag{4.11}$$

where $\bar{v}$ is the mean longshore-current velocity, and $u_{bm}$ the maximum velocity under a breaking wave. The value of $u_{bm}$ is calculated by

$$u_{bm} = \left(2\bar{E}_b/\rho h_b\right)^{1/2}. \tag{4.12}$$

Based on the calculation of Longuet-Higgins (1970), $\bar{v}$ is given by

$$\bar{v} = (5\pi/4)(\tan\beta/f_w) u_{bm} \sin\alpha_b \tag{4.13}$$

under the assumption that horizontal mixing is negligible, where $\tan\beta$ is the bottom slope. Substitution of Equation (4.13) into Equation (4.11) and comparison with Equation (4.8) necessitates the conclusion,

$$\tan\beta/f_w = \text{constant}. \qquad (4.14)$$

However, in the theory of Longuet-Higgins (1970), the mean longshore-current velocity depends on the parameter $P=2\pi N\tan\beta/(\gamma f_w)$ which describes the importance of lateral mixing relative to bottom friction, $N$ and $\gamma$ being constants. In order to make clear the above results on the basis of a theoretical treatment, it is essential to clarify the dynamics in the nearshore area, especially inside the surf zone.

The above formulas, Equations (4.8) and (4.11), are applicable for the total alongshore sediment-transport rate inside the breaker line. In recent years, a great effort has been devoted to measuring the distribution of sediment-transport rate across the surf zone by, for example, using sand traps in a wave basin with a movable bed (Sawaragi & Deguchi 1978), or by taking a large number of core samples during dyed-sand tracer experiments in the field (N. C. Kraus, personal communication, 1979). This kind of approach, especially in the field, is truly laborious, but holds promise for advancing our knowledge of longshore sediment transport.

There is a standing debate on the subject of the dominant sediment-transport mode comprising longshore sediment transport. That is to say, which is more significant, suspended transport or bed-load transport? The impression that the suspended load is much more important within the surf zone has been supported traditionally (Dean 1973). On the other hand, Komar (1976) roughly estimated the suspended load and concluded that the suspended load comprises no more than approximately one fifth of the total transport. At present, it is fairly difficult to determine precisely the ratio between the suspended and bed transport rates. Therefore it may be a more suitable approach to focus on the total longshore transport rate. Dyed-sand experiments such as initiated by Komar & Inman (1970) are being actively performed by the Nearshore Sediment Transport Study Group (Seymour & Duane 1978) and in the writer's research group. Detailed analysis of core samples can yield important information on distribution of the moving-layer thickness and of the sand advection velocity of sediment particles across the surf zone. These data can be correlated with the wave and current characteristics at each sampling site.

## 5 SIMULATION OF SHORE PROCESSES

In order to treat coastal processes, as stated previously, we commonly separate the phenomena into two parts, namely onshore-offshore processes and alongshore processes. Beach-profile changes are considered to have a seasonal variation, that is to say, beach profiles seem to follow an approximately one-year cycle. In contrast, the alongshore change in beach topography is caused mainly by the local balance of alongshore sediment transport; variation in the longshore transport rate along the shoreline is the principal mechanism governing erosion and deposition of beach sediment. From this standpoint, Iwagaki (1966) formulated the following equation:

$$\partial\bar{h}/\partial t=\left(1-\bar{h}/h_{\mathrm{i}}\right)\partial h_{\mathrm{i}}/\partial t+\left(\partial Q_y/\partial y\right)/(1-\lambda)B \tag{5.1}$$

where $B$ is the width of the littoral zone, $Q_y$ the longshore transport rate, $\bar{h}$ the mean water depth in the littoral zone, $h_{\mathrm{i}}$ the water depth which determines the offshore limit of the littoral zone, $\lambda$ the bottom sediment porosity, and the $y$ axis is taken in the alongshore direction. Equation (5.1) indicates that coastal change has two contributions, the local variation of longshore transport, $\partial Q_y/\partial y$, and the time variation of $h_{\mathrm{i}}$, which is determined by the time history of the incoming-wave characteristics. It is a consequence that

1. Even on a coast where $\partial Q_y/\partial y=0$, beach erosion can occur with increasing wave height, because $\partial h_{\mathrm{i}}/\partial t>0$ in this situation and it follows that $\partial\bar{h}/\partial t>0$.
2. When $\partial h_{\mathrm{i}}/\partial t=0$, erosion or deposition will completely depend on the sign of $\partial Q_y/\partial y$. That is to say, if $\partial Q_y/\partial y>0$, erosion will occur because $\partial\bar{h}/\partial t>0$, while if $\partial Q_y/\partial y<0$, deposition will occur because $\partial\bar{h}/\partial t<0$.

It can be said in general that the long-term variation in coastal topography is caused by local variation in the longshore transport rate. Through use of the above rules, if the rate of sediment transport along a beach can be evaluated, then equilibrium, erosional, or depositional regions along the coast can be determined as demonstrated in Figure 4. The prediction of shoreline changes occurring due to the presence of coastal structures can be predicted by this procedure as well. This approach has been developed and applied to practical problems.

The mathematical treatment of shoreline change was initiated by Pelnard-Considère (1956), who used a simple model as shown in Figure 5*a*. This concept was applied to the practical problems of predicting shoreline change due to the construction of groin. Price, Tomlinson & Willis (1972) developed a numerical model and checked the validity of its prediction by comparison with laboratory results. Hashimoto (1974) treated shoreline changes caused by the construction of a detached breakwater. Sasaki & Sakuramoto (1978) extended the one-line theory to a situation where the incident-wave direction varied with time, and

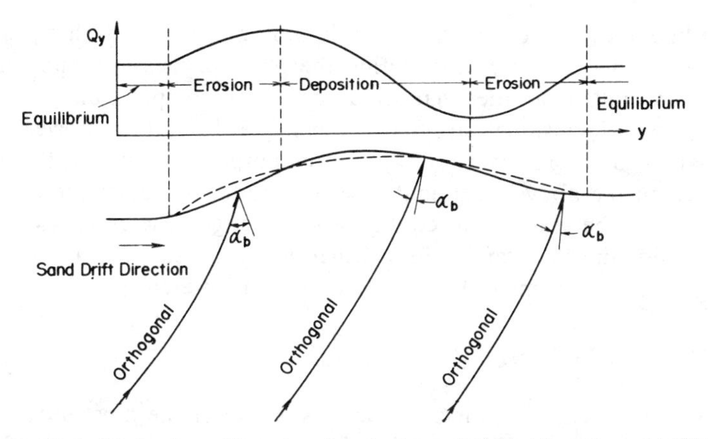


*Figure 4* Mechanism of long-term beach change (adapted from Iwagaki 1966).

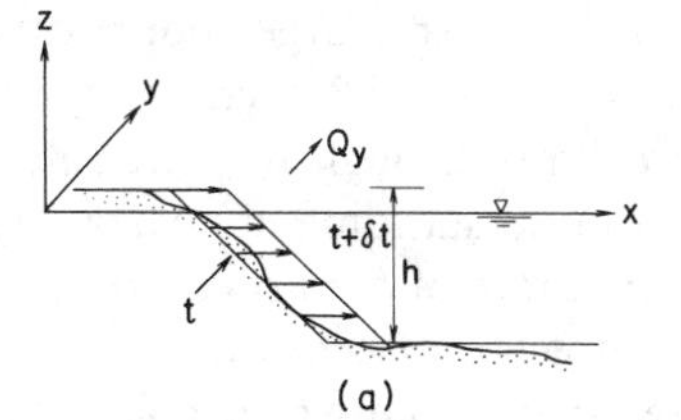


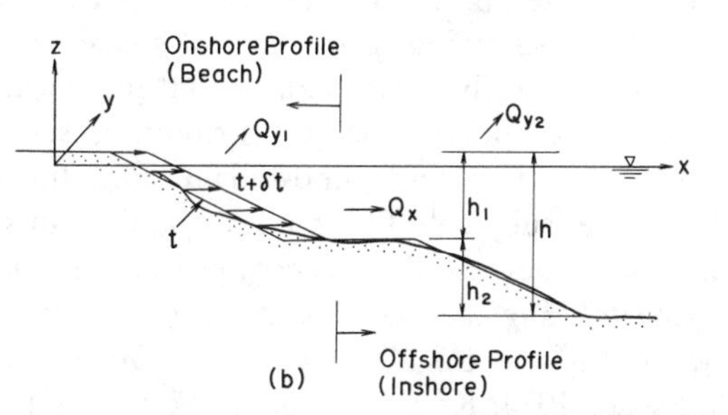


*Figure 5* Schematic diagram for (*a*) the one-line theory and (*b*) the two-line theory [adapted from Pelnard-Considère (1956) for (*a*) and Bakker, Klein Breteler & Roos (1970) for (*b*)].

compared computed results obtained under certain reasonable assumptions with field-observation data. They confirmed that the agreement seemed to be fairly good for practical purposes.

In order to improve the modeling of beach-profile change, Bakker, Klein Breteler & Roos (1970) proposed the two-line theory as shown in Figure 5b, in which the onshore-offshore sediment transport is partially treated. More generally, the two-line theory could be extended to a multi-line theory. However, it should not be forgotten that further studies are needed to clearly define the physical meaning and limitations of various quantities that are used in such computations.

The two-dimensional approach seems to be rather powerful in evaluating shore processes for engineering purposes. However, the models used in the above treatments are too crude to predict the detailed structure of the sea-bottom configuration change. Therefore, as a goal we should aim to clarify three-dimensional coastal processes, even though the goal seems to be far away from our present position.

## 6 CONCLUDING REMARKS

The study of coastal sediment processes is one of the most difficult in coastal engineering. This field has been treated by scientists and engineers during the last few decades without any remarkable advancement. However, in recent years several large cooperative research groups have been organized in various countries such as Japan and the US with the strong intention of breaking through the difficult barriers obstructing the development of our understanding of coastal sediment processes. Therefore, the time when the state of the art can be completely reviewed will hopefully come in the not too distant future.

In preparing this review article, the writer collected and consulted a large amount of material. However, only a portion of the relevant literature could be cited owing to the lack of space. In addition to this, the subject matter had to be restricted, so that numerous related topics, such as the mass-transport velocity distribution, suspended-sediment concentration distribution, and wind-blown sand transport, were not covered. For these subjects, the reader should consult appropriate reference materials such as the *Proceedings of the International Conferences on Coastal Engineering*, *Coastal Engineering in Japan*, *Shore Protection Manual of the US Army Corps of Engineers*, or appropriate textbooks (Silvester 1974, Komar 1976, Horikawa 1978a).

### ACKNOWLEDGMENTS

The writer would like to thank Dr. N. C. Kraus and Mr. T. Shibayama for their helpful comments during the preparation of the manuscript.

### *Literature Cited*

Abou-Seida, M. M. 1965. Bed load function due to wave action. *Univ. Calif. Hydraul. Engrg. Lab., Wave Res. Proj. HEL 2–11*. 78 pp.

Arthur, R. S. 1962. A note on the dynamics of rip currents. *J. Geophys. Res.* 67:277–79.

Bagnold, R. A. 1946. Motion of waves in shallow waters, interaction between waves and sand bottom. *Proc. R. Soc.*

*London, Ser. A* 187:1–12.

Bagnold, R. A. 1963. Mechanics of marine sedimentation. In *The Sea*, vol. 3, ed. M. N. Hill, pp. 507–28. New York: Wiley. 963 pp.

Bakker, W. T., Klein Breteler, E. H. J., Roos, A. 1970. The dynamics of a coast with a groyne system. *Proc. 12th Coastal Engrg. Conf., Washington, DC*, pp. 1001–20

Bowen, A. J. 1969a. The generation of longshore currents on a plane beach. *J. Mar. Res.* 27:206–15

Bowen, A. J. 1969b. Rip currents. I. Theoretical investigations. *J. Geophys. Res.* 74:5467–78

Bowen, A. J., Inman, D. L., Simmons, V. P. 1968. Wave "set-down" and "set-up," *J. Geophys. Res.* 73:2569–77

Caldwell, J. M. 1956. Wave action and sand movement near Anaheim Bay. *US Army Corps Engrs. Beach Erosion Board. Tech. Memo 68.* 21 pp.

Carstens, M. R., Neilson, F. M., Altinbilek, H. O. 1969. Bed forms generated in the laboratory under an oscillatory flow: analytical and experimental study. *US Army Corps Engrs., Coastal Eng. Res. Cent. Tech. Memo. 28.* 105 pp.

Dean, R. C. 1973. Heuristic models of sand transport in the surf zone. *Proc. Conf. Engrg. Dynamics in the Surf Zone, Sydney*, pp. 208–14

Dingler, J. R., Inman, D. L. 1976. Wave-formed ripples in nearshore sands. *Proc. 15th Coastal Engrg. Conf., Honolulu*, pp. 2109–26

Galvin, C., Vitale, P. 1976. Longshore transport prediction-SPM 1974 equation. *Proc. 15th Coastal Engrg. Conf., Honolulu*, pp. 1133–48

Hashimoto, H. 1974. Simulation model for estimating the effect of an offshore breakwater on the adjacent coast. *Proc. 21st Engrg. Conf. in Japan*, pp. 181–85 (in Japanese)

Hom-ma, M., Horikawa, K. 1962. Suspended sediment due to wave action. *Proc. 8th Coastal Engrg. Conf., Mexico City*, pp. 168–93

Horikawa, K. 1970. Advanced treatise on coastal sedimentation. *Summer Seminar Hydraul. Engrg., Jpn. Soc. Civil Engrs.* pp. 501–34 (in Japanese)

Horikawa, K. 1978a. *Coastal Engineering—An Introduction To Ocean Engineering.* Univ. Tokyo Press and A. Haltsted Press. 402 pp.

Horikawa, K. 1978b. Present state of coastal sediment studies. *Mitteilungen, Leichtweiss-Inst. f. Wasserbau, Tech. Univ. Braunschweig* 56:77–197

Horikawa, K., Harikai, S., Kraus, N. C. 1979. A physical and numerical modeling of waves, currents and sediment transport near a breakwater. *Ann. Rep. Engrg. Res. Inst., Fac. Engrg., Univ. Tokyo* 38:41–48

Horikawa, K., Sasaki, T. 1970. Tables for the critical water depth for the inception of sand grains due to wave action. *Proc. Jpn. Soc. Civil Engrs. 55* (5):58–63 (in Japanese)

Horikawa, K., Sunamura, T., Shibayama, T. 1977. Laboratory study on the two-dimensional beach profile change—devices for measuring sand transport rate in the offshore zone. *Proc. 24th Coastal Engrg. Conf. in Japan*, pp. 170–74 (in Japanese)

Horikawa, K., Watanabe, A. 1967. A study on sand movement due to wave action. *Coastal Engrg. Jpn.* 10:39–57

Hotta, S., Mizuguchi, M. 1978. Field observation of waves inside the surf zone. *Proc. 25th Coastal Engrg. Conf. in Japan*, pp. 151–54 (in Japanese)

Inman, D. L. 1957. Wave generated ripples in nearshore sands. *US Army Corps Engrs., Beach Erosion Board. Tech. Memo. 100.* 42 pp.

Inman, D. L., Bagnold, R. A. 1963. Littoral processes. In *The Sea* III, ed. M. N. Hill, pp. 529–53. New York: Wiley. 963 pp.

Inman, D. L., Bowen, A. J. 1963. Flume experiments on sand transport by waves and currents. *Proc. 8th Coastal Engrg. Conf., Mexico City*, pp. 137–50

Iwagaki, Y. 1966. A treatise on beach erosion. *Summer Seminar Hydraul. Engrg. Jpn. Soc. Civil Engrs.* 17:1–17 (in Japanese)

Jonsson, I. G. 1966. Wave boundary layers and friction factors. *Proc. 10th Coastal Engrg. Conf., Tokyo*, pp. 127–48

Jonsson, I. G. 1976. Discussion of "Friction factor under oscillatory waves" by Kamphuis, J. W.. *J. Waterways, Harbors Coastal Engrg. Div.* 102:108–9

Kajiura, K. 1964. On the bottom friction in an oscillatory current. *Bull. Earthquake Res. Inst. Univ. Tokyo* 42:147–74

Kajiura, K. 1968. A model of the bottom boundary layer in water waves. *Bull. Earthquake Res. Inst. Univ. Tokyo* 46:75–123

Kalkanis, G. 1964. Transportation of bed material due to wave action. *US Army Corps Engrs., Coastal Engrg. Res. Center. Tech. Memo. 3.* 68 pp.

Kamphuis, J. W. 1975. Friction factor under oscillatory waves. *J. Waterways, Harbors Coastal Engrg. Div.* 101:135–44
Kana, T. W. 1978. Surf zone measurements of suspended sediment. *Proc. 16th Coastal Engrg. Conf., Hamburg*, pp. 1725–43
Komar, P. D. 1976. *Beach Processes and Sedimentation*. Englewood Cliffs, NJ: Prentice-Hall. 429 pp.
Komar, P. D., Inman, D. L. 1970. Longshore sand transport on beaches. *J. Geophys. Res.* 75:5914–27
Kraus, N. C., Sasaki, T. 1979. Influence of wave angle and lateral mixing on the longshore current. *Mar. Sci. Commun.* 5(2):91–126
Liu, P. L., Dalrymple, R. A. 1978. Bottom friction stress and longshore currents due to waves with large angles of incidence. *J. Mar. Res.* 36:357–75
Longuet-Higgins, M. S. 1970. Longshore currents generated by obliquely incident sea waves. I, II. *J. Geophys. Res.* 75:6778–89, 6790–801
Longuet-Higgins, M. S., Stewart, R. W. 1960. Changes in the form of short gravity waves on long waves and tidal currents. *J. Fluid Mech.* 8:565–83
Longuet-Higgins, M. S., Stewart, R. W. 1962. Radiation stress and mass transport in gravity waves, with application to "surf beats." *J. Fluid Mech.* 13:481–504
Lundgren, H. 1963. Wave thrust and wave energy level. *Proc. 10th Congr. Int. Assoc. Hydraul. Res., London*, vol. 1, pp. 147–51
Madsen, O. S., Grant, W. D. 1976. Quantitative description of sediment transport by waves. *Proc. 15th Coastal Engrg. Conf., Honolulu*, pp. 1093–112
Manohar, M. 1955. Mechanics of bottom sediment movement due to wave action. *US Army Corps Engrs., Beach Erosion Board. Tech. Memo 75.* 121 pp.
Mizuguchi, M., Tsujioka, K., Horikawa, K. 1978. A consideration on wave height change inside the surf zone. *Proc. 25th Coastal Engrg. Conf. in Japan*, pp. 155–59 (in Japanese)
Mogridge, G. R., Kamphuis, J. W. 1973. Experiments on bed form generation by wave action. *Proc. 13th Coastal Engrg. Conf., Vancouver*, pp. 1123–42
Nielsen, P., Svendsen, I. A., Staub, C. 1978. Onshore-offshore sediment movement on a beach. *Proc. 16th Coastal Engrg. Conf., Hamburg*, pp. 1475–92
Noda, E. K. 1974. Wave-induced nearshore circulation. *J. Geophys. Res.* 79:4097–106
Pelnard-Considère, R. 1956. Essai de théorie de l'évolution des formes de rivages en plages de sable et de galets. *IVmes J. de l'Hydraulique, Paris*, vol. 1, pp. 289–298
Phillips, O. M. 1977. *The Dynamics of the Upper Ocean*. Cambridge Univ Press. 336 pp. (2nd ed.)
Price, W. A., Tomlinson, D. W., Willis, D. H. 1972. Predicting the changes in plan shape of beaches. *Proc. 13th Coastal Engrg. Conf., Vancouver*, pp. 1321–29
Sasaki, T. 1975. Simulation on shoreline and nearshore current. *Proc. Specialty Conf. on Civil Engrg. in the Ocean, ASCE*, pp. 176–96
Sasaki, T., Sakuramoto, H. 1978. Field verification of a shoreline simulation model. *Int. Seminar on Water Resources, AIT, Bangkok*, pp. 501–18
Sawamoto, M., Yamaguchi, S. 1979. Theoretical modeling on wave entrainment of sand particles from rippled bed. *Trans. Jpn. Soc. Civil Engrs.* 288:107–13
Sawaragi, T., Deguchi, I. 1978. Distribution of sand transport rate across a surf zone. *Proc. 16th Coastal Engrg. Conf., Hamburg*, pp. 1596–613
Seymour, R. J., Duane, D. B. 1978. The nearshore sediment transport study, *Proc. 16th Coastal Engrg. Conf., Hamburg*, pp. 1555–62
Shibayama, T., Horikawa, K. 1980. Laboratory study on sediment transport mechanism due to wave action. *Proc. Jpn. Soc. Civil Engrs.* 296:131–41
Silvester, R. 1974. *Coastal Engineering*, Vols. 1, 2. Amsterdam: Elsevier. 457 pp., 338 pp.
Sleath, J. F. A. 1978. Measurements of bed load in oscillatory flow. *J. Waterways, Harbors, Coastal Engrg. Div.* 104:291–307
Sunamura, T., Bando, K., Horikawa, K. 1978. Laboratory study on the sediment transport mechanism and rate by waves over asymmetrical ripples. *Proc. 25th Coastal Engrg. Conf. in Japan*, pp. 250–54 (in Japanese)
Tunstall, E. B., Inman, D. L. 1975. Vortex generation by oscillatory flow over rippled surfaces. *J. Geophys. Res.* 80:3475–84
Watanabe, A., Riho, Y., Horikawa, K. 1979. Two-dimensional beach profile change and onshore-offshore sediment transport rate distribution. *Proc. 26th Coastal Engrg. Conf. in Japan*, pp. 172–76 (in Japanese)

*Ann. Rev. Fluid Mech. 1981. 13:33–55*


# SOME FLUID-DYNAMICAL PROBLEMS IN GALAXIES

*8168

*C. C. Lin*
Department of Mathematics, Massachusetts Institute of Technology, Cambridge, Massachusetts 02139

*William W. Roberts, Jr.*[1]
Institute for Computer Applications in Science and Engineering, NASA Langley Research Center, Hampton, Virginia 23665

## I. INTRODUCTION

There are many interesting fluid-dynamical problems in astrophysics. The universe, viewed as a whole in studies of its origin and expansion, is usually modeled as a fluid. In this article, we deal with certain aspects of astrophysical problems, specifically the fluid dynamics associated with individual galaxies.

Galaxies were first classified into three types—ellipticals, normal spirals, and barred spirals—by Hubble as early as 1926. Figure 1 shows Hubble's "tuning fork" classification scheme for galaxies. For the spirals, one of the central problems is to understand and explain the spiral structure (and the bar structure in the barred spirals). In Figure 2 we show a photograph of an Sc galaxy, NGC 5364. This normal spiral is one striking example where the spiral arms are not only long and regular but also very narrow. Figure 3 shows the barred spiral NGC 1300. This galaxy is cited by Sandage (1961) as the prototype of the pure SBb(s) system. The bar is distinct and smooth in texture. Two straight and narrow dust lanes emerging on opposite sides of the nucleus can be traced to the ends of the bar, where they turn sharply and follow the inside of the spiral arms. These two galaxies, despite their apparent intricacies and complexities on the small scale (as in all galaxies), serve as good examples of the degree of order on the large

[1]On sabbatical leave from the Department of Applied Mathematics and Computer Science, University of Virginia, Charlottesville, Virginia 22901.



0066-4189/81/0115-0033$01.00

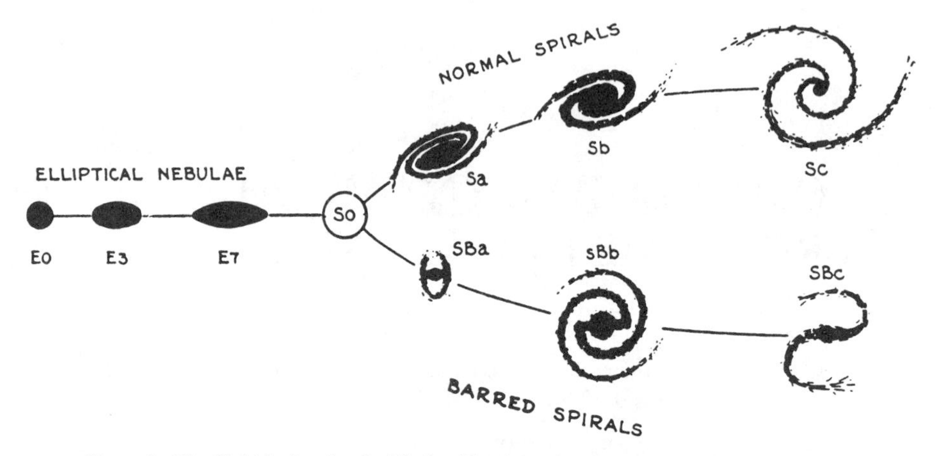


*Figure 1* The Hubble "tuning fork" classification scheme for galaxies (Hubble 1926).

*Figure 2* Photograph of the Sc galaxy NGC 5364, one of the most regular normal spiral galaxies in the sky. Its spiral arms are long and regular and very narrow. Faint dust lanes can be traced along the inside edges of the two spiral arms (Hale Observatories).

*Figure 3* Photograph of the barred spiral galaxy NGC 1300 (Hale Observatories). The dark, narrow dust lanes are thought to be tracers of large-scale galactic shock waves in the gaseous component. The nature of such shocks and their associated dust lanes reveals a great deal about the structure and underlying dynamics of a galaxy.

scale in such systems exhibited through luminous spiral arms, the bar structure, and dark, narrow dust lanes.

In the early 1960s, a theory was in its initial stages of development by which the spiral structure (and perhaps even the bar structure in barred spirals) could be viewed as a density-wave manifestation (Lin & Shu 1964, 1966; also see Lin 1971). At that time there was already the suspicion that the dark, narrow dust lanes along the spiral arms and along the bar structure in many galaxies may be related to galactic shock waves formed within the gaseous component in these systems (Prendergast 1962). In this theory, now known as the density-wave theory, the luminosity of a spiral arm is believed to originate primarily from the very young, newly formed stars whose ages are only about one one-thousandth of the age of the galaxy and whose mass makes up only a very small percentage of the bulk of galactic mass; and the spiral arm itself is believed to be a spiral wave that is capable of triggering from the gas, via a galactic shock, the formation of the young stars selectively along the wave crest (Fujimoto 1968, Roberts 1969, Roberts & Yuan 1970, Shu et al 1972, Shu, Milione & Roberts 1973).

Galaxies, although complex, can be viewed through two primary components: gas and stars. For those intriguing galaxies classified as normal spirals and barred spirals, the gaseous and stellar components are found to be concentrated in a thin disk; often a prominent nuclear bulge is present in the inner parts. Figure 4 illustrates such a disk and nuclear bulge as evident in the galaxy NGC 4565, which is oriented edge-on to the line of sight of earth-based telescopes. The spheroidal nuclear bulge is composed of stars with large velocity dispersion, whereas the disk contains stars with relatively small velocity dispersion, both perpendicular to the disk and in the plane of the disk. The gas lies essentially in the disk. Most of the gas has such a low temperature that

*Figure 4* **Photograph of the galaxy NGC 4565, seen edge-on. The galactic matter is concentrated in a thin disk and nuclear bulge (Hale Observatories).**

its molecular velocity is on the order of only 1 km $s^{-1}$ or less. Even its turbulent motion has a velocity only on the order of 10 km $s^{-1}$, which is still much smaller than the typical velocity dispersion of disk stars.

The spiral-like or bar-like gravitational field in the disk, due to the underlying distribution of matter in a normal spiral or barred spiral galaxy, will influence the gaseous component much more strongly than the stellar component, perhaps more by an order of magnitude since the gaseous component has an equivalent pressure (due to turbulent motion) that is about 10 times smaller than that in the disk stars. Thus, even when the gravitational field is so small that the behavior of the stellar component can be treated on the basis of a linear theory, the behavior of the gas must be treated on the basis of a nonlinear theory, perhaps with shock waves included. The dynamical problem governing the behavior of galactic material is therefore essentially very complicated. Fortunately, the gaseous component usually has such a small mass that we may first neglect its effect in an approximate treatment of the stellar component. In a second step, we may consider the nonlinear behavior of the gas under the influence of the stellar gravitational field.

Section II of this article deals with the gaseous disk of matter, the interstellar medium. We consider the large-scale motion of the interstellar medium in the presence of the collective gravitational field of the more massive stellar component in normal and barred spiral galaxies. In Sections III and IV we turn to the stellar component, which we also view from a fluid-dynamical perspective. Section III considers the theory of stellar density-wave modes in spiral galaxies. Section IV concentrates on the asymptotic theory, dynamical mechanisms, and modal maintainence. Section V provides other aspects of these areas of fluid dynamics on galactic scales and our concluding remarks.

## II. MOTION OF THE INTERSTELLAR MEDIUM IN NORMAL AND BARRED SPIRAL GALAXIES

We now know that many phenomena occurring in galaxies can be viewed through the dynamics of the gaseous component in response to the background gravitational field of the more massive stellar component. The dynamics of the gaseous disk can be considered through the hydrodynamic equations (written with respect to a rotating frame with angular velocity $\mathbf{\Omega}_\mathrm{p}$):

$$\frac{\partial \rho}{\partial t} + \nabla \cdot (\rho \mathbf{u}) = 0, \tag{1}$$

$$\frac{\partial \mathbf{u}}{\partial t} + \nabla\left(\tfrac{1}{2}u^2\right) + (\nabla \times \mathbf{u}) \times \mathbf{u} = -\nabla\left(\Psi - \frac{1}{2}\Omega_\mathrm{p}^2 r^2\right) - 2\mathbf{\Omega}_\mathrm{p} \times \mathbf{u} - \rho^{-1} \nabla P. \tag{2}$$

$P$, $\rho$, and $\mathbf{u}$ are the gaseous pressure, density, and velocity. $\Psi(r, \theta, z, t)$, the gravitational potential underlying the galaxy, is composed of two parts: $\Psi_0(r, z)$, an equilibrium potential field corresponding to the axisymmetric mass distribution; and $\Psi'(r, \theta, z, t)$, a perturbing potential field that is spiral-like in the outer parts for a normal spiral and bar-like in the inner parts for a barred spiral. $\Omega_p$ is the constant angular pattern speed of the perturbing potential, and $(r, \theta, z)$ are the cylindrical coordinates for the galactic disk. The results reviewed in this article are for an isothermal gas with a mean equivalent dispersion speed $c = (\partial P/\partial \rho)^{1/2}$, which may be viewed as partly kinetic, partly due to turbulence, and partly due to cosmic-ray particles.

We consider first the "steady-state" flow of gas in normal spirals with tightly wound spiral arms. In these systems the velocity variations from an equilibrium state of purely circular rotation are only moderate, and it is convenient to view the physical quantities of pressure, density, and velocity in terms of their equilibrium and perturbed parts, the latter calculated as the response to the perturbing potential. For a tightly wound normal spiral, an asymptotic approximation can be made that neglects the small variation in the perturbed quantities in the direction parallel to spiral equipotential contours (e.g. along an induced shock) with respect to their variation in the direction normal to such contours (e.g. normal to such a shock). This asymptotic approximation simplifies considerably the (nonlinear) calculation of the "steady-state" flow of gas in such systems.

### *Galactic Shocks and Implications for Star Formation*

The gas is found to respond quite strongly to a perturbing potential of even mild amplitude, with the formation of large-scale galactic shock waves in the gas. Figure 5 illustrates the calculated distribution of gas density along a typical streamline. The nonlinear density response of the gas has a rather narrow peak induced by the shock wave that forms in the potential well of the background stellar density-wave arm. Galactic shocks are thought to provide a possible triggering mechanism for the gravitational collapse of gas clouds, leading to star formation along spiral arms. Gas flows into this shock and compression region from left to right in Figure 5. Before reaching the shock, some of the large clouds and cloud complexes may be on the verge of gravitational collapse. A sudden compression of the clouds in the shock could conceivably trigger the gravitational collapse of some of the largest gas clouds. As the gas leaves the shock region, it is rather quickly decompressed and star formation ceases.

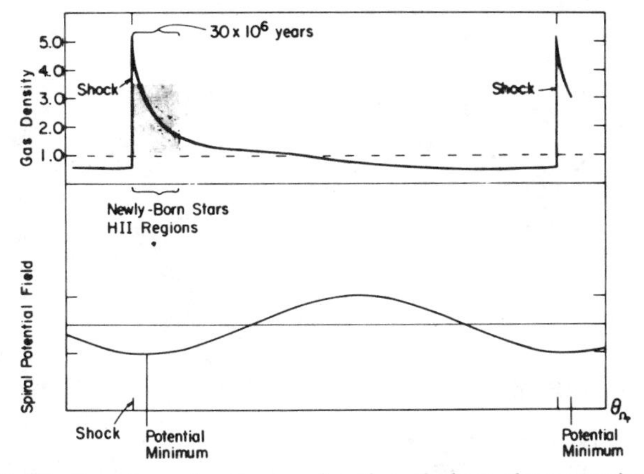


*Figure 5* Distribution of gas density as a function of phase along a typical streamline, relative to an underlying (stellar) gravitational perturbing potential of 10% amplitude (Roberts 1969).

Certainly the real picture of star formation in galaxies is very complex, and there may be other competing mechanisms that influence the formation of stars in addition to the large-scale shock mechanism. Because the interstellar medium is so irregular and perturbed, strong large-scale shocks most certainly trigger irregularly distributed star formation and supernova activity, perhaps in quite a stochastic manner (cf Gerola & Seiden 1978). Thus, even for galaxies with the strongest large-scale shocks, a stochastic star formation process may also enter and act to decrease the degree of order that shocks might otherwise provide on the large scale. Superposition of several wave patterns would also tend to give rise to irregular features.

## *Comparison of Theory and Observations for a Representative Normal Spiral Galaxy*

NGC 3031 (M 81) is a representative galaxy for which blue-light photographs show well-developed, narrow spiral arms in the luminous star distribution. These are viewed to be a consequence of strong shocks. Rots & Shane (1975) show that the observed surface-density distribution of gaseous neutral hydrogen (H I) in M 81 also exhibits similarly well-developed spiral structure, extending to large distances from the center of the galaxy, as expected if strong shocks are present.

Visser (1978) constructs a density-wave model for M 81 and calculates its gas dynamics following Roberts (1969) and Shu et al (1973) in the use of the asymptotic equations for the "steady-state" gas flow.

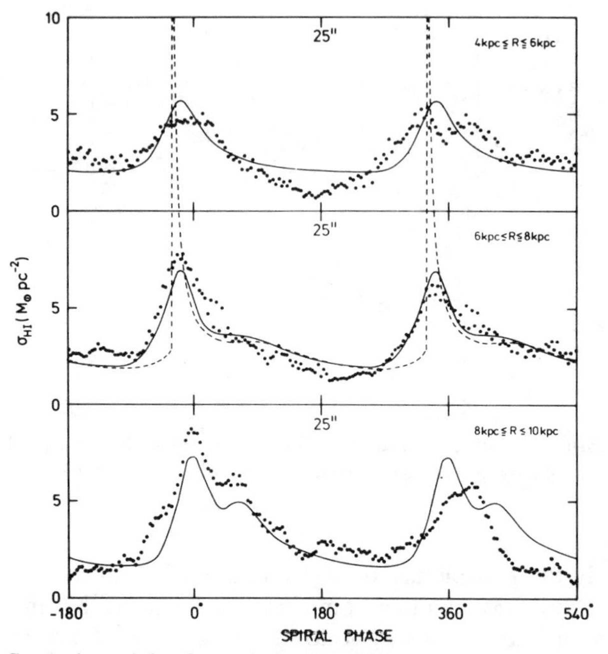


*Figure 6* Comparison of the observed neutral hydrogen (H I) surface-density distribution in the normal spiral galaxy M 81 (dots, from Rots 1975) and the theoretically calculated gas-density distribution in Visser's density-wave model (solid lines). From Visser (1978).

Figure 6 shows a comparison between the observed H I surface-density distribution (dots), plotted versus spiral phase in three circular bands around M 81, and the density distribution calculated for the gas in the density-wave model (solid lines). Effects of smoothing due to the telescope beam were included for the display of the theoretically calculated gas-density distribution (solid lines) just as if it were derived through observations with the telescope. It is important to note that the observed arm peaks (dots) are substantially narrower than the observed interarm troughs (dots), altogether indicative of nonlinear wave phenomena. Some observed peaks even show the skewed character of steep rise followed by more gradual fall off, characteristic of a strong shock. Because of the good agreement between the theoretically calculated gas-density distribution and the observed data, the physical understanding of these observed characteristics provided by the density-wave theory seems to be strongly supported.

## *Barred Spirals: The Bar Structure and Large Noncircular Motions*

Velocity variations from purely circular rotation in barred and open-armed normal spirals are observed to be quite large, in some cases of the same order of magnitude as the mean rotational velocity. In addition, for these systems an asymptotic approximation like the one discussed for tightly wound spirals is expected to be a good approximation only for the regime of flow near shocks. Accordingly, Roberts, Huntley & van Albada (1979) consider the "steady-state" flow of gas in such systems by an analysis that enables the "two-dimensional" flow to be broken into two physical regimes: regime I near and within the bar (and spiral arms) where the flow is subsonic and moderately supersonic and the effect of pressure is important in the formation of a large-scale shock, and regime II over the remainder of the disk where the flow is highly supersonic and pressure effects are of secondary importance. The flow in regime I is determined through the asymptotic approximation that neglects the small variation of the physical quantities of pressure, density, and velocity along a shock with respect to their variation normal to the shock, whereas the flow in regime II is determined through a different asymptotic approximation that neglects secondary terms proportional to the square of the dispersion speed, such as the transverse gradient of pressure along a streamline.

The composite picture calculated by this analysis for the "steady-state" flow of gas is illustrated in Figure 7 for one case with a weak bar-like, spiral-like perturbing potential. The arrowed streamlines trace the "steady" flow of gas about the disk. The response of the gas is found to be strongly nonlinear with a shock wave (heavy line) forming near the minimum (dashed line) of the perturbing potential. It is in the region near the maximum radius of the streamline that the "piling up" of gas is

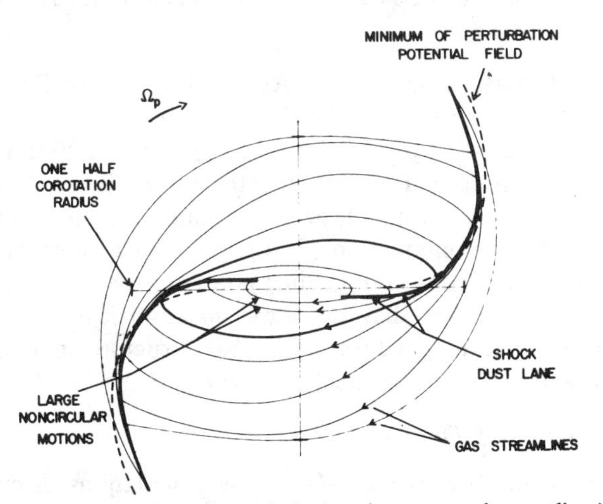


*Figure 7* Oval, "steady-state" gas flow (clockwise, along arrowed streamlines) driven by a 9% bar-like, spiral-like perturbing potential. Noncircular motions are present with radial velocities as high as 100 km $s^{-1}$ along the inner streamlines (Roberts, Huntley & van Albada 1979).

*Figure 8* Photographic simulation of the "steady-state" gas-density distribution corresponding to the flow along the streamlines in Figure 7. A strong shock together with a sharp, narrow gas-density ridge (dark lane) lies along the bar structure, offset toward its leading edge, and along the (inside edge of the) spiral arms (Roberts, Huntley & van Albada 1979).

so pronounced that a shock occurs. As gas flows out of this region, its component of motion normal to the shock is subsonic.

Figure 8 shows a photographic simulation of the resulting gas-density distribution for this case. The highest intensity levels are darkened (by the computer) in order to emphasize the location of the shock and the narrow ridge of gas compression (dark lane) and to note how narrow this region really is. The resulting "dust lanes" in this simulation help to underline the conclusion that the dark, narrow dust lanes observed offset toward the leading edges of the bar structure in many barred spirals may be tracers of such shocks in these systems.

## *Bar-Driven Spiral Density Waves*

In addition to the "steady-state" gas-flow calculations discussed thus far, it is possible to consider the gas flow from the standpoint of an initial-value problem and carry out time-dependent, two-dimensional,

*Figure 9* Photographic simulation of the gas-density distribution (a snapshot at one point in time) in response to a mild stellar-driving bar, computed in a time-dependent, two-dimensional, numerical hydrodynamical calculation (Huntley, Sanders & Roberts 1978).

numerical hydrodynamical calculations using Equations (1) and (2) but with no asymptotic approximation. This approach is followed by a number of authors (see Sanders & Huntley 1976, Sorensen, Matsuda & Fujimoto 1976, Huntley 1977, 1980, Huntley, Sanders & Roberts 1978, Liebovitch 1978, and Sanders & Tubbs 1980) for the time-evolutionary response of the gas to bar-like perturbing potentials. Even without a spiral-like contribution to the perturbing potential, the time-evolutionary gas response still shows "open" trailing spiral density waves rotating with the angular speed of the bar-like perturbation. Figure 9, from Huntley, Sanders & Roberts (1978), for a model with a mild driving bar emphasizes the prominence of the gaseous response in the form of "open" trailing spiral gaseous density-wave arms and the central gas bar. These gas-wave arms appear as if they are driven from the ends of the stellar bar. The overall gas circulation here resembles the "steady-state" flow illustrated in Figure 7; in both cases the gas undergoes a perennial drift inward during each circulation of the disk.

# III. THEORY OF SPIRAL MODES

We now turn to the stellar component in galaxies. Although the formulation to be given below applies to both normal and barred spirals, existing studies have focused mostly on the former. We shall describe the theory of unstable spiral modes from a fluid-dynamical perspective. To be sure, the mass of a galaxy is largely in the form of stars, and therefore a stellar dynamical theory ought to be developed. This has been done (see Bertin 1980, and the references cited there). However, it is much more complicated; and the fluid-dynamical theory does bring out the essential mechanisms. In particular, it shows how the spiral modes are generated and maintained, and explains why *trailing* patterns are preferred over *leading* ones.

In the fluid theory, the state of the stellar component in a galaxy is characterized by 1. a density distribution $\rho_\star(r,\theta,z,t)$, 2. a pressure distribution $P_\star(r,\theta,z,t)$ or equivalently an enthalpy distribution $h(r,\theta,z,t)$, 3. a velocity field $v(r,\theta,z,t)$, and 4. a gravitational potential $\Psi(r,\theta,z,t)$. These physical variables are governed by (*a*) the equation of continuity, (*b*) the equations of motion, (*c*) the thermodynamic equation of state (which we take to be barotropic), and (*d*) the Poisson equation for gravitational potential. For a thin stellar disk, we speak of a surface density $\sigma(r,\theta,t)$; the corresponding gravitational potential $\Psi(r,\theta,z,t)$ then satisfies the Poisson equation

$$\nabla^2\Psi=4\pi G\rho_\star=4\pi G\sigma\delta_0(z), \tag{3}$$

where $G$ is the universal constant of gravitation and $\delta_0(z)$ is the delta function of Dirac. We shall assume the fluid, which simulates the stellar component, to have a barotropic relationship between its pressure $P_\star$ and density $\rho_\star$, with a mean equivalent dispersion speed $a$ given by $a^2=\rho_\star(dh/d\rho_\star)=\sigma(dh/d\sigma)$. In the solar vicinity, $a$ is about 30 km s$^{-1}$.

The state of equilibrium has rotational symmetry. The variables 1., 2., and 4. in this state will carry a subscript zero. Let us denote the angular velocity by $\Omega(r)$. A more important parameter is the epicyclic frequency $\kappa$ defined by

$$\kappa^2=(2\Omega)^2\left\{1-\frac{s}{2}\right\},\ s=-\frac{d\ln\Omega}{d\ln r}.$$

In the case of uniform rotation, $s=0$ and $\kappa$ reduces to the usual Coriolis parameter. Other important parameters are the density $\sigma_0$, the acoustic velocity $a_0$, the length scales $\delta=\pi G\sigma_0/\kappa^2$ and $k_0^{-1}=a_0^2/(\pi G\sigma_0)$, and the dimensionless parameter $Q=a_0/\kappa\delta=(k_0\delta)^{-1/2}$. The parameters $\Omega(r)$, $\sigma_0(r)$, and $Q(r)$ are given in Figure 10 for a typical galaxy.

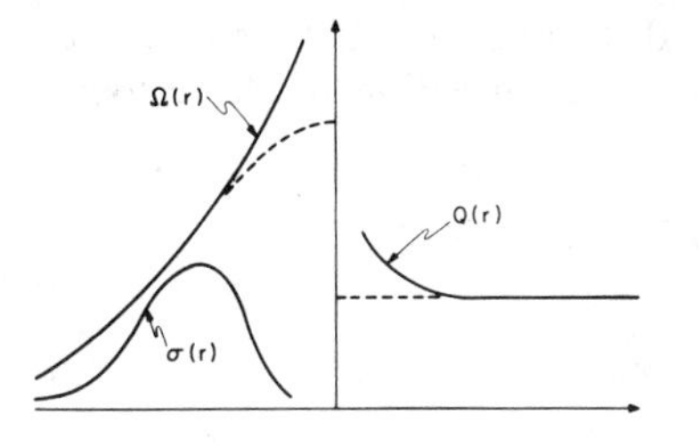


*Figure 10* Typical distributions of rotational velocity, surface mass density, and velocity dispersion (as measured by $Q$) in the class of galaxies considered (Lin & Lau 1979).

We now consider small disturbances from the state of equilibrium; the amplitudes of the variables 1., 2., and 4. in the perturbed state, as functions of $r$, will carry the subscript one. If we neglect nonlinear terms of second order from the equations of motion and continuity, the resultant linear equations are cyclic in the variables $t$ and $\theta$, and hence we may seek solutions in the form of normal modes, with a typical quantity $q$ in the form

$$q(r,\theta,t)=q_0(r)+q_1(r)e^{i(\omega t-m\theta)}, \tag{4}$$

where $z=0$ is introduced in $\Psi_1(r,\theta,z,t)$ for a thin disk, $m$ is an integer corresponding to the number of spiral arms, and the frequency parameter $\omega$ may be conveniently written in the form $\omega=m\Omega_p-\mathrm{i}\gamma$, so that the pattern speed (angular velocity) is $\Omega_p$ and the growth rate is $\gamma$. The spiral nature of the disturbance becomes especially apparent if $q_1(r)$ is written in the form $\hat{q}_1(r)\exp\{\mathrm{i}\int k dr\}$ and if $k$ is of the form of $\Lambda f(r)$, where $\Lambda$ is a large parameter, and $f(r)$, as well as $\hat{q}_1(r)$, is a slowly varying function of the radial distance $r$, i.e. $d\ln f(r)/d\ln r=O(1)$. As one can easily verify, the combination

$$q_1(r)\exp\{\mathrm{i}(\omega t-m\theta)\}=\hat{q}_1(r)\exp\{\gamma t+\mathrm{i}\phi(r,\theta,t)\},$$

where

$$\phi(r,\theta,t)=m(\Omega_p t-\theta)+\int k(r)\,dr,$$

represents spiral waves whose crests are located approximately along the curves $\phi=0(\mathrm{mod}\,2\pi)$. Thus, the spiral form of an equipotential curve (for example), at any instant $t$, is given by

$$m(\theta-\theta_0)=\int_{r_0}^{r}k(r)\,dr,$$

and $|k|$ is the wave number. The waves are short, since $|k|r\gg 1$, when $\Lambda$ is large. The waves are called *trailing* if $k<0$ and *leading* if $k>0$. The spiral structure is inclined at a *pitch angle* $i$ with respect to the circular direction, where $i$ is given by

$$|\tan i|=|m/kr|.$$

The set of linear equations for small disturbances can be reduced to the following integro-differential system by elimination of the velocity variables:

$$\psi_1(r) = -2\pi G \int_0^\infty K(r, r')\sigma_1(r')dr', \tag{5}$$

$$\mathcal{L}(h_1 + \psi_1) = -Ch_1, \tag{6}$$

where $h_1$ is related to $\sigma_1$ by

$$h_1 = a_0^2 \sigma_1 / \sigma_0 \tag{7}$$

and $\mathcal{L}$ is a second-order differential operator

$$\mathcal{L} = \frac{d^2}{dr^2} + A\frac{d}{dr} + B. \tag{8}$$

The coefficient functions $A$, $B$, and $C$ are defined as

$$A = -\frac{1}{r}\frac{d\ln \mathcal{Q}}{d\ln r}, \qquad \mathcal{Q} = \frac{\kappa^2(1-\nu^2)}{\sigma_0 r},$$

$$B = -\frac{m^2}{r^2} - \frac{4m\Omega(r\nu')}{\kappa r^2(1-\nu^2)} + \frac{2\Omega m}{r^2 \kappa \nu}\frac{d\ln(\kappa^2/\sigma_0\Omega)}{d\ln r},$$

$$C = -\frac{\kappa^2(1-\nu^2)}{a_0^2}.$$

The kernel function $K$ is the usual one in potential theory; it has a logarithmic singularity at $r = r'$. In the coefficient functions $A$, $B$, and $C$, the frequency $\omega$ appears only through the dimensionless parameter

$$\nu = \frac{\omega - m\Omega}{\kappa} \qquad (\nu' \text{ means } d\nu/dr), \tag{9}$$

where the real and imaginary parts of $\nu$ are

$$\nu_r = \frac{m(\Omega_p - \Omega)}{\kappa} \quad \text{and} \quad \nu_i = -\frac{\gamma}{\kappa}. \tag{10}$$

In the optically prominent inner part of a galaxy, it turns out that $\nu_r$ is usually negative. Thus, $-\nu_r$ is the dimensionless measure of the frequency at which the material, traveling at angular velocity $\Omega$, experiences the periodic effect of the spiral pattern of $m$ arms, traveling at angular velocity $\Omega_p$. Thus we may call $-\nu_r$ the *dimensionless frequency of encounter* of the material with the wave pattern, and $-\nu_i$ the *dimensionless rate of growth*.

The importance of the parameter $\nu$ can be seen from the fact that the coefficients of the differential equation (6) have singularities and other special features at

$$\nu = 0, +1, -1. \tag{11}$$

These are usually designated as the *corotation resonance*, the *outer Lindblad resonance*, and the *inner Lindblad resonance*. At the Lindblad resonances, the encounter frequency is equal to the epicyclic frequency $\kappa$.

Equations (5), (6), and (7) constitute the basic equations, from which $\sigma_1(r)$ can be eliminated. Thus, we have one integral equation and one differential equation of the second order for the two variables $\psi_1(r)$ and $h_1(r)$. The complete formulation of the problem of spiral modes must still depend on $(a)$ the proper formulation of the boundary conditions, and $(b)$ the proper understanding of the behavior of the solution in the neighborhood of the singularities at $\nu = 0, \pm 1$.

$(a)$ *Boundary conditions*. There should certainly be 1. a condition of *regularity* at the center. At infinity, the correct boundary condition for spiral modes turns out to be 2. a *radiation condition* of *outgoing* (trailing) spiral waves. *This latter condition is crucial* if the fluid-dynamical model is to simulate a stellar dynamical system.

$(b)$ *Singularities at resonances*. In common with many other problems of instability theory, the governing equations are expected to hold for the *unstable modes* along the real axis of the independent variable (regarded as complex). The singularities $\nu = 0, \pm 1$ are points off the real axis. In this case, the explicit proof has been given by Feldman & Lin (1973). "Stable modes" are unimportant in any instability theory; in the solution of the problem by the initial-value approach, they are always dominated by the contributions associated with the initial conditions. In the modal approach, they are dominated by the contributions from a continuous spectrum of neutral solutions with discontinuities at the singular points.

There is a deeper problem in our special case: Can we expect the fluid-dynamical theory to model the stellar dynamical system properly at these resonances? The answer is *no* if we take the equations at face value. The proper treatment of these singularities can only be understood by comparison with results obtained from the stellar dynamical theory. Fortunately, these are available through the work of Mark (1976a).

The system of Equations (5), (6), and (7) is still quite complicated to solve. Even after the solution of the boundary-value problem is obtained, we still have to face the difficulty of understanding the mechanism by which such modes are maintained. Historically, in order to reach physical understanding, one made use of the fact that the spiral waves have rather small pitch angles. This allows us to develop the

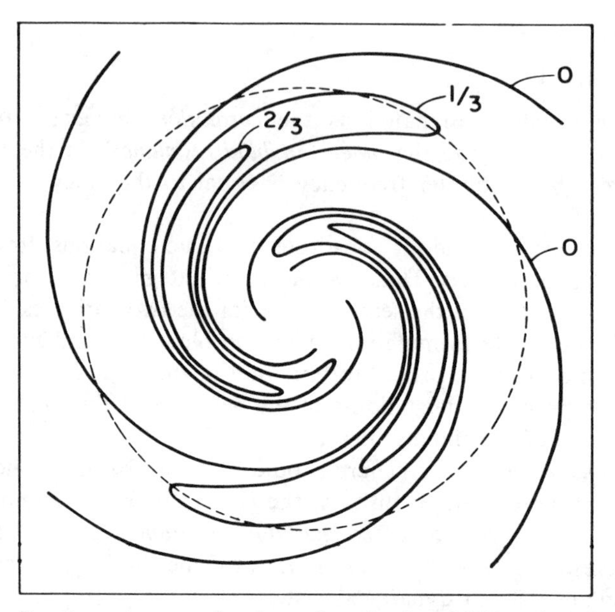


*Figure 11* Density contour map for the $n=0$ mode of model I based on the asymptotic theory. The numbers denote the fraction of maximum perturbation density on the contours. The dashed circle is the corotation circle (Lau & Bertin 1978).

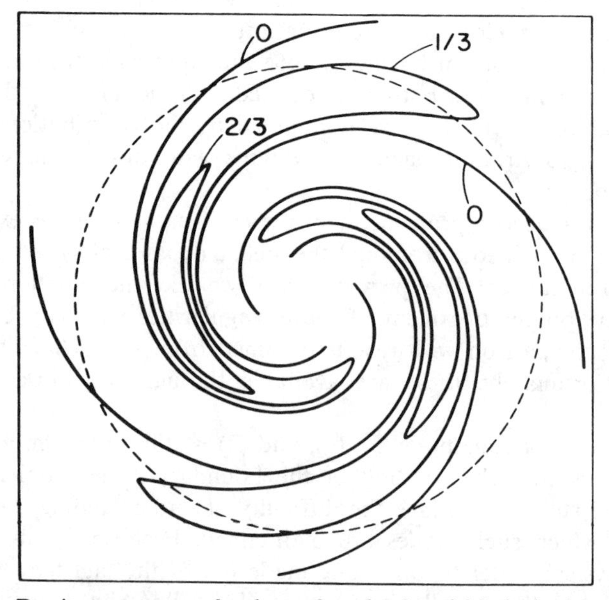


*Figure 12* Density contour map for the $n=0$ mode of model I obtained from the exact theory. Same explanation and scale as in Figure 11 (Pannatoni 1979).

asymptotic theory (to be discussed in Section IV), which more easily conveys the various physical concepts. Recently, however, it has also been possible to solve the above-formulated problem directly, by a fairly elaborate numerical procedure (Pannatoni 1979). In Figures 11 and 12, we note the good agreement obtained by Pannatoni & Lau (1979) in the comparison of the asymptotic approach and the exact formulation for a certain galactic model I.

## IV. ASYMPTOTIC THEORY OF SPIRAL DENSITY WAVES: MECHANISMS

In the previous section, we outlined the formulation of the basic equations and the boundary conditions for determining the spiral wave patterns over a galactic disk. We now describe an asymptotic method for simplifying this mathematical problem and for bringing out the essential physical mechanisms. As we shall see below, we can arrive at a simple *quantum condition* in the form of a phase integral for the determination of the eigenfrequency.

As expected, in the lowest approximation the asymptotic theory leads to a dispersion relationship. In this case, it is found to be

$$(\omega - m\Omega)^2 = \kappa^2 + (ka_0)^2 - 2\pi G\sigma_0|k|, \tag{12}$$

where the symbols are those defined in Section III. If we solve for the wave number $k$, we get

$$k = -(k_0 \pm k_3) = -k_0\left\{1 \pm \sqrt{1 - Q^2(1-\nu^2)}\right\} \tag{13a}$$

in the case of trailing waves ($k<0$), or

$$k = +(k_0 \pm k_3) = +k_0\left\{1 \pm \sqrt{1 - Q^2(1-\nu^2)}\right\} \tag{13b}$$

in the case of leading waves. There are thus *four* sets of waves. A calculation of the group velocity $c_g = -\partial\omega/\partial k$ shows that

$$c_g = -\frac{a_0^2}{\kappa}\frac{k+k_0}{\nu}. \tag{14}$$

Thus, we have the situation summarized in Table 1.

Since there are trailing (leading) waves propagating in opposite directions, we may expect to be able to construct a mode out of trailing (leading) waves alone, provided that there are suitable turning points (radii) in the system, where trailing (leading) waves beget trailing (leading) waves. This is indeed found to be the case. In order to show this, we obtain differential equations as approximations for the original

**Table 1** Group propagation for four types of density waves

| Waves for | | |
|---|---|---|
| $k<0$: | Short trailing | Long trailing |
| $k>0$: | Long leading | Short leading |
| Propagation: | Departing corotation | Approaching corotation |

integro-differential system. The final equation obtained is

$$\frac{d^2u}{dr^2}+\hat{k}_3^2u=0, \tag{15}$$

where $u$ is related to $h_1$ by

$$h_1=\left\{\frac{\kappa^2(1-\nu^2)}{\sigma_0 r}\right\}^{1/2}u\exp\left\{-i\int k_0 dr\right\}, \tag{16}$$

and

$$\hat{k}_3^2=k_3^2+\tfrac{1}{4}\mathcal{J}^2Q^2\cdot\left(\frac{\kappa}{a_0}\right)^2 \tag{17}$$

and the dimensionless parameter $\mathcal{J}$ is related to several of the quantities defined in Section III as $\mathcal{J}=2m(\delta/r)(2\Omega/\kappa)\sqrt{s}$ . In the lowest approximation, we may neglect the term $\frac{1}{4}\mathcal{J}^2Q^2$ in (17) in view of the small factor $\delta/r$ in $\mathcal{J}$. Equation (15) is then simplified to

$$\frac{d^2u}{dr^2}+k_3^2u=0. \tag{18}$$

This equation is clearly compatible with the dispersion relationship (13a) given above whenever $k_3^2\neq 0$.

A typical form of the distribution of $k_3^2$ is given in Figure 13. There are two turning points, designated as $r_{ce}$ and $r_{co}$, where $k_3^2$ vanishes. The waves, both trailing, propagate back and forth between these turning points. The waves are evanescent to the left of $r_{ce}$. To the right of $r_{co}$, the waves are propagating.

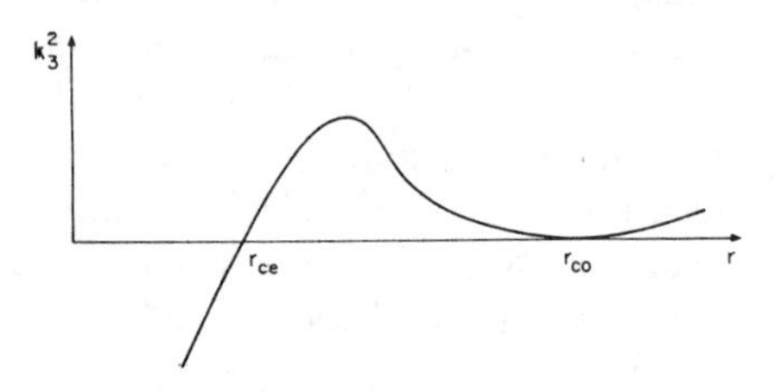


*Figure 13* Distribution of the square of the reduced wave number (Lin & Lau 1979).

To set up a discrete mode of trailing waves, we impose the boundary condition of an *outward* propagating wave. This is to simulate the *absorption* of these waves at outer Lindblad resonance [according to the mechanism studied by Mark (1971)]. This is a crucial step. It results in a *complex* quantum condition that leads to a *complex* eigenvalue that implies an exponential growth of the discrete mode. The condition is

$$\int_{r_{ce}}^{r_{co}} k_3 dr = \left(n+\tfrac{1}{2}\right)\pi + \tfrac{1}{4} i \ln 2, \qquad n = 0, 1, \ldots \tag{19}$$

whose real and imaginary parts are approximately

$$\int_{r_{ce}}^{r_{co}} k_3 dr = \left(n+\tfrac{1}{2}\right)\pi \qquad (\omega = \omega_r \text{ in } k_3) \tag{20}$$

and

$$\gamma\tau = \tfrac{1}{4}\ln 2, \tag{21}$$

where $\gamma$ is the growth rate, and

$$\tau = \int_{r_{ce}}^{r_{co}} \frac{dr}{|c_g|} \tag{22}$$

is the time of propagation of the waves between $r_{ce}$ and $r_{co}$. Thus, the growth of these waves over one round trip ($r_{ce} \rightarrow r_{co} \rightarrow r_{ce}$) is

$$e^{2\gamma\tau} = \sqrt{2}\,. \tag{23}$$

More specifically, the outgoing wave is turned back at $r_{co}$ with a reinforcement (because there is another wave leaving $r_{co}$ to go to infinity) that corresponds to a *doubling* of energy. The reinforcement results because the outgoing wave (for $r > r_{co}$) carries angular momentum (or energy) of a sign *opposite* to that of the proper part of the wave pattern inside of the corotation circle $r = r_{co}$.

A careful examination of the nature of the outgoing wave shows that this mechanism is possible only when we are dealing with short trailing waves outside of the corotation circle. We thus have an *instability of trailing-wave patterns*. There is no corresponding mechanism for leading-wave patterns. This growth mechanism is known as the WASER mechanism—wave amplification by stimulation of emission of radiation (see Mark 1976b, Lau & Bertin 1978). Through this mechanism, an unstable system tends to emit waves on its own and is expected to do so when stimulated. The WASER mechanism is even stronger when the minimum in the curve of $k_3^2$ (in Figure 13) is above the real axis. This is found to be the situation when the asymptotic analysis is carried out to a higher approximation (see Lau & Bertin 1978, who also give a physical interpretation of this extra enhancement).

## V. CONCLUDING REMARKS

Over the past fifteen years, we have come a long way towards the understanding of spiral patterns in galaxies through the introduction of the density-wave theory. Especially well understood are galactic spirals with small pitch angle. But there are still large areas of challenging theoretical investigations to be made and many observational studies to be carried out. We wish to discuss some of the theoretical points briefly.

In Section II, we presented some of the implications of the density-wave theory and their observational confirmation. We have seen the gradual emergence of a classification scheme from a dynamical point of view. We may now pose this question: Can one classify all galaxies into types according to their dynamical characteristics (primarily their mass distributions) and establish a correspondence with their empirical classifications? Specifically, what is the difference between normal spirals and barred spirals from this point of view? Speculations on this point have been given by Lin & Lau (1979) to yield a dynamical classification scheme that corresponds to de Vaucouleur's (1964) classification of galaxies into SA, SAB, and SB. Observational support for the dynamical classification of galaxies into SAa, SAb, and SAc has already been presented by Roberts, Roberts & Shu (1975).

Another point we wish to emphasize is the intrinsic interest of the study of a galaxy as a *dynamical system*. It is a system similar in nature to an electromagnetic plasma. In both galaxies and plasmas, we may adopt either a description of the physical system in the continuum model, or a description in terms of a system of discrete particles. In the latter case, we may either deal with the distribution function in phase space or treat the motion of a large number of particles by direct machine calculation. Large-scale numerical experiments have yielded much useful information and will continue to do so in the future. In contrast, in the present article, we adopt the continuum model as the primary basis, and our treatment is largely analytical.

The relative merits of the various models are discussed briefly by Lin & Lau (1979). Here, we comment on one aspect of the dynamical problem, namely, the description of the evolutionary process as an initial-value problem. In the "numerical experiments," the evolutionary process is directly followed in the course of time. In the analytical approach, we have a choice. We may follow the evolutionary process either directly or by a superposition of normal modes. This is of course a general point in all studies of stability problems, and it is well known that the two procedures can be shown to be equivalent in the end. However, the two approaches can lead to very different emphases on the various dynamical processes. Indeed, the first approach tends to give the impression that the system (and the spiral structure, in particular) will continue to evolve and will not settle down into a quasi-stationary mode, even if it actually does. It may also give the impression of great instability when no unstable modes exist, because the growth of the disturbances could be transient [see the classical problem of hydrodynamic stability discussed by Orr (1907)]. The second approach tends to give the opposite impression. One must be reminded of the fact that several (perhaps as few as three) competing spiral modes of comparable growth rates may lead to a final state that is fairly transitory and quite irregular in appearance. The existence of unstable spiral modes must not be construed, by itself, to be a guarantee for an eventual attainment of a regular stationary spiral structure, even though it is a hopeful sign.

In those cases where several modes of sufficiently high instability are present, the final pattern will be similar to that of *turbulent* flow in ordinary hydrodynamics. Such a flow is often dominated by a small number of interacting modes at large scales, with superposed motions at smaller scales. In turbulent jets, the larger-scale motions can be relatively easily detected; in turbulent boundary layers over a flat plate or a convex surface, it is difficult to discern clear patterns. In spiral galaxies, we may also expect to find a variety of situations. For example, one may imagine the existence of a slowly evolving spiral pattern on a certain general length scale, as a result of the coexistence of two competing patterns on similar scales. At any given moment, there is a grand design that may be evolving at a rather slow rate on the order of the "beat frequency" of the two patterns.

The amplitude of a fully developed turbulent motion is determined by a rather complicated process of energy balance involving nonlinear processes and energy dissipation. It is difficult to trace this process in detail, but a general outline is known. In a galaxy, similar work should perhaps be pursued.

On the other hand, it is well known that, even when discrete unstable modes do not exist in a dynamical system, a regular wavy pattern can persist for a long time and show transient growth, even though the disturbances will eventually die away. A classical example is the simple uniform shear flow studied by Orr (1907). Studies using Orr's approach have been carried out by Lau (1979) who extended Orr's work to the case of hydromagnetic waves and to the case of density waves. Thus, in those cases where no discrete unstable modes exist, a spiral pattern may still be in existence for a long time after the system is disturbed by an outside agency. The least unstable mode may be most noticeable. For example, consider a drum head hit with a baton; the mode excited eventually dies away since there is no instability.

We have attempted in this review to introduce the fluid dynamicist to some interesting and as yet unsolved problems in the dynamics of galaxies. We have not discussed some other classical aspects of gravito-fluid dynamics. Two standard references on these topics have been published by Chandrasekhar (1961, 1969). The problem of convection in the interior of stars is a very challenging one. The dynamics of the interplanetary medium, the interstellar medium, and the intergalactic medium are all very complicated. To attempt to present them all might give the unfortunate picture that the subject of astrophysical fluid dynamics is broad and intractable. We hope that the limited approach adopted in this article will encourage research workers in fluid dynamics to tackle these problems on the huge scales of astronomical objects.

ACKNOWLEDGMENTS

Portions of the work discussed in this article were supported under grants from the National Science Foundation (MCS-7513878, CCL; AST-7909935, WWR); WWR also acknowledges support by NASA while in residence at ICASE (NAS1-15810).

*Literature Cited*

Bertin, G. 1980. On the density wave theory for normal spiral galaxies. *Phys. Rep.* Submitted for publication

Chandrasekhar, S. 1961. *Hydrodynamic and Hydromagnetic Stability*. Oxford: Clarendon. 652 pp.

Chandrasekhar, S. 1969. *Ellipsoidal Figures of Equilibrium*. New Haven: Yale Univ. Press. 252 pp.

de Vaucouleurs, G. 1964. *Reference Catalogue of Bright Galaxies*. Austin: Univ. Texas Press. 268 pp.

Feldman, S. I., Lin, C. C. 1973. A forcing mechanism for spiral density waves in galaxies. *Stud. Appl. Math.* 52:1

Fujimoto, M. 1968. Gas flow through a model spiral arm. *IAU Symp. No. 29*, p. 453

Gerola, H., Seiden, P. E. 1978. Stochastic star formation and spiral structure of galaxies. *Astrophys. J.* 223:129

Hubble, E. 1926. Extra-galactic nebulae. *Astrophys. J.* 64:321

Huntley, J. M. 1977. *Gas response to perturbations in disk galaxies*. PhD thesis. Univ. Va., Charlottesville

Huntley, J. M. 1980. Self-gravitating gas flow in barred spiral galaxies. *Astrophys. J.* Submitted for publication

Huntley, J. M., Sanders, R. H., Roberts, W. W. 1978. Bar-driven spiral waves in disk galaxies. *Astrophys. J.* 221:521

Lau, Y. Y. 1979. Transient growth in a current-carrying plasma. *Phys. Rev. Lett.* 42:779

Lau, Y. Y., Bertin, G. 1978. Discrete spiral modes, spiral waves, and the local dispersion relationship. *Astrophys. J.* 226:508

Liebovitch, L. S. 1978. *Two dimensional calculation of gas flow in barred spiral galaxies*. PhD thesis. Harvard Univ., Cambridge

Lin, C. C. 1971. Theory of Spiral Structure. *Highlights Astron.* 2:88

Lin, C. C., Lau, Y. Y. 1979. Density wave theory of spiral structure of galaxies. *Stud. Appl. Math.* 60:97

Lin, C. C., Shu, F. H. 1964. On the spiral structure of disk galaxies. *Astrophys. J.* 140:646

Lin, C. C., Shu, F. H. 1966. On the spiral structure of disk galaxies, II. Outline of a theory of density waves. *Proc. Natl. Acad.*

*Sci. USA* 55:229

Mark, J. W.-K. 1971. The spiral wave of our galaxy near inner Lindblad resonance. *Proc. Natl. Acad. Sci. USA* 68:2095

Mark, J. W.-K. 1976a. On density waves in galaxies. II. The turning point problem at the corotation region. *Astrophys. J.* 203:81

Mark, J. W-K. 1976b. On density waves in galaxies. III. Wave amplification by stimulated emission. *Astrophys. J.* 205:363

Orr, W. McF. 1907. The stability or instability of the steady motions of a liquid. *Proc. R. Irish Acad. Sect. A* 27:69

Pannatoni, R. F. 1979. *A general fluid dynamical theory of unstable spiral modes in disk-shaped galaxies*. PhD thesis. MIT, Cambridge

Pannatoni, R. F., Lau, Y. Y. 1979. Unstable spiral modes in disk-shaped galaxies: a general fluid dynamical theory with some preliminary results. *Proc. Natl. Acad. Sci. USA* 76:4

Prendergast, K. H. 1962. The motion of gas in barred spiral galaxies. In *Interstellar Matter in Galaxies*, ed. L. Woltjer, p. 217. New York: Benjamin

Roberts, W. W. 1969. Large-scale shock formation in spiral galaxies and its implications on star formation. *Astrophys. J.* 158:123

Roberts, W. W., Huntley, J. M., van Albada, G. D. 1979. Gas dynamics in barred spirals: Gaseous density waves and galatic shocks. *Astrophys. J.* 233:67

Roberts, W. W., Roberts, M. S., Shu, F. H. 1975. Density wave theory and the classification of spiral galaxies. *Astrophys. J.* 196:381

Roberts, W. W., Yuan, C. 1970. Application of the density wave theory to the spiral structure of the Milky Way system. III. Magnetic field: large-scale hydromagnetic shock formation. *Astrophys. J.* 161:877

Rots, A. H. 1975. Distribution and kinematics of neutral hydrogen in the spiral galaxy M81. II. Analysis. *Astron. Astrophys.* 45:43

Rots, A. H., Shane, W. W. 1975. Distribution and kinematics of neutral hydrogen in the spiral galaxy M81. I. Observations. *Astron. Astrophys.* 45:25

Sandage, A. 1961. *The Hubble Atlas of Galaxies*. Carnegie Inst. Washington

Sanders, R. H., Huntley, J. M. 1976. Gas response to oval distortions in disk galaxies. *Astrophys. J.* 209:53

Sanders, R. H., Tubbs, A. D. 1980. Gas as a tracer of barred spiral dynamics. *Astrophys. J.* 235:803

Shu, F. H., Milione, V., Gebel, W., Yuan, C., Goldsmith, D. W., Roberts, W. W. 1972. Galactic shocks in an interstellar medium with two stable phases. *Astrophys. J.* 173:557

Shu, F. H., Milione, V., Roberts, W. W. 1973. Nonlinear gaseous density waves and galactic shocks. *Astrophys. J.* 183:819

Sorensen, S. A., Matsuda, T., Fujimoto, M. 1976. On the formation of large scale shock waves in barred galaxies. *Astrophys. Space Sci.* 43:491

Visser, H. C. D. 1978. *The dynamics of the spiral galaxy M81*. PhD thesis. Univ. Groningen, Holland

*Ann. Rev. Fluid Mech. 1981. 13:57–77*


# DEBRIS FLOW

*8169

*Tamotsu Takahashi*
Disaster Prevention Research Institute, Kyoto University, Kyoto, Japan

## INTRODUCTION

The phenomenon of debris flow, as the agent forming alluvial cones in the mouths of mountain canyons, has attracted the attention of physiographers for more than a century. Debris flows are also of concern to engineers who are responsible for human life and property. Although various kinds of countermeasures have been invented, debris flow is still one of the most threatening natural phenomena in some regions in the world. For example, about 90 lives a year on the average are lost by debris flows in Japan. Just as subaerial debris flows have attracted the attention of physiographers and engineers, subaqueous sediment gravity flows have drawn geologists' attention regarding the geneses of some sorts of sedimentary rocks and submarine canyons. Analytical study of the mechanism of debris flow, however, had not been carried out before 1954, when Bagnold published his concept of dispersive stress. Since then, except for many speculative descriptions based on limited circumstantial evidence, relatively insignificant progress has been made in the analysis of the mechanism of debris flow.

One of the reasons for this situation is presumably the difficulty of observing the actual phenomena. Debris flows are episodic events; they occur within a very short time, particularly with very heavy rainfall in isolated areas. Recently, some observation systems have been set up on the basins in Japan where debris flows occur several times a year. Because of these systems and efforts of researchers, we can now observe on film and videotape debris flows as they are flowing. The writer believes that studies of the mechanism of debris flow will thus be promoted.

### *Debris Flow as a Type of Massive Subaerial Sediment Motion*

We define massive sediment motion as the falling, sliding, or flowing of conglomerate or the dispersion of sediment, in which all particles as well



0066-4189/81/0115-0057$01.00

as the interstitial fluid are moved by gravity, so that the relative velocity between the solid phase and fluid in the direction of displacement of mass plays only a minor role. By contrast, in fluid flow, lift and drag forces due to relative velocity are essential for individual particle transport.

Four types of massive subaerial sediment motion, as shown in Table 1, may be distinguished on the basis of the mechanism of motion, properties of interstitial fluid, velocity of displacement, and travel distance: (*a*) *fall, landslide, creep*, in which all individual particles move downward with little internal deformation, (*b*) *sturzstrom*, defined by Hsü (1975) for catastrophic landslides, in which disintegration occurs in the initial stage of movement and the debris flows along a nearly horizontal valley floor with tremendous speed, (*c*) *pyroclastic flow*, a very rapid flow of a mixture of hot ash and gas (sometimes water) ejected explosively from a volcanic crater, in which the sediment could be supported by the upward flow of fluid due to the rapid expansion of infolded air or emitted gas, (*d*) *debris flow* (that occurring at the time of volcanic eruptions is sometimes called *lahar*), in which the grains are dispersed in a water or clay slurry (hot water in the case of lahar) with the concentration a little thinner than in stable sediment accumulation.

The motions classified in (*b*), (*c*), and (*d*) are more or less the phenomena of flowage, and they could be inclusively called *sediment gravity flow*. For the sediment gravity flow to be continuous, some kind

**Table 1** Classification of subaerial massive sediment motions

| Type of flow | Sediment support mechanism | Interstitial fluid | Velocity | Travel distance | Deposit |
|---|---|---|---|---|---|
| Fall, landslide, creep | Fall, jump, roll, slide | Air and water | Free fall∼ 2,3 mm/y | ∼2·fall height | Talus |
| Sturzstrom | Grain interaction? Fluidization? | Air (Vacuum) | <50 m/s | 200 m∼ 10 km | Sturzstrom deposit |
| Pyroclastic flow | Fluidization Grain interaction? | Hot air Volcanic gas (Water) | <50 m/s | $\lesssim$ 60 km | Pyroclastic deposit |
| Debris flow (lahar) | Grain interaction Matrix strength Buoyancy | Water Clay slurry (Hot water) | 20 m/s∼ 0.5 m/s | 200 m∼ 10 km | Debris cone |

of upward sediment-supporting forces are necessary. As mentioned above, fluidization could be the agent in the case of pyroclastic flow. But what is the main sediment-supporting force in debris flow? Turbulence in the interstitial fluid would be too weak to suspend the large boulders. Bagnold (1954) proved the existence of a dispersive pressure resulting from the exchange of momentum between the grains in neighboring layers. When the voids are filled by dense clay slurry, large stones can be dispersed under rather small dispersive pressure, helped by buoyancy in the fluid phase. This is one approach to the mechanics of debris flow. Another approach considers the plastic strength of interstitial clay slurry as the grain-supporting force, and regards the grain-to-grain interactions as trivial. Middleton & Hampton (1976) distinguish *debris flow* from *grain flow*. They attribute the main sediment-supporting force to the plastic strength of the fluid phase in the former, and to grain-to-grain interaction in the latter. In this paper, however, we define debris flow in a broader sense, which embraces both debris and grain flows as defined by Middleton & Hampton.

In a typical case of sturzstrom, there is neither buoyancy nor plastic strength in the pore fluid. Moreover, sturzstrom can occur even on the moon. Therefore, grain-to-grain interaction could be the main mechanism. But why can the sturzstrom flow on a plane that is flatter than the angle of repose? The mechanics of sturzstrom is for the time being an enigma.

Actual flows exist throughout a continuum between these typical conceptual classifications. It is often difficult to classify a particular event as, say, landslide or debris flow, landslide or sturzstrom, pyroclastic flow or lahar. Only thorough understanding of the mechanism of each phenomenon in its manifold features can help further progress.

We begin by reviewing the characteristics of real debris flow. Next we discuss the mechanism of its occurrence, the mechanics of steady movement, and the process of deposition. A comparison of several models of flowage confirms Bagnold's dilatant fluid model.

Because of limitations of space, highly practical problems such as the delineation of hazardous areas for debris flow, the evaluation of various countermeasures, and the study of dynamic forces on structures are not discussed, nor are geological, geographical, and hydrological factors, or methods of forecasting.

## GENERAL CHARACTERISTICS

Figure 1 shows a debris flow passing through a debris dam. The four photographs were taken by Okuda et al (1977, Okuda 1978) at the gully Kamikamihori, Mt. Yake, Japan, on 3 August 1976. As is clear from

Photo 1. The front is coming. Note the scanty discharge in advance of the debris flow.

Photo 2. 2 seconds later than Photo 1. The big stone is 2.5 m in diameter.

*Figure 1* Debris flow passing through a debris dam.

Photo 3. 2 seconds later than Photo 2.

Photo 4. 2 seconds later than Photo 3.

these photos, the front of the debris flow swells and brings together the larger boulders. The rear part contains more water, and grain sizes in this part are smaller than in the front. The velocity of the debris flow at this point was 3.7 m $s^{-1}$. The gradient of the gully at this point is 18°.

These are typical features for debris flow, and the records of witnesses so far, for example Schlumberger (1882), Blackwelder (1928), Sharp & Nobles (1953), Curry (1966), and Johnson (1970), all give similar results. The velocities of debris flows vary widely, however, due to differences not only in the character of the debris such as grain concentration and grain size distribution, but also in the shape of the course of passage such as its width, slope, etc. The observed velocities are between about 0.5 m $s^{-1}$ and 20 m $s^{-1}$. Bulk densities also vary case by case. Although there exists doubt on the accuracy of measurement, densities of from 1400 $kgm^{-3}$ (Okuda et al 1977) to 2530 $kgm^{-3}$ (Curry 1966) have been reported. These bulk densities are equivalent to a volume concentration of solid material from about 25% to 70 or 80% respectively.

Debris flow is sometimes known as mud flow. But the ordinary debris flow shown in the photos contains a smaller amount of fine particles. A few records report the results of mechanical analysis of sediment samples, which show that less than 20% of particles are finer than silt size, and only a few per cent are finer than clay (Curry 1966, Sharp & Nobles 1953, Okuda et al 1977). On the other hand, often surprisingly huge stones are transported. In Japan, there is a famous stone weighing about 3000 tons that was transported several kilometers by a debris flow.

Debris flow makes a deposit at the mouth of a canyon, and it builds a geographical feature that is known as an alluvial cone or debris cone. This is steeper than the usual alluvial fan and consists of a poorly sorted mixture of particles. Sometimes the front of a debris flow comes to rest without spreading to the full width of the cone or the valley. In such a case, the distal end of the deposition retains its lobate form just as if the flow had been suddenly frozen (Curry 1966, Johnson 1970).

A debris flow exerts enormous impact forces on obstacles in its way. Okuda et al (1977) measured a force of up to 60 kN on an area 15 cm square. However, after the flow issues from the canyon and spreads out upon the plain, its power is quickly lost. It is not uncommon that a house is buried to the eaves by debris without serious damage (see, for example, Sharp & Nobles 1953). This probably suggests that the flow is very thin and slow at the moment of the stoppage.

The erosivity of debris flow in the source area is severe, and one can see that the bottom of a canyon is thoroughly scoured to the bedrock; but fully freighted debris flow has little erosive effect, and one can sometimes see that the paving of roads has not been damaged on passage of the flow.

The data so far reveal that debris flows occur on sediment beds in mountain canyons whose slopes are steeper than 15°, and they come to rest on plains or in canyons whose slopes are steeper than 3°. Finer particles, however, may be transported to flatter places as bed load and suspended load by the surface stream squeezed out by the deposition of debris.

## INITIATION OF DEBRIS FLOW

As defined above, debris flow is a rapid flowage of debris with sufficient water to disperse grains uniformly throughout the whole depth. Mechanical interpretation of the initiation of debris flow should answer how water is supplied and mingled with grains just after the commencement of motion of debris mass; otherwise the discussion would involve nothing but the mechanism of slides and slumps, an area extensively studied in soil mechanics.

The initiation of debris flow has been attributed to the following causes: (*a*) under some conditions a landslide turns into debris flow, (*b*) a naturally built dam that checks a gully may collapse into debris flow, (*c*) with the appearance of a surface water stream in heavy rainfall, an accumulation of debris may become unstable and turn into a debris flow. Among these causes, the third (*c*) is most common, and its mechanism is most easily formulated. Takahashi (1978a) analyzed the initiation of debris flow under these circumstances.

Imagine a thick uniform layer of loosely packed noncohesive grains, whose slope angle is $\theta$. It is assumed that at the moment when the surface flow of water of depth $h_o$ appears, the pore spaces among grains are saturated and a parallel seepage flow without any excessive pore pressure occurs. The characteristic distribution of shear stress in the bed should be like one of those shown in Figure 2, in which $\tau$ is the applied tangential stress and $\tau_L$ the internal resistive stress.

Case 1 in Figure 2 occurs under the condition

$$\tan\theta \geqslant \frac{c_*(\sigma-\rho)}{c_*(\sigma-\rho)+\rho}\tan\phi \tag{1}$$

in which $c_*$ is the grain concentration by volume in the static debris bed, $\sigma$ and $\rho$ are the densities of grains and fluid respectively, and $\phi$ is the internal friction angle. When Case 2 occurs, the following equation should be satisfied:

$$\tan\theta = \frac{c_*(\sigma-\rho)}{c_*(\sigma-\rho)+\rho\left(1+h_o a_L^{-1}\right)}\tan\phi, \tag{2}$$

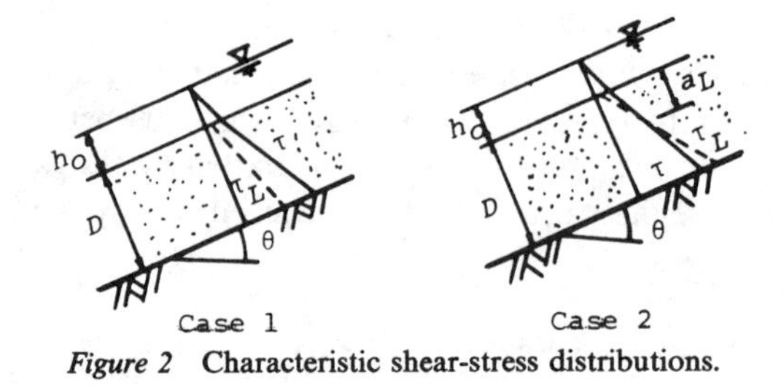


*Figure 2* Characteristic shear-stress distributions.

in which $a_L$ is the depth where $\tau$ and $\tau_L$ coincide.

The whole bed in Case 1 and the part above the depth $a_L$ in Case 2 will begin to flow as soon as the surface flow appears. This type of instability in the bed is due not to the dynamic force of fluid flow but to static disequilibrium, so that the flow should be called sediment gravity flow.

The condition for the occurrence of sediment gravity flow is, therefore, $a_L \geqslant d$, in which $d$ is the diameter of grains. Substitute this condition into Equation (2), and we obtain

$$\tan\theta \geqslant \frac{c_*(\sigma-\rho)}{c_*(\sigma-\rho)+\rho(1+h_o d^{-1})}\tan\phi. \tag{3}$$

Even though the condition $a_L \geqslant d$ is satisfied, when $a_L$ is far less than $h_o$ grains cannot be uniformly dispersed throughout the whole depth of the flow due to the rather small colliding dispersibility. Therefore, a sediment gravity flow that is appropriately called a debris flow should meet the condition $a_L \geqslant \kappa h_o$, in which $\kappa$ is a numerical coefficient, determined from experiments to be about 0.7. Substitution of the condition $a_L \geqslant \kappa h_o$ into Equation (2) gives

$$\tan\theta \geqslant \frac{c_*(\sigma-\rho)}{c_*(\sigma-\rho)+\rho(1+\kappa^{-1})}\tan\phi. \tag{4}$$

Debris flow occurs when Equations (3) and (4) are simultaneously satisfied. Attention must be paid to Case 1 where Equations (3) and (4) are satisfied. A grain bed that satisfies Equation (1) might slip even though the seepage flow did not reach the surface. In such a case, in the neighborhood of the surface layer, $\tau_L$ would be larger than $\tau$, and there the structure of the bed would be maintained. This phenomenon should be called landslide rather than debris flow. Therefore, on such a steep permeable bed, unless the supply of water is very abrupt, say, because of the collapse of a natural dam, a landslide would occur before the increment of the stage of seepage flow reaches the surface of the bed. A pile of debris caused by a landslide usually has a smaller gradient than it originally had on the mountain slope, so that it may be moved as debris flow with the appearance of a surface water flow. The time lag between the occurrence of a landslide and the removal of its deposit by the generation of debris flow would depend on the water supply, and it sometimes could be so short that one would hardly recognize the transition between the two phenomena.

As shown in Figure 3, the domains of occurrence of various types of sediment transportation are defined by Equations (1), (2), (3), (4), the condition for fall ($\theta=\phi$), and the equation of critical tractive force on a steep channel (Ashida et al 1973),

$$\frac{\rho u_{*c}^2}{(\sigma-\rho)gd}=0.034\cos\theta\left[\tan\phi-\frac{\sigma}{(\sigma-\rho)}\tan\theta\right]\times 10^{0.32(d/h_c)} \tag{5}$$

in which $u_{*c}$ = critical shear velocity $[=(gh_o\sin\theta)^{0.5}]$ and $g$ = acceleration due to gravity. The domain labeled *1* is that of no particle movement; *2*

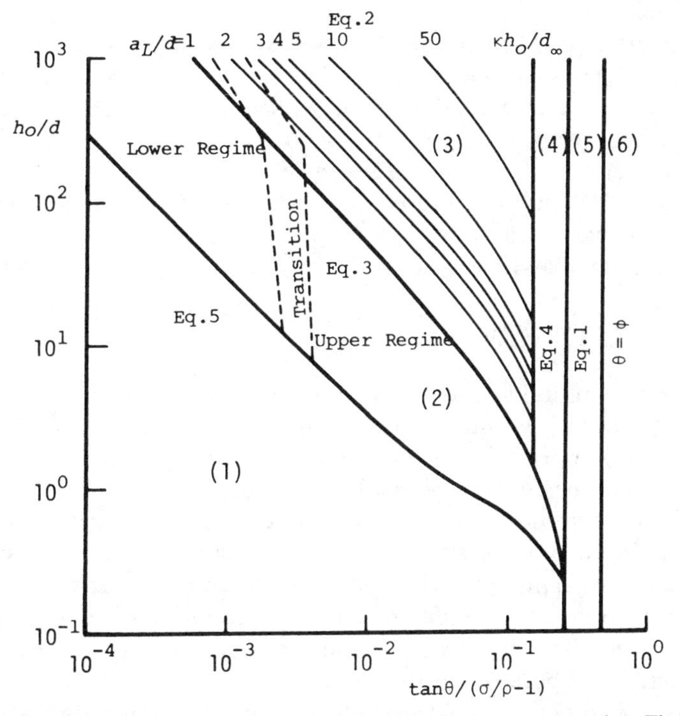


*Figure 3* Criteria for occurrence of various types of sediment transportation. The curves are obtained under the condition that $c_*=0.7$, $\sigma=2.65$ gcm$^{-3}$, $\rho=1.0$ gcm$^{-3}$, $\kappa=0.7$, and $\tan\phi=0.8$.

is the domain of individual particle movement due to the dynamic force of fluid flow, i.e. bed load transport; *3* is the domain of sediment gravity flow, in which the effect of dynamic force of fluid flow coexists and in the flow there is a clear water layer over a dense mixture of grain and water. Numbers attached to the curves in this domain correspond to the thickness of the moving layer of grains. The effect of dynamic action should decrease for increasing thickness of the moving layer. Note that the domains of the transition and the upper regime in the bed form so far proposed (Ashida & Michiue 1972) contain both domains of individual particle movement and sediment gravity flow. *4* is the domain of debris flow in which the grains are dispersed in the whole layer; *5* is the domain of the occurrence of both landslide and debris flow; and in *6* the sediment bed is unstable under no fluid flow.

The usual debris flow consists of grains whose sizes are very widely distributed. What is the representative sediment size? Experiments using materials whose standard deviation $(d_{84}/d_{16})^{0.5}$ varied from 1.0 to 3.46 reveal that we can adopt the median diameter as the representative size of the mixture.

Takahashi (1978b) extended the theory to a cohesive sediment bed and determined that the slightest cohesive strength in the material can considerably increase the stability of the bed. Residual cohesive strength has been one of the main difficulties in understanding the mechanism of the formation of mud flow (Hutchinson & Bhandari 1971). Recently, Vallejo (1979) studied the process of the widening of fissures in clay structures due to the effect of evaporation and frost action. In his analysis small clay blocks behave like grains in noncohesive materials.

## PROCESS OF GROWTH

Consider a uniform bed whose slope angle satisfies Equation (3) and (4) but does not satisfy Equation (1), i.e. domain *4* in Figure 3. As soon as surface flow appears at the upstream end of the bed, debris above the depth $a_L$ will begin to move, and these grains mingled with water will form a bore, whose height is $h$ and whose grain concentration is $c_d$ by volume. In the next instant the bore front will tumble over the surface of the downstream bed. Then if the concentration of the bore front is thinner than the fully freighting value, the bed beneath the front will be scoured, and the front will swell and the concentration increase. Thus, the front of the bore will grow until it reaches a steady state. Takahashi (1978a) analyzed the process of growth. According to him, the equilibrium concentration $c_{d\infty}$ is

$$c_{d\infty}=\frac{\rho\tan\theta}{(\sigma-\rho)(\tan\phi-\tan\theta)}. \tag{6}$$

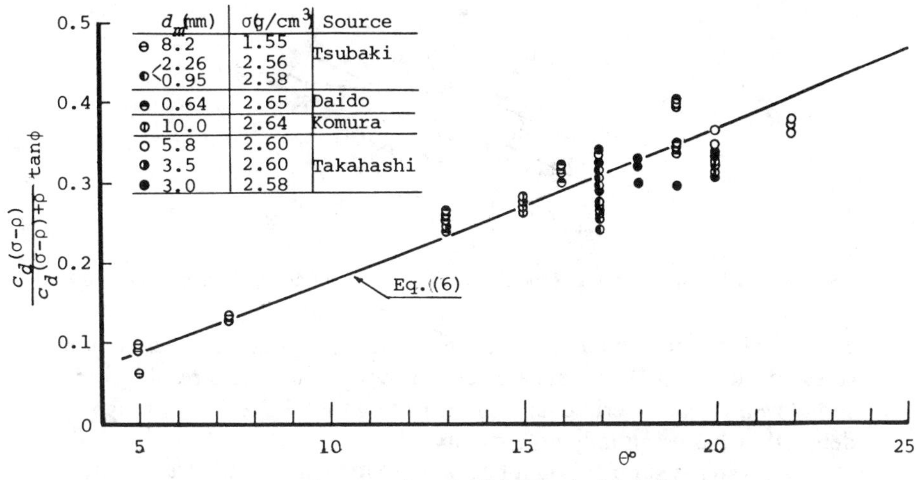


*Figure 4* Concentration of solid phase in debris flows.

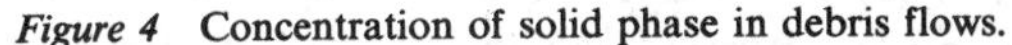

Clearly, $c_d$ does not exceed $c_*$ since the angle of the bed does not satisfy Equation (1). Figure 4 compares the experimental concentration with that obtained by the calculation of Equation (6).

After the passage of the front, the bed continues to be scoured as long as water is supplied. Takahashi (1977) analyzed this process assuming that the concentration of the rear part of the debris flow is given by

$$c_d = \frac{\rho(\tan\theta - \partial y/\partial x)}{(\sigma-\rho)(\tan\phi - \tan\theta + \partial y/\partial x)}, \tag{7}$$

in which $x$ is the position and $y$ the distance from the original bed surface to the bottom of the debris flow. Using Equation (7) and the equation of continuity, and neglecting smaller terms, he obtained an equation of diffusion type, whose solution at time $t$ is

$$\frac{y}{2(Kt)^{0.5}} = -M \cdot \mathrm{ierfc}\left[\frac{x}{2(Kt)^{0.5}}\right] \tag{8}$$

in which

$$M = \tan\theta - \left\{c_*(\sigma-\rho)\tan\phi / \left[c_*(\sigma-\rho) + \rho(1+\kappa^{-1})\right]\right\},$$

$$K = \frac{\sigma-\rho}{\rho} \frac{c_{d\infty}^2 c_* q_o}{\left\{c_* - \left[s + (1-s)c_*\right]c_{d\infty}\right\}^2} \frac{\tan\phi}{\tan^2\theta},$$

and $s$ is the degree of saturation in the bed in advance of the arrival of the debris flow. Equation (8) implies that the longitudinal profile of the bottom of the debris flow upstream of the essentially steady uniform part is nearly linear. Thus, the longitudinal profile of the debris flow at

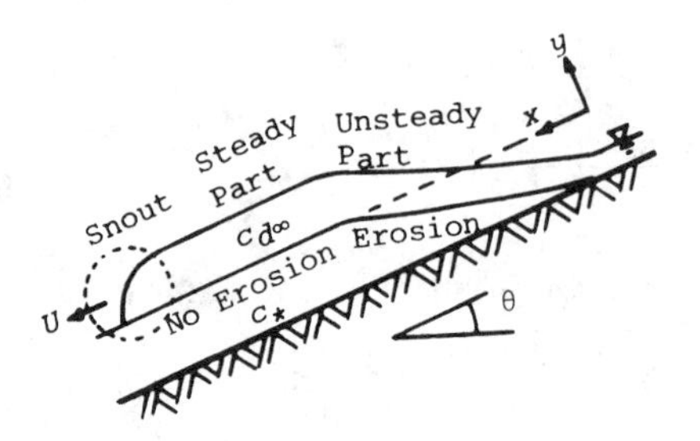


*Figure 5* Schematic diagram of the shape of the profile of an equilibrium debris flow.

any moment would be as shown schematically in Figure 5. The region of steady uniform flow elongates as it flows down. These results correspond well to the experiments, demonstrating that the real fully freighted debris flow has essentially no erosivity.

When water is supplied on a bed whose slope angle satisfies Equation (1), and if the bed is saturated as soon as the bore front arrives, then no stable bed should remain and the front height would continuously grow downstream. But, in fact, appreciable time is necessary to infiltrate the bed, and there is a tendency of the bore to stop. The same situation occurs for a flatter slope if the bed is not saturated in advance.

## MECHANICS OF STEADY FLOW

### *Dilatant Fluid Model*

Bagnold (1954) sheared a dispersion of neutrally buoyant particles in a Newtonian fluid in the annular space between two concentric drums. The particles dilated to the extent of exerting pressure on the vertical walls perpendicular to the main flow. He reasoned that this dispersive pressure is the result of momentum exchange associated with grain encounters, and he found that the dispersive pressure is proportional to the shear stress. When the applied shear strain $du/dy$ is small the resulting shear stress is a mixed one due to the effect of fluid viscosity as modified by the presence of grains, whereas when the applied shear strain is large the viscosity of the interstitial fluid is insignificant, and the resulting shear stress is essentially due to grain interaction. For the latter case, Bagnold gives the formulas

$$\left.\begin{aligned} P &= a\,\sigma\left[\left(c_*/c_d\right)^{1/3}-1\right]^{-2} d^2 (du/dy)^2 \cos\alpha \\ T &= P\tan\alpha \end{aligned}\right\} \tag{9}$$

in which $P$ is the dispersive pressure, $T$ the shear stress, $\alpha$ a dynamic angle of internal friction, and $a$ a numerical constant equal to 0.042 for

$c_d < 0.81c_*$. Bagnold's experiments show that the fully inertial condition is satisfied when $G^2 = \sigma d^2 T[(c_*/c_d)^{1/3} - 1]\mu^{-2}$ is larger than 3000, where $\mu$ is the viscosity of the interstitial fluid. This condition should be easily met in a real debris flow (Takahashi 1980).

Application of Equation (9) to steady uniform flow in a wide open channel leads to the velocity distribution

$$\frac{u_s - u}{u_s} = \left(1 - \frac{y}{h}\right)^{3/2} \tag{10}$$

in which $u_s$ is the surface velocity, $u$ the velocity at height $y$, and $h$ the depth of flow.

The cross-sectional mean velocity $U$ is

$$U = \frac{2}{5d}\left\{\frac{g}{a\sin\alpha}\left[c_d + (1-c_d)\frac{\rho}{\sigma}\right]\right\}^{1/2}\left[\left(\frac{c_*}{c_d}\right)^{1/3} - 1\right] h^{3/2}(\sin\theta)^{1/2}. \tag{11}$$

The relationship between the velocity and the discharge $q_0$ of water supplied at the upstream end is (Takahashi 1978a)

$$\frac{U(gd)^{-0.5}}{\left[q_0^2(gd^3)^{-1}\right]^{0.3}} = \frac{0.693\sin^{0.2}\theta}{(a\sin\alpha)^{0.2}} \frac{\left[\left(c_* c_d^{-1}\right)^{1/3} - 1\right]^{0.4}\left[c_d + (1-c_d)\rho\sigma^{-1}\right]^{0.2}}{\left\{1 - \left(c_d c_*^{-1}\right)\left[s + (1-s)c_*\right]\right\}^{0.6}}. \tag{12}$$

Elimination of $U$ from Equations (11) and (12) predicts the depth of steady uniform flow.

Takahashi (1980) discussed the lobate longitudinal profile of the snout using Equation (11) and the condition that in the equilibrium state no erosion under the snout should occur. The result is

$$h + h_\infty \ln\frac{|h - h_\infty|}{h_\infty} = \tan\theta x' \tag{13}$$

in which $h_\infty$ is the asymptotic depth of the steady flow and $h$ is the depth at $x'$, measured upstream from the tip of the snout. This corresponds well to experiment in which uniform gravel is used (Figure 6).

Bagnold (1954, 1968) reasons qualitatively that when grains of mixed sizes are sheared together the larger grains tend to drift towards the free surface, because for a given shear strain the dispersive stress appears to increase as the square of the size. Since the flow surface necessarily moves fastest. the large material should drift towards the front of the

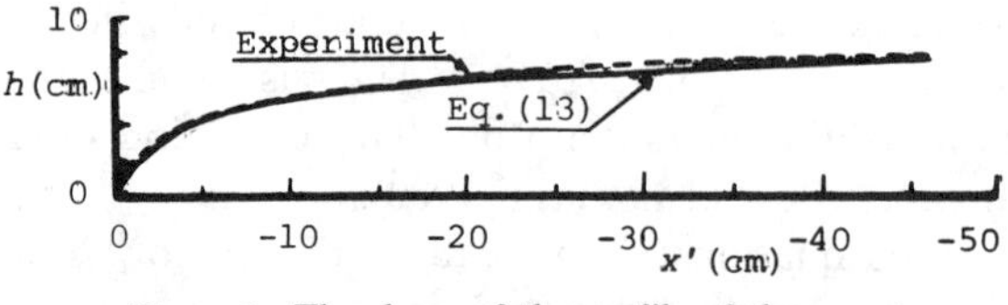


*Figure 6* The shape of the profile of the snout.

flow, as shown in the photos. Using Bagnold's theory, Takahashi (1980) discussed quantitatively the motion of a particle whose size is different from that of the other uniform particles that determine the structure of the flow. A particle whose diameter is larger tends to drift upward with a velocity faster than that of a smaller particle drifting downward. Experiments revealed that the drag coefficient of the upward-drifting particle is very large ($C_D \simeq 2000$), even though the flow itself is in the fully inertial range.

Owing to the accumulation of larger particles in the snout, the translational speed of the front should tend to decrease, as seen in Equation (11). This could be the reason why the front part of a real debris flow swells.

## *Plastico-Viscous Rheological Models*

The surfaces of some debris flows seem comparatively placid and very viscous, and sometimes it seems that boulders float on the surface of mud. To explain these characteristics, Yano & Daido (1965) and Johnson (1970) proposed a Bingham plastic-fluid model. The flow of clay slurry is well modeled as a Bingham fluid.

The stress-strain relationship in a Bingham fluid is

$$\tau = \tau_y + \mu \frac{du}{dy}, \tag{14}$$

where $\tau_y$ is the yield strength. From Equation (14) we obtain the following formula for the velocity distribution on a wide open channel:

$$\frac{u_s - u}{u_s} = \left(1 - \frac{h}{H}\frac{y}{h}\right)^2 \tag{15}$$

where $H$ is the height from the bottom of the flow to the point where the applied shear stress is equal to the yield strength. As the shear stress above $H$ is less than $\tau_y$, this part of the flow should be rigid as if it were a raft on laminar flow. This part is often called the "plug," and the thickness $(h - H)$ is

$$(h - H) = \frac{\tau_y}{[\sigma c_d + (1 - c_d)\rho] g \sin\theta}. \tag{16}$$

When $H$ is equal to $h$, the fluid is simple Newtonian.

The cross-sectional mean velocity is given by

$$U = \frac{(Hh^{-1})^2\left[1 - H(3h)^{-1}\right]}{2\mu}\left[\sigma c_d + (1 - c_d)\rho\right] g h^2 \sin\theta. \tag{17}$$

Prediction of velocity by this model requires values of $\tau_y$ and $\mu$. However, no method has been proposed for obtaining them except for pure clay slurries. From his observation of real debris flow at Wrightwood, California, Johnson (1970) found that $\tau_y = 602$ Nm$^{-2}$ and $\mu = 76$ Pa·s. These values greatly exceed, say by three powers of ten, those for ordinary very dense clay slurries. This would suggest that the effect of grain interaction is significant in real debris flows.

It is possible for a Bingham fluid to flow in a channel of very low slope provided $h$ becomes large. This is not in accordance with real debris-flow phenomena. To avoid this contradiction Johnson (1970) proposed his Coulomb-viscous model, in which the stress-strain relationship is

$$\tau = c + \sigma_n \tan\phi + \mu\frac{du}{dy}, \tag{18}$$

where $c$ is the cohesion and $\sigma_n$ the internal normal stress. In this model the yield strength is a function of not only the properties of the materials but also the thickness of the flow. But $c$ and $\mu$ are again unknowns.

## Other Models

Tsubaki et al (1972) used experimental flumes, whose slopes were $\sin\theta = 0.383$ and 0.285. The sediment bed was a mixture of $d_{50} = 8$ mm and 0.32 mm with a mixing ratio of 1 : 1 by volume. They claim that the velocities of the front of the debris flows and those of the bores of pure water are both given by

$$U = 2.5(gh\sin\theta)^{0.5}. \tag{19}$$

Syanozhetsky et al (1973) treated their general equation of motion of debris flow by neglecting the body-force term and the resistance term, and obtained the following formulas to predict the velocities of the front.

For the case of a "turbulent mud-stream"

$$U = 2\left[gh\cos\theta(1 + 1.5\sin\theta)\right]^{0.5}. \tag{20}$$

For the case of a "structural mud-stream"

$$U = 1.5\left[gh\cos\theta(1 + 1.5\sin\theta)\right]^{0.5}. \tag{21}$$

Classification of debris flow into "turbulent" and "structural" seems to be generally accepted in the Soviet Union. The former is rich with water and the latter contains little water. The effects of viscosity dominate in the latter. Syanozhetsky et al claim that the two kinds of debris flow can be divided by the value $c_d = 0.47$.

In the Crimean district of the Soviet Union, the following formula of Sribniy's is used (Gol'din & Lyubashevsky 1966):

$$U = 6.5h^{2/3}(\sin\theta)^{1/4}\left[\frac{\sigma}{\rho}\frac{(\sigma-\rho)c_d}{(\sigma+\rho)(1-c_d)}+1\right]^{-1/2}. \tag{22}$$

## *Comparison of the Models*

Figure 7 shows the velocity distribution in debris flow with practically uniform gravel. It will be understood that the dilatant-fluid model interprets the experimental results better than the Bingham-fluid model (in the experimental case, no clay was contained, so that the model should be the Newtonian fluid model).

The objection might arise that, if clay was contained, the Bingham-fluid model would fit better. But according to Michaels & Bolger (1962) and Daido (1970) the yield strength and viscosity of clay slurry are at most 6.5 $\mathrm{Nm}^{-2}$ and 0.04 Pa·s respectively. Therefore, the thickness of the plug is at most 0.15 cm for flow on the slope $\theta = 18°$. This is enough information to understand that the velocity distribution in the debris flow would be little affected by clay content, which would usually be less than 20%.

The plastico-viscous rheological model attributes the transportation of large boulders to their buoyancy and the strength in interstitial

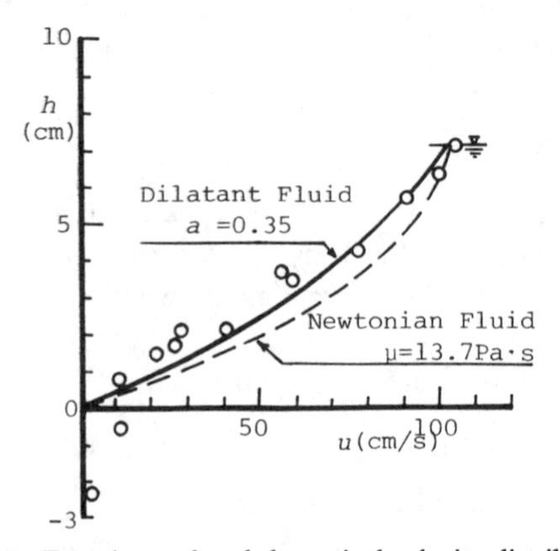


*Figure 7* Experimental and theoretical velocity distributions.

clay-water slurry. Hampton (1975, 1979) obtained from his experiments the relationships between the clay contents in clay-water slurry and the competences to float grains. His results show that sand-sized particles can be floated but larger ones can not.

It should be noted as shown in Figure 7 that very large viscosity must be given to roughly fit the Bingham-fluid model even though the flow completely lacks clay or silt-sized particles. Thus, the apparent high viscosity and high competence to transport larger particles are nothing but the results of grain interactions. A dilatant-fluid model has been introduced by considering the effect of grain interaction, and this model describes various aspects of debris flow as discussed above. This suggests that the dilatant-fluid model is at present the better one.

When a debris flow is about to stop, however, the flow is by no means fully inertial, and the viscosity of interstitial clay-water slurry plays an important role. Further, the mud flows that sometimes occur from the toe of an earth creep contain much clay, and they could be well modeled as a Bingham fluid.

Figure 8 compares the experimental velocities obtained by Takahashi (1977) with values calculated from various velocity-predicting formulas. One should be aware that Syanozhetsky's formula has the tendency to

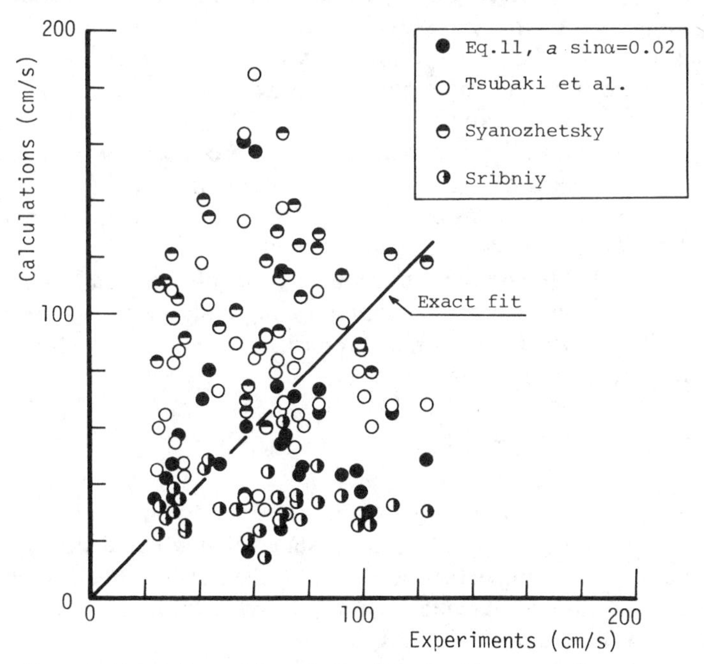


*Figure 8* Calculations of the velocity of the front by various formulae.

predict velocities that are too high, and that of Sribniy velocities that are too low. The experimental formula of Tsubaki et al and Equation (11) have a similar tendency. A reason for this coincidence could be the similarity in experimental conditions.

There is appreciable scatter in the experimental data. It is likely that in experimental work the measurement of depth, velocity, and concentration inherently involves considerable error. A further cause for scatter is probably the nature of the sediment bed in advance of the passage of flow: the bed was in one case rather dry and in the other case saturated. Infiltration of water into the bed from the front of the debris flow could influence the value of $a$, whereas the constant value of 0.042 obtained by Bagnold was used in the calculation. We find evidence of this in Figure 7, where the value $a=0.35$ must be used to fit the experimental velocity distribution.

## DEPOSITION OF DEBRIS FLOW

The mechanics of the stoppage and deposition of debris flow have been little studied, notwithstanding its importance from the point of view of prevention of hazard.

Hooke (1967) carried out experiments in which small alluvial cones were built in a working area by several repetitions of an episode consisting of a debris flow followed by a water flow, where the debris flow was made by mixing a slurry of water and sediment. Laboratory cone slopes were usually between 4° and 8°, values that are about average for natural cones with a high percentage of cobbles and boulders. Hooke aimed to answer physiographical questions, and his interpretations of the mechanics of deposition were qualitative and based on the Bingham-fluid model.

Johnson (1970) discussed the lobate form of the distal end of the deposition of debris flow. Assuming a simple plastic substance, whose surface is everywhere in critical equilibrium, he obtained the form of the curved part of the snout as

$$x=-\frac{\tau_y}{\gamma_d}\ln\left[\cos\left(\frac{\gamma_d}{2\tau_y}y\right)\right] \tag{23}$$

where $\gamma_d$ is the bulk density of the plastic substance. This equation shows a bulbous form, and Johnson stated that it could be applied to the snout of a moving debris flow. But for the moving snout the shape is given by Equation (13), which is more slowly asymptotic to the uniform depth than Equation (23).

Takahashi & Yoshida (1979) studied the deposition of debris flow at a position where the slope of the channel abruptly changes to a flatter value without expansion of width. They considered the momentum change in the snout in the flatter channel downstream, and obtained the condition that the snout would stop as

$$\tan\theta < \frac{(\sigma-\rho)c_{du}\tan\alpha}{(\sigma-\rho)c_{du}+\rho} \tag{24}$$

where $c_{du}$ is the concentration in the steeper channel upstream. This equation shows that the denser the interstitial fluid and the thinner the concentration of debris flow, the flatter the slope of the channel in which the debris flow arrives. When $c_{du}$ satisfies Equation (6), Equation (24) is equivalent to

$$\tan\theta < \frac{\tan\alpha}{\tan\phi}\tan\theta_u, \tag{25}$$

in which $\theta_u$ is the angle of the upstream channel. Since $\tan\alpha$ is generally smaller than $\tan\phi$, the debris flow can arrive at a flatter place even though it freights its full capacity in the upstream steeper channel.

When the flow stops, the concentration of grains should change from $c_{du}$ to $c_*$, which means that the excess pore water would be squeezed out of the deposition and flow on the surface of the pile of sediment. Therefore, if the slope of the surface of the deposition just before stoppage is greater than the critical slope for the occurrence of sediment gravity flow, the grains at the surface cannot secure their position and they will continue to flow until they form a slope that is equal to the critical one. Thus, the stable slope of the deposition should be

$$\tan\gamma = \frac{c_*(\sigma-\rho)}{c_*(\sigma-\rho)+\rho(1+h_o d^{-1})}\tan\phi \tag{26}$$

where $h_o$ is the depth of the surface flow. Equation (26) was verified by experiments.

A schematic presentation of the depositing process is given in Figure 9. The shape of the profile at any particular time should be calculated applying the mass-conservation equation. The height, H′, of the jump at the upstream end of the deposition must be predicted in advance of the calculation. Some experiments have shown that there is negligible energy loss in the jump process.

In the case where $\theta$ is larger than $\gamma$, the surface slope of the deposition should always be larger than $\gamma$, so that the sediment gravity flow would always occur at least at the neighborhood of the surface

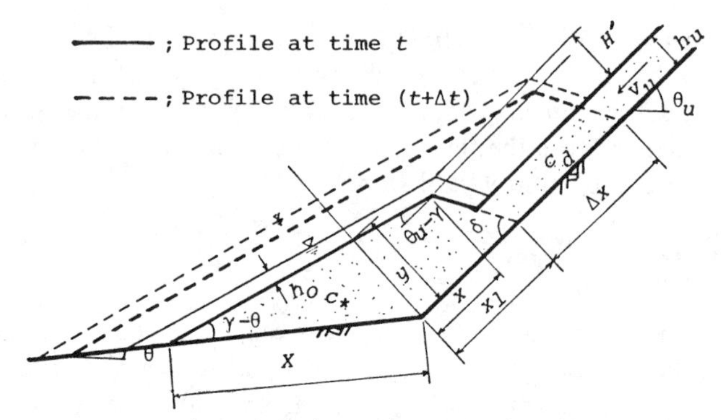


*Figure 9* A schematical presentation of the depositing process.

during the process of deposition. Experiments have revealed that the surface slope of the deposition in such a case is approximately the mean of the bed slopes of the upstream and downstream channels.

## CONCLUDING REMARKS

We have reviewed the existing mechanical investigations of the processes of occurrence, flowage, and deposition of debris flow. As far as the principal characteristics of debris flow are concerned, they have been quantitatively interpreted by the simple idealized theories. Natural debris flows, however, show a very wide variety of characteristics under various complex natural conditions. Much room remains for improvement of theories. Some examples of basic themes to be investigated are as follows: how the equation of flowage should be modified when the complex properties of natural debris flow are considered, such as rotation of particles, nonhomogeneity in concentration, mixing of a wide range of particle sizes, instability of flow on very steep channels, etc.; how much buoyancy the boulders receive from the dense suspension that is composed of a rather continuous distribution of fine particles; how debris flows behave in an irregular channel; what the mechanism of formation of alluvial cones is. Innumerable problems remain, also, in the field of practical engineering.

### ACKNOWLEDGMENTS

I am sincerely grateful to Dr. S. Okuda and his colleagues in the field observation study of debris flow for their kindness in permitting me to use their valuable photographs.

*Literature Cited*

Ashida, K., Daido, A., Takahashi, T., Mizuyama, T. 1973. Study on the resistance law and the initiation of motion of bed particles in a steep slope channel.

*Ann. Disaster Prev. Res. Inst. Kyoto Univ.* 16B:481–94

Ashida, K., Michiue, M. 1972. Study on hydraulic resistance and bed-load transport rate in alluvial streams. *Proc. Jpn. Soc. Civil Engrg.* 206:59–69

Bagnold, R. A. 1954. Experiments on a gravity-free dispersion of large solid spheres in a Newtonian Fluid under shear. *Proc. R. Soc. London Ser. A.* 225:49–63

Bagnold, R. A. 1968. Deposition in the process of hydraulic transport. *Sedimentology* 10:45–56

Blackwelder, E. 1928. Mudflow as a geologic agent in semiarid mountains. *Bull. Geol. Soc. Am.* 39:465–84

Curry, R. R. 1966. Observation of Alpine mudflows in the Tenmile Range, Central Colorado. *Bull. Geol. Soc. Am.* 77:771–76

Daido, A. 1970. *Dosekiryu ni kansuru Kisoteki Kenkyu*. Dr. Engrg. thesis. Kyoto Univ., Japan. 178 pp.

Gol'din, B. M., Lyubashevskiy, L. S. 1966. Computation of the velocity of mudflows for Crimean rivers. *Sov. Hydrol.* 2:179–81

Hampton, M. A. 1975. Competence of fine-grained debris flows. *J. Sediment. Petrol.* 45:834–44

Hampton, M. A. 1979. Buoyancy in debris flows. *J. Sediment. Petrol.* 49:753–58

Hooke, R. L. 1967. Processes on arid-region alluvial fans. *J. Geol.* 75:438–60

Hsü, K. J. 1975. Catastrophic debris streams (sturzstroms) generated by rockfalls. *Bull. Geol. Soc. Am.* 86:129–40

Hutchinson, J. N., Bhandari, R. K. 1971. Undrained loading, a fundamental mechanism of mudflows and other mass movements. *Géotechnique* 21:353–58

Johnson, A. M. 1970. *Physical Processes in Geology*, pp. 433–534. San Francisco: Freeman. 577 pp.

Michaels, A. S., Bolger, J. C. 1962. The plastic behavior of flocculated kaolin suspensions. *J. Ind. Engrg. Chem. Fundam.* 1:153–62

Middleton, G. V., Hampton, M. A. 1976. Subaqueous sediment transport and deposition by sediment gravity flows. In *Marine Sediment Transport and Environmental Management*, ed. D. J. Stanley, D. J. P. Swift. 11:197–218. New York: Wiley

Okuda, S., Suwa, H., Okunishi, K., Nakano, M., Yokoyama, K. 1977. Synthetic observation on debris flow, Part 3, observation at Valley Kamikamihorizawa of Mt. Yakedake in 1976. *Ann. Disaster Prev. Res. Inst. Kyoto Univ.* 20B-1:237–63

Okuda, S. 1978. *Observation of the motion of debris flow and its geomorphological effects*. Presented at Paris Symp. Int. Geograph. Union, Comm. Field Exp. Geomorph.

Schlumberger. 1882. Uber den Muhrgang am 13 August 1876 im Wildbache von Faucon bei Barcelonnte (Niederalpen). In *Studien über die Arbeiten der Wildebewaldung und Barausung der Gebirge*, ed. P. Demontzey, pp. 289–99.

Sharp, R. P., Nobles, L. H. 1953. Mudflow of 1941 at Wrightwood, Southern California. *Bull. Geol. Soc. Am.* 64:547–60

Syanozhetsky, T. G. V., Beruchashvili, G. M., Kereselidze, N. B. 1973. Hydraulics of rapid turbulent and quasilaminar (structural) mud-streams in deformed bed with abrupt slopes. *Proc. Istanbul Conf. IAHR* 1:507–15

Takahashi, T. 1977. A mechanism of occurrence of mud-debris flows and their characteristics in motion. *Ann. Disaster Prev. Res. Inst. Kyoto Univ.* 20B-2:405–35

Takahashi, T. 1978a. Mechanical characteristics of debris flow. *J. Hydraul. Div., ASCE.* 104:1153–69

Takahashi, T. 1978b. The occurrence and flow mechanism of debris flow. *Tsuchi to Kiso*. 26:45–50

Takahashi, T., Yoshida, H. 1979. Study on the deposition of debris flows (1)—Deposition due to abrupt change of bed slope. *Ann. Disaster Prev. Res. Inst. Kyoto Univ.* 22B-2:315–28

Takahashi, T. 1980. Debris flow on prismatic open channel. *J. Hydraul. Div., ASCE* 106:381–96

Tsubaki, T., Hirano, M., Uchimura, K. 1972. *Keikoku Taiseki Dosha no Ryudo*. Presented at the 27th Jpn. Natl. Congr. Civil Engrg.

Vallejo, L. E. 1979. An explanation for mudflows. *Géotechnique* 24:351–54

Yano, K., Daido, A. 1965. Fundamental study of mudflow. *Bull. Disaster Prev. Res. Inst. Kyoto Univ.* 14:69–83

*Ann. Rev. Fluid Mech. 1981. 13:79–95*


# EXISTENCE THEOREMS FOR THE EQUATIONS OF MOTION OF A COMPRESSIBLE VISCOUS FLUID



*V. A. Solonnikov*
Steklov Institute of Mathematics, Leningrad, USSR

*A. V. Kazhikhov*
Institute of Hydrodynamics, Novosibirsk, USSR

## 1. INTRODUCTION

One of the fundamental questions that should be answered concerning any problem of mathematical physics is whether it is well set, that is whether solutions actually exist and whether they are unique. This question is usually answered by existence and uniqueness theorems, which are results of a rigorous mathematical investigation of the problem. It is the aim of this paper to present some theorems of this type for the equations of motion of a compressible viscous fluid (Serrin 1959),

$$\begin{aligned} &\frac{d\rho}{dt}+\rho\nabla\cdot\mathbf{v}=0,\\ &\rho\frac{d\mathbf{v}}{dt}=\rho\mathbf{f}+\nabla\cdot\mathbf{T}, \qquad \mathbf{T}=(-p+\lambda\nabla\cdot v)\mathbf{I}+2\mu\mathbf{D}, \qquad (1)\\ &\rho\frac{dE}{dt}=\nabla\cdot\kappa\nabla\theta+\mathbf{T}:\mathbf{D}. \end{aligned}$$

Here $\rho=\rho(x,t)$ is the density, $\mathbf{v}(x,t)=(v_1,v_2,v_3)$ is the velocity of the liquid at time $t$ at the point $x=(x_1,x_2,x_3)$, $\mathbf{f}(x,t)$ is a vector of external forces, and $\theta(x,t)$ is the temperature. The coordinates $x_1, x_2, x_3$ (so-called Eulerian coordinates) refer to a fixed Cartesian coordinate system. The pressure $p$, the specific internal energy $E$, the coefficients of viscosity $\lambda$ and $\mu$, and the coefficient of heat conduction $\kappa$ are given



0066-4189/81/0115-0079$01.00

functions of the variables $\rho, \theta$ satisfying the conditions $\kappa, \mu > 0$, $2\mu + 3\lambda \geqslant 0$. By $\nabla$ we mean the operator of differentiation with respect to variables $x_i$: $\nabla = (\partial/\partial x_1, \partial/\partial x_2, \partial/\partial x_3)$; for any function $a(x)$, vector $\mathbf{b}(x)$, and matrix $\mathbf{A}(x)$ with elements $a_{jk}(x)$, $j, k = 1,2,3$ we have

$$\nabla a = \operatorname{grad} a = \left(\frac{\partial a}{\partial x_1}, \frac{\partial a}{\partial x_2}, \frac{\partial a}{\partial x_3}\right), \qquad \nabla \cdot \mathbf{b} = \operatorname{div} \mathbf{b} = \frac{\partial b_1}{\partial x_1} + \frac{\partial b_2}{\partial x_2} + \frac{\partial b_3}{\partial x_3},$$

$$\nabla \times \mathbf{b} = \operatorname{curl} \mathbf{b}, \qquad \nabla^2 a = \nabla \cdot (\nabla a) = \frac{\partial^2 a}{\partial x_1^2} + \frac{\partial^2 a}{\partial x_2^2} + \frac{\partial^2 a}{\partial x_3^2},$$

$$(\mathbf{b} \cdot \nabla) a = b_1 \frac{\partial a}{\partial x_1} + b_2 \frac{\partial a}{\partial x_2} + b_3 \frac{\partial a}{\partial x_3},$$

$$\nabla \cdot \mathbf{A} = \mathbf{g}, \qquad g_i = \frac{\partial a_{1i}}{\partial x_1} + \frac{\partial a_{2i}}{\partial x_2} + \frac{\partial a_{3i}}{\partial x_3}, \qquad i = 1,2,3.$$

The material derivative $d/dt$ is defined by $d/dt = \partial/\partial t + (\mathbf{v} \cdot \nabla)$. Finally, $\mathbf{I}$ is the unit matrix of the third order, $\mathbf{D} = \mathbf{D}(\mathbf{v})$ is the deformation tensor, i.e. a matrix with elements $\frac{1}{2}(\partial v_i/\partial x_j + \partial v_j/\partial x_i) = D_{ij}$, $\mathbf{T}$ is the stress tensor, and

$$\mathbf{T} : \mathbf{D} = \sum_{i,j=1}^{3} T_{ij} D_{ij}.$$

As

$$\frac{dE}{dt} = E_\theta \frac{d\theta}{dt} + E_\rho \frac{d\rho}{dt} = c_v \frac{d\theta}{dt} - \rho E_\rho \nabla \cdot \mathbf{v},$$

the Equations (1) can be written in the form

$$\frac{d\rho}{dt} + \rho \nabla \cdot \mathbf{v} = 0, \tag{2a}$$

$$\rho \frac{d\mathbf{v}}{dt} = \rho f + \nabla(\lambda \nabla \cdot \mathbf{v}) + 2\nabla \cdot \mu \mathbf{D} - \nabla p, \tag{2b}$$

$$\rho c_v \frac{d\theta}{dt} = \nabla \cdot \kappa \nabla \theta + \lambda (\nabla \cdot \mathbf{v})^2 + 2\mu \mathbf{D} : \mathbf{D} + \left(\rho^2 E_\rho - p\right) \nabla \cdot \mathbf{v}, \tag{2c}$$

where $c_v = E_\theta = \partial E(\rho, \theta)/\partial \theta$ is the specific heat of the liquid which is a positive function of $\rho$ and $\theta$, and $E_\rho = \partial E(\rho, \theta)/\partial \rho$. It is possible to introduce in (2c) the specific entropy $S(\rho, \theta)$ instead of $E(\rho, \theta)$, since

$$c_v = \theta \partial S/\partial \theta, \qquad \rho^2 E_\rho - p = \rho^2 \theta \partial S/\partial \theta.$$

We suppose that the functions $\rho, \mathbf{v}, \theta$ are specified at the moment $t = 0$:

$$\rho(x,0) = \rho_0(x), \ \mathbf{v}(x,0) = \mathbf{v}_0(x), \ \theta(x,0) = \theta_0(x), \tag{3}$$

the density $\rho_0$ and the temperature $\theta_0$ being strictly positive. On the boundary of the domain occupied by the liquid we suppose the usual no-slip condition

$$\mathbf{v}(x,t)=0 \tag{4}$$

as well as one of the classical boundary conditions for the temperature,

$$\theta=\theta_1(x,t), \tag{5a}$$

$$\frac{\partial\theta}{\partial n}=g(x,t), \tag{5b}$$

$$\frac{\partial\theta}{\partial n}-k(\theta-\theta_2(x,t))=0, \tag{5c}$$

to be satisfied. They mean that the temperature of the liquid $\theta_1$ or the heat flux $g$ are prescribed at the walls of the container of the liquid, or the heat flux is proportional to the difference of the temperature of the liquid at the wall and the temperature $\theta_2$ of the wall itself.

We shall be concerned with the following problems for the Equations (2):

*Cauchy problem*: To find functions $\rho, \mathbf{v}, \theta$ in a strip $\Pi_{t_0}=R^3\times(0,t_0): x \in R^3$, $t\in(0,t_0)$ satisfying equations (2) and initial conditions (3) when $f(x,t), \rho_0(x), \mathbf{v}_0, \theta_0$ are given.

*Initial-boundary value problem*: To find functions $\rho, \mathbf{v}, \theta$ in $Q_{t_0}=\Omega\times(0,t_0): x_3\in\Omega\subset R^3$, $t\in(0,t_0)$ satisfying Equations (2) in $Q_{t_0}$, initial conditions (3) in $\Omega$, boundary condition (4) and one of the conditions (5a), (5b), (5c) on $\Gamma_{t_0}=S\times(0,t_0): x\in S$, $t\in(0,t_0)$, where $S$ is the boundary of $\Omega$. Some other problems are mentioned in Section 3.

The system (2) is of a mixed type, since the Equation (2a) is hyperbolic and the others form a parabolic system, provided $\rho>0$, $c_v>0$, $\mu>0$, $2\mu+\lambda>0$. At this point the system (2) differs from the Navier-Stokes system for incompressible fluid, which is not parabolic. It is natural that the investigation of Equations (2) rests to a considerable extent upon the theories of the Cauchy problem and of the initial-boundary-value problems for parabolic systems of differential equations which were constructed in the 1950s and early 1960s (see Eidel'man 1964, Friedman 1964, Solonnikov 1965).

Let us now define the basic spaces of functions that are generally used in the theory of parabolic initial-boundary-value problems and that are needed for the formulation of the main existence theorem. Let $G$ be a domain in $R^n$, $Q_T=G\times(0,T): x=(x_1,\ldots,x_n)\in G$, $t\in(0,T)$ and let $l$ be a positive noninteger. By $C^l(G)$ we mean the set of all functions $u(x)$ that are defined in $G$, bounded, continuous, $[l]$ times continuously differentiable ($[l]$ is the integer part of $l$) and whose derivatives of the order $[l]$ satisfy the Hölder condition with the exponent $\lambda = l - [l]$. By $C^{\alpha,\alpha/2}(Q_T)$, $\alpha \in (0,1)$, we mean the set of all functions $u(x,t)$, defined in $Q_T$, bounded and satisfying the Hölder condition with the exponent $\alpha$ with respect to $x$ and with the exponent $\alpha/2$ with respect to $t$. Finally, by $C^{2+\alpha,1+\alpha/2}(Q_T)$ we mean the set of functions $u(x,t)$ defined in $Q_T$ and having the following differentiability properties: $(a)$ they are bounded, continuous, and have continuous derivatives $\partial u/\partial x_i$, $\partial^2 u/\partial x_i \partial x_j$, $\partial u/\partial t$, $(b)$ $\partial u/\partial t, \partial^2 u/\partial x_i \partial x_j \in C^{\alpha,\alpha/2}(Q_T)$, $\partial u/\partial x_j$ satisfy the Hölder condition with the exponent $1/2+\alpha/2$ with respect to $t$.

All these sets of functions are Banach spaces. The norms in $C^{\alpha,\alpha/2}(Q_T)$ and $C^{2+\alpha,1+\alpha/2}(Q_T)$ are defined by

$$|u|_{Q_T}^{(\alpha)} = \sup_{x,t \in Q_T} |u(x,t)| + \sup_{x,y,t} |x-y|^{-\alpha} |u(x,t) - u(y,t)| + \sup_{x,t,\tau} |t-\tau|^{-\alpha/2} |u(x,t) - u(x,\tau)|,$$

$$|u|_{Q_T}^{(2+\alpha)} = \sup_{Q_T} |u(x,t)| + \sum_{i=1}^{n} \sup_{Q_T} \left| \frac{\partial u}{\partial x_i} \right| + \left| \frac{\partial u}{\partial t} \right|_{Q_T}^{(\alpha)} + \sum_{i,j=1}^{n} \left| \frac{\partial^2 u}{\partial x_i \partial x_j} \right|_{Q_T}^{(\alpha)} + \sum_{i=1}^{n} \sup_{x,t,\tau} |t-\tau|^{-1/2-\alpha/2} \left| \frac{\partial u(x,t)}{\partial x_i} - \frac{\partial u(x,\tau)}{\partial x_i} \right|.$$

Using local coordinates, it is not difficult to define such spaces for functions that are prescribed on manifolds, in particular, on boundaries of domains in $R^n$. The same notations will be used for the spaces of vector functions, the norm of a vector supposed to be equal to the sum of norms of all its components.

## 2. LOCAL EXISTENCE THEOREMS

The main existence theorem for Equations (2) is the theorem asserting that the Cauchy problem [(2), (3)] and initial-boundary-value problem [(2)–(5)] are solvable locally with respect to time, i.e. they have a solution in a certain time interval $(0, t_1)$, depending on the data $(\mathbf{f}, \rho_0, \mathbf{v}_0, \theta_0, \text{etc})$. For the Cauchy problem this theorem was proved by Nash (1962), Itaya (1970), and by Vol'pert & Khudyaev (1972) who considered more general systems of equations of a mixed type. Solonnikov (1976) has proved the local solvability of the initial-boundary-value problem for the system (2a), (2b) with $p=p(\rho)$, $\lambda=\text{const}$, $\mu=\text{const}$, and with no-slip condition (4) at the boundary. The general system (2) was considered in Tani (1977). The main result of this paper is as follows.

*Theorem*: Suppose that $\Omega\subset R^3$ is a bounded or unbounded domain with a boundary belonging to $C^{2+\alpha}$, $\alpha\in(0,1)$. Let the functions $\mathbf{f}$, $\partial\mathbf{f}/\partial x_i \in C^{\alpha,\alpha/2}(Q_{t_0})$, $\rho\in C^{1+\alpha}(\Omega)$, $\mathbf{v}_0\in C^{2+\alpha}(\Omega)$, $\theta_1\in C^{2+\alpha,1+\alpha/2}(\Gamma_{t_0})$ satisfy the conditions

$$0<\theta'\leqslant\theta_0(x)\leqslant\theta'', \qquad \theta'\leqslant\theta_1(x)\leqslant\theta'', \qquad 0<\rho'\leqslant\rho_0(x)\leqslant\rho'' \tag{6}$$

and the compatibility conditions

$$\mathbf{v}_0(x)=0, \qquad \theta_0(x)=\theta_1(x,0),$$

$$\rho_0(x)\mathbf{f}(x,0)+\nabla(\lambda(\rho_0,\theta_0)\nabla\cdot\mathbf{v}_0)+2\nabla\cdot\mu(\rho_0,\theta_0)\mathbf{D}(\mathbf{v}_0)-\nabla p(\rho_0,\theta_0)=0,$$

$$\nabla\cdot\kappa(\rho_0,\theta_0)\nabla\theta_0+\lambda(\rho_0,\theta_0)(\nabla\cdot\mathbf{v}_0)^2+2\mu(\rho_0,\theta_0)\mathbf{D}(\mathbf{v}_0):\mathbf{D}(\mathbf{v}_0)$$

$$+\left(\rho_0^2E_{\rho_0}(\rho_0,\theta_0)-p(\rho_0,\theta_0)\right)\nabla\cdot\mathbf{v}_0=\rho_0c_v(\rho_0,\theta_0)\left.\frac{\partial\theta_1}{\partial t}\right|_{t=0}$$

for $x\in S$. Suppose that $\lambda(\rho,\theta)$, $\mu(\rho,\theta)$, $p(\rho,\theta)$, $E(\rho,\theta)$ are defined for

$$\beta\rho'\leqslant\rho\leqslant\beta^{-1}\rho'', \qquad \beta\theta'\leqslant\theta\leqslant\beta^{-1}\theta'', \qquad \beta\in(0,1) \tag{7}$$

and belong to $C^{2+\alpha}$. Then the problem (2)–(5a) has a unique solution $(\mathbf{v},\rho,\theta)$ defined in $Q_{t_1}$, where $t_1\leqslant t_0$ depends on the data, $\mathbf{v}\in C^{2+\alpha,1+\alpha/2}(Q_{t_1})$, $\theta\in C^{2+\alpha,1+\alpha/2}(Q_{t_1})$, $\rho$, $\partial\rho/\partial x_i$, $\partial\rho/\partial t\in C^{\alpha,\alpha/2}(Q_{t_1})$ and the conditions (7) are satisfied.

This theorem as well as similar results of the authors mentioned above follows actually from the contraction-mapping principle, since both the Cauchy problem and initial-boundary-value problems for the Equations (2) can be reduced to the equation of the form $V=\mathfrak{B}V$ in a certain Banach space and the operator $\mathfrak{B}$ is a contraction operator, provided $t_1$ is small enough. It should be noted, however, that some of the authors prefer to use a more direct procedure of successive approximations. To illustrate the basic ideas of the proof, we consider the Cauchy problem (2), (3) where we take $\mathbf{f}=0$ for the sake of simplicity. First of all, it is convenient to change from Eulerian to Lagrangian coordinates in (2). We recall that the Lagrangian coordinates of the particle located at time $t$ at the point $x$ are coordinates of the point $\xi\in R^3$ where the particle was at the time $t=0$. The Eulerian and Lagrangian coordinates are connected by the relation [see, for instance, Rozhdestvenskii & Yanenko (1978)]

$$x=\xi+\int_0^t\mathbf{u}(\xi,\tau)\,d\tau=X(\xi,t),$$

where $\mathbf{u}(\xi, \tau)$ is the velocity at the time $t$ of the particle located at $t=0$ at the point $\xi$. Hence, the system (2) in the coordinates $(\xi, t)$ takes the form

$$\frac{\partial r}{\partial t} + r\nabla_u \cdot \mathbf{u} = 0, \tag{8a}$$

$$r\frac{\partial \mathbf{u}}{\partial t} - \nabla_u(\lambda \nabla_u \cdot \mathbf{u}) - 2\nabla_u \cdot \mu \mathbf{D}_u(\mathbf{u}) + \nabla_u p = 0, \tag{8b}$$

$$rc_v \frac{\partial T}{\partial t} - \nabla_u \cdot \kappa \nabla_u T - 2\mu \mathbf{D}_u(\mathbf{u}) : \mathbf{D}_u(\mathbf{u}) - \lambda(\nabla_u \cdot \mathbf{u})^2 - F(r,T)\nabla_u \cdot \mathbf{u} = 0. \tag{8c}$$

Here $r(\xi, t) = \rho(X(\xi, t), t)$ and $T(\xi, t) = \theta(X(\xi, t), t)$ are the density and the temperature in the Lagrangian coordinates, $F(r,T) = r^2\, \partial E(r,T)/\partial r - p(r,T), \mathbf{D}_u(\mathbf{u})$ is a matrix with elements

$$\sum_{m=1}^{3} (a_{jm}\partial u_i/\partial \xi_m + a_{im}\partial u_j/\partial \xi_m), \qquad i, j, = 1,2,3$$

$$\nabla_u = \mathbf{A}\nabla = \left( \sum_{m=1}^{3} a_{1m}\partial/\partial \xi_m, \sum_{m=1}^{3} a_{2m}\partial/\partial \xi_m, \sum_{m=1}^{3} a_{3m}\partial/\partial \xi_m \right),$$

$$\mathbf{A} = (\mathbf{J}_X^{-1})^*.$$

$\mathbf{J}_X$ is the Jacobi matrix of the transformation $X$ whose elements are $\partial X_i/\partial \xi_j = \delta_{ij} + \int_0^t \partial u_i/\partial \xi_j d\tau$. Integrating Equation (8a) we can reduce our problem to the Cauchy problem for the parabolic system (8b), (8c) with $r(\xi, t) = \rho_0(\xi) \exp(-\int_0^t \nabla_u \cdot \mathbf{u}(\xi, \tau)\, d\tau)$ and with initial conditions

$$\mathbf{u}(\xi, 0) = \mathbf{v}_0(\xi), \qquad T(\xi, 0) = \theta_0(\xi). \tag{9}$$

Let $\mathbf{u}', T'$ be functions defined in $\Pi_\tau = R^3 \times (0, \tau)$ and satisfying the conditions (9) , and let

$$X'(\xi, t) = \xi + \int_0^t \mathbf{u}'(\xi, \tau)\, d\tau, \qquad \mathbf{A}' = (\mathbf{J}_{X'}^{-1})^*,$$

$$r' = \rho_0 \exp\left( -\int_0^t \nabla_{u'} \cdot \mathbf{u}'\, d\tau \right).$$

We consider an auxiliary linear Cauchy problem

$$r'\frac{\partial \mathbf{u}}{\partial t} - \nabla_{u'}(\lambda' \nabla_{u'} \cdot \mathbf{u}) - 2\nabla_{u'} \cdot \mu' \mathbf{D}_{u'}(\mathbf{u}) + p'_{T'}\nabla_{u'} T = -p'_{r'}\nabla_{u'} r', \tag{10}$$

$$r'c'_v \frac{\partial T}{\partial t} - \nabla_{u'} \cdot \kappa' \nabla_{u'} T - \lambda(\nabla_{u'} \cdot \mathbf{u}')(\nabla_{u'} \cdot \mathbf{u})$$

$$-2\mu' \mathbf{D}_{u'}(u') : \mathbf{D}_{u'}(\mathbf{u}) - F'\nabla_{u'} \cdot \mathbf{u} = 0,$$

$$\mathbf{u}(\xi, 0) = \mathbf{v}_0(\xi), \qquad T(\xi, 0) = \theta_0(\xi), \tag{11}$$

where $\lambda' = \lambda(r', T')$, $p' = p(r', T')$ etc, $\mathbf{D}_{u'}(\mathbf{u})$ is a matrix with elements

$$\sum_{m=1}^{3} (a'_{jm}\partial u_i/\partial \xi_m + a'_{im}\partial u_j/\partial \xi_m).$$

The relations (10), (11) can be written in a shorter form

$$\mathfrak{L}(U')U = H(U'), \quad U|_{t=0} = U_0$$

where $U = (\mathbf{u}, T)$, $U' = (\mathbf{u}', T')$, $H(U') = (-p'_{r'}\nabla_{u'}r', 0)$, $U_0 = (\mathbf{v}_0, \theta_0)$ are four-component vectors and $\mathfrak{L}(U')$ is a matrix differential operator in the left-hand side of (10). The nonlinear problem (8b), (8c), (9) corresponds to the case $U' = U$, i.e.

$$\mathfrak{L}(U)U = H(U), \qquad U|_{t=0} = U_0. \tag{12}$$

Let us reduce this problem to the equation $V = \mathfrak{B}V$ in the Banach space of four-component vectors $V \in C^{2+\alpha, 1+\alpha/2}(\Pi_\tau)$ vanishing for $t = 0$. To this end we define (at least for small $\tau$) the vector $U_1$ as the solution of the Cauchy problem in $\Pi_\tau$

$$\mathfrak{L}(U_0)U_1 = H(U_0), \quad U_1|_{t=0} = U_0 \tag{13}$$

and introduce a new unknown vector $V = U - U_1$. Subtracting (13) from (12) we obtain

$$\begin{aligned} \mathfrak{L}(U_1)V = {} & [\mathfrak{L}(U_1) - \mathfrak{L}(U_1 + V)](U_1 + V) + [H(U_1 + V) - H(U_1)] \\ & + [\mathfrak{L}(U_0) - \mathfrak{L}(U_1)]U_1 + [H(U_1) - H(U_0)], \\ V|_{t=0} = {} & 0. \end{aligned}$$

These relations are equivalent to the equation

$$\begin{aligned} V = {} & \mathfrak{R}(\tau)\{[\mathfrak{L}(U_1) - \mathfrak{L}(U_1 + V)](U_1 + V) + [H(U_1 + V) - H(U_1)]\} \\ & + \mathfrak{R}(\tau)\{[\mathfrak{L}(U_0) - \mathfrak{L}(U_1)]U_1 + [H(U_1) - H(U_0)]\} = \mathfrak{B}_\tau V \end{aligned} \tag{14}$$

where $\mathfrak{R}(\tau)$ is a linear operator which is defined on the space of vectors $F \in C^{\alpha, \alpha/2}(\Pi_\tau)$ by letting $W = \mathfrak{R}(\tau)F \in C^{2+\alpha, 1+\alpha/2}(\Pi_\tau)$ be a solution of the following Cauchy problem for a linear parabolic system: $\mathfrak{L}(U_1)W = F$, $W|_{t=0} = 0$ (see Eidelman 1964). If $\tau$ is small enough, more exactly, if the condition

$$(\tau + \tau^{1/2-\alpha/2})|U|_{\Pi_\tau}^{(2+\alpha)} \leqslant \delta \tag{15}$$

is satisfied for a small $\delta \geqslant 0$, the vector $H(U)$ and the coefficients of the operator $\mathfrak{L}(U)$ are smooth functions of the arguments $U$, $\partial U/\partial \xi$, $r = \rho_0(\xi)\exp(-\int_0^t \nabla_u \cdot \mathbf{u}(\xi, t')\,dt')$ and $\partial r/\partial \xi_k$. Hence, if $U_1$, $U_1 + V$ and

$U_1+W$ satisfy (15) and $V(\xi,0)=W(\xi,0)=0$, then

$$|\mathfrak{B}_\tau V|_{\Pi_\tau}^{(2+\alpha)} \leqslant \tau^\gamma \|\mathfrak{R}(\tau)\| \left(c_1|V|_{\Pi_\tau}^{(2+\alpha)}+c_2|U_0-U_1|_{\Pi_\tau}^{(2+\alpha)}\right),$$

$$|\mathfrak{B}_\tau V-\mathfrak{B}_\tau W|_{\Pi_\tau}^{(2+\alpha)} \leqslant c_3\tau^\gamma \|\mathfrak{R}(\tau)\| |V-W|_{\Pi_\tau}^{(2+\alpha)} \tag{16}$$

for a certain $\gamma>0$. The constants $c_1, c_2, c_3$ do not depend on $\tau, V, W$.

Now we choose the numbers $t_1>0$ and $\eta>0$ satisfying the inequalities

$$\left(t_1+t_1^{1/2-\alpha/2}\right)|U_0|_{\Pi_{t_1}}^{(2+\alpha)} \leqslant \delta, \qquad \left(t_1+t_1^{1/2-\alpha/2}\right)|U_1|_{\Pi_{t_1}}^{(2+\alpha)} \leqslant \frac{\delta}{2},$$

$$c_3 t_1^\gamma \|\mathfrak{R}(t_1)\| < 1, \qquad c_1 t_1^\gamma \|\mathfrak{R}(t_1)\| \leqslant \frac{1}{2}, \tag{17}$$

$$2c_2 t_1^\gamma \|\mathfrak{R}(t_1)\| |U_1-U_0|_{\Pi_{t_1}}^{(2+\alpha)} \leqslant \eta \leqslant \frac{1}{2}\delta\left(t_1+t_1^{1/2-\alpha/2}\right)^{-1}.$$

Let $K_\eta$ be the set of all vectors $V \in C^{2+\alpha,1+\alpha/2}(\Pi_{t_1})$, satisfying the conditions $V(\xi,0)=0$, $|V|_{\Pi_{t_1}}^{(2+\alpha)} \leqslant \eta$. The inequalities (16) and (17) entail the following consequences: (*a*) the vectors $U_0$ and $U_1+V$ with any $V \in K_\eta$ satisfy (15) and the norm of $\mathfrak{R}(t_1)$ is bounded; (*b*) the operator $\mathfrak{B}_{t_1}$ maps $K_\eta$ into itself and is a contraction operator; i.e. for any $V, W \in K_\eta$ we have $|\mathfrak{B}_{t_1}V-\mathfrak{B}_{t_1}W|_{\Pi_{t_1}}^{(2+\alpha)} \leqslant c_4|V-W|_{\Pi_{t_1}}^{(2+\alpha)}$, $c_4<1$. Hence, the Equation (14) has a unique solution in $K_\eta$.

The solution of the problem (12) obtained in this way is unique also in the class of vectors in $C^{2+\alpha,1+\alpha/2}(\Pi_{t_1})$ such that $\beta\theta' \leqslant T \leqslant \beta^{-1}\theta''$, $\beta\rho' \leqslant r \leqslant \beta^{-1}\rho''$. The uniqueness of solutions of initial-boundary-value problems for the Equations (2a) in the class of functions with a finite energy integral was first proved in Serrin (1959).

The question of the solvability of the above problems on an arbitrary time interval $(0, t_1)$ independent of the data is still open. It rests on a priori estimates of their solutions in this interval, which would enable us to construct the solutions for $t \in (0, t_1)$ by means of a repeated application of the local existence theorems.

## 3. GLOBAL EXISTENCE THEOREMS IN THE ONE-DIMENSIONAL CASE

In this section we consider one-dimensional viscous compressible flows, i.e. we assume that the vector of velocity is directed along the $x_1$-axis and that $\rho$, $\theta$, and $v_1$ depend only on $t$ and $x_1=x$. Moreover, we suppose that there are no external forces and that the moving continuum is a perfect and polytropic gas, i.e. that

$$p=R\rho\theta, \qquad E=c_v\theta$$

where $R$ and $c_v$ are positive constants. The coefficients of viscosity and of heat conduction are also supposed to be constants.

In the one-dimensional case it is convenient to use the material Lagrangian coordinates $(q, t)$ where $q=\int_0^\xi \rho(\lambda,0)\,d\lambda$ and $\xi$ is the Lagrangian coordinate defined in the preceding section (we suppose that the gas occupies an interval in $R^1$ containing the point $\xi=0$). Furthermore, we introduce instead of density $\rho$ the specific volume $v=\rho^{-1}$. Then the system (2) will assume the form (Rozhdestvenskii & Yanenko 1978)

$$\frac{\partial v}{\partial t}-\frac{\partial u}{\partial q}=0,$$

$$\frac{\partial u}{\partial t}=\mu_1\frac{\partial}{\partial q}\left(\frac{1}{v}\frac{\partial u}{\partial q}\right)-R\frac{\partial}{\partial q}\frac{T}{v}, \tag{18}$$

$$c_v\frac{\partial T}{\partial t}=\kappa\frac{\partial}{\partial q}\left(\frac{1}{v}\frac{\partial T}{\partial q}\right)+\frac{\mu_1}{v}\left(\frac{\partial u}{\partial q}\right)^2-\frac{RT}{v}\frac{\partial u}{\partial q},$$

where $\mu_1=\lambda+2\mu>0$ and $v(q,t)$, $u(q,t)$, and $T(q,t)$ are the specific volume, velocity, and temperature.

We consider also a model system of two equations

$$\frac{\partial v}{\partial t}-\frac{\partial u}{\partial q}=0, \tag{19a}$$

$$\frac{\partial u}{\partial t}+\frac{\partial p(v)}{\partial q}=\mu_1\frac{\partial}{\partial q}\left(\frac{1}{v}\frac{\partial u}{\partial q}\right). \tag{19b}$$

For these systems global existence theorems, i.e. theorems concerning their solvability in an arbitrary time interval, are proved.

The Cauchy problem for the system (19) with initial conditions

$$u(q,0)=u_0(q), \qquad v(q,0)=v_0(q) \tag{20}$$

was first studied apparently in Kanel' (1968) where its global solvability was established under the following assumptions on the data:

(*a*) $p(v)$, $v>0$ is a monotonically decreasing function that is smooth everywhere except perhaps for the point $v=0$,

(*b*) $u_0\in C^{2+\alpha}(R^1)$, $v_0\in C^{1+\alpha}(R^1)$, $v_0$ is positive and tends to a constant $v_\infty$ as $q\to\pm\infty$; the functions $u_0$, $v_0-v_\infty$ and their derivatives are square integrable in $R^1$,

(*c*) the function $\psi(q)=\int_{v_\infty}^{v_0(q)}[p(v_\infty)-p(s)]\,ds$ is integrable in $R^1$.

With other assumptions on the data the same result was obtained in Itaya (1976) in the case $p(v)=\text{const}\cdot v^{-1}$.

Kazhikhov & Nikolaev (1979) have extented Kanel's result to the case of a non-monotonic function $p(v)$ with the following properties:

($a$) there exists at least one number $v_*>0$ (in particular, $v_*=v_\infty$) such that $[p(v)-p(v_*)][v-v_*]\leqslant 0$ for all $v>0$,

($b$) if $p'(v)>0$, then $p'(v)<\text{const}\, v^{-1}$.

The global solvability of the initial-boundary-value problem for the system (19) in the domain $Q_{t_1}=(0,1)\times(0,t_1)$ with boundary conditions

$$u(0,t)=u(1,t)=0 \tag{21}$$

was established in Kazhikhov (1975) for the case $p(v)=\text{const}\, v^{-\gamma}$, $\gamma\geqslant 1$, and in Kazhikhov & Nikolaev (1979) for non-monotonic $p(v)$ satisfying the above restrictions. It is necessary to mention also the works of Itaya (1974), Tani (1974), and Kazhikhov (1976b) where analogous results have been obtained for generalized Burgers' equations, i.e. for the system (19) without the term $\partial p/\partial q$ in (19b).

The study of non-monotonic $p(v)$ is aimed at the investigation of the equations of motion of a continuum for which the van der Waals relation

$$p=\frac{RT}{v-b}-\frac{a}{v^2} \tag{22}$$

is valid. Since the function on the right-hand side is defined only for $v>b$, the a priori estimate of the type $v(q,t)>b$ is needed in the treatment of these equations. This estimate is announced in Kazhikhov (1979a) for the solutions of the problem (19)–(21) under the condition that $p(v)$ is defined and subjected to ($a$), ($b$) for $v>b$ and $\lim_{v\to b}(v-b)p(v)>0$. The function (22) with any fixed $T>0$ possesses these properties.

As pointed out in Section 2, the proof of the solvability of the above problems in the interval $(0,t_1)$ rests on a priori estimates for their solutions. Let us give an idea of these estimates, following mainly Kazhikhov (1975). We consider the problem (19)–(21) where we take for simplicity $p(v)=v^{-1}$.

One of the basic estimates is the one for $v(q,t)$:

$$0<N_1\leqslant v(q,t)\leqslant N_2 \tag{23}$$

($N_i$ are positive constants depending only on $t$, $v_0$, $u_0$). In virtue of (19),

$$\frac{1}{2}\frac{\partial u^2}{\partial t}+\frac{\partial(v-\ln v)}{\partial t}+\frac{\mu_1}{v}\left(\frac{\partial u}{\partial q}\right)^2=\frac{\partial}{\partial q}\left(\frac{\mu_1}{v}u\frac{\partial u}{\partial q}+u-\frac{u}{v}\right), \tag{24}$$

$$\frac{\partial^2}{\partial q\partial t}\left[\int_0^q u(s,t)\,ds-\mu_1\ln v(q,t)+\int_0^t\frac{d\tau}{v(q,\tau)}\right]=0. \tag{25}$$

Integrating Equations (19a) and (24) with respect to $q$ and $t$ we obtain

$$\int_0^1 v(q,t)\,dq=\int_0^1 v_0(q)\,dq=N_3,$$

$$\int_0^1\left\{\frac{1}{2}u^2(q,t)+[v(q,t)-\ln v(q,t)]\right\}dq+\int_0^1\int_0^t\frac{\mu_1}{v}\left(\frac{\partial u}{\partial q}\right)^2 dq\,d\tau\leqslant N_4. \tag{26}$$

Furthermore (25) implies

$$\int_0^q u(s,t)\,ds-\mu_1\ln v(q,t)+\int_0^t\frac{d\tau}{v(q,\tau)}=a(q)+b(t). \tag{27}$$

It can be assumed that $b(0)=0$, therefore

$$a(q)=\int_0^q u_0(s)\,ds-\mu_1\ln v_0(q),$$

$$b(t)=\int_0^1 dq\left[\int_0^q u(s,t)\,ds-\mu_1\ln v(q,t)+\int_0^t\frac{d\tau}{v(q,\tau)}-a(q)\right]$$

and consequently

$$|a(q)|\leqslant N_5,\qquad b(t)\geqslant -N_6. \tag{28}$$

The formula (27) is equivalent to

$$\frac{1}{\mu_1 v}\exp\left(\int_0^t\frac{d\tau}{\mu_1 v(q,\tau)}\right)=\frac{\partial}{\partial t}\exp\left(\int_0^t\frac{d\tau}{\mu_1 v(q,\tau)}\right)=A(q,t)B(t) \tag{27a}$$

with functions $A(q,t)=\mu_1^{-1}\exp(\mu_1^{-1}a(q)-\mu_1^{-1}\int_0^q u(s,t)\,ds)$, $B(t)=\exp\mu_1^{-1}b(t)$. In virtue of (26), (28) they satisfy the inequalities

$$0<N_7\leqslant A(q,t)\leqslant N_8,\qquad B(t)\geqslant\exp(-\mu_1^{-1}N_6),$$

$$\int_0^1\left(\frac{\partial A(q,t)}{\partial q}\right)^2 dq\leqslant N_9. \tag{29}$$

Integrating (27a) with respect to $t$, we can express $\exp(\int_0^t\mu_1^{-1}v^{-1}(q,\tau)\,d\tau)$ in terms of $A$ and $B$ and finally obtain the relation

$$\mu_1 v(q,t)B(t)=A^{-1}(q,t)\left[1+\int_0^t A(q,\tau)B(\tau)\,d\tau\right]. \tag{30}$$

Hence,

$$\mu_1 N_3\,B(t)=\int_0^1 A^{-1}(q,t)\left[1+\int_0^t A(q,\tau)B(\tau)\,d\tau\right]dq$$

$$\leqslant N_7^{-1}\left(1+N_8\int_0^t B(\tau)\,d\tau\right),$$

which implies $B(t) \leqslant (\mu_1 N_3 N_7)^{-1} \exp[(\mu_1 N_3)^{-1} N_8 t_1] = N_{10}$. In turn, this inequality as well as (29) and (30) imply the estimates (23) and

$$\int_0^1 \left( \frac{\partial v(q,\tau)}{\partial q} \right)^2 dq \leqslant N_{11}. \tag{31}$$

After the inequalities (23) and (31) are proved we can estimate the derivatives $\partial u/\partial q$, $\partial^2 u/\partial q^2$, $\partial u/\partial t$ by means of methods of the theory of parabolic equations. First of all, multiplying (19b) by $\partial^2 u/\partial^2 q$ and integrating over $Q_\tau$ with an arbitrary $\tau \leqslant t_1$ we obtain after simple calculations the estimate

$$\sup_{\tau \leqslant t_1} \int_0^1 \left( \frac{\partial u(q,\tau)}{\partial q} \right)^2 dq + \int_0^{t_1} d\tau \int_0^1 \left( \frac{\partial^2 u}{\partial q^2} \right)^2 dq \leqslant N_{12}$$

and then from (19) we conclude that

$$\sup_{\tau \leqslant t} \int_0^1 \left( \frac{\partial v(q,\tau)}{\partial \tau} \right)^2 dq + \int_0^{t_1} d\tau \int_0^1 \left( \frac{\partial u}{\partial \tau} \right)^2 dq \leqslant N_{13}.$$

With all these inequalities we are able to estimate Hölder norms of $u, v$ and of their derivatives by means of well-known methods.

Let us go on to the discussion of the system (18). Kazhikhov (1976a, 1977) proved the global solvability of the problem describing the process of the flowing out of a finite mass of gas into a vacuum. It is the initial-boundary-value problem for the Equations (18) in the domain $Q_{t_1}$ with the initial and boundary conditions

$$u(q,0) = u_0(q), \qquad v(q,0) = v_0(q), \qquad T(q,0) = T_0(q), \tag{32}$$

$$\mu_1 \frac{\partial u}{\partial q} - RT = 0, \qquad \frac{\partial T}{\partial q} = 0 \qquad (q=0, q=1). \tag{33}$$

The conditions (33) mean that the stress and the heat flux vanish on the boundary. In Eulerian coordinates this problem is a free-boundary problem.

Kazhikhov & Shelukhin (1977) obtained the same results for this problem with another boundary condition,

$$u = 0, \qquad \frac{\partial T}{\partial q} = 0, \qquad (q=0, q=1). \tag{34}$$

In both papers the a priori estimates of solutions are obtained by means of the methods discussed above. But the calculations are considerably more complicated, since the estimates of the type (23) are needed both for the specific volume and for the temperature. Recently Kanel' (1979) proved the global solvability of the Cauchy problem (18), (32) under the assumption that $u_0(q), v_0(q), T_0(q)$ tend to constant limits $u_\infty, v_\infty, T_\infty$ as $q \to \pm\infty$ and are close to these constants in a certain sense.

The behavior of solutions of Equations (18) (19) for large $t>0$ is investigated in Kanel' (1968, 1979) and Kazhikhov (1979b). Kanel' proved that solutions of Cauchy problems (19), (20) and (18), (32) tend to $u_\infty, v_\infty, T_\infty$ as $t \to +\infty$. Kazhikhov (1979b) considered the initial-boundary-value problem (19)–(21) and proved that its solutions also tend to the constant limits exponentially.

Finally, we briefly describe the recent results of V. V. Shelukhin concerning the solvability of Equations (18) and (19) with a periodic or quasi-periodic external force $f$, i.e. of equations

$$\frac{\partial v}{\partial t} - \frac{\partial u}{\partial q} = 0, \qquad (35)$$

$$\frac{\partial u}{\partial t} + \frac{\partial p(v)}{\partial q} = \mu_1 \frac{\partial}{\partial q}\left(\frac{1}{v}\frac{\partial u}{\partial q}\right) + f\left(\int_0^q v(s,t)\,ds, t\right),$$

and

$$\frac{\partial v}{\partial t} - \frac{\partial u}{\partial q} = 0,$$

$$\frac{\partial u}{\partial t} + R\frac{\partial}{\partial q}\left(\frac{T}{v}\right) = \mu_1 \frac{\partial}{\partial q}\left(\frac{1}{v}\frac{\partial u}{\partial q}\right) + f\left(\int_0^q v(s,t)\,ds, t\right), \qquad (36)$$

$$c_v \frac{\partial T}{\partial t} = \kappa \frac{\partial}{\partial q}\left(\frac{1}{v}\frac{\partial u}{\partial q}\right) + \frac{\mu_1}{v}\left(\frac{\partial u}{\partial q}\right)^2 - \frac{RT}{v}\frac{\partial u}{\partial q}.$$

Shelukhin (1979, 1980) proves that the system (35) has a smooth, periodic with respect to time, or almost-periodic, or bounded solution defined in a strip $q \in (0,1)$, $-\infty < t < \infty$, and satisfying the boundary conditions (21), provided the function $f$ is periodic, almost-periodic, or bounded. On the other hand, the system (36) cannot have periodic solutions, satisfying the conditions (34) for any periodic $f(x,t)$ if $f$ is not a function of the space variable $x$ only.

## 4. EXISTENCE THEOREMS FOR THE EQUATIONS OF MOTION OF MULTI-COMPONENT AND MICROPOLAR FLUIDS

In this section we consider systems of differential equations whose solvability can be established by means of the same methods as for the Equations (2).

The motion of a homogeneous mixture of $n$ viscous compressible components is often described (see Ishii 1975) by a model system of equations

$$\frac{\partial \rho_i}{\partial t}+\nabla\cdot\rho_i\mathbf{v}_i=0, \tag{37}$$

$$\rho_i\left[\frac{\partial \mathbf{v}_i}{\partial t}+(\mathbf{v}_i\cdot\nabla)\mathbf{v}_i\right]-\mu_i\nabla^2\mathbf{v}_i-\lambda_i\nabla(\nabla\cdot\mathbf{v}_i)+\nabla p_i=\rho_i\mathbf{f}_i+\sum_{j=1}^{n}\mathbf{h}_{ij},$$

$$i=1,\ldots,n,$$

where $\rho_i, \mathbf{v}_i, p_i(\rho_i)$ are the density, the velocity, and the pressure of the $i$th component, $\mathbf{h}_{ij}$ is the intensity of the impulse exchange between the $i$th and $j$th components, which is equal to $K_{ij}(\mathbf{v}_i-\mathbf{v}_j)$, $K_{ij}=K_{ji}=\text{const}$, $\lambda_i$, $\mu_i$ are the constant coefficients of viscosity of an $i$th component. Förste (1978a) proved for $n=2$ the local solvability of the initial-boundary-value problem for the system (37) in a bounded domain $\Omega$ with initial and boundary conditions

$$\mathbf{v}_i(x,0)=\mathbf{v}_{0i}(x), \qquad \rho_i(x,0)=\rho_{0i}(x)>0, \qquad x\in\Omega$$

$$\mathbf{v}_i(x,t)=0, \qquad x\in S,$$

where $S$ is a boundary of $\Omega$. The method of proof is close to the arguments in Solonnikov (1976) where the case $n=1$ is considered. In the one-dimensional case Kazhikhov & Petrov (1978) established the global solvability of this problem for arbitrary $n$ under the condition $p_i'(\rho_i)>0$.

The flow of a two-phase heterogeneous mixture, consisting of a continuous and a dispersible phase, is considered in Förste (1979). The flow is characterized by the density $\rho$, the velocity $\mathbf{v}$ of the mixture, and the concentration $\alpha$ of the dispersible phase. These functions are connected by a model system of equations:

$$\frac{d\rho}{dt}+\rho\nabla\cdot\mathbf{v}=0,$$

$$\rho\frac{d\mathbf{v}}{dt}-\mu\nabla^2\mathbf{v}-\lambda\nabla(\nabla\cdot\mathbf{v})+h\nabla\alpha+\nabla p(\rho)=\rho\mathbf{f},$$

$$\frac{\partial\alpha}{\partial t}+\nabla\cdot\left[\alpha(\mathbf{v}+\mathbf{v}_\infty)\right]=D\nabla^2\alpha,$$

where $\mu$, $\lambda$, $h$, $D>0$ are constants, $\mathbf{v}_\infty$ is a certain constant velocity that is characteristic for the mixture. J. Förste proved the local solvability of the initial-boundary-value problem for these equations with initial and boundary conditions:

$$\rho(x,0)=\rho_0(x)>0,\ \alpha(x,0)=\alpha_0(x),\ \mathbf{v}(x,0)=\mathbf{v}_0(x),$$

$$\mathbf{v}=0,\ \frac{\partial\alpha}{\partial n}=0,\ x\in S$$

(the last condition means that the boundary is impenetrable for the dispersible phase).

In 1966 A. Eringen introduced a class of fluids possessing a deformable microstructure which he called micropolar fluids. The movement of isotropic micropolar fluids is characterized, except for $\rho$ and $\mathbf{v}$, by the microrotation vector $\boldsymbol{\nu}$ and the microinertia $j$. These functions are connected by the system of equations of motion:

$$\frac{d\rho}{dt}+\rho\nabla\cdot\mathbf{v}=0,\ \frac{dj}{dt}=0,$$

$$\rho\frac{d\mathbf{v}}{dt}-(\mu+k)\nabla^2\mathbf{v}-\lambda\nabla(\nabla\cdot\mathbf{v})-k\nabla\times\boldsymbol{\nu}+\nabla p=\rho\mathbf{f},$$

$$\rho j\frac{d\boldsymbol{\nu}}{dt}-\gamma\nabla^2\boldsymbol{\nu}-b\nabla(\nabla\cdot\boldsymbol{\nu})-k\nabla\times\mathbf{v}+2k\boldsymbol{\nu}=\rho\mathbf{l},$$

where $\mathbf{l}$ is a specified vector of body couple, $k, \gamma, b$ are given positive constants.

In Förste (1978b) the local existence theorem is proved for the initial-boundary-value problem for these equations, in which the initial and boundary conditions are

$$\rho(x,0)=\rho_0(x), j(x,0)=j_0(x),\ \mathbf{v}(x,0)=\mathbf{v}_0(x),\ \boldsymbol{\nu}(x,0)=\boldsymbol{\nu}_0(x),$$

$$\mathbf{v}=0,\ \boldsymbol{\nu}=0,\ x\in S.$$

The functions $\rho_0$ and $j_0$ are strictly positive.

## 5. CONCLUDING REMARKS

As seen from the statements above the mathematical analysis of the equations of motion of a viscous compressible fluid is far from complete, even as far as the existence theorems are concerned. In two- and three-dimensional cases only local existence theorems are established. One of the most important unsolved problems is whether or not there exist global solutions of the problems discussed or at least whether one can prove their global solvability under additional restrictions on the data, as is done for the Navier-Stokes equations (see the review of Ladyzhenskaya 1975). Matsumura & Nishida (1980) considered the Cauchy problem for the equations of motion of a viscous perfect polytropic heat-conductive gas in three-dimensional space and proved that it is solvable "in the large" with respect to time, provided the initial data are close to constants.

In the one-dimensional case the most interesting unsolved problems, in the opinion of the authors, are (*a*) the proof of the global solvability of the complete system of equations of motion of a viscous compressible gas with a non-monotonic function $p(v, T)$ and in the first line with a function $p(v, T)$ defined by (22), (*b*) the investigation of initial-boundary-value problems with more general initial data and with non-homogeneous boundary conditions, in particular, of problems describing the flow through the interval $\Omega$, in which it is difficult to use Lagrangian coordinates, (*c*) the analysis of the behavior of solutions when the coefficient of viscosity tends to zero.

*Literature Cited*

Èidel'man, S. D. 1964. *Parabolicheskiye Sistemy*. Moscow: Nauka (In Russian) 443 pp.

Eringen, A. C. 1966. Theory of micropolar fluids. *J. Math. Mech*. 16:1–18

Förste, J. 1978a. Über die Lösbarkeit der Feldgleichungen des Zweiflüssigkeitsmodells für ein staubbeladenes Gas. *Z. Angew. Math. Mech*. 58:189–98

Förste, J. 1978b. Zur Theorie kompressibler micropolarer Flüssigkeiten. *Z. Angew. Math. Mech*. 58:464–66.

Förste, J. 1979. Über das Driftmodell für disperse Zweiphasenströmungen. *Math. Nachr*. 89:57–69

Friedman, A. 1964. *Partial Differential Equations of Parabolic Type*. New York: Prentice-Hall. 347 pp.

Ishii, M. 1975. *Thermo-Fluid Dynamic Theory of Two-Phase Flow*. Paris: Eyrolles

Itaya, N. 1970. The existence and uniqueness of the solution of the equations describing compressible viscous fluid flow. *Proc. Jpn. Acad*. 46:379–82

Itaya, N. 1974. On the temporally global problem of the generalized Burgers' equation. *J. Math. Kyoto Univ*. 14:129–77

Itaya, N. 1976. A survey on the generalized Burgers' equation with a pressure model term. *J. Math. Kyoto Univ*. 16:223–40

Kanel', Ya. I. 1968. On a model system of equations of one-dimensional movement of gas. *Diff. Uravn*. 4:721–34 (In Russian) Translation in *Diff. Equations* 4:374–80

Kanel', Ya. I. 1979. On the Cauchy problem for the equations of gas dynamics with viscosity. *Sib. Mat. Zh*. 20:293–306 (In Russian) Translation in *Sib. Math. J*. 20:208–18

Kazhikhov, A. V. 1975. Correctness "in the large" of initial-boundary value problems for model system of equations of a viscous gas. *Din. Sploshnoi Sredy* 21:18–47 (In Russian)

Kazhikhov, A. V. 1976a. On global solvability of one-dimensional boundary value problems for equations of motion of a viscous heat-conducting gas. *Din. Sploshnoi Sredy* 24:45–61 (In Russian)

Kazhikhov, A. V. 1976b. On boundary value problems for Burgers' equation for the compressible flow in domains with moving boundaries. *Din. Sploshnoi Sredy* 26:60–76 (In Russian)

Kazhikhov, A. V. 1977. Sur la solubilité globale des problèmes monodimensionnels aux valeurs initiales-limites pour les équations d'un gas visqueux et calorifère. *C. R. Acad. Sci. Paris Ser. A* 284:317–20

Kazhikhov, A. V. 1979a. Some problems of the theory of the Navier-Stokes equations for a compressible fluid. *Din. Sploshnoi Sredy* 38:33–47 (In Russian)

Kazhikhov, A. V. 1979b. On the stabilization of solutions of the initial-boundary value problem for the equations of motion of a barotropic viscous fluid. *Diff. Uravn*. 15:662–67 (In Russian) Translation in *Diff. Equations* 15:463–67

Kazhikhov, A. V., Nikolaev, V. B 1979. On the correctness of boundary value problems for the equations of a viscous gas with a non-monotonic function of state.

*Chislennye Metody Mekh. Sploshnoi Sredy* 10:77–84 (In Russian)

Kazhikhov, A. V., Petrov, A. N. 1978. On the correctness of an initial-boundary value problem for a model system of equations for a multi-component mixture. *Din. Sploshnoi Sredy* 35:61–73 (In Russian)

Kazhikhov, A. V., Shelukhin, V. V. 1977. The unique solvability "in the large" with respect to time of initial-boundary value problems for one-dimensional equations of a viscous gas. *Prikl. Mat. Mekh.* 41:282–91 (In Russian) Translation in *J. Appl. Math. Mech.* 41, no. 2:273–82

Ladyzhenskaya, O. A. 1975. Mathematical analysis of Navier-Stokes equations for incompressible fluids. *Ann. Rev. Fluid Mech.* 7:249–72

Matsumura, A., Nishida, T. 1980. The initial value problem for the equations of motion of viscous and heat-conductive gases. *J. Math. Kyoto Univ.* 20:67–104

Nash, J. 1962. Le problème de Cauchy pour les équations différentielles d'un fluide général. *Bull. Soc. Math. France* 90:487–97

Rozhdestvenskii, B. L., Yanenko, N. N. 1978. *Sistemy Kvasilineinykh Uravnenii i Ikh Prilozheniya k Gasovoi Dinamike.* Moscow: Nauka. 687 pp.

Serrin, J. 1959. On the uniqueness of compressible fluid motion. *Arch. Ration. Mech. Anal.* 3:271–88

Shelukhin, V. V. 1979. Periodic flow of viscous gas. *Din. Sploshnoi Sredy* 42:80–102 (In Russian)

Shelukhin, V. V. 1980. On bounded and quasi-periodic solutions of the equations of a viscous gas. *Din. Sploshnoi Sredy* 44:147–63 (In Russian)

Solonnikov, V. A. 1965. On boundary value problems for linear parabolic systems of differential equations of a general type. *Tr. Mat. Inst. Steklov* 83:3–163 (In Russian) Translation in *Proc. Steklov Inst. Math. No.* 83. Providence, R.I.: Am. Math. Soc. iv + 184 pp.

Solonnikov, V. A. 1976. On the solvability of an initial-boundary-value problem for the equations of motion of a compressible viscous fluid. *Zap. Nauch. Semin. Leningr. Otd. Mat. Inst. Steklov* 56:128–42 (In Russian)

Tani, A. 1974. On the first initial-boundary value problem of the generalized Burgers' equation. *Publ. Res. Inst. Math. Sci. Kyoto Univ.* 10:209–33

Tani, A. 1977. On the first initial-boundary value problem of compressible viscous fluid motion. *Publ. Res. Inst. Math. Sci. Kyoto Univ.* 13:193–253

Vol'pert, A. I., Khudyaev, S. I. 1972. On the Cauchy problem for composite systems of non-linear differential equations. *Mat. Sb.* 87:504–28 (In Russian) Translation in *Math. USSR-Sbornik* 16:517–44

Ann. Rev. Fluid Mech. 1981. 13:97–129


# TURBULENCE IN AND ABOVE PLANT CANOPIES*



*M. R. Raupach*
Division of Environmental Mechanics, CSIRO, Canberra City, A.C.T. 2601, Australia

*A. S. Thom*
Department of Meteorology, University of Edinburgh, EH9 3JZ, Scotland

## 1 INTRODUCTION

Traditionally, the main motive for studying turbulent flow in the plant environment has been to understand the processes governing momentum, heat, and mass exchange between the atmosphere and the biologically active canopy. This exchange regulates the microclimate in which plants grow, provides them with carbon dioxide for photosynthesis, and removes the water vapor produced in transpiration. An understanding of its mechanisms is essential for a variety of applications in biology, hydrology, agriculture, and forestry, as well as being relevant to wider questions concerning the global balances of carbon dioxide and nitrogen. Because of these strong and diverse practical motives, most research into turbulence in the plant environment has been empirical and observational; several decades of effort by many workers have produced no general and successful theory.

Most theoretical work has assumed that fluxes within a plant canopy are governed by the local diffusion equation

$$Q_S = -\rho K_S \, \partial\bar{s}/\partial z \tag{1}$$

where $Q_S$ is the vertical flux density of a property with mean concentration $\bar{s}$ per unit mass, and where the turbulent diffusivity $K_S$ is specifiable in terms of local flow or canopy parameters. However, the assumptions

*In memory of Alasdair Strang Thom, who died shortly after this review was completed.



0066-4189/81/0115-0097$01.00

underlying this equation have long been known to be questionable. Therefore, recent research has moved towards identifying the limitations of local-diffusion models and providing alternatives of greater physical reality. One promising approach is the application within the canopy region of higher-order closure techniques developed for solving the Reynolds equations in the planetary boundary layer (e.g. Donaldson 1973).

Although the diffusion approach has fallen into disfavor among micrometeorologists and turbulence workers in recent years, it remains the foundation for a great deal of work on the plant microclimate. It has given approximate but useful insight into the way in which physical and biological factors combine to govern the transpiration and photosynthesis rates of a plant canopy, and has been used to parameterize turbulent transport in several computer models of the physical and biological processes in crops. These developments are reviewed in the book edited by Monteith (1975), especially in the chapters by P. E. Waggoner (on physical-biological modeling) and A. S. Thom (on the application of the diffusion approach to turbulent transport in the plant environment).

Here, we bypass such applied aspects of the subject. Rather, we explore current limits to knowledge about the physics of turbulence in canopies, concentrating on momentum transport but mentioning heat and mass transport where appropriate.

## 2 MECHANICS OF CANOPY TURBULENCE

### *2.1 The Situation under Consideration*

A plant canopy consists of numerous elements such as leaves, stems, and branches, aggregated into complex structures. When this rough surface interacts with the airflow above and within it, the following important aerodynamic processes occur:

1. Momentum is absorbed from the flow by both form and skin-friction drag on elements.
2. Heat, and other scalar properties like water vapor and carbon dioxide, are exchanged between the flow and the elements.
3. Momentum and scalar properties are transported vertically by turbulent diffusion, and possibly by dispersion as well. (We later define precisely the term "dispersion" in this context.)
4. The elements of the canopy generate turbulent wakes, which convert the mean kinetic energy of the flow into turbulent kinetic energy at length scales characteristic of the elements. (For economy of expression, we henceforth refer to the mean kinetic energy as MKE and the turbulent kinetic energy as TKE.)

5. Most plants wave in the wind, thereby storing MKE as strain potential energy, to release it as TKE half a waving cycle later.

To understand these processes in detail, it is convenient to restrict attention to a level, horizontally homogeneous, extensive canopy over which the mean wind is steady and unidirectional. Over the canopy there forms a boundary layer in streamwise equilibrium; it includes a surface layer in which vertical fluxes above the canopy are approximately constant with height, and in which the mean streamwise pressure gradient is negligible. The surface layer must be considered in two parts: the upper part, the *inertial sublayer* (Tennekes 1973), is a region where effective height is the only length scale in adiabatic conditions, where profiles obey semi-logarithmic laws or their diabatic extensions, and where the mean flow can be described one-dimensionally using surface-layer similarity theory (e.g. Wyngaard 1973). (In admitting diabatic influences we are using Tennekes' terminology loosely, as he defined the inertial sublayer for adiabatic flow only.) The lower part of the surface layer is the region close to and within the canopy itself, where the mean flow is three-dimensional because it is mechanically and thermally influenced by nearby canopy elements. We call this region the *roughness sublayer*, but the terms "transition layer" and "interfacial layer" are also used. Despite its three-dimensional nature, the roughness sublayer can be described using a single vertical axis if flow and canopy properties are horizontally averaged. However, this operation requires care, because the resulting one-dimensional description must correctly parameterize the several essentially three-dimensional processes mentioned above.

## 2.2 Conservation Equations

At any point in the roughness sublayer, the airflow obeys three-dimensional, time-averaged conservation equations for mass, momentum, and a passive property $S$ with specific concentration $s$:

$$\partial \bar{u}_i / \partial x_i = 0 \tag{2}$$

$$\frac{\partial \bar{u}_i}{\partial t} + \bar{u}_j \frac{\partial \bar{u}_i}{\partial x_j} + \frac{\partial}{\partial x_j} \overline{u_i' u_j'} = -\frac{1}{\rho} \frac{\partial \bar{p}}{\partial x_i} + \frac{g \bar{\theta}}{T_0} \delta_{i3} + \nu \nabla^2 \bar{u}_i, \tag{3}$$

$$\frac{\partial \bar{s}}{\partial t} + \bar{u}_j \frac{\partial \bar{s}}{\partial x_j} + \frac{\partial}{\partial x_j} \overline{u_j' s'} = \kappa_S \nabla^2 \bar{s}. \tag{4}$$

Here $u_i$ and $x_i$ are velocity and position vectors, respectively, with $i=1$, 2, and 3 denoting the longitudinal, lateral, and vertical directions relative to the mean wind; $t$ is time, $\rho$ air density, $p$ pressure without hydrostatic component, $g$ acceleration due to gravity, $\theta$ deviation from a reference temperature that decreases adiabatically with height, $T_0$ an average absolute temperature, $\nu$ kinematic viscosity of air, and $\kappa_S$ molecular diffusivity for the property $S$. The summation convention applies for indices repeated within terms, and $\delta_{ij}$ is the Kronecker delta. Overbars and primes denote time averages and fluctuations, respectively, so that $u_i=\bar{u}_i+u_i'$ and $s=\bar{s}+s'$. The time-averaging procedure has introduced the Reynolds-stress tensor $\overline{u_i'u_j'}$ and the vectorial turbulent flux $\overline{u_j's'}$ of $S$. The momentum-conservation equation (3) includes no Coriolis-force term because it is usually negligible in the surface layer (Lumley & Panofsky 1964).

The effects of the canopy do not appear explicitly in the conservation equations until a horizontal average is taken (Wilson & Shaw 1977). This is best done by horizontally averaging the time-averaged equations (2), (3), and (4) over an area large enough to eliminate variation caused by individual canopy elements.[1] Let angle brackets denote a horizontal average, and double primes a departure of a time-averaged quantity therefrom; thus, $\bar{u}_i=\langle\bar{u}_i\rangle+\bar{u}_i''$ and $\bar{s}=\langle\bar{s}\rangle+\bar{s}''$. Formally, if $\psi(x_1,x_2,x_3)$ is a scalar field defined in the air but not at points occupied by canopy elements,

$$\langle\bar{\psi}\rangle=\frac{1}{A}\iint_R\bar{\psi}(x_1,x_2,x_3)\,dx_1\,dx_2$$

where $A$ is the area of the region $R$ of the $x_1x_2$ plane. In the free atmosphere, the operation of horizontal averaging commutes with spatial differentiation, as is the case for time averaging. However, this is not always so in the canopy, where the region $R$ is multiply connected because it is intersected by plant parts. Full analysis shows that if $\bar{\psi}$ is constant at the air-element interfaces, then averaging and differentiation commute, so that $\langle\partial\bar{\psi}/\partial x_i\rangle=\partial\langle\bar{\psi}\rangle/\partial x_i$. Otherwise, they do not commute; in particular, $\langle\partial\bar{\psi}''/\partial x_i\rangle\neq 0$. A simple example is provided by the pressure field for the flow over and about a series of fences lying across a wind in the $x_1$ direction. A pressure differential exists across each fence because form drag takes place there; therefore, in the space between adjacent fences, $\partial\bar{p}/\partial x_1=\partial\bar{p}''/\partial x_1>0$. Clearly, a horizontal average gives $\langle\partial\bar{p}''/\partial x_1\rangle>0$. However, $\partial\langle\bar{p}''\rangle/\partial x_1=0$ by definition, so the operators are noncommutative.

When this horizontal-averaging operation is applied to Equations (2), (3), and (4), considerable simplification is gained (and no physics is lost) by considering only a horizontally homogeneous canopy subject to

[1]Our discussion of horizontal averaging within canopies is condensed from an analysis by M. R. Raupach and R. H. Shaw, to be published.

stationary flow. Using the meteorological notation $u_i=(u,v,w)$ and $x_i=(x,y,z)$, with the mean wind vector in the surface layer aligned with the $x$ axis, the conservation equations for momentum and the property $S$ become

$$\left.\begin{aligned}\frac{\partial}{\partial z}\langle\bar{u}''\bar{w}''\rangle+\frac{\partial}{\partial z}\langle\overline{u'w'}\rangle&=-\frac{1}{\rho}\left\langle\frac{\partial\bar{p}''}{\partial x}\right\rangle+\nu\langle\nabla^2\bar{u}''\rangle\\&=f_D+f_V\end{aligned}\right\}\tag{5}$$

$$\left.\begin{aligned}\frac{\partial}{\partial z}\langle\bar{w}''\bar{s}''\rangle+\frac{\partial}{\partial z}\langle\overline{w's'}\rangle&=\kappa_S\langle\nabla^2\bar{s}''\rangle\\&=q_S\end{aligned}\right\}\tag{6}$$

We assume no mean streamwise pressure gradient, and negligible vertical molecular transport in comparison with vertical turbulent transport. The terms that account for effects induced specifically by the canopy are the first term on the left of each equation, and all the terms on the right. These terms are absent in the inertial sublayer, where the equations simply state that vertical fluxes are constant with height.

The terms on the right have all arisen because of the noncommutativity of horizontal averaging and spatial differentiation. In Equation (5) there are two such terms, respectively equal to the form drag $f_D$ and the viscous drag $f_V$ per unit mass of air. The above example of flow through a series of fences has indicated how form drag may be identified with $-\langle\partial\bar{p}''/\partial x\rangle/\rho$; a similar argument can be used to identify $\nu\langle\nabla^2\bar{u}''\rangle$ with the viscous drag. Likewise, the term on the right of Equation (6) can be identified as the specific creation density $q_S$, or emission rate per unit mass of air, of the property $S$ by canopy elements—for example, the horizontally averaged emission rate of sensible heat, water vapor, or carbon dioxide. It is important to note that these drag and creation-density terms emerge from a proper averaging of the conservation equations in the canopy, as first shown by Wilson & Shaw (1977). The alternative procedure is to add the drag forces arbitrarily into the equation of motion, where they appear as body forces. Although this approach has a long history, it is conceptually in error: body forces suppress TKE and dampen turbulence levels, whereas form drag converts MKE to TKE in the element wakes, thus contributing to the very high turbulence levels found inside real canopies. This is further discussed below.

The first term on the left of Equations (5) and (6) is a dispersive flux, arising from the spatial correlation of regions of mean updraft or downdraft with regions where $\bar{u}$ or $\bar{s}$ differs from its spatial mean. It is not a truly turbulent flux, since it could also arise if the flow were laminar. It combines with the turbulent flux $\overline{u'w'}$ to produce a total horizontally averaged vertical flux $\langle \bar{u}''\bar{w}''+\overline{u'w'}\rangle$, and similarly for the property $S$. The dispersive flux is potentially significant not only within the canopy, but also in that part of the roughness sublayer lying above the canopy, where constancy of flux with height applies to the sum $\langle \bar{u}''\bar{w}''+\overline{u'w'}\rangle$, not to the turbulent component $\overline{u'w'}$ alone. In the roughness sublayer over laboratory rough surfaces in wind tunnels, measurements have shown that $|\overline{u'w'}|$ decreases as the surface is approached (Antonia & Luxton 1971, Mulhearn & Finnigan 1978, Raupach et al 1980). Antonia & Luxton attributed this to a stress contribution from $\langle \bar{u}''\bar{w}''\rangle$, although attempts at direct measurement have been unsuccessful (Mulhearn 1978). It is not yet known how significant dispersive fluxes may be in real canopies. However, there is evidence that the covariance $\overline{w's'}$ successfully measures property fluxes for heat and water vapor in the roughness sublayer above forest canopies (Thompson 1979), suggesting that the dispersive mechanism is not significant for scalar properties above the canopy.

## 2.3 *Second-Moment Equations*

For canopy flow, as in the free atmosphere, the equations for the second moments of the flow properties provide considerable understanding of the nature of the turbulence, and also open possibilities for modeling that do not exist if attention is restricted to the conservation equations alone. As a useful example of the behavior of second moments in the canopy, we consider the kinetic energy per unit mass, $\frac{1}{2}\,\overline{u_i u_i}=\frac{1}{2}(\overline{u^2}+\overline{v^2}+\overline{w^2})$. This is usually expressed as the sum of a mean component $\frac{1}{2}\,\bar{u}_i\bar{u}_i$, the MKE, and a turbulent component $\frac{1}{2}\,\overline{u_i'u_i'}$, the TKE. The MKE and TKE budgets may be derived from the conservation equations (e.g. Lumley & Panofsky 1964). Ignoring buoyancy forces for simplicity, the three-dimensional, time-averaged TKE budget at a single point is

$$\frac{1}{2}\left(\frac{\partial}{\partial t}+\bar{u}_j\frac{\partial}{\partial x_j}\right)\overline{u_i'u_i'}=-\overline{u_i'u_j'}\,\frac{\partial \bar{u}_i}{\partial x_j}-\frac{\partial}{\partial x_j}\left(\frac{\overline{p'u_i'}}{\rho}\delta_{ij}+\frac{\overline{u_j'u_i'u_i'}}{2}\right)-\varepsilon \quad (7)$$

where molecular contributions to transport are assumed negligible, and where $\varepsilon=\nu\overline{(\partial u_i'/\partial x_j)(\partial u_i'/\partial x_j)}$ is the dissipation rate. The significance of the terms in this equation is well known: time rate-of-change and advection terms appear on the left, and production, pressure transport, inertial transport, and dissipation terms on the right, the transport terms being here bracketed together. The production term describes the conversion of MKE to TKE, and the dissipation term accounts for the ultimate conversion of TKE to heat.

As with the conservation equations, the operation of horizontal averaging makes certain canopy effects explicit. Using the same rules as before, and again considering stationary flow and a horizontally homogeneous canopy, Equation (7) becomes

$$
\begin{aligned}
0 = & -\langle \overline{u'w'} \rangle \frac{\partial \langle \bar{u} \rangle}{\partial z} - \left\langle \overline{u_i'u_j'}'' \frac{\partial \bar{u}_i''}{\partial x_j} \right\rangle \\
& \qquad\quad \text{I} \qquad\qquad\qquad \text{II} \\
& - \frac{\partial}{\partial z}\left( \frac{\langle \overline{p'w'} \rangle}{\rho} + \frac{1}{2}\langle \overline{w'u_i'u_i'} \rangle + \frac{1}{2}\langle \bar{w}''\, \overline{u_i'u_i'}'' \rangle \right) - \langle \varepsilon \rangle \qquad (8) \\
& \qquad\qquad \text{III} \qquad\qquad \text{IV} \qquad\qquad\qquad \text{V} \qquad\qquad \text{VI}
\end{aligned}
$$

Here, terms I, III, IV, and VI are analogous to the corresponding terms in Equation (7). The remaining terms are unique to the canopy environment, and may be interpreted as follows: term V represents transport of TKE by a dispersive flux equivalent to the dispersive momentum flux $\langle \bar{u}''\bar{w}'' \rangle$. As with momentum, its importance is unknown. Term II is of much more certain significance: it accounts for the production of TKE from MKE in the wakes of canopy elements, and hence may be labeled the "wake production" term. In a typical wake, the local Reynolds stress $\overline{u_i'u_j'}$ is directed down the local gradient $\partial \bar{u}_i/\partial x_j$, so that, after horizontal averaging, the wake- and shear-production terms will be of the same sign—both inducing a gain in TKE. The energy converted by wake production from MKE to TKE is equal to the work done by the flow against form drag, $\langle \bar{u} \rangle f_D$. However, wake production occurs throughout the wake region of an element, unlike the form drag, which is localized at the element itself. Hence, the wake-production term can be equated to $\langle \bar{u} \rangle f_D$ only after it has been averaged over the region of influence of the wake. In particular, some wake production will occur in that part of the roughness sublayer lying above the canopy, at levels where no form drag occurs.

The turbulence generated by wake production has a length scale equal to that of the canopy elements, which is much smaller than the characteristic length scales of the shear-generated turbulence. Wake turbulence is therefore quickly dissipated to heat (the dissipation rate being of order $e^3/\ell$, where $e$ and $\ell$ are turbulence velocity and length scales, respectively) and so is short-lived in the canopy environment. The difference between the length scales of the two types of turbulence has another important consequence: form drag extracts energy not only from the mean flow but also from the large, shear-generated eddies, converting this energy directly to small-scale turbulence by wake production. The normal eddy-cascade process for the decay of large eddies is thereby short-circuited, and their dissipation rate enhanced. This mechanism is not represented directly in Equation (8), which does not discriminate between turbulent motion of differing scales.

The flow performs work not only against the form drag but also against the viscous drag, so that the *total* rate of working of the mean flow is $\langle\bar{u}\rangle(f_D+f_V)$. When the MKE budget within the canopy is constructed, the portion $\langle\bar{u}\rangle f_D$ appears as a wake-production term similar to that in the TKE budget, and the portion $\langle\bar{u}\rangle f_V$ appears as a viscous-dissipation term for MKE. Viscous destruction of MKE is negligible in the free-surface layer, but is important in the viscous sublayers surrounding individual elements. It is a significant mechanism for kinetic-energy destruction within the canopy whenever viscous drag is significant in comparison with form drag.

## *2.4 Parameterization of Drag Forces and Property-Creation Densities*

If a single canopy element is exposed to a uniform wind of speed $u$, then the drag force $F$ upon the element is

$$F=\rho C_M A u^2 \tag{9}$$

where $A$ is the area of the element and $C_M$ is a drag coefficient of order 1/2. (In common with meteorological convention, no factor 1/2 appears explicitly.) If the element wake is laminar, then $C_M \propto \mathrm{Re}^{-\frac{1}{2}}$ (where $\mathrm{Re}=uL/\nu$ is the Reynolds number, $L$ being the element's streamwise length scale in this case) and $A$ is the wetted surface area (Schlichting 1968, p. 128). If the element wake is turbulent and form drag is the principal drag mechanism, then $A$ is the frontal surface area and $C_M$ is approximately independent of a Reynolds number based on a cross-sectional length scale. Typically, $C_M \approx 0.5$ when Re is between $10^3$ and $10^5$ (Goldstein 1938, p. 419). Above $\mathrm{Re} \approx 10^5$, the "drag crisis" leads to a sudden decrease in $C_M$, but such high Reynolds numbers are not found for individual elements in canopies. In practice, the drag on an element is usually a mixture of form and viscous; for example, Thom (1968) found that for bean leaves at typical angles of incidence, the ratio of form to viscous drag was 3 to 1. The result is a wind-speed dependence $C_M \propto u^{-n}$, where $n$ lies between 0 and 1/2.

A similar transfer equation applies to the flux $Q_S$ of the property $S$ from an element in a uniform wind:

$$Q_S=\rho C_S A u(s_0-s) \tag{10}$$

where $C_S$ is the transfer coefficient, $s_0$ is the concentration at the element surface, and $A$ is the wetted surface area. The transfer of properties other than momentum occurs by molecular diffusion alone;

therefore, it is usually less efficient than momentum transfer (Thom 1972).

When element drag and transfer coefficients are measured in situ within a canopy, the resulting values are usually several times smaller than those obtained for an element in an equivalent uniform wind. This was termed the "shelter effect" by Thom (1971). The "equivalent uniform wind" is a uniform flow with speed and property concentration equal to the values of $\bar{u}$ and $\bar{s}$ in those spaces between canopy elements where measurements are possible—usually, the spaces between plants. Although these values are not unique, spatial deviations in $\bar{u}$ and $\bar{s}$ are insufficient to account for the shelter effect (Seginer et al 1976). To quantify the effect, shelter factors $p_M$ and $p_S$ (for momentum and property, respectively) can be defined as

$$p_M \, (\text{or } p_S) = \frac{F \, (\text{or } Q_S) \text{ in equivalent uniform wind}}{F \, (\text{or } Q_S) \text{ in situ}}. \tag{11}$$

In a wind-tunnel study of momentum and water-vapor exchange from spruce shoots and their constituent needles, Landsberg & Thom (1971) found that $p_M \approx p_S$. Unfortunately, this convenient equality is only likely to hold in situations where form drag is not significant. Field values of $p_M$ are typically between 3 and 4, as found both by Thom (1971) for a bean crop, and Stewart & Thom (1973) for a pine forest.

Wind-tunnel results show that the shelter effect is complex. In a canopy of cylindrical rods with Re≈60, Thom (1971) found that $p_M = 1$. However, in a similar canopy with Re≈1000, Seginer et al (1976) found that $p_M \approx 3$. This suggests the importance of a turbulent-wake field with length scales tuned to the element dimensions, a view substantiated by several sets of results on the variation of element form drag with the intensity and length scale of turbulence in the incident wind (e.g. Lee 1976) and on the mutual aerodynamic interference of proximate identical elements (e.g. Zdravkovich & Pridden 1977).

A further complication is that of element movement in the wind. Schuepp (1972) showed that property-transfer coefficients are 40% higher, on average, for fluttering single leaves than for equivalent stationary leaves.

# 3 MOMENTUM TRANSPORT WITHIN THE CANOPY

## *3.1 The Canopy Wind Profile*

The mean wind profile is one of the most easily measured aerodynamic properties of a plant canopy: a vertical array of cup anemometers, sited between plants, is usually used. Figure 1 shows examples of canopy

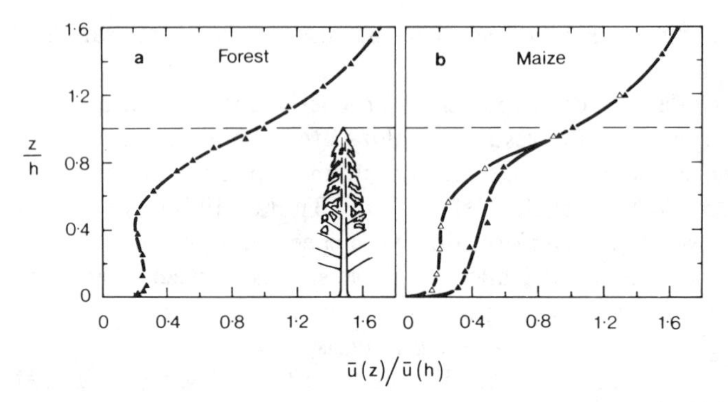


*Figure 1* (*a*) Profile of mean wind speed in a pine forest canopy ($h=16$ m); data averaged from 18 very-near-neutral one-hour runs. (*b*) Profiles of mean wind speed in a maize canopy ($h=2.1$ m) during periods of light wind [$\bar{u}(h)=0.88$ m s$^{-1}$, ▲] and strong wind [$\bar{u}(h)=2.66$ m s$^{-1}$, △].

wind profiles measured in a pine forest[2] and in a maize canopy.[3] It is convenient to present the profiles in dimensionless form, using $h$ and the mean wind speed $\bar{u}(h)$ at the top of the canopy as normalizers. Characteristically, the mean wind shear is high in the upper part of the canopy and very low to negligible in the lower part, where there is often a stem or trunk space largely free from leaves or branches. Another highly sheared layer occurs close to the ground.

The forest canopy wind profile in Figure 1*a* shows a reversal of the wind gradient $\partial\bar{u}/\partial z$ in the lower part of the canopy, resulting in a secondary maximum or "bulge" near $z/h=0.1$. Similar behavior has been observed in numerous other forest and agricultural canopies (Shaw 1977). In a one-dimensional situation, such as must exist in a forest with sufficient homogeneous upwind fetch, the momentum flux throughout the canopy must be downward because there are no sources of mean momentum anywhere in the canopy. Therefore, counter-gradient momentum transport is occurring in regions where $\partial\bar{u}/\partial z<0$. A preliminary warning is thus sounded that conventional diffusion theory

[2] E. F. Bradley, unpublished data. A detailed forest turbulence experiment has been carried out at Uriarra, A.C.T., Australia, over a uniform, level forest of Ponderosa pine with a fetch of about 1 km. Simultaneous measurements were made of the wind, temperature, and humidity fluctuations at two different heights, and of wind, temperature, and humidity profiles above and within the canopy. Results will be published in full by E. F. Bradley, O.T. Denmead, and G. W. Thurtell, as "Measurements of turbulence and heat and moisture transport in a forest canopy" (in preparation). We will refer to this experiment again, as the "Uriarra forest experiment."

[3] O. T. Denmead, unpublished data from an experiment near Griffith, N.S.W., Australia, in March 1979.

may be seriously in error in the canopy environment. To draw this conclusion, one must discount several possible gradient-reversal mechanisms associated with departure of the flow from horizontal homogeneity, such as "blow-through" from the leading edge of a plant stand, slope effects, or the influence of gaps in the upper storey of the canopy. Also, allowance must be made for the fact that a cup anemometer measures the total scalar wind $(u^2+v^2)^{\frac{1}{2}}$, whereas diffusion theory considers the streamwise mean wind component, $\bar{u}$. However, none of these mechanisms satisfactorily accounts for all the observations (Shaw 1977), so counter-gradient momentum transport must be accepted as a real feature of many canopy flows.

In practice, the shape of the canopy wind profile is influenced strongly by prevailing meteorological conditions. The maize-canopy wind profiles in Figure 1*b* were measured on warm, sunny days with comparable insolations but with different mean wind speeds, and therefore with different degrees of atmospheric instability. There is relatively more scalar wind in the lower parts of the canopy in light wind conditions, possibly because the associated increase in free convective activity causes the development of isolated, long-lived thermal plumes that draw air from the lower parts of the canopy. Structures of this kind have been observed using smoke trails (Oliver 1973). However, the problem of the effect of stability on the canopy wind regime is still far from resolved.

## *3.2 Practical Application of the Momentum Equation*

By combining the momentum equation (5) with the parameterization (9) for the total drag force $F$ on an individual element in a canopy, one obtains

$$\partial\tau/\partial z = c_M(U)a(z)U^2(z) \tag{12}$$

where $\tau(z) = -\langle\overline{u'w'}\rangle$ is the average kinematic shear stress at height $z$ in the canopy, $U=\langle\bar{u}\rangle$ is a short notation for the averaged wind velocity, $a(z)$ is the area density (defined as the frontal area of individual elements per unit volume of canopy), and $c_M(U)=C_M(U)/p_M$ is the *effective* drag coefficient of an element subject to a shelter factor $p_M$ within the canopy, defined from Equation (11) by assuming that the "uniform equivalent wind" is of speed $U$. The dispersive flux term in Equation (5) has been ignored here.

Integration of Equation (12) enables a shear-stress profile $\tau(z)$ to be calculated from the canopy wind profile. If we let $c_M(U)=c_1U^{-n}$, where $n$ varies from 0 to $1/2$ according to the ratio of form to viscous drag experienced by canopy elements, then Equation (12) integrates to

$$\tau(z)-\tau(0)=c_1\int_0^z a(z)U(z)^{2-n}\,dz. \tag{13}$$

The area density profile $a(z)$ can be found from measurements of the canopy geometry. For most canopies, the shear stress at the ground, $\tau(0)$, is negligible in comparison with $\tau(h)$, so that an approximate drag profile is

$$\frac{\tau(z)}{\tau(h)}=\frac{\int_0^z a(z)U(z)^{2-n}\,dz}{\int_0^h a(z)U(z)^{2-n}\,dz}. \tag{14}$$

In forming the ratio $\tau(z)/\tau(h)$, we have eliminated the proportionality factor $c_1$, which is usually unknown because it involves an unknown shelter factor $p_M$. The only restriction upon $p_M$ required by Equation (14) is that it be constant throughout the canopy, which requires, in effect, that $p_M$ be independent of foliage density.

Although Equation (14) has been used by many authors (e.g. Uchijima & Wright 1964, Thom 1971, Inoue & Uchijima 1979), there are very few direct tests of it because of the scarcity of direct observations of $\tau(z)$ within canopies. Figure 2 shows a test using one of the very few available data sets, measured in the wind-tunnel flow through a canopy of cylindrical rods with $h=19$ cm (Seginer et al 1976). There is good

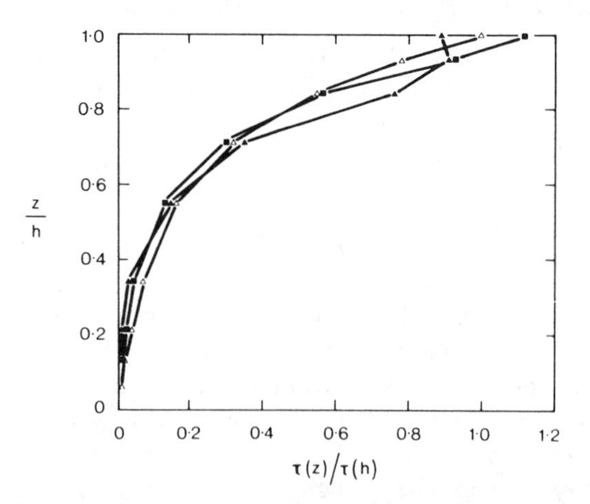


*Figure 2* Profiles of measured shear stress and shear stress calculated from Equation (14), for a canopy of cylindrical rods (Seginer et al 1976). Normalizing parameters: $\bar{u}(h)=9.0$ m s$^{-1}$, $h=19$ cm, $\tau(h)=3.3$ m$^2$ s$^{-2}$. ▲, $\overline{u'w'}$ measured out of wake; ■, $\overline{u'w'}$ measured in wake; △, calculated $\tau(z)$.

agreement between the measured values of $\tau(z)$ and the profile calculated from Equation (14) with a wind-speed-independent drag coefficient ($n=0$).

Equation (13) can also be used to find the shelter factor $p_M$, by calculating from it a stress $\tau'(h)$ with the assumption $p_M=1$. Then, if the true stress $\tau(h)$ is known independently, $p_M=\tau'(h)/\tau(h)$. For example, application of this method to the data of Figure 2 gives $p_M \approx 3$ for the canopy of Seginer et al, when a free-stream element drag coefficient $C_M=0.5$ is assumed. The canopy-shelter factors quoted in Section 2.4 were obtained in this way.

# 4 MODELS FOR TRANSPORT WITHIN THE CANOPY

## *4.1 Local-Diffusion Models*

A primary aim of the study of canopy turbulence is to be able to compute important characteristics of the plant microclimate (especially the vertical fluxes of heat, water vapor, and carbon dioxide, and their variation with height) from a suitable specification of the physical and biological properties of the vegetation. This involves solving the horizontally averaged conservation equations (5) and (6), after parameterizing the drag force and creation-density terms. As always happens when modeling turbulent boundary layers, the averaging process introduces a closure problem, which must be circumvented by making additional assumptions. The simplest response, almost universally used hitherto, is to close the conservation equations at first order by using the diffusion equation (1), together with a plausible assumption about the diffusivity $K_S$. Here, we discuss this approach in detail for momentum transport.

If we let

$$\tau(z)=K_M\frac{\partial U}{\partial z} \tag{15}$$

and eliminate $\tau(z)$ between Equations (12) and (15), the result is

$$\frac{\partial K_M}{\partial z}\frac{\partial U}{\partial z}+K_M\frac{\partial^2 U}{\partial z^2}=c_M(U)a(z)U^2. \tag{16}$$

To solve this second-order differential equation in $U(z)$, one must specify $c_M(U)$, $a(z)$, and two boundary conditions for $U$, usually $U(h)$ and $U(0)$ (which is zero by the no-slip condition). Also, $K_M$ must be plausibly related to the other variables in the equation. There are two general ways of doing this, which we illustrate by showing how each leads to an analytical wind profile in the special case of a uniform canopy [$a(z)$=constant] and a velocity-independent drag coefficient [$c_M(U)$=constant].

The *mixing-length approach* is based upon Prandtl–von Kármán mixing-length theory for a turbulent boundary layer, which postulates

$$K_M=\ell^2\frac{\partial U}{\partial z} \tag{17}$$

where the mixing length $\ell$ is most simply taken as constant within the canopy. Then, one solution of Equation (16) is

$$U(z)/U(h)=\exp[\alpha_1(z/h-1)], \tag{18}$$

in which the attenuation coefficient $\alpha_1$ is equal to $(\frac{1}{2}c_M a h^3 \ell^{-2})^{1/3}$ (Inoue 1963, Cionco 1965). This is the exponential wind profile. There are too many assumptions in its derivation for it to be regarded as any more than a single-parameter empirical fit, but it is nevertheless widely used, largely because it describes the upper part of most canopy wind profiles quite well when $\alpha_1$ is suitably chosen. Cionco (1972b) has summarized the best values of $\alpha_1$ for numerous canopies; it usually lies between 2 and 3.

In the *diffusivity approach*, $K_M$ is constructed directly from known variables without recourse to a mixing-length assumption. Cowan (1968) postulated that $K_M \propto U$, from which he derived the wind profile

$$\frac{U(z)}{U(h)}=\left(\frac{\sinh(\alpha_2 z/h)}{\sinh \alpha_2}\right)^{\frac{1}{2}} \tag{19}$$

by solving Equation (16). Another possible assumption is $K_M$=constant, from which Landsberg & James (1971) and Thom (1971) generated the profile

$$U(z)/U(h)=[1+\alpha_3(1-z/h)]^{-2}. \tag{20}$$

As with the exponential profile, both $\alpha_2$ and $\alpha_3$ can be expressed in terms of the parameters governing Equation (16) and $K_M$, but are more appropriately regarded as empirical coefficients.

Note that only the Cowan profile (19) satisfies boundary conditions at both $z=0$ and $z=h$. The exponential and constant $K_M$ profiles, (18) and (20), both fail to vanish at $z=0$; hence, although analytically and empirically convenient, they are mathematically arbitrary because they are solutions of a second-order differential equation constrained by only one boundary condition.

Equations (18), (19), and (20) all give rather similar wind profiles inside the canopy. Figure 3 shows the three analytical profiles in comparison with measured wind speeds within the model canopy of

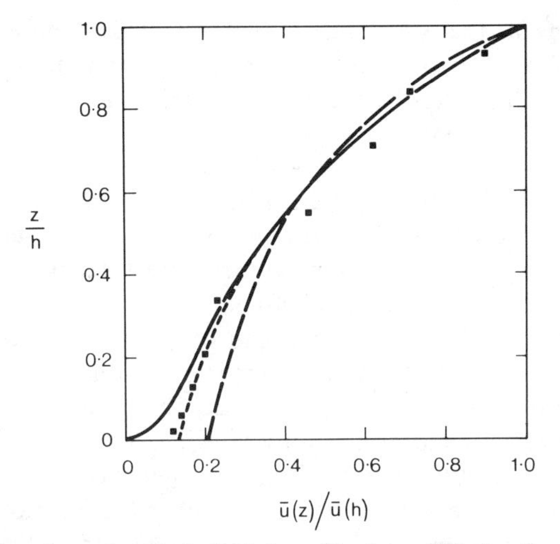


*Figure 3* Comparison of analytical wind profiles from diffusion theory with measurements (squares) of Seginer et al (1976) in a uniform canopy of cylindrical rods. Dotted line, exponential: Equation (18) with $\alpha_1 = 2$; solid line, Equation (19) with $\alpha_2 = 4$; dashed line, constant $K_M$: Equation (20) with $\alpha_3 = 1.2$.

Seginer et al (1976), which has a uniform $a(z)$ and hence provides a fair test of the theories. In spite of the similarity of $U(z)$ profiles, the associated $K_M$ profiles differ widely: for Equations (18) and (19), the $K_M$ profile is identical in shape to the $U$ profile and so decreases sharply with decreasing $z$, whereas for Equation (20), $K_M$ is constant with height. The insensitivity of $U(z)$ to $K_M$ is such that any reasonable assumption about $K_M$ generates a plausible-looking wind profile.

Diffusion theories of this general type—often involving computer solution of Equation (16) in order to incorporate realistic $a(z)$ profiles—have been used extensively to calculate canopy wind profiles. Likewise, profiles of temperature and humidity have been computed by models involving a local-diffusion assumption, the net radiation at each level in the canopy being partitioned into sensible heat and water-vapor fluxes by a combination approach (e.g. Monteith 1975). This avoids the need to know the foliage surface temperatures and humidities that would otherwise be required to calculate property emission rates from Equation (10) for use in Equation (6). These models have had some successes in simulating the profiles, and related processes such as the photosynthesis and transpiration rates. Nevertheless, there are important weaknesses in the local-diffusion assumption. That we now explore these weaknesses in detail, together with more sophisticated models that may overcome them, should not be interpreted as an undervaluation of the successes of simple models based on local-diffusion theories. On the contrary, it is important to maximize the proper use of such models by identifying clearly the areas in which they are significantly deficient.

One symptom of the inadequacy of local-diffusion theories is their inability to cope with counter-gradient fluxes of momentum (see Section 3.1) and other properties; for example, strong counter-gradient heat fluxes have been observed in the Uriarra forest experiment. The underlying reason for the failure is twofold. First, turbulence is inherently nonlocal. A local-diffusion equation can only properly describe transport if the length scales of the flux-carrying motions are much smaller than the scales over which average gradients change appreciably. Such a condition is often violated, but canopy flow is a more extreme case than most: observations show (Section 6) that much of the flux-carrying turbulence close to and within the canopy takes the form of large-scale, intermittent downsweeps of high-velocity fluid emanating from the overlying boundary layer. The second problem is the multiplicity of length and velocity scales governing canopy flow. In any local-diffusion theory, the turbulent diffusivity $K_M$ or $K_S$ must be guessed on dimensional grounds; this can only be done with some hope of success when these scales are unique (Tennekes & Lumley 1972, p. 57). This is why local-diffusion theories work in the inertial sublayer, but it is also one reason why they fail in the canopy.

## 4.2 *Second-Order Closure Models*

One way of avoiding a simple local-diffusion assumption is to close the equations describing turbulent flow at second rather than at first order. The technique is to write not only the dynamic equations (3) and (4) for the mean velocity $\bar{u}_i$ and the property concentration $\bar{s}$, but also the equations for the second moments $\overline{u_i'u_j'}$, $\overline{u_i's'}$, and $\overline{s'^2}$. The closure problem causes the resulting set of equations to contain extra variables introduced by the averaging process, such as the triple-correlation and pressure-correlation terms in the TKE equation, (7). (The TKE equation is actually half the sum of equations for $\overline{u'^2}$, $\overline{v'^2}$, and $\overline{w'^2}$, which are used separately in most second-order closure models.) These extra variables are parameterized to produce a closed set of equations, which is then solved numerically with appropriate boundary conditions. This method has been used extensively for modeling the planetary boundary layer (e.g. Donaldson 1973, Mellor 1973) and has succeeded in explaining the counter-gradient heat flux that usually occurs throughout its upper portion in convective conditions (Deardorff 1966).

Second-order closure models introduce an element of nonlocality into the mathematical description of turbulence by permitting second moments, including the turbulent stresses or fluxes, to be influenced by events in layers adjacent to the one being considered (Donaldson 1973, p. 363). Local-diffusion assumptions still occur in the theory (e.g. Mellor 1973) since they are used to achieve second-order closure in much the same way as for the first-order closure schemes discussed in Section 4.1. However, it is taken on faith that the increased physical content of second-order models makes their predictions more accurate: "if a crude assumption for second moments predicts first moments adequately, then perhaps a crude assumption for third moments will predict second moments adequately" (Lumley & Khajeh-Nouri 1974, p. 171).

A second-order closure model for momentum transport in neutral canopy flow has been constructed by Wilson & Shaw (1977). Their model consists of horizontally averaged dynamic equations for five variables: the mean momentum $\langle\bar{u}\rangle = U$, the Reynolds stress $\langle\overline{u'w'}\rangle = -\tau$, and the three components of turbulent kinetic energy, $\langle\overline{u'^2}\rangle$, $\langle\overline{v'^2}\rangle$, and $\langle\overline{w'^2}\rangle$. The closure parameterizations of higher-order terms introduce several length scales that are assumed proportional, with constants determined by forcing the model to reproduce the observed flow characteristics of the free, neutral atmospheric surface layer. The model was applied to a corn canopy for which detailed observations were available; Figure 4 shows some computed and measured aerodynamic characteristics of the canopy. The computed wind profile includes a realistic-looking secondary maximum close to the ground, confirming that the model can describe counter-gradient momentum transport (cf Section 3.1). The computed stress profile is also realistic.

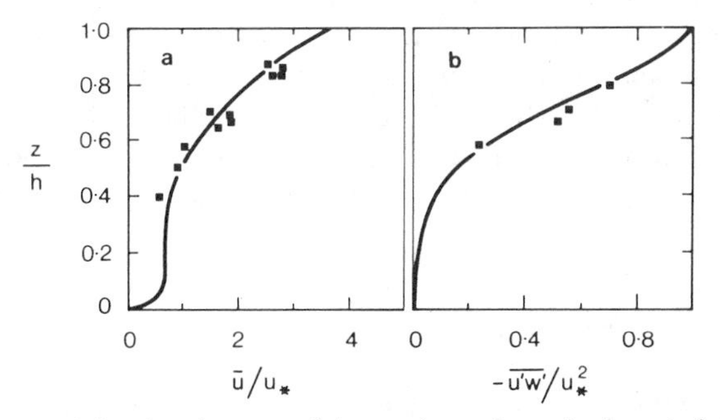


*Figure 4* Computed (lines) and measured (squares) aerodynamic characteristics of a corn canopy (Wilson & Shaw 1977): (*a*) wind profile; (*b*) stress profile. The friction velocity $u_*$ is defined by $\tau(h)=u_*^2$.

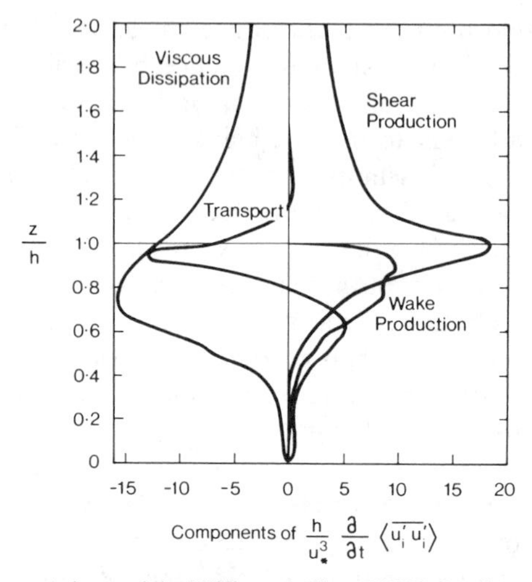


*Figure 5* Predictions of the model of Wilson & Shaw (1977) for the relative magnitude of terms in the turbulent kinetic energy budget, Equation (8), within a corn canopy.

An important part of the model is the incorporation of the wake-production term in the TKE equation (8), which accounts for TKE production by working of the mean flow against form drag. Figure 5 shows the model predictions for the relative magnitudes of the terms in Equation (8). Shear production of TKE is largest in the highly sheared flow near the top of the canopy, whence the turbulence transports TKE predominantly downwards. In the lowest 60% of the canopy, transport from higher levels is the largest single contributor to locally dissipated TKE. The wake-production term accounts for significant TKE production, exceeding shear production in all but the top 20% of the canopy.

Measurements with which these predictions can be compared are practically nonexistent. Wind-tunnel data (Mulhearn & Finnigan 1978) show that in the flow just above a rough stone surface, local loss of TKE by downward turbulent transport is about 15% of the production term. Maitani (1978) has observed significant downward TKE fluxes over wheat and rice canopies.

## 5 THE FLOW ABOVE THE CANOPY

As pointed out earlier, the influence upon the flow of a canopy of height $h$ is felt throughout a roughness sublayer that extends considerably above $z=h$. At its upper limit, the roughness sublayer merges with an inertial sublayer in which the following local flux-gradient relationships are known to apply:

$$\partial \bar{u}/\partial \ln(z-d)=u_*\phi_M/k, \tag{21}$$

$$\partial \bar{\theta}/\partial \ln(z-d)=-\theta_*\phi_H/k, \tag{22}$$

$$\partial \bar{q}/\partial \ln(z-d)=-q_*\phi_E/k. \tag{23}$$

Here, $\theta$ and $q$ are the potential temperature and specific humidity, respectively, $k$ is the von Kármán constant, $d$ is the zero-plane displacement, and $u_*$ the friction velocity. The inertial-sublayer scaling parameters $u_*$, $\theta_*$, and $q_*$ are defined from the (kinematic) shear stress $\tau$, the heat flux $H$, and the water vapor flux $E$ by the equations

$$\left.\begin{aligned} \tau &= -\overline{u'w'} = u_*^2 \\ H &= \rho c_p \overline{w'\theta'} = \rho c_p u_* \theta_* \\ E &= \rho \overline{w'q'} = \rho u_* q_* \end{aligned}\right\} \tag{24}$$

where $c_p$ is the specific heat of air. The influence functions $\phi_M$, $\phi_H$, and $\phi_E$ express the dependence of the flux-gradient relationships upon ambient or site conditions; in the inertial sublayer, they are functions only of the dimensionless stability parameter $(z-d)/L=-kg\theta_*(z-d)/(T_0u_*^2)$, where $L$ is the Monin-Obukhov length. This functional dependence is reasonably well known, although some uncertainty persists (Yaglom 1977). In a neutral inertial sublayer, $\phi_M$ is 1 by definition and Equation (21) integrates to give the semi-logarithmic wind profile

$$\bar{u}(z)=\frac{u_*}{k}\ln\left(\frac{z-d}{z_0}\right) \tag{25}$$

where $z_0$ is the roughness length.

A proper understanding of Equations (21) to (23) is crucial to the very important practical problem of inferring vertical fluxes from measurements of property gradients above a canopy. Two factors complicate the issue: the first is $d$, which has always created problems for micrometeorologists, and the second is the roughness sublayer.

The zero-plane displacement is usually thought of as the level to which the effective surface must be raised to make the neutral wind profile in the inertial sublayer obey the semi-logarithmic law (25). In practice, when only wind-profile data are available, this definition is very hard to apply. Not only is it notoriously insensitive because it involves a three-parameter fit in $u_*$, $d$, and $z_0$ to Equation (25) (Bradley & Finnigan 1973), but also there is no guarantee that the wind profiles used to determine $d$ have been measured entirely in the inertial sublayer

—thereby excluding observations from the roughness sublayer (Raupach et al 1980). The "curve-fitting" connotations of this approach can be removed, and a better physical understanding of $d$ may be gained, by defining $d$ as the *mean level of momentum absorption* by the canopy (Thom 1971), which can be calculated if the drag profile within the canopy is known. Thom noted initially that the two definitions of $d$ coincided for a model canopy in a wind tunnel. Recently, Jackson (1981) has shown that the concept of $d$ as the mean level of momentum absorption is implicit in the usual derivation of the semi-logarithmic law by asymptotic matching arguments. This places $d$ on a secure theoretical foundation, but does not ease the problem of measuring it unless an accurate drag profile within the canopy can be deduced. A further problem is the assumption in Equations (21) to (23) that $d$ is identical for all properties; a contrary assumption has been advanced by Hicks et al (1979), who postulated that additional zero-plane displacments $d_H$, for heat, and $d_E$, for water vapor, should be used in Equations (22) and

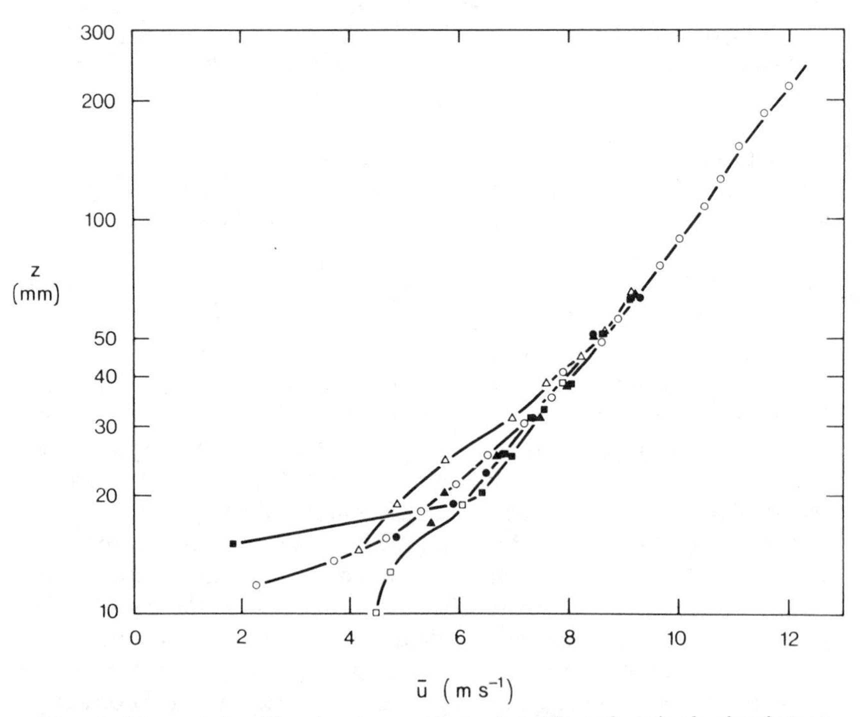


*Figure 6* Mean wind profiles at various points over a surface of randomly placed stones with average diameter 15 mm and average spacing 50 mm (Mulhearn & Finnigan 1978). Note that no zero-plane displacement has been used in the ordinate.

(23), respectively. However, such a course involves complications, and we shall avoid it here.

Once $d$ has been ascertained, more substantial issues arise: what is the depth of the roughness sublayer, and how do flux-gradient relationships within it differ from those in an inertial sublayer? We first consider momentum. Wind-tunnel studies of the flow close to rough surfaces have identified two direct surface influences on the mean wind field: the first is horizontal inhomogeneity, dramatically demonstrated by wind profiles measured at several points over an array of stones (Mulhearn & Finnigan 1978) and shown in Figure 6. This effect extends up to about $z=h+D$, where $D$ is the mean inter-element spacing (Schlichting 1936, Raupach et al 1980). Second, there is a perturbation on the (horizontally averaged) dimensionless mean wind gradient $(1/u_*)\partial\langle\bar{u}\rangle/\partial\ln(z-d)$, such that the gradient in the roughness sublayer is less than in the inertial sublayer—or, equivalently, that $\phi_M<1$ in the roughness sublayer. This has been attributed to a "wake diffusion" effect whereby element wakes augment the turbulent diffusivity in the roughness sublayer (O'Loughlin & Annambhotla 1969, Thom et al 1975). Raupach et al (1980) studied the flow in a wind tunnel over cylinders in regular arrays of varying density, and concluded that the height $z_w$ of the region of wake influence is about $h+1.5\ell_t$ where $\ell_t$ is the tranverse-element length scale. The wake-diffusion effect is demonstrated, in Figure 7, by the curvature near the surface of wind profiles from that experiment (which will be referred to again as the "Edinburgh wind-tunnel experiment").

For heat and water vapor, information about the roughness sublayer has only been obtained in the field; as an example, Figure 8 shows flux-gradient data from the extensively instrumented forest at Thetford, U.K. (Raupach 1979) and from the Uriarra forest experiment (Footnote 2, Section 3.1). In both cases, above-canopy gradients and eddy-correlation flux values were combined to give $\phi_M$, $\phi_H$, and $\phi_E$ from Equations (21) to (24). Both forests were of height $h\approx 16$ m, with mean tree spacings $D$ between 2 and 3 m. Gradients were calculated at a reference height $z_R=21$ m, about 5 m above the canopy top, with an assumed value for $d$ of 0.75 $h$ or 12 m (Thom et al 1975). The results are plotted as $\phi_{M,H,E}^{-1}$ [which are proportional to the turbulent diffusivities, since $K_{M,H,E}=ku_*(z-d)\phi_{M,H,E}^{-1}$] against the stability parameter $(z_R-d)/L$. Curves representing typical inertial-sublayer behavior (Dyer 1974) are also shown. Over both forests, the behavior of $\phi_M^{-1}$ or $K_M$ is fairly close to that expected in an inertial sublayer, which is consistent with the wind-tunnel result of Raupach et al (1980) for the depth of the layer of wake influence for momentum. However, $\phi_H^{-1}$ and $\phi_E^{-1}$, and hence

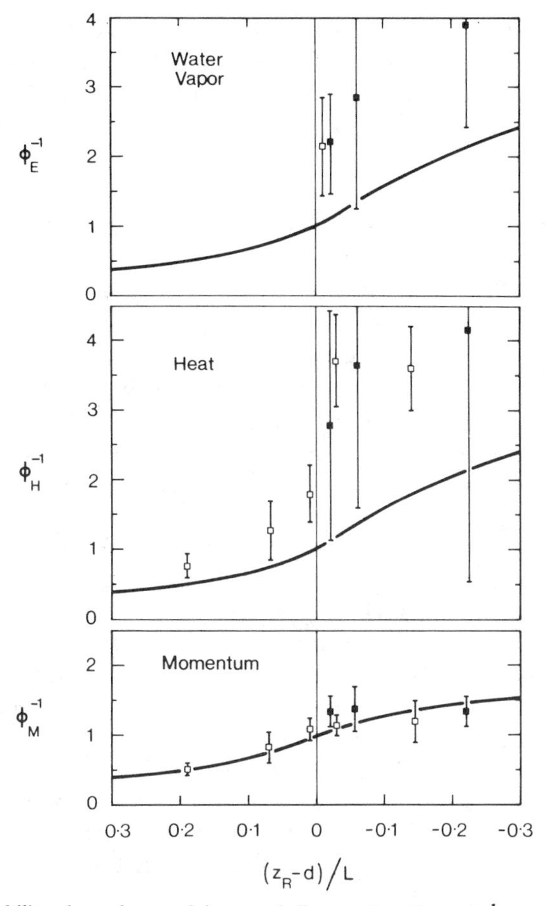


*Figure 8* Stability dependence of inverse influence functions $\phi_{M,H,E}^{-1}$, measured over Thetford forest (□; Raupach 1979) and Uriarra forest (■). The smooth curves represent typical inertial-sublayer behavior.

found that both $K_H$ and $K_M$ were greater than their inertial-sublayer values by a factor between 1.5 and 2.

Differences like this are not surprising. The flow just above a rough surface is strongly influenced by the geometric and thermal character of the surface, not only through "wake diffusion" effects, but also through the spatial distributions of sources and sinks for heat and water vapor. We should not expect to describe all these variations by simple scaling considerations. More fundamentally, the local diffusion framework used in this section is likely to be subject to the same flaws above the canopy as within (Section 4.1), so that comprehensive models of the flow in the roughness sublayer will require second-order closure techniques. Nevertheless, the simplicity of the local-diffusion approach means that, in practice, it will continue to be used wherever possible, perhaps with semi-empirical adjustments to account for "anomalies" caused by surface influences.

# 6 THE TURBULENCE FIELD

## *6.1 Turbulence Intensities*

The longitudinal turbulence intensity $i_u = \sigma_u/\bar{u}$, the simplest indicator of "turbulence strength," is always higher within canopies than above them. Cionco (1972a) summarized numerous observations and concluded that, in a gross way at least, $i_u$ increases with canopy density: typically, $i_u \approx 0.4$ in agronomic crops, $i_u \approx 0.6$ in temperate forests, and $0.7 \lesssim i_u \lesssim 1.2$ in tropical forests. He and Inoue et al (1975) also showed that $i_u$ is constant with height in canopies with a height-invariant area density $a(z)$, such as wheat, rice, barley, and various wind-tunnel model canopies. However, when plant waving is significant, $i_u$ increases with height in such canopies (Finnigan 1979a; see also Section 6.5). In strongly stratified canopies, profiles of $i_u$ tend broadly to follow the area-density profile. Mean wind speed and stability within the canopy also influence $i_u$ (see, for example, Inoue & Uchijima 1979), causing $i_u$ values calculated from (say) one-hour runs to be highly scattered.

The lateral and vertical turbulence intensities, $i_v = \sigma_v/\bar{u}$ and $i_w = \sigma_w/\bar{u}$, respectively, are usually roughly proportional to the longitudinal component, such that $i_w < i_v < i_u$ (Shaw et al 1974a). Close to the ground, $i_w$ usually decreases rapidly because vertical fluctuations are constrained there (Seginer et al 1976).

## *6.2 Spectra and Cospectra*

Measurements of turbulence spectra and cospectra within canopies (e.g. Uchijima & Wright 1964, Allen 1968, Isobe 1972, Shaw et al 1974b, Seginer et al 1976, Desjardins et al 1978, Inoue & Uchijima 1979, Finnigan 1979a) show that the spectral properties of canopy turbulence differ substantially from those of turbulence in an inertial sublayer. Work over the last ten years (e.g. Busch 1973) has shown that inertial-sublayer turbulence over *short* vegetation is characterized by generally hump-shaped spectra and cospectra for velocity components and property concentrations, which are universal with respect to height and wind speed when plotted against a dimensionless frequency $f = nz/\bar{u}$, where $n$ is the frequency in Hertz. The peak dimensionless frequency $f_p = n_p z/\bar{u}$ depends on the property (typically, in near-neutral conditions, it is of order $10^{-1}$ for the $w$ spectrum, $10^{-2}$ for $uw$ or $ws$ cospectra, and $10^{-3}$ for $u$ or $s$ spectra, where $s$ is a scalar concentration) and decreases with increasing instability. Hence, inertial-sublayer turbulence has a peak length scale $\ell_p = \bar{u}/n_p = z/f_p$ which is proportional to the height. On the high-frequency side of the peak, spectra obey the Kolmogorov $-5/3$ power law and cospectra a $-7/3$ power law.

In the canopy environment, this picture alters in two main ways. The first concerns scaling. Above a vegetation canopy of significant height, spectra and cospectra scale with a dimensionless frequency $f = n(z - d)/\bar{u}$ calculated with respect to the effective height $(z-d)$, rather than the height $z$ above ground (Shaw et al 1974b). This situation parallels that for the flux-gradient relationships (21) to (23). Within the canopy, on the other hand, spectra and cospectra do not scale with any dimensionless frequency; rather, their peak frequency $n_p$ tends to be independent of height (Allen 1968, Seginer et al 1976). The constancy with depth of a time scale rather than a length scale (cf. Silversides 1974) is a significant result.

Second, turbulence in the canopy environment is due to both shear and wake production, and in the case of waving crops is significantly altered by a third mechanism, the resonant coupling at the waving frequencies between the crop and the airflow (Section 6.5). Both wake production and the waving phenomenon introduce high-frequency peaks into the velocity-component spectra, which may prevent the $-5/3$ law from being observed in the usual range of roughly 1 to 10 Hz. Isobe (1972) and Baines (1972) noted eddies in agricultural crops at frequencies corresponding to vortex shedding from stems. Shaw et al (1974b) found no such peaks in either corn or forest, but instead observed an increased "peakiness" in the $w$ spectrum and all cospectra, which they attributed to the wakes of entire plants. Seginer et al (1976) observed high-frequency spectral peaks due to vortex shedding from the vertical cylinders of their model canopy, but noted that the mechanism accounted for less than 10% of the turbulent kinetic energy. Finnigan (1979a) has observed significant high-frequency peaks over waving wheat, due to resonance.

## 6.3 *Two-Point Correlations of Velocity Components*

Correlation statistics measured at two vertically separated points, rather than at one point only, are vital for answering the important questions of the degree and nature of the coupling between the airflow above and within the canopy. However, little information is presently available: only Seginer & Mulhearn (1978), Inoue & Uchijima (1979), and Finnigan (1979b) present two-point correlation statistics. Figure 9 shows the

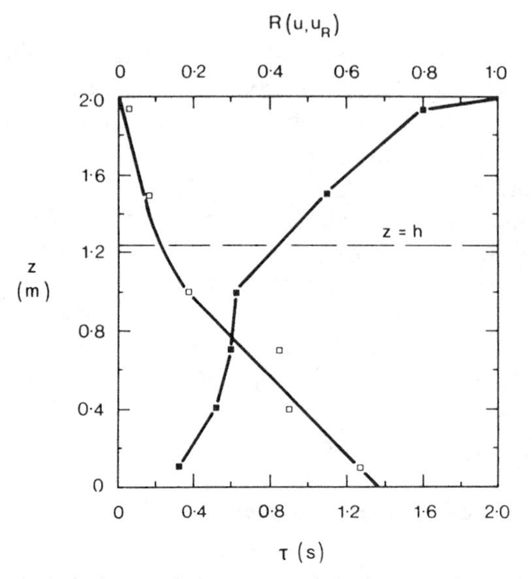


*Figure 9* Two-point velocity correlation over waving wheat: ■, the maximum correlation coefficient $R(u, u_R)$ between a reference velocity $u_R$ (at height 2 m) and the velocity $u(z)$; □, the time delay $\tau$ at which $R$ is maximum (Finnigan 1979b).

correlation $R(u, u_R)$ between longitudinal velocity fluctuations $u'_R$ at a reference height $z_R=2$ m above a waving wheat canopy, and similar fluctuations $u'$ at various lower heights $z$ (Finnigan 1979b). These correlations have been calculated at the time delay $\tau$ at which $R$ is a maximum; that is,

$$R(u, u_R)=\max\left[\overline{u'(t+\tau)u'_R(t)}\left(\overline{u'^2}\right)^{-1}\left(\overline{u_R'^2}\right)^{-1}\right]. \tag{26}$$

The values of $\tau$ used to calculate $R$ are also shown. As expected, $R$ diminishes with increasing separation, but not rapidly within the canopy, suggesting substantial coherence between velocity fluctuations within the waving canopy. In the absence of waving, such strong coherence is not observed (Inoue & Uchijima 1979). The time lag $\tau$ is such that longitudinal velocity disturbances propagate downwards into the canopy at a convection speed $dz/d\tau$ which is approximately constant within the canopy, and increases with height above the canopy (Finnigan 1979b). Seginer & Mulhearn (1978) presented similar results for $\tau$.

## *6.4 Turbulence Structures and Their Interaction with the Surface*

Our earlier discussion of the nonlocality of canopy turbulence has shown the desirability of examining the structures responsible for turbulent transport, and the manner in which they contribute to it. The statistics already mentioned contribute diagnostically to our knowledge of the turbulence structures, but they may be powerfully augmented by a range of *conditional-sampling techniques* developed originally for the study of turbulence in laboratory boundary layers (see, for example, R. A. Antonia's review in this volume). Together with flow visualization and cross-correlation studies, these techniques have shown that within the apparent chaos of a turbulent boundary layer, there exist large, coherent structures distributed randomly in time and in the horizontal plane, with finite lifetimes and with length scales ranging from some surface-defined scale $\ell_s$ to the boundary-layer thickness $\delta$. They may be visualized as inclined horseshoe vortices, leaning with the shear (Hinze 1975, pp. 560 and 660), or as "double roller" eddies (Townsend 1976, p. 120). These structures undoubtedly also occur in the atmospheric boundary layer, and their interaction with the underlying surface is central to the nature of canopy turbulence.

A simple conditional-sampling technique, which can provide some structural information from a single-point time history, is the quadrant representation of shear stress (e.g. Willmarth 1975). The time-averaged value of $\overline{u'w'}$ at a single point is written as the sum of four contributions, one from each quadrant $i$ of the $(u', w')$ plane. Each contribution accounts for a stress fraction $S_i$, defined thus:

$$S_i = \left(\overline{u'w'}\right)^{-1} \lim_{T\to\infty} \frac{1}{T} \int_0^T u'(t)w'(t)I_i\,dt$$

where the indicator function $I_i$ is 1 when $(u', w')$ is in quadrant $i$, and 0 otherwise. Clearly, $S_1 + S_2 + S_3 + S_4 = 1$. The four quadrants correspond to the following events:

$i = 1(u' > 0, w' > 0)$: outward interaction,

$i = 2(u' < 0, w' > 0)$: ejection,

$i = 3(u' < 0, w' < 0)$: inward interaction,

$i = 4(u' > 0, w' < 0)$: sweep or gust.

Stress is transported downwards by ejections and sweeps, and upwards by the two interaction events, so that $S_2 > 0$, $S_4 > 0$, $S_1 < 0$, $S_3 < 0$ (since the overall stress is downward). Studies using this type of analysis have shown that most momentum transport close to or within a rough surface occurs during sweep events, in which high-velocity fluid from above moves downwards towards the surface. In contrast, ejections and sweeps are of comparable importance for momentum transport in the inertial sublayer (e.g. Raupach 1981). The near-surface dominance of sweeps is shown by Figure 10, in which profiles of $S_i$ are plotted for the inertial and roughness sublayers over the roughest surface of the Edinburgh wind-tunnel experiment.

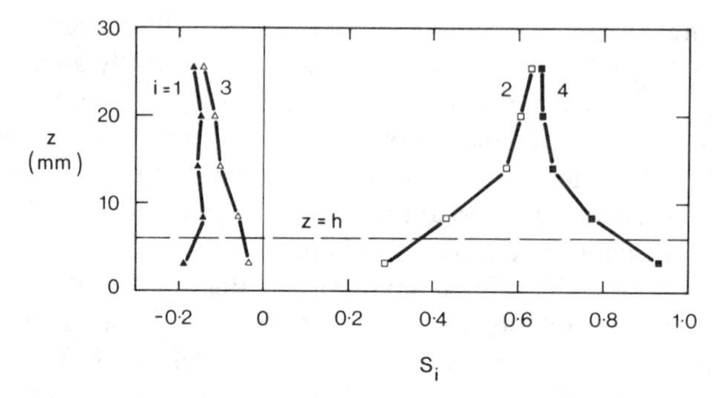


*Figure 10* Stress fractions $S_i$ over a rough surface of cylindrical elements of height 6 mm (surface *F*, Figure 7): ▲, outward interaction ($i=1$); □, ejection ($i=2$); △, inward interaction ($i=3$); ■, sweep ($i=4$). Data from Raupach (1981).

The stress fractions $S_i$ are closely related to the third moments $\overline{u'u'u'}$, $\overline{u'u'w'}$, $\overline{u'w'w'}$, and $\overline{w'w'w'}$ of $u$ and $w$. For example, the skewnesses $Sk(u)=\overline{u'^3}/\sigma_u^3$ and $Sk(w)=\overline{w'^3}/\sigma_w^3$, and the difference $\Delta S=S_4-S_2$ between the stress fractions due to sweeps and ejections, were found to be approximately proportional, such that

$$\Delta S \approx 0.4 Sk(u), -\Delta S \approx 0.7 Sk(w) \qquad (27)$$

everywhere in the boundary layers of the Edinburgh wind-tunnel experiment, including the roughness sublayers (Raupach 1981). Figure 11 illustrates this. A similar relationship applies in the field, as shown by data from a waving wheat canopy (Finnigan 1979b), presented in Figure 12. Note that a proportionality relationship between $\Delta S$ and the skewnesses can be derived theoretically if the joint probability-density function has no cumulants of higher than third order, and if the ratios

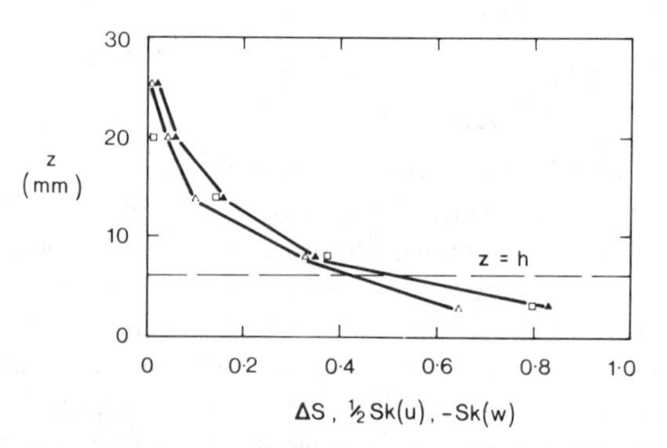


*Figure 11* Skewnesses $Sk(u)$, ▲, and $Sk(w)$, □, of $u$ and $w$, and the difference $\Delta S$, △, between stress fractions due to sweeps and ejections, over the same rough surface as in Figure 10.

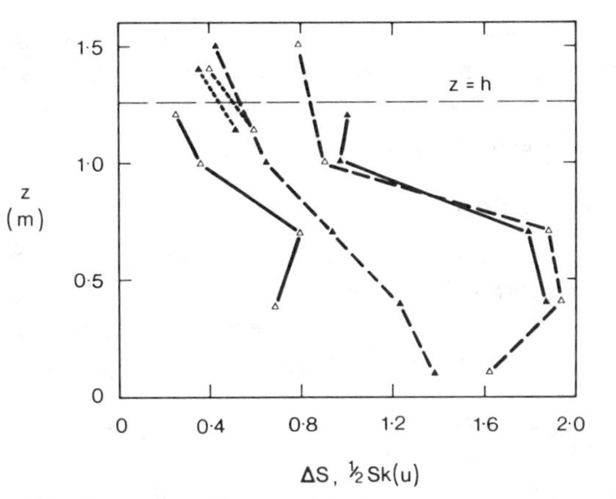


*Figure 12* Skewnesses $Sk(u)$, ▲, of $u$, and $\Delta S$, △, over a waving wheat surface (Finnigan 1979a, Figure 4; 1979b, Figure 6). Data from 3 runs: R2 (dashed lines), R4 (solid lines), R5 (dotted lines).

between third-order cumulants are constant (Raupach 1981). Most boundary layers appear to approximate these conditions well enough for relationships like (27) to be useful.

Figure 11 shows that $\Delta S$, $Sk(u)$, and $Sk(w)$ are all of large magnitude near $z=h$, and all decrease with height to a value close to zero in the inertial sublayer ($z \gtrsim 15$ mm in the Edinburgh wind-tunnel experiment; see Figure 7). This suggests that the skewnesses can be used to determine the height of the roughness sublayer, and hence to show where local diffusion theory may be safely applied. From Figure 11, the criterion $|Sk(w)| \lesssim 0.2$ should ensure the applicability of inertial-sublayer flux-gradient relationships, at least for momentum in near-neutral conditions.

The relationships between $\Delta S$ and the third moments also show the link between the dominance of sweeps in momentum transport close to a canopy and the behavior of the turbulent flux of TKE, $\frac{1}{2}\overline{w'u_i'u_i'} = \frac{1}{2}\overline{w'(u'^2+v'^2+w'^2)}$. For order-of-magnitude calculations, we may assume that $\overline{w'u'^2} \approx \overline{w'v'^2} \approx \overline{w'^3}$ (e.g. Raupach (1981) and $\sigma_w \approx u_*$. Equation (27) then implies that $\frac{1}{2}\overline{w'u_i'u_i'} \approx (3/2)\overline{w'^3} \approx -2\Delta S \sigma_w^3 \approx -2\Delta S u_*^3$. Since $\Delta S$ is at least 0.5 near $z=h$, from Figures 11 and 12, the turbulent TKE flux is at least $-u_*^3$ close to the canopy, and directed downward.

The penetration of sweeps or gusts into the canopy was investigated by Finnigan (1979b), who conditionally sampled the velocity profile within a wheat canopy using the instantaneous velocity above the canopy as a trigger. He found that the short-term mean canopy-wind profile was different during lulls and gusts (or sweeps); gusts produced a much higher shear in the upper part of the canopy than did lulls, and were responsible for a disproportionately high share of the shear stress —as suggested also by quadrant analysis.

## 6.5 *Plant Waving*

Plant waving, or aeroelasticity, is an important phenomenon that is only now beginning to receive the attention it merits. That waving is dynamically significant in some circumstances can be illustrated by writing the drag effective force $F$ per unit length of a waving wheat stalk:

$$F=\rho c_M A\left(u-\frac{dx}{dt}\right)\left|u-\frac{dx}{dt}\right| \tag{28}$$

where $A$ is the area of the stalk element and $x$ its streamwise position coordinate. The modulus sign ensures that the drag force always opposes the local velocity relative to the stalk, even if that velocity momentarily reverses because of the waving or the turbulence field. Finnigan & Mulhearn (1978a, b) showed that, after time averaging, the contributions to the total drag from the waving terms in Equation (28) are comparable to the other terms, in typical conditions for waving wheat. They also showed, with a mathematical model and with a properly scaled aeroelastic model canopy in a wind tunnel, that significant coupling occurs between the airflow in the canopy and the coherently waving stalks, such that sharply tuned oscillations in pressure and velocity components appear at the natural vibration frequency of the stalks.

The phenomenon of "honami" (Inoue 1955) is the wavelike traveling disturbance seen in cereal or grassy canopies on windy days. Recent observations (Finnigan & Mulhearn 1978a, Finnigan 1979a, Maitani 1979) yield the following picture of honami in wheat: a gust or sweep (Section 6.4) bends over a patch of the crop as it passes, storing downwardly transported kinetic energy as strain-potential energy. After passage of the gust, the stalks bounce back and vibrate with their own natural frequency, releasing this stored energy into the flow by drag coupling and so producing pressure and velocity fluctuations at the waving frequency. As the gust travels over the crop, a small phase difference is created between adjacent patches, giving a wavelike impression. Finnigan (1979a), by examining $u$ and $w$ spectra and cospectra during five-minute intervals over a wheat crop when honami was occurring, was able to identify peaks corresponding to the gust frequency (which was significant in the $uw$ cospectrum), the waving frequency, and twice the waving frequency (which occurred primarily in the $w$ spectrum). A "Doppler effect" caused these impressed motions to shift to higher and lower frequencies, respectively, as they were convected upwards or downwards, through the sheared flow, from their level of generation, which was centered near the top of the canopy.

# 7 SUMMARY AND CONCLUSIONS

We have stressed two related themes: that local-diffusion models of transport are seriously deficient in the canopy environment, and that organized structures in the overlying boundary layer are the principal agents determining the nature of canopy turbulence. To define properly the limits of local-diffusion theory, and to replace it where necessary, second-order closure models for momentum and property transport provide the best hope. Their predictions will need to be checked with carefully designed field and wind-tunnel experiments, in which measurements are made of as many terms as possible in the second-moment equations. It may prove possible to incorporate aeroelasticity into some of these models.

In studying the role of organized structures in the field and the wind tunnel, using conditional-sampling techniques, canopy-turbulence workers can draw upon an extensive body of knowledge relating to smooth wall flows (e.g. Willmarth 1975), but such information is much sparser for rough wall flows. In this area at least, canopy-turbulence research has contributions to make to fundamental knowledge about turbulent boundary layers.

Finally, a warning is necessary about oversimplifications. The micrometeorological custom of searching for scaling schemes using nondimensional variables, evident here in Figure 8, is not likely to lead to correct results in canopy-flow studies, except in very simple cases. For this reason, the subject will probably remain partially empirical for some time to come.

ACKNOWLEDGMENTS

We are indebted to Dr. E. F. Bradley for making available the extensive data set from the Uriarra forest experiment, and to Dr. O. T. Denmead for use of his unpublished data for maize. Dr. Bradley, and Drs. A. R. G. Lang and R. Leuning, provided valuable criticism of a draft of this review.

*Literature Cited*

Allen, L. H. 1968. Turbulence and wind speed spectra within a Japanese larch plantation. *J. Appl. Meteorol.* 7:73–78

Antonia, R. A., Luxton, R. E. 1971. The response of a turbulent boundary layer to a step change in surface roughness. Part 1. Smooth to rough. *J. Fluid Mech.* 48:721–61

Baines, G. B. K. 1972. Turbulence in a wheat crop. *Agric. Meteorol.* 10:93–105

Bradley, E. F., Finnigan, J. J. 1973. Heat and mass transfer in the plant-air continuum. *Proc. Australasian Conf. Heat & Mass Transfer, 1st, Melbourne, 1973, Reviews*, pp. 57–78

Busch, N. E. 1973. The surface boundary layer (part 1). *Boundary-Layer Meteorol.* 4:213–40

Cionco, R. M. 1965. A mathematical model for air flow in a vegetative canopy. *J. Appl. Meteorol.* 4:517–22

Cionco, R. M. 1972a. Intensity of turbulence within canopies with simple and complex roughness elements. *Boundary-Layer Meteorol.* 2:453–65

Cionco, R. M. 1972b. A wind-profile index for canopy flow. *Boundary-Layer Meteorol.* 3:255–63

Cowan, I. R. 1968. Mass, heat and momentum exchange between stands of plants and their atmospheric environment. *Q. J. R. Meteorol. Soc.* 94:318–32

Deardorff, J. W. 1966. The counter-gradient heat flux in the lower atmosphere and in the laboratory. *J. Atmos. Sci.* 23:503–6

Desjardins, R. L., Allen, L. H., Lemon, E. R. 1978. Variations of carbon dioxide, air temperature, and horizontal wind within and above a maize crop. *Boundary-Layer Meteorol.* 14:369–80

Donaldson, C. duP. 1973. Construction of a dynamic model of the production of atmospheric turbulence and the dispersal of atmospheric pollutants. In *Workshop on Micrometeorology*, ed. D. A. Haugen, pp. 313–92. Boston: Am. Meteorol. Soc. 392 pp.

Dyer, A. J. 1974. A review of flux-profile relationships. *Boundary-Layer Meteorol.* 7:363–72

Finnigan, J. J. 1979a. Turbulence in waving wheat. I. Mean statistics and honami. *Boundary-Layer Meteorol.* 16:181–211

Finnigan, J. J. 1979b. Turbulence in waving wheat. II. Structure of momentum transfer. *Boundary-Layer Meteorol.* 16:213–36

Finnigan, J. J., Mulhearn, P. J. 1978a. Modelling waving crops in a wind tunnel. *Boundary-Layer Meteorol.* 14:253–77

Finnigan, J. J., Mulhearn, P. J. 1978b. A simple mathematical model of airflow in waving plant canopies. *Boundary-Layer Meteorol.* 14:415–31

Garratt, J. R. 1978. Flux profile relations above tall vegetation. *Q. J. R. Meteorol. Soc.* 104:199–212

Goldstein, S., ed. 1938. *Modern Developments in Fluid Dynamics*. Oxford: Clarendon. 702 pp.

Hicks, B. B., Hess, G. D., Wesely, M. L. 1979. Analysis of flux-profile relationships above tall vegetation—an alternative view. *Q. J. R. Meteorol. Soc.* 105:1074–77

Hinze, J. O. 1975. *Turbulence*. New York:McGraw-Hill. 790 pp. 2nd ed.

Inoue, E. 1955. Studies of the phenonema of waving plants ("Honami") caused by wind. Part 1. Mechanism and characteristics of waving plants phenomena. *J. Agric. Meteorol. (Jpn)* 11:18–22

Inoue, E. 1963. On the turbulent structure of airflow within crop canopies. *J. Meteorol. Soc. (Jpn).* 41:317–26

Inoue, K., Uchijima, Z. 1979. Experimental study of microstructure of wind turbulence in rice and maize canopies. *Bull. Natl. Inst. Agric. Sci. (Jpn.), Ser. A* 26:1–88

Inoue, K., Uchijima, Z., Horie, T., Iwakiri, S. 1975. Studies of energy and gas exchange within crop canopies (10) Structure of turbulence in rice crop. *J. Agric. Meteorol. (Jpn.)* 31:71–82

Isobe, S. 1972. A spectral analysis of turbulence in a corn canopy. *Bull. Natl. Inst. Agric. Sci. (Jpn.), Ser. A* 19:101–12

Jackson, P. S. 1981. On the displacement height in the logarithmic velocity profile. *J. Fluid Mech.* In press

Landsberg, J. J., James, G. B. 1971. Wind profiles in plant canopies: studies on an analytical model. *J. Appl. Ecol.* 8:729–41

Landsberg, J. J., Thom, A. S. 1971. Aerodynamic properties of a plant of complex structure. *Q. J. R. Meteorol. Soc.* 97:565–70

Lee, B. E. 1976. Some effects of turbulence scale on the mean forces on a bluff body. *J. Industrial Aerodynamics* 1:361–70

Lumley, J. L., Khajeh-Nouri, B. 1974. Computational modelling of turbulent transport. *Adv. Geophys.* 18A:169–92

Lumley, J. L., Panofsky, H. A. 1964. *The Structure of Atmospheric Turbulence.* New York: Interscience. 239 pp.

Maitani, T. 1978. On the downward transport of turbulent kinetic energy in the surface layer over plant canopies. *Boundary-Layer Meteorol.* 14:571–84

Maitani, T. 1979. An observational study of wind-induced waving of plants. *Boundary-Layer Meteorol.* 16:49–65

Mellor, G. 1973. Analytic prediction of the properties of stratified planetary surface layers. *J. Atmos. Sci.* 30:1061–69

Monteith, J. L., ed. 1975. *Vegetation and the Atmosphere*, Vols. 1, 2. London: Academic. 278 pp, 439 pp.

Mulhearn, P. J. 1978. Turbulent flow over a periodic rough surface. *Phys. Fluids* 21:1113–15

Mulhearn, P. J., Finnigan, J. J. 1978. Turbulent flow over a very rough, random surface. *Boundary-Layer Meteorol.* 15:109–32

Oliver, H. R. 1973. Smoke trails in a pine forest. *Weather* 28:345–47

O'Loughlin, E. M., Annambhotla, V. S. S. 1969. Flow phenomena near rough boundaries. *J. Hydraul. Res.* 7:231–50

Raupach, M. R. 1979. Anomalies in flux-gradient relationships over forest. *Boundary-Layer Meteorol.* 16:467–86

Raupach, M. R. 1981. Conditional statistics of Reynolds stress in smooth and rough wall turbulent boundary layers. *J. Fluid Mech.* In press

Raupach, M. R., Thom, A. S., Edwards, I. 1980. A wind-tunnel study of turbulent flow close to regularly arrayed rough surfaces. *Boundary-Layer Meteorol.* 18:373–97

Schlichting, H. 1936. Experimentelle Untersuchungen zum Rauhigkeitsproblem. *Ing-Arch.* 7:1–34

Schlichting, H. 1968. *Boundary Layer Theory*. New York: McGraw-Hill. 747 pp. 6th ed.

Schuepp, P. H. 1972. Studies of forced-convection heat and mass transfer of fluttering realistic leaf models. *Boundary-Layer Meteorol.* 2:263–74

Seginer, I., Mulhearn, P. J. 1978. A note on vertical coherence of streamwise turbulence inside and above a model plant canopy. *Boundary-Layer Meteorol.* 14:515–23

Seginer, I., Mulhearn, P. J., Bradley, E. F., Finnigan, J. J. 1976. Turbulent flow in a model plant canopy. *Boundary-Layer Meteorol.* 10:423–53

Shaw, R. H. 1977. Secondary wind speed maxima inside plant canopies. *J. Appl. Meteorol.* 16:514–21

Shaw, R. H., den Hartog, G., King, K. M., Thurtell, G. W. 1974a. Measurements of mean wind flow and three-dimensional turbulence intensity within a mature corn canopy. *Agric. Meteorol.* 13:419–25

Shaw, R. H., Silversides, R. H., Thurtell, G. W. 1974b. Some observations of turbulence and turbulent transport within and above plant canopies. *Boundary-Layer Meteorol.* 5:429–49

Silversides, R. H. 1974. On scaling parameters for turbulence spectra within plant canopies. *Agric. Meteorol.* 13:203–11

Stewart, J. B., Thom, A. S. 1973. Energy budgets in pine forest. *Q. J. R. Meteorol. Soc.* 99:154–70

Tennekes, H. 1973. Similarity laws and scale relations in planetary boundary layers. See Donaldson 1973, pp. 177–216

Tennekes, H., Lumley, J. L. 1972. *A First Course in Turbulence*. Cambridge, Mass: MIT Press. 300 pp.

Thom, A. S. 1968. The exchange of momentum, mass and heat between an artificial leaf and airflow in a wind tunnel. *Q. J. R. Meteorol. Soc.* 94:44–55

Thom, A. S. 1971. Momentum absorption by vegetation. *Q. J. R. Meteorol. Soc.* 97:414–28

Thom, A. S. 1972. Momentum, mass and heat exchange of vegetation. *Q. J. R. Meteorol. Soc.* 98:124–34

Thom, A. S., Stewart, J. B., Oliver, H. R., Gash, J. H. C. 1975. Comparison of aerodynamic and energy budget estimates of fluxes over a pine forest. *Q. J. R. Meteorol. Soc.* 101:93–105

Thompson, N. 1979. Turbulence measurements above a pine forest. *Boundary-Layer Meteorol.* 16:293–310

Townsend, A. A. 1976. *The Structure of Turbulent Shear Flow.* Cambridge Univ. Press. 429 pp. 2nd ed.

Uchijima, Z., Wright, J. L. 1964. An experimental study of air flow in a corn plant–air layer. *Bull. Natl. Inst. Agric. Sci. (Jpn), Ser. A* 11:19–65

Willmarth, W. W. 1975. Structure of turbulence in boundary layers. *Adv. Appl. Mech.* 15:159–254

Wilson, N. R., Shaw, R. H. 1977. A higher-order closure model for canopy flow. *J. Appl. Meteorol.* 16:1198–205

Wyngaard, J. L. 1973. On surface-layer turbulence. See Donaldson 1973, pp. 101–49

Yaglom, A. M. 1977. Comments on wind and temperature flux-profile relationships. *Boundary-Layer Meteorol.* 11:89–102

Zdravkovich, M. M., Pridden, D. L. 1977. Interference between two circular cylinders; series of unexpected discontinuities. *J. Industrial Aerodynamics* 2:255–70

*Ann. Rev. Fluid Mech. 1981. 13:131–56*


# CONDITIONAL SAMPLING IN TURBULENCE MEASUREMENT



*R. A. Antonia*
Department of Mechanical Engineering, University of Newcastle, N.S.W., 2308 Australia

## 1 INTRODUCTION AND DEFINITIONS

The technique of conditional sampling and averaging can be described, in general terms, as a means to distinguish and provide quantitative information about interesting regions of a turbulent flow. It has been fairly widely used for the experimental study of turbulent shear flows. Specifically, it has been applied to the study of

(*a*) Shear flows which exhibit a turbulent-nonturbulent interface;
(*b*) Shear layers perturbed by interaction with another field of turbulence;
(*c*) Quasi-periodic or periodic flows (these include flows behind passing blades in rotating machinery, boundary layers over a solid or liquid surface with stationary or progressive periodic waves, and more generally flows subjected to either internal or external periodic perturbations);
(*d*) Coherent structures in different shear flows (these structures are currently receiving close attention by the turbulence community in view of their importance to the flow dynamics).

Conditional sampling and averaging have been used mainly to provide quantitative information needed to complement usually qualitative but interesting information obtained from flow-visualization experiments. This is particularly true for flows in (*a*) and (*d*). Corrsin (1943) first observed the clear demarcation between turbulent and nonturbulent regions at a relatively thin interface or viscous superlayer. Using dye injected at the wall, Hama (see Corrsin 1957) first observed, in the sublayer, an intermittent streamwise streaky structure later visualized



0066-4189/81/0115-0131$01.00

and studied in detail by Kline et al (1967), Corino & Brodkey (1969), and others. More recently, the visual recognition of quasi-deterministic coherent vortex structures, often completely embedded in a turbulent environment, has received a great deal of attention. The information obtained from these flow-visualization experiments has already been reviewed by Davies & Yule (1975), Roshko (1976), and Laufer (1975). The established existence of the turbulent-nonturbulent interface first prompted Townsend (1949) to make use of conditional averaging and obtain averaged values of the normal velocity in the turbulent region only of the heated wake of a circular cylinder. The technique did not, however, receive widespread usage until much later when advances in signal-processing techniques and the emergence of fast analog-to-digital conversion followed by further digital processing rendered its application more attractive. While both analog and digital conditional sampling measurements have been made, the digital approach (Van Atta 1974) appears to be particularly suited to the study of (*d*).

It should be noted that various aspects of conditional sampling of turbulent flows have already been discussed in the literature (e.g. Bradshaw 1972, Kaplan 1973, Van Atta 1974, 1980, Willmarth 1975, Kovasznay 1979) and some of these will not be considered here. In particular, periodic and quasi-periodic sampling (category *c*) will not be considered as it has already been treated by Van Atta (1974) and Kovasznay (1979).

It is appropriate to define (e.g. Blackwelder 1977) a conditional average as a special type of generalized cross-correlation. We first consider the correlation

$$R(\mathbf{x}, \Delta\mathbf{x}, \tau_j) = \lim_{N\to\infty} \frac{1}{N} \sum_{i=1}^{N} c(\mathbf{x}, t_i) f(\mathbf{x}, \Delta\mathbf{x}, t_i + \tau_j) \tag{1}$$

between a digital conditioning function $c$, derived from one or more turbulence signals, and a digital function $f$, which may also be derived from one or more turbulence signals. It should be noted that the signals used to obtain $c$ and $f$ may be identical or completely different (in general, they may be generated at different positions in space). $N$ is the number of digital points to be averaged over, while the time delay $\tau_j$ may be either positive or negative.

The function $c$, which defines the condition under which averaging is to occur, is usually generated by a detection function $d$ (Figure 1) which should, for practical reasons, be as simple as possible. The algorithm used to detect the turbulent-nonturbulent interface may include linear (e.g. differentiation, smoothing), as well as nonlinear operations (e.g. rectification) and combinations of components (e.g. sum of squares).

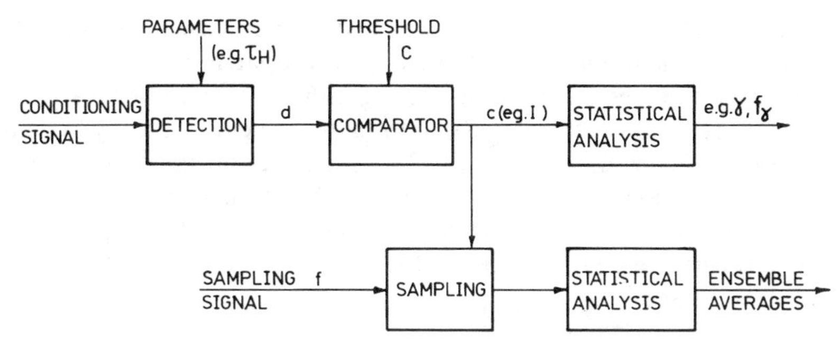


*Figure 1* General conditional sampling and averaging scheme.

A simple type of conditional average is the zone average, obtained as the accumulation of values of $f$ during those intervals for which the flow at a given point is turbulent or nonturbulent. With the intermittency function $I(\mathbf{x}, t_i)$ defined as

$$I = \begin{pmatrix} 1, & \text{for turbulent flow} \\ 0, & \text{for nonturbulent flow} \end{pmatrix} \tag{2}$$

the turbulent and nonturbulent zone averages $\bar{f}_t$ and $\bar{f}_n$ can be deduced from (1) when $c$ is identified with $I$ and $1-I$ respectively. With $\Delta\mathbf{x}$ and $\tau_j$ set equal to zero,

$$\bar{f}_t = \lim_{N\to\infty} \sum_{i=1}^{N} If / \sum_{i=1}^{N} I, \tag{3}$$

and

$$\bar{f}_n = \lim_{N\to\infty} \sum_{i=1}^{N} (1-I)f / \sum_{i=1}^{N} (1-I). \tag{4}$$

The conventional average of $f$ is given by $\bar{f} = \lim_{N\to\infty} \Sigma_{i=1}^{N} f/N$. The intermittency factor $\gamma$, or fraction of the time for which the flow at a position in space is turbulent, is equal to $\lim_{N\to\infty} \Sigma_{i=1}^{N} I/N$. The frequency of occurrence of turbulent or nonturbulent regions is $f_\gamma$. Another relatively simple type of average is the point average when $f$ is averaged only at the location of the front (or back) of a turbulent region. The accuracy of determining a point average is lower than that of a zone average for a given duration of the digital time series $f$. Antonia & Bradshaw (1971) and Antonia (1972) presented point averages throughout a turbulent (or nonturbulent) region as a function of the time from the front of the region normalized by the duration of that region. A similar form of averaging was used by LaRue & Libby (1974a, 1978) who referred to it as range-conditioned point averaging.

Another type of average is that obtained over an ensemble of events or structures mentioned in $(d)$. Again the output $c$ of the detection algorithm (the types of inputs to the algorithm may vary a great deal and may be derived from visual information such as the hydrogen-bubble coordinates used by Grass 1971) is a series of time points, each of these corresponding to a particular phase of that structure. Once these points are identified, ensemble averages can be obtained relative to a particular reference point of the structure. A common form of ensemble average is

$$\langle f(\mathbf{x}, \tau_j)\rangle = \sum_{m=1}^{M} f(\mathbf{x}, t_m+\tau_j)/M \tag{5}$$

where $t_m$ are the points in time when detection occurs and $M$ is the maximum number of events detected. When turbulent and nonturbulent regions are the events under study, the values of $\langle f\rangle$, obtained when $t_m$ are the positions of the fronts and backs of turbulent regions, are denoted here by $\langle f_{\mathrm{f}}\rangle$ and $\langle f_{\mathrm{b}}\rangle$ and represent the point averages mentioned previously.

The overall and rather simplified scheme of Figure 1 is sufficiently general to include most, if not all, conditional sampling and averaging procedures used in practice. Of course, such generality means that it is of little use in guiding us in individual cases, especially with respect to the choice of parameters, threshold $C$, and operations required to form $d$. A commonly used parameter is $\tau_{\mathrm{H}}$, a smoothing or hold-time which should not be much larger than the smallest time scale of interest for the event under study ($\tau_{\mathrm{H}}$ is sometimes applied after comparison with $C$). An option (not included in Figure 1) available to the experimenter is to adjust one or several of the parameters and/or $C$ after comparing $c$ with the turbulence signal(s) in the detection scheme. Note also that $c$ can be used in a more practical sense than suggested by Figure 1 (it can for example, be applied to flow visualization).

## 2 TURBULENT-NONTURBULENT INTERFACE DETECTION AND RESULTS

The generation of the intermittency function $I$ suffers from a certain amount of arbitrariness due to the inevitable subjectivity inherent in any detection scheme. Most of the detection functions used for the earlier conditional measurements have been used on velocity fluctuations. Tabulations of these functions may be found in Hedley & Keffer (1974a), Schon & Charnay (1975). Bradshaw & Murlis (1974) noted three major difficulties that arise in the determination of $I$ by analog or digital analysis of velocity fluctuations:

(*a*) The physical difficulty that the interface is highly re-entrant (smoke photographs of Fiedler & Head 1966 and Falco 1977 reveal the highly contorted three-dimensional behavior of the interface);
(*b*) The mathematical difficulty that the criterion function, supposed to be positive within the turbulence, can have occasional zeroes;
(*c*) The operational difficulty that this criterion function may respond to high wave-number intermittency as well as interfacial intermittency.

These difficulties lead to the prediction of short but not necessarily spurious "nonturbulent" indications within "turbulent" regions, a situation that has been described as the "false alarm" or sometimes "drop out" or "hole" problem. The practice of rejecting all short regions is associated by Bradshaw & Murlis with the generation of "wholesale" intermittency, as distinct from "retail" intermittency which retains the difficulty contained in (*a*). Most of the analog (and digital) intermittency meters based on velocity inputs have relied essentially on setting $C$ and $\tau_H$ to generate "wholesale" intermittency. Kibens et al (1974) and Antonia & Atkinson (1974) used pseudo-turbulent signals to calibrate analog intermittency circuits with an aim to optimize $C$ and $\tau_H$. The latter authors used a pseudo-turbulent signal derived by random switching between either two signals of similar spectra simulated on an analog computer or two real signals obtained in the turbulent ($\gamma \simeq 1$) and irrotational regions respectively of a jet. Although there is clearly an element of subjectivity in the construction of artificial signals, the calibration procedure was found to be helpful in arriving at satisfactory settings of $C$ and $\tau_H$.

Use of a passive scalar (usually temperature) to mark the position of the interface has proved to be operationally superior to the use of velocity or a component of vorticity. To generate $I$ in a slightly heated circular jet, Antonia et al (1975) were able to dispense with $\tau_H$ and simply compare the instantaneous temperature in the jet with the ambient air temperature. The variations of $\gamma$ and $f_\gamma$ with threshold $C$ showed that little ambiguity, except perhaps when $\gamma$ approached unity, occurred in the selection of $C$. Errors in zone averages, such as $\overline{(uv)}_t$, $\overline{(u\theta)}_t$, and $\overline{(v\theta)}_t$ ($u$, $v$ are the streamwise and normal velocity fluctuations, $\theta$ is the temperature fluctuation; $U$, $V$, and $T$ denote corresponding instantaneous quantities), were found to be small at all values of $\gamma$. Kovasznay & Ali (1974) and LaRue (1974) also used $\theta$ to generate $I$ but retained a nonzero hold time $\tau_H$; LaRue showed that the statistics of $I$ were nearly independent of the values of the detection parameters.

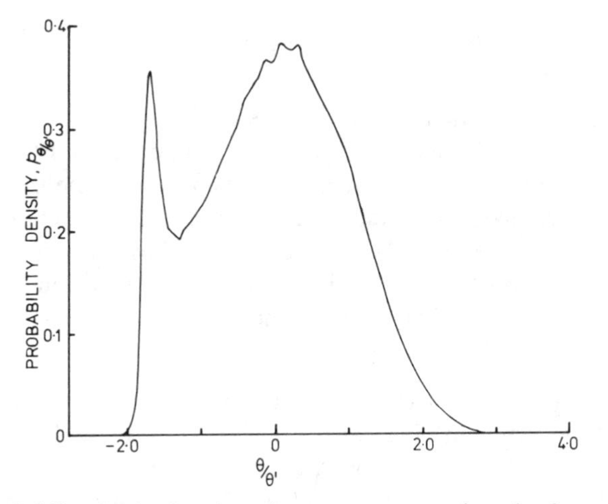


*Figure 2* Probability density function of temperature near the axis of an axisymmetric heated jet in a co-flowing stream. $\gamma = 0.91$. (Prime denotes rms value.)

That the temperature signal offers good discrimination between turbulent (usually marked) and nonturbulent regions can be quickly ascertained from the often bimodal shape of the probability density function (pdf) $p_\theta$ of temperature (Figure 2). The narrow spike in $p_\theta$ is associated (Bilger et al 1976) mainly with the presence of free-stream temperature fluctuations (often low-frequency fluctuations associated with drafts in the laboratory) and noise in the electronics. Bilger et al observed that a Gaussian curve closely fitted this spike and that the area under this Gaussian represented a good estimate of $(1-\gamma)$, the accuracy of this measurement being limited by the signal-to-noise ratio. This method was used by Sreenivasan et al (1978) to distinguish between turbulent and nonturbulent (both "cold" and "hot") regions of the axisymmetric mixing layer of a round jet where $p_\theta$ can exhibit a trimodal shape. With the assumption that the turbulent and nonturbulent parts of the flow are mutually exclusive, the probability density function of any quantity $f$ may be written as

$$p_f = \gamma p_f^{\mathrm{T}} + (1-\gamma) p_f^{\mathrm{N}}. \tag{6}$$

The switching between turbulent and nonturbulent flow regions may not be quite independent of statistics of these regions, e.g. large values of $(\bar{U}_{\mathrm{t}} - \bar{U}_{\mathrm{n}})$ are in general associated with large durations of turbulent regions while $\bar{U}_{\mathrm{t}} \simeq \bar{U}_{\mathrm{n}}$ for small durations. Using (6), the error in detecting nonturbulent fluid in a turbulent region may be expressed as an analytic function of the threshold. Chen & Blackwelder (1978) considered $p_\theta$ to show that there is no region where estimated values of $\gamma$ and turbulent zone averages are independent of $C$. More recently, Haverbeke et al (1979) examined in some detail the overlap of contributions fom the turbulent and nonturbulent regions to $p_\theta$, and suggested that this overlap is due primarily to transitions associated with the viscous superlayer. An approximate upper bound was given for the uncertainty in measuring $\gamma$ and the optimum choice for $C$ for conditioned measurements was found to be the same as that for $\gamma$.

Conditional measurements have been made in a number of different flows (apart from one exception, marked with an asterisk, the maximum temperature difference, $\bar{T}_0$ say, across all the heated flows referred to below is such that temperature may be treated as a passive marker).

*Boundary layer*: unheated (Kaplan & Laufer 1968, Imaki 1968, Kovasznay et al 1970, Antonia 1972, Hedley & Keffer 1974b, Murlis 1975) and heated (Chen 1975, Subramanian & Antonia 1978);
*Wake of a circular cylinder*: unheated (Thomas 1973); and heated (LaRue & Libby 1974a, b, Fabris 1974, 1979, Barsoum et al 1978);
*Wake of a heated flat plate*: Ali & Kovasznay (1975);
*Wake of an axisymmetric disk*: Oswald & Kibens (1971);
*Two-dimensional mixing layer*: unheated (Wygnanski & Fiedler 1970); and heated (Fiedler 1974);
*Plane jet*: unheated (Gutmark & Wygnanski 1976); and heated (Davies et al 1975, Jenkins & Goldschmidt 1974, Moum et al 1979); helium jet (Anderson et al 1979);
*The wall jet*: unheated (Paizis & Schwarz, 1975); and heated (Alcaraz et al 1977, Hopfinger* 1978);
*Circular jet*: heated (Antonia et al 1975, Chevray & Tutu 1978a, b).

It may be useful to indicate briefly some of the results obtained in a few of these investigations. In the intermittent region of a turbulent boundary layer, it has been firmly established that while $\bar{U}_n$ is larger than $\bar{U}_t$, measurements of the normal component of velocity indicate that $\bar{V}_t$ and $\bar{V}_n$ are positive and negative respectively, with respect to the conventional velocity $\bar{V}$, throughout a turbulent bulge. The values of $\bar{V}_t$ and $\bar{V}_n$ obtained by Hedley & Keffer (1974b) are in good agreement with those of Kovasznay et al (1970). There is reasonable agreement between the point-averaged normal velocities measured by Hedley & Keffer (1974b) and those of Kovasznay et al (1970). With respect to $\bar{V}_n$, $\langle V_f \rangle$ and $\langle V_b \rangle$ are always negative and positive respectively, emphasizing the vortical behavior of the turbulent bulge. This behavior was demonstrated by Blackwelder & Kovasznay (1972) who used their point-averaged data to construct the velocity field associated with an "average" turbulent bulge moving at a constant convection velocity ($\simeq 0.93 U_\infty$). Measurements of the Reynolds shear stress in a turbulent region have confirmed the expectation that $\overline{(uv)}_t = \overline{uv}/\gamma$, with negligible contribution provided by the irrotational flow. The previously noted quantitative agreement between different measurements of $\langle V_f \rangle$ and $\langle V_b \rangle$ does not extend to point-averaged streamwise velocities, perhaps as a result of the less pronounced changes in $U$, in comparison to $V$ or $T$ at the back of the bulge. The bulk of the available data seems to be consistent with the observation that $\langle U_f \rangle$ is generally slightly larger than $\langle U_b \rangle$.

Our expectation that for a fluid of molecular Prandtl number near unity, the vorticity and thermal interfaces are nearly coincident has been supported by a number of observations. Dumas et al (1972) found that boundary-layer distributions of $\gamma$, obtained using discriminating procedures based either on velocity or temperature, were essentially coincident. Jenkins & Goldschmidt (1974) found that in a plane jet, distributions of $\gamma$ and $f_\gamma$, based on either velocity or temperature, were identical to within the uncertainty of measurement. These authors further confirmed the coincident termination at or near the interfaces of point averages of velocity and temperature, although the jump across the interface is undoubtedly sharper for the temperature than for the velocity. Davies et al (1975) verify, in a heated plane turbulent jet, that the spread rates of the thermal and momentum fields, when considered in terms of the turbulent fluid only, were nearly coincident.

Antonia et al (1975) found that the zone-averaged mean temperature and temperature-fluctuation intensity were nearly homogeneously distributed in the outer turbulent region of a slightly heated circular jet with a co-flowing stream. The ratio $(\overline{T}_t - \overline{T}_n)/\overline{T}_0$ is considerably larger than corresponding ratios for either $U$ or $V$. This result is certainly true in other flows [e.g. a mixing layer (Fiedler 1974) or the wake (Kovasznay & Ali 1974)] and is not surprising in view of the good distinction between turbulent and nonturbulent states provided by $\theta$ (cf. Figure 2). Antonia et al (1975) also noted that the Reynolds shear stress $\overline{(uv)}_t$ decreased in the outer turbulent-flow region, the radial heat flux $\overline{(v\theta)}_t$ continued to increase, with a subsequent decrease in the turbulent Prandtl number (vis-à-vis the conventional turbulent Prandtl number). Jenkins & Goldschmidt (1974) found that the turbulent Prandtl number was essentially constant and equal to 0.4 across the fully turbulent region of a plane jet.

Fiedler (1974) obtained a nearly homogeneous mean temperature and an approximately constant temperature variance in the turbulent region of a mixing layer, and ascribed this behavior to the large-scale vortical motion. Fluid from the "hot" or "cold" sides of a heated axisymmetric mixing layer is found (Sreenivasan et al 1978) on the opposite side, consistent with bulk transport of nonturbulent fluid by a large-scale motion comparable to the width of the flow. Density fluctuations (Roshko 1976) in the mixing layer between parallel streams of helium and nitrogen showed deep excursions of each gas into the other.

The similarity of conventional mean-velocity and temperature profiles in self-preserving shear flows can obviously be extended to turbulent zone-averaged distributions. Davies et al (1975) found that the similarity exhibited by turbulent zone-averaged mean velocity and temperature in a plane jet was certainly as good as that demonstrated by the conventional profiles, a result that seems consistent with the good similarity exhibited by $\gamma$ profiles obtained at different streamwise stations in the flow. Equivalent observations have been made by LaRue & Libby (1974a, b) in a wake[1] and Subramanian & Antonia (1979) who measured turbulent zone-averaged mean-temperature profiles in a boundary layer at different Reynolds numbers. Jenkins & Goldschmidt's (1974) point-averaged velocity and temperature profiles, measured relative to the interface at different values of $\gamma$ and different streamwise stations, also exhibited good similarity. LaRue & Libby (1978) reported that similarity within the turbulent fluid exists in even greater detail than the previous measurements suggest. They found, using the range-conditioning technique, that the mean and rms temperature distributions within bulges (or "structures" in their terminology) of essentially the same duration exhibit similarity in the central intermittent region of the wake. Bulges of relatively large duration ($l_c/2U_\infty$, where $l_c$ is an appropriate characteristic length) have an exponential ramplike behavior with the maximum temperature occurring near the back of the bulge. LaRue & Libby identified three regions for these bulges: a central region where gradients of the temperature moments are relatively small and two interfacial regions, typically of thickness $10\eta$ ($\eta$ is the Kolmogorov microscale), associated with large gradients. Recently, Sreenivasan et al (1979) used this range-conditioning technique in the intermittent region of a round jet for bulges of relatively large duration (these bulges were identified, with reservation, with signatures of the large structure of the flow). The temperature distribution bears close similarity to that of LaRue & Libby (1978), except of course that the maximum temperature now occurs at the front instead of the back of the bulge. The front and back of the bulge have positive and negative normal velocities, emphasizing the large-scale vortical motion within the

[1]Although Kovasznay & Ali (1974) reported a lack of self-preservation of $\bar{U}_t$ and $\bar{T}_t$ profiles in the wake of a flat plate they also mentioned that these averages are likely to be in error in the outer part of the wake because of an incorrect setting of $C$.

bulge. It is perhaps worth mentioning here that this range-conditioning technique enables a separate study of the large-scale signature and of the superposed fluctuations, accepting that such a separation may be overly simplifying. While the large-scale signature appears to be responsible for the breakdown in local isotropy (as evidenced for example by the nonzero skewness of the temperature derivative), the superposed fluctuations are consistent with local isotropy and may be responsible for a significant part of the turbulent momentum and heat transport. Probability densities of these fluctuations are a reasonable approximation to the Gaussian whereas conventional and "turbulent" pdf's deviate significantly from it.

Several statistics related to the average shape and size of a bulge have been obtained. The topography information of the interface has been studied by several investigators using multiple probes. Paizis & Schwarz (1974) found that overhangs occurred in about 40% of interfaces in a wall jet. LaRue & Libby (1976) obtained only 15% in a two-dimensional wake but noted that about 25% of the interfaces at the front of a bulge exhibit overhangs, as compared with only about 6% at the back. Mulej & Goldschmidt (1975) also noted, for a plane jet, that folding is more likely to be detected on the front of a bulge and towards the centerline of the jet. Kawall & Keffer (1979) observed that, in the near wake of a cylinder, folding is not as significant as for the far-wake observations of LaRue & Libby (1976). Subjecting the wake to a strain field flattens out the interface. Information related to the spatial extent of an average bulge has been obtained by auto- and cross-correlations of the intermittency function $I$, i.e. by computing (1) with $f$ replaced by $I$. From the autocorrelation of $I$, an average streamwise length can be inferred by using Taylor's hypothesis. The Fourier transform of this autocorrelation has been found to agree with earlier boundary-layer measurements of Corrsin & Kistler (1955). The spectral density of $I$ does not suggest any periodicity of the interface. In a plane jet, Moum et al (1979) found that the extent of the bulge is of the same order of magnitude in the three coordinate directions, being tilted backward, with no spanwise yaw at about 26° to the lateral axis. In a wake (Barsoum et al 1978), the spanwise extent is smaller and larger than the streamwise and lateral extents respectively.

## 3 CONDITIONAL-SAMPLING QUADRANT ANALYSES

A conditional-sampling and averaging technique introduced by Wallace et al (1972) and Willmarth & Lu (1972) to quantify the contribution to the Reynolds shear stress during the cycle of events observed in the wall region of turbulent-boundary-layer and pipe flows has often been used in both wall-bounded and free turbulent shear flows. The cycle of events includes the burst [a three-stage process described by Kim et al (1971), by the formation of a low-speed streak, its liftup and oscillation, and finally its breakup] and the sweep (Corino & Brodkey 1969, also Offen & Kline 1974). Essentially, the technique sorts contributions to $\overline{uv}$ into quadrants of the $u$–$v$ plane. At a given **x**, the conditioning function in (1) is, in the notation of Lu & Willmarth (1973),

$$h_j(t_i)=\begin{cases} 1, & \text{if the point } (u,v) \text{ is in the } j\text{th quadrant of the } u\text{–}v \text{ plane} \\ 0, & \text{otherwise.} \end{cases} \tag{7}$$

With the sampling function $f \equiv uv$, contributions to the conventional shear stress $\overline{uv}$ from ejection ($u<0$, $v>0$), sweep ($u>0$, $v<0$), and interaction quadrants at different positions in the flow were reported by Wallace et al (1972); Brodkey et al (1974); Sabot & Comte-Bellot (1976); Ueda & Mizushima (1977); Badri Narayanan et al (1977), and others. The dominant contribution to $\overline{uv}$ in the viscous (linear) sublayer ($y^+ = yU_\tau/\nu \leqslant 10$, where $U_\tau$ is the friction velocity and $\nu$ the kinematic viscosity of the fluid) is associated with the sweep. Lu & Willmarth (1973) modified $h_i$ to identify more accurately contributions from violent ejections and sweeps. Specifically, $h_i$ was replaced by $S_i$ defined as

$$S_j(H,t_i)=\begin{cases} 1, & \text{if } |uv|>Hu'v' \text{ and point } (u,v) \text{ is in the } j\text{th quadrant of } u\text{–}v \text{ plane} \\ 0, & \text{otherwise.} \end{cases} \tag{8}$$

An objective choice of the hole size $H$ is not simple. Lu & Willmarth identified large or violent ejections ("bursts" in their terminology) with $H \simeq 4.5$ (corresponding to $|uv| \simeq 10\ \overline{uv}$) since contributions from other quadrants are negligible at this value of $H$. A violent sweep was identified with $H \simeq 2.5$, when contributions from interaction quadrants disappear. The mean period between violent ejections $\bar{t}_e$ was found to be the same as that between violent sweeps and, when normalized by $U_\infty$ and the displacement thickness $\delta^*$, approximately independent of either position across a layer or Reynolds number. Perhaps coincidentally, the numerical value for this period (which does depend on $H$) was in good agreement with $\bar{t}_B$, the mean period obtained by Rao et al (1971) for high-frequency pulses [identifiable with the intermittent fine structure of turbulence, e.g. Badri Narayanan et al (1977)]. Comte-Bellot et al (1979) carried out an extensive study of the effect of $H$ on the period between ejections or sweeps, the occupancy time or "intermittency" factor $\bar{S}_j$, and contributions to $\overline{uv}$ from different quadrants. They suggested that Lu & Willmarth (1973) singled out for study events that are too violent and that a more appropriate selection of $H$ for events in a particular quadrant may simply be based on the percentage contribution to $\overline{uv}$ from that quadrant. Values of $H$ obtained for the ejection and sweep quadrants were about 1 and 0.7 respectively, significantly smaller than the values determined by Lu & Willmarth. The period $\bar{t}_e$ was found to be correspondingly smaller than that obtained by Lu & Willmarth but, surprisingly, was inferred to be in good agreement with $\bar{t}_B$. This inference was based on a comparison of the ratio $\bar{t}_e/\bar{t}_0$, where $\bar{t}_0$ is the period between zero crossings of an unfiltered $u$ signal, with the ratio $\bar{t}_B/\bar{t}_0$ obtained by Badri Narayanan et al (1977). While the relationship between $\bar{t}_B$ and the period between zero crossings of an appropriately filtered $u$ signal may be reasonable, the above inconsistency is perhaps due to the difficulty in determining $\bar{t}_0$ precisely. It should perhaps be added that further conditions can be added to (8) as was done by Comte-Bellot et al (1979) who imposed the extra constraints $|u| \geqslant h_u u'$ or $|v| \geqslant h_v v'$, when analysing amplitudes of $u$ and $v$ during ejections and sweeps.

It should be noted that the quadrant analysis studies described above were accompanied by studies (e.g. Gupta & Kaplan 1972, Lu & Willmarth 1973) of the statistics of the product $uv$ and especially of the joint pdf of $u$ and $v$. Antonia & Atkinson (1973) used a cumulant-discard method to predict third- and fourth-order moments and the pdf of $uv$. Nakagawa & Nezu (1977) predicted the magnitude of the contribution to $\overline{uv}$ from ejections and sweeps by considering conditional pdf's of $uv$, derived by application of a cumulant-discard method to the Gram-Charlier joint pdf $p(u, v)$ of $u$ and $v$. Wallace & Brodkey (1977) presented measurements of $p(u, v)$ and of the weighted pdf $uvp(u, v)$ at different positions across a fully developed turbulent channel flow. Comparison of $p(u, v)$ and $uvp(u, v)$ indicated that the most probable pairs of velocities did not coincide with the pairs associated with the largest contribution to $\overline{uv}$. Perry & Hoffman (1976) also presented measurements of $uvp(u, v)$ in a slightly heated turbulent boundary layer and compared these with measurements of $v\theta p(v, \theta)$. Ejections and sweeps in $uv$ were accompanied, with high probability, by corresponding events in $v\theta$. However, the mean period, using Lu & Willmarth's (1973) setting for $H$, between $v\theta$ ejections was found to be 2.3 times as large as that between $uv$ ejections, indicating the need for a more critical look at the choice of $H$ before this approach can provide a meaningful quantitative assessment of the similarity of the instantaneous products $uv$ and $v\theta$.

The quadrant analysis (and hole-size) technique has been used (Takeuchi et al 1977) to show that the various events are more vigorous in a boundary layer over a large-amplitude wave of fixed frequency than over a flatter surface. Chambers & Antonia (1979) found that, in the marine surface layer, the period between ejections depends on the ratio $C/U_\tau$, $C$ being the phase velocity associated with the dominant wave frequency. It should finally be mentioned that the technique has been used in the core region of a pipe flow by Sabot & Comte-Bellot (1976) to identify ejections from opposite halves of the flow. It has also been used in a slightly heated circular jet (Sreenivasan & Antonia 1979) where the quadrants ($u>0$, $v>0$) and ($u<0$, $v<0$) were identified with ejections of high-momentum fluid and sweeps of relatively low-momentum fluid. At a radial position where the production of turbulent energy is maximum, quadrant occupancy times and contributions from ejections and sweeps are consistent with results of Brodkey et al (1974) near $y^+=15$. Measurements of $p(u,v)$ and $p(v,\theta)$ in the turbulent region only of the flow indicated that the departure from Gaussianity is less pronounced for the outward than for the inward radial motion (Antonia & Sreenivasan 1977).

Sweeps and ejections have also been studied by Zaric (1974, 1975) with a different conditional-averaging technique, the conditioning function being based on the product of $u$ and $\partial u/\partial t$. Little usage has been made of conditional sampling for the study of the Reynolds-number-dependent, intermittent fine-structure regions of turbulence. Kuo & Corrsin (1972) attempted to identify the geometrical character of such regions, dispersed in nearly isotropic grid turbulence, using two-point coincidence functions of an intermittency function (see Kuo & Corrsin 1971 for details) set equal to 1 and 0 depending on when the hot wire is in a fine-structure region and when it is not. The regions were found to be more nearly like rods than blobs or slabs. Coincidence measurements, with a spatial array of wires, may, with appropriate conditional averaging, yield interesting information about the geometry of the fine structure in a turbulent shear flow and its possible interaction with the large structure.

## 4 INTERACTING LAYERS

Extensive use has been made of conditional sampling, using temperature as a passive marker, to quantify the interaction of a turbulent shear layer and another field of turbulence. Tagging one of the interacting fields with temperature allows a distinction to be made between hot-zone and cold-zone contributions. The three main classes of interacting flows considered by Bradshaw (1977a) will be discussed briefly.

### *Interacting Shear Layers*

Dean & Bradshaw's (1976) measurements of interacting boundary layers (one heated) in a rectangular duct indicate that, after the layers merge, the interaction region near the centerline is best described by the time-sharing between hot and cold fluids. The turbulence structure is little affected by the interaction so that superposition of the turbulence fields, strictly not allowed by the nonlinear Navier-Stokes equations, yields (Bradshaw et al 1973) good predictions of a fully developed duct flow. Conditionally sampled measurements (Weir & Bradshaw 1977) in two merging mixing layers (one heated) of a plane jet showed that the turbulence structure of each layer is significantly altered by interaction with the other. Conditionally sampled measurements in the symmetric and asymmetric near wake of a thin aerofoil (Andreopoulos 1979) reveal an inner wake, where fine-grain mixing is dominant, and two outer unmixed layers with only minor changes in turbulence structure. Fabris (1974) used conditional sampling to study the interaction of the wakes of two parallel circular cylinders mounted in the same plane (the lower one being slightly warm). The interaction was found to increase the lateral spreading rate in the upper half of the double wake as the enhanced normal velocity of the heated lumps in the interaction region was more than twice as large as the corresponding normal velocity in the single wake.

### *Internal Layers*

Johnson (1959) first observed the temperature intermittency of the internal layer that grows within a boundary layer subjected to a step change in surface temperature. Antonia et al (1977) examined the response of a turbulent boundary layer to a sudden increase in surface heat flux (zero heat flux upstream) and presented statistics of the thermal interface and conditional measurements with respect to this interface. Near the step, the hot-zone-averaged Reynolds shear stress and heat flux are considerably larger than conventional averages (when the thermal intermittency factor is small) as temperature acts as an effective marker of fluid ejected from the wall region. Zone- and point-averaged measurements, with respect to both inner and outer thermal interfaces, were made by Charnay et al (1979) in a turbulent boundary layer at one station downstream of a sudden decrease in wall temperature.

### *Boundary Layers With a Turbulent Free Stream*

Charnay et al (1976) obtained temperature-conditioned averages inside and outside bulges of a heated boundary layer subjected to an external grid-generated turbulence whose intensity and length scale decrease and increase, respectively, with distance from the grid. The standard deviation (and frequency) of the thermal interface increased as the free-stream turbulence level increased, a result supported by photographs (e.g. Bradshaw 1977b) of a smoke-filled boundary layer. The importance of the external-turbulence length scale was demonstrated by Hancock (1978) who also used temperature-conditioned sampling in a heated boundary layer subjected to an external turbulent free stream with a relatively wide range of length scales.

It seems appropriate to also mention the thermal "mixing layer" study of Keffer et al (1977) who introduced a step change in temperature in a decaying homogeneous turbulent field. The transfer of heat across the thermal interface took place by turbulent diffusion rather than molecular diffusion.

## 5 CONDITIONAL SAMPLING AND AVERAGING OF COHERENT STRUCTURES

Conditional sampling has been extensively used to recognise and yield phase- or ensemble-averaged information related to organized coherent structures. These may range from the large organized quasi-deterministic structures observed (Brown & Roshko 1974) in the mixing layer or the perhaps less organized large-scale motion in the outer part of a boundary layer to the smaller-scale burst in the wall region of bounded shear flows. The difficulties in recognizing these often three-dimensional vortical structures are numerous but not necessarily insurmountable. The structures usually occur randomly in space and time while their three-dimensional geometry and convection speed can exhibit a large amount of jitter. These difficulties may be overcome partly by the possibility (e.g. Blackwelder 1979) of increased use of mathematical tools of pattern recognition and partly by the use of spatial arrays of sensors (e.g. Wygnanski 1979) to experimentally capture a significant volume of the structure. While most arrays have been essentially one dimensional, two- and three-dimensional arrangements are being tried [e.g. to study flow patterns of large eddies, Townsend (1979) arranged eight hot wires in three parallel lines with two sensors in each outer row and four in the central row]. Ideally, we require the detection algorithm to recognize and lock onto some phase relationship that exists between the conditioning and sampling functions. The randomness of phase caused by the random environment can cause serious degradation of the conditional averages. The possibility of correcting for phase-scrambling effects has been considered (e.g. Blackwelder 1977 and Yule 1979). Such corrections or palliatives can facilitate the interpretation of measurements usually obtained in an Eulerian reference frame in terms of information derived (as in flow visualization) in a quasi-Lagrangian frame moving with the structures.

To detect the burst near the wall of a turbulent boundary layer, Blackwelder & Kaplan (1976) used a relatively simple detection algorithm to identify regions, at the edge of the sublayer, exhibiting a high degree of fluctuation. A variable-interval time-averaging (VITA) technique was employed (the averaging time must be of order of the time scale of the event under study). Reference times for each burst were obtained whenever the VITA variance signal exceeded a certain threshold value. Note that this type of detection (and the use of threshold and hold times) is not very different from that used in Section 2. Once the reference times were identified, conditional sampling was performed following the procedure outlined in Section 1. Using a rake of 10 hot wires, Blackwelder & Kaplan obtained conditionally averaged velocity profiles in the region $0<y^+<100$ with the detection criterion applied to the signal at $y^+=15$. An inflexional profile (observed earlier by Kim et al 1971 and Corino & Brodkey 1969) was evident immediately before detection. The conditionally averaged Reynolds shear stress was approximately an order of magnitude greater than its conventionally averaged value and decayed slowly downstream after correcting for the random-convection velocity. Using the criterion that the filtered velocity at two side-by-side points ($y^+\simeq 16$) should be simultaneously low and decreasing, Willmarth & Lu (1972) identified individual contributions to $\overline{uv}$ as large as $62\overline{uv}$. Recently, Blackwelder & Eckelmann (1979) used quadrant-probability analysis and conditional-sampling techniques in the vicinity of the wall to infer that pairs of counter-rotating streamwise vortices are responsible for the ejection and formation of streaks of low-speed fluid. The VITA technique had been earlier applied to signals from an array of hot wires, mounted very near the surface in a spanwise direction, by Gupta et al (1971). It was established that the characteristic wavelength $\lambda$ of the streaky sublayer structure is given by $\lambda^+=100$, consistent with the results of Kline et al (1967). For $y^+<100$, Blackwelder & Kaplan (1976) found that their detection frequency was in reasonable agreement with the "pulse" frequency of Rao et al (1971) and scaled on outer-flow variables. This detection frequency was also found (Blackwelder & Woo 1974) to be independent of the frequency of a periodic pressure perturbation applied in the irrotational region outside the turbulent-nonturbulent interface.

Wallace et al (1977) developed a scheme to detect and ensemble-average patterns that occur repeatedly in the $u$ signal and are characterized by a relatively weak deceleration followed by a strong acceleration. In the region $10<y^+<30$, this pattern occurred over 65% of the total sampling time and was interpreted as the $u$ signature of a burst. The scheme also normalized each pattern to an arbitrarily chosen time interval (the maximum and minimum slopes of the pattern were fixed at two predetermined points along this interval). Always using $u$ to detect the pattern, conditional averages were obtained (see Eckelmann et al 1977), over the same interval, for a number of signals including the instantaneous spanwise vorticity and the instantaneous turbulence production. The $v$ signal was on average 180° out-of-phase with the $u$ signal (as noted by Blackwelder & Kaplan 1976) and the ejection was found to be the principal producer of turbulent energy. Eckelmann et al (1977) suggested that the interesting dynamics occur during the acceleration phase of the pattern. It should also be noted that conditional sampling based on quadrant analysis (Section 3) can be used to provide reference times (when the threshold level is exceeded in the relevant quadrant) for the event. Comte-Bellot et al (1979) have used such an approach to obtain $\langle u \rangle$, $\langle v \rangle$, and $\langle uv \rangle$ signatures associated with ejections and sweeps. The quadrant analysis approach may be more effective for the detection of coherent structures if applied to signals (not necessarily different) simultaneously measured at different points in space.

Brown & Thomas (1977) used an array of hot wires and wall-shear-stress probes to establish the presence of an organized large-scale motion in a turbulent boundary layer, working on the premise that the existence of a large structure implies significant correlation between the motion throughout the structure. The large-scale motion produces a slowly varying component in the wall shear stress with a high-frequency large-amplitude fluctuation, described as the characteristic response of the wall region to the large structure, occurring slightly before the maximum of the slowly varying component. Similar characteristics of the wall-shear-stress signal were observed, in a turbulent duct flow, by Rajagopalan & Antonia (1979a). The averages $\langle u \rangle$, $\langle v \rangle$, and $\langle uv \rangle$ were obtained by Thomas (1977) (also, Thomas & Brown 1977) who first low-pass filtered $u$ before forming a high-frequency component by subtracting the filtered signal from the total signal. The output of the detection scheme consisted of the times when the rectified and smoothed high-frequency component exceeded a predetermined threshold level. While there was no indication that the detection was independent of either this level or the filter cutoff, ensemble averages depended only weakly on these settings.[2] The detection scheme essentially focuses on the back of the structure where Brown & Thomas observed a "sharp"

[2]A similar observation was found by van de Ven (reported by Beljaars 1979) who used an analogous detection method.

increase in $u$. Ensemble averages were enhanced by a procedure consisting of first calculating the average using every member of the ensemble then correlating each member with the previously formed average to determine the optimum time shift for each member (negative correlation at zero time delay indicated a false detection with subsequent rejection of that member). After realignment of the surviving members, the average is recomputed. Each iteration sharpens the definition of the detection criterion and only a few iterations are necessary for convergence. This type of procedure has been used, with minor variations, by various investigators wishing to improve the eduction of a structure. For example, Zilberman et al (1977) used a similar form of enhancement to educe the signature of a turbulent spot artificially introduced in a turbulent boundary layer.

Chen (1975) (also Chen & Blackwelder 1978) made use of temperature as a passive marker of the turbulent region of a boundary layer to identify, with the use of a 10 cold-wire rake, the existence of a sharp internal (but highly three-dimensional) temperature front, characterized by a rapid decrease in temperature (corresponding to the above-mentioned increase in $u$), throughout the entire boundary layer. While in the outer region the front is identifiable with the back of the bulge it is found to be related to the sharp acceleration observed outside the sublayer. Both Chen & Blackwelder (1978) and Thomas (1977) suggested that this front may be associated with an internal shear layer between accelerated and decelerated fluid regions (e.g. Nychas et al 1973).

Ensemble averages of $u$ and $v$ (based on a VITA variance of $\theta$) indicate that the temperature front is associated with an internal shear layer. The inclination of this front to the wall seems to be in the range 30–40° (as inferred from Falco's 1974, 1979 smoke measurements, Chen's 1975 coincidence-correlation measurements, or Subramanian & Antonia's 1979 measurements of the temperature front over a wide range of Reynolds numbers). It should be mentioned that the spatially coherent characteristic ramp-like feature of $\theta$ observed in the laboratory boundary layer has also been observed in the atmospheric surface layer (e.g. Taylor 1958). Phong-anant et al (1980) obtained $\langle u \rangle$, $\langle v \rangle$, $\langle \theta \rangle$, $\langle uv \rangle$, and $\langle v\theta \rangle$ by identifying visually the time points at which $\theta$ is maximum and, after alignment of these points to an arbitrary common origin ($t_1$ say), averaging over a sufficient time interval on either side of $t_1$. Visual identification was not too cumbersome as only about 20 ramps (simultaneously observed at four different heights) were noted over a 30-min record but the possibility of degradation (this applies also to ensemble averages generated by the previous iterative procedure) of the averages with increasing time from $t_1$ should not be overlooked.

Ensemble averages of $uv$ indicated that for nearly neutral conditions, the contribution to $\overline{uv}$ was less than 10%. This differs from the result of Thomas & Brown (1977) who found that the large structure of a laboratory boundary layer makes very large contributions to the Reynolds shear stress.

Use has been made of combined flow visualization–hot wire (or film) experiments to either generate new conditional averages or compare different types of conditional-sampling and averaging techniques. Offen & Kline (1973), with the simultaneous usage of hot film and dye visualization in a boundary layer, compared five kinds of conditional-sampling methods (including the method of Blackwelder & Kaplan 1976 and the quadrant analysis approach) using conditional sampling with their visual studies. While the comparison revealed that all methods correlated positively with each other, the magnitude of the correlation was not such as to indicate true correspondence between any two methods. In particular, none of the methods indicated that the number of events was independent of the threshold setting. While the study of Offen & Kline need not be taken as definitive (for example, Blackwelder & Kaplan mention that Offen & Kline's setting of the threshold was considerably lower than that used in their own study), comparative studies of this kind seem important before $\langle u \rangle$, $\langle v \rangle$, $\langle \theta \rangle$, etc, and contributions from the large structure to $\overline{uv}$ and $\overline{v\theta}$ can be trusted on a quantitative basis. Testing any detection method with a pseudo-turbulent and "structureless" signal(s) produced by a random-noise generator to ensure that a structure is not educed from total disorder is a useful, if not essential, procedure. Of course, such a procedure is not sufficient to ensure the accuracy of ensemble averages educed from the flow signals. Falco (1977) conditionally sampled hot-wire signals to specific flow features of a boundary layer marked by smoke. Large-scale motions, whose upper boundary coincides with a bulge and whose lower boundaries are associated with sharp concentration gradients extending deep into the layer, were found to contribute significantly to $\overline{uv}$ in the logarithmic region. A "typical eddy" motion formed on the back of the large-scale motion (perhaps a consequence of the internal shear layer) was the major contributor to $\overline{uv}$ in the outer half of the layer at $R_\theta \simeq 1200$. Bandyopadhyay (1978) found strong $u$ correlations between two hot wires along a line (at about 40° to the surface for $R_\theta \simeq 2000$) corresponding to the back of the large-scale motion.

Average signatures of the coherent structure in a mixing layer have been determined, using conditional sampling, by various investigators. The reader is referred to the papers by Winant & Browand (1974), Browand & Weidman (1976), Lau & Fisher (1975), Bruun (1977), and Yule (1978). The last three investigations used the spikelike appearance of $u$ near the core and near the low-speed side of the mixing layer of a circular jet to identify the passage of the structure. Browand & Weidman (1976) used combined flow visualization and hot-film measurements to obtain $\langle u \rangle$, $\langle v \rangle$, $\langle uv \rangle$, and isovorticity contours, during and after pairing. Significant contributions to $\overline{uv}$ occurred during pairing. Rajagopalan & Antonia (1979b) obtained $\langle u \rangle$, $\langle v \rangle$, $\langle \theta \rangle$, $\langle uv \rangle$, and $\langle v\theta \rangle$ using a detection criterion based on $\theta$ in the mixing layer of a slightly heated plane jet.

A worthwhile objective of turbulence research is the synthesis of a relatively complex flow situation by progressive relaxation of restrictions on an idealized turbulent flow. The notion of a transitional turbulent spot (Emmons 1951) as an orderly structure which may be considered as the basic module or building block of a turbulent boundary layer is certainly attractive (e.g. Wygnanski 1978). Conditional sampling has been used (e.g. Wygnanski et al 1976, and Cantwell et al 1978) to study the flow associated with the spot. The sampling is strictly periodical since the spot is usually generated by a periodic disturbance but the slight jitter in shape and convection speed of the spot may require the identification of the front and/or back of each spot prior to subsequent realignment to enhance ensemble averages.

## 6 CONCLUDING REMARKS

There is little doubt that the conditional-sampling technique has helped to focus on the more interesting regions of a turbulent flow and has provided quantitative information supplementing (usually but not always) qualitative data obtained from flow-visualization experiments. The data have ranged from, perhaps in order of increasing difficulty and decreasing accuracy, zone and point averages obtained in turbulent -flow regions bounded by a turbulent-nonturbulent interface to point averages obtained relative to the occurrence of structures. The number of events in the ensemble is usually small (often due to their strong three-dimensionality and random occurrence) with a correspondingly small accuracy for the ensemble average. Also, the nonuniform trajectories of the structures and jitter in their convection speed and spatial shape make it difficult to phase lock the detection output to the occurrence of the event. Despite these difficulties, conditional sampling and averaging has helped to close the gap between data collected in an Eulerian frame of reference with information in a Lagrangian frame, as provided by flow-visualization experiments. While quantitative differences exist between results produced by different detection methods supposedly focusing on the same event, the onus is on the experimenter to at least demonstrate that the event recognised is not an artifact of the detection method used. There is a clear need for a critical comparison of quantitative data obtained by different detection methods.

Studies that may benefit from the application of conditional sampling are the influence of Reynolds number on the large structure, the spatial geometry of the Reynolds-number-dependent fine structure and its interaction with the large structure, the connection and interaction between different events at different times in their "life" or similar events (e.g. interaction between large-scale motions), and the magnitude of contributions made by the coherent structure to $\overline{uv}$ and $\overline{v\theta}$.

No mention was made in this review of the theoretical attempts at conditioning the conservation equations of fluid dynamics. Using a model equation for the conditioning functions $I$, properties of conditioned quantities have been predicted and compared with measurements in different flows (e.g. Libby 1975, 1976, Chevray & Tutu 1978a, Dopazo & O'Brien 1978). The prediction of conditional averages associated with the events may also be possible provided a suitable model equation for the conditioning function can be written.

## ACKNOWLEDGMENTS

The author is grateful to the Australian Research Grants Committee for supporting his work and to his colleagues Drs. A. J. Chambers, N. Phan-Thien, and S. Rajagopalan for their helpful comments on an earlier draft. A useful discussion with Dr. J. D. Atkinson is also acknowledged.

*Literature Cited*

Alcaraz, E., Charnay, G., Mathieu, J. 1977. Measurements in a wall jet over a convex surface. *Phys. Fluids* 20:203–210

Ali, S. F., Kovasznay, L. S. G. 1975. Structure of the turbulence in the plane wake behind a heated flat plate. *Tech. Rep. 75-2.* The Johns Hopkins University, Baltimore, Md.

Anderson, P., LaRue, J. C., Libby, P. A. 1979. Preferential entrainment in a two-dimensional turbulent jet in a moving stream. *Phys. Fluids* 22:1857–61

Andreopoulos, J. 1979. Turbulent near wake of a thin aerofoil. *Proc. Symp. Turbulent Shear Flows, 2nd, London*, pp. 2.1–2.5

Antonia, R. A. 1972. Conditionally sampled measurements near the outer edge of a turbulent boundary layer. *J. Fluid Mech.* 56:1–18

Antonia, R. A., Atkinson, J. D. 1973. High-order moments of Reynolds shear stress fluctuations in a turbulent boundary layer. *J. Fluid Mech.* 58:581–93

Antonia, R. A., Atkinson, J. D. 1974. Use of a pseudo-turbulent signal to calibrate an intermittency measuring circuit. *J. Fluid Mech.* 64:679–99

Antonia, R. A., Bradshaw, P. 1971. Conditional sampling of turbulent shear flows. *I.C. Aero Rep. 71-04.* Imperial College, London

Antonia, R. A., Danh, H. Q., Prabhu, A. 1977. Response of a turbulent boundary layer to a step change in surface heat flux. *J. Fluid Mech.* 80:153–77

Antonia, R. A., Prabhu, A., Stephenson, S. E. 1975. Conditionally sampled measurements in a heated turbulent jet. *J. Fluid Mech.* 72:455–80

Antonia, R. A., Sreenivasan, K. R. 1977. Conditional probability densities in a turbulent heated round jet. *Proc. Aust. Hydraulics & Fluid Mech. Conf., 6th, Adelaide*, pp. 411–14

Badri Narayanan, M. A., Rajagopalan, S., Narasimha, R. 1977. Experiments on the fine structure of turbulence. *J. Fluid Mech.* 80:251–57

Bandyopadhyay, P. 1978. Combined smoke-flow visualization and hot-wire anemometry in turbulent boundary layers. In *Lecture Notes in Physics—Structure and Mechanisms of Turbulence I*, ed. H. Fiedler, pp. 206–16. Berlin:Springer

Barsoum, M. L., Kawall, J. G., Keffer, J. F. 1978. Spanwise structure of the plane turbulent wake. *Phys. Fluids* 21:157–61

Beljaars, A. 1979. *A model for turbulent exchange in boundary layers*. PhD thesis. Eindhoven Univ. Tech., Holland

Bilger, R. W., Antonia, R. A., Sreenivasan, K. R. 1976. Determination of intermittency from the probability density function of a passive scalar. *Phys. Fluids* 19:1471–74

Blackwelder, R. F. 1977. On the role of phase information in conditional sampling. *Phys. Fluids* 20:S232–42

Blackwelder, R. F. 1979. Pattern recognition of coherent eddies. *Proc. Dynamic Flow Conf.—Dynamic Measurements in Unsteady Flows, 1978, Marseille, Baltimore*, pp. 173–90

Blackwelder, R. F., Eckelmann, H. 1979. Streamwise vortices associated with the bursting phenomenon. *J. Fluid Mech.* 94:577–94

Blackwelder, R. F., Kaplan, R. E. 1976. On the wall structure of the turbulent boundary layer. *J. Fluid Mech.* 76:89–112

Blackwelder, R. F., Kovasznay, L. S. G. 1972. Time scales and correlations in a turbulent boundary layer. *Phys. Fluids* 15:1545–54

Blackwelder, R. F., Woo, H-H. W. 1974. Pressure perturbation of a turbulent boundary layer. *Phys. Fluids* 17:515–19

Bradshaw, P. 1972. An introduction to conditional sampling of turbulent flow. *I.C. Aero Rep. 72–18*. Imperial College, London

Bradshaw, P. 1977a. Interactions between adjacent turbulent flows. See Antonia & Sreenivasan 1977, pp. 399–402

Bradshaw, P. 1977b. Interacting shear layers in turbomachines and diffusers. In *Turbulence in Internal Flows*, ed. S. N. B. Murthy, pp. 35–66. Hemisphere

Bradshaw, P., Dean, P. B., McEligot, D. M. 1973. Calculation of interacting turbulent shear layers:duct flow. *J. Fluids Engng*. 213–19

Bradshaw, P., Murlis, J. 1974. On the measurement of intermittency in turbulent flow. *I.C. Aero Rep. 74-04*. Imperial College, London

Brodkey, R. S., Wallace, J. M., Eckelmann, H. 1974. Some properties of truncated turbulence signals in bounded shear flows. *J. Fluid Mech.* 63:209–24

Browand, F. K., Weidman, P. D. 1976. Large scales in the developing mixing layers. *J. Fluid Mech.* 76:127–44

Brown, G. L., Roshko, A. 1974. On density effects and large structure in turbulent mixing layers. *J. Fluid Mech.* 64:775–816

Brown, G. L., Thomas, A. S. W. 1977. Large structure in a turbulent boundary layer. *Phys. Fluids* 20:S243–52

Bruun, H. H. 1977. A time-domain analysis of the large-scale flow structure in a circular jet. Part 1. Moderate Reynolds number. *J. Fluid Mech.* 83:641–71

Cantwell, B., Coles, D., Dimotakis, P. 1978. Structure and entrainment in the place of symmetry of a turbulent spot. *J. Fluid Mech.* 87:641–72

Chambers, A. J., Antonia, R. A. 1979. Wave-induced effect on the similarity of Reynolds shear stress and heat flux in the marine surface layer. *Rep. TN FM 43*. Dept. Mech. Engrg., Univ. Newcastle, N.S.W., Australia

Charnay, G., Mathieu, J., Comte-Bellot, G. 1976. Response of a turbulent boundary layer to random fluctuations in the external stream. *Phys. Fluids* 19:1261–72

Charnay, G., Schon, J. P., Alcaraz, E., Mathieu, J. 1979. Thermal characteristics of a turbulent boundary layer with an inversion of wall heat flux. In *Turbulent Shear Flows I*, pp. 104–18. Berlin:Springer

Chen, C-H. P. 1975. *The large scale motion in a turbulent boundary layer: A study using temperature contamination*. PhD thesis. Univ. S. Calif., Los Angeles

Chen, C-H. P., Blackwelder, R. F. 1978. Large-scale motion in a turbulent boundary layer: a study using temperature contamination. *J. Fluid. Mech.* 89:1–31

Chevray, R., Tutu, K. M. 1978a. Intermittency and preferential transport of heat in a round jet. *J. Fluid Mech.* 88:133–60

Chevray, R., Tutu, N. K. 1978b. Conditional measurements in a heated turbulent jet. See Bandyopadhyay 1978, pp. 73–84

Comte-Bellot, G., Sabot, J., Saleh, I. 1979. Detection of intermittent events maintaining Reynolds stress. See Blackwelder 1979, pp. 213–29

Corino, E. R., Brodkey, R. S. 1969. A visual investigation of the wall region of a turbulent flow. *J. Fluid Mech*. 37:1–30

Corrsin, S. 1943. Investigation of flow in an axially symmetrical heated jet of air. *NACA WR W-94*

Corrsin, S. 1957. Some current problems in turbulent shear flows. *Naval Hydrodyn. Publ. 515*, Ch. XV, pp. 373–400

Corrsin, S., Kistler, A. L. 1955. Free stream boundaries of turbulent flows. *NACA Rep. 1244*

Davies, A. E., Keffer, J. F., Baines, W. D. 1975. Spread of a heated plane turbulent jet. *Phys. Fluids* 18:770–75

Davies, P. O. A. L., Yule, A. J. 1975. Coherent structures in turbulence. *J. Fluid Mech*. 69:513–57

Dean, R. B., Bradshaw, P. 1976. Measurements of interacting turbulent shear layers in a duct. *J. Fluid Mech*. 78:641–76

Dopazo, C., O'Brien, E. E. 1978. Intermittency in free turbulent shear flows. See Charnay et al 1979, pp. 6–23

Dumas, R., Fulachier, L., Arzoumanian, E. 1972. Facteurs d'intermittence et de dyssimétrie des fluctuations de température et de vitesse dans une couche limite turbulente. *C. R. Acad. Sci. Paris, Ser. A* 274:267–70

Eckelmann, H., Nychas, S. G., Brodkey, R. S., Wallace, J. M. 1977. Vorticity and turbulence production in pattern recognized turbulent flow structures. *Phys. Fluids* 20:S225–31

Emmons, H. W. 1951. The laminar-turbulent transition in a boundary layer. Part I. *J. Aero. Sci*. 18:490–98

Fabris, G. 1974. *Conditionally sampled turbulent thermal and velocity fields in the wake of a warm cylinder and its interaction with an equal cool wake*. PhD thesis. Ill. Inst. Tech., Chicago

Fabris, G. 1979. Conditional sampling study of the turbulent wake of a cylinder. Part I. *J. Fluid Mech*. 94:673–709

Falco, R. E. 1974. Some comments on turbulent boundary layer structure inferred from the movements of a passive contaminant. *AIAA Pap. No. 74-99*

Falco, R. E. 1977. Coherent motions in the outer region of turbulent boundary layers. *Phys. Fluids* 20:S124–32

Falco, R. E. 1979. Comments on "large structure in a turbulent boundary layer". *Phys. Fluids* 22:2042

Fiedler, H. E. 1974. Transport of heat across a plane turbulent mixing layer. *Adv. Geophys*. 18A:93–109

Fiedler, H., Head, M. R. 1966. Intermittency measurements in the turbulent boundary layer. *J. Fluid Mech*. 25:719–35

Grass, A. J. 1971. Structural features of turbulent flow over smooth and rough boundaries. *J. Fluid Mech*. 50:233–55

Gupta, A. K., Kaplan, R. E. 1972. Statistical characteristics of Reynolds stress in a turbulent boundary layer. *Phys. Fluids* 15:981–85

Gupta, A. K., Laufer, J., Kaplan, R. E. 1971. Spatial structure in the viscous sublayer. *J. Fluid Mech*. 50:493–517

Gutmark, E., Wygnanski, I. 1976. The planar turbulent jet. *J. Fluid Mech*. 73:465–95

Hancock, P. E. 1978. *Effect of free-stream turbulence on turbulent boundary layers*. PhD thesis. Imperial College, London

Haverbeke, A., Wood, D. W., Smits, A. J. 1979. Uncertainties and errors in conditional sampling. See Andreopoulos 1979, pp. 11.1–11.6

Hedley, T. B., Keffer, J. F. 1974a. Turbulent/non-turbulent decisions in an intermittent flow. *J. Fluid Mech*. 64:625–44

Hedley, T. B., Keffer, J. F. 1974b. Some turbulent/non-turbulent properties of the outer intermittent region of a boundary layer. *J. Fluid Mech*. 64:645–78

Hopfinger, E. J. 1978. Buoyancy effects on the large scale structure of free turbulent shear flows. See Bandyopadhyay 1978, pp. 65–85

Imaki, K. 1968. Structure of the superlayer in the turbulent boundary layer I & II. *Inst. Space & Aeronaut. Sci*., Univ. Tokyo 4:348–67; 4:536–63

Jenkins, P. E., Goldschmidt, V. W. 1974. A study of the intermittent region of a heated two-dimensional plane jet. *Rep. HL 74-45*. Purdue Univ., Lafayette, Ind.

Johnson, D. S. 1959. Velocity and temperature fluctuation measurements in a turbulent boundary layer downstream of a stepwise discontinuity in wall temperature. *J. Appl. Mech*. 26:325–36

Kaplan, R. E. 1973. Conditional sampling techniques. In *Turbulence in Liquids*, ed.

G. K. Patterson, J. L. Zakin, pp. 274–83. Univ. Missouri—Rolla

Kaplan, R. E., Laufer, J. 1968. The intermittently turbulent region of the boundary layer. *Rep. USC AE 110*. Univ. S. Calif., Los Angeles

Kawall, J. G., Keffer, J. F. 1979. Interface statistics of a uniformly distorted heated turbulent wake. *Phys. Fluids* 22:31–9

Keffer, J. F., Olsen, G. J., Kawall, J. G. 1977. Intermittency in a thermal mixing layer. *J. Fluid Mech*. 79:595–607

Kibens, V., Kovasznay, L. S. G., Oswald, L. J. 1974. Turbulent–non-turbulent interface detector. *Rev. Sci. Instrum*. 45:1138–44

Kim, H. T., Kline, S. J., Reynolds, W. C. 1971. The production of turbulence near a smooth wall in a turbulent boundary layer. *J. Fluid Mech*. 50:133–60

Kline, S. J., Reynolds, W. C., Schraub, F. A., Runstadler, P. W. 1967. The structure of turbulent boundary layers. *J. Fluid Mech*. 30:741–73

Kovasznay, L. S. G. 1979. Measurement in intermittent and periodic flow. See Blackwelder 1979, pp. 133–59

Kovasznay, L. S. G., Ali, S. F. 1974. Structure of the turbulence in the wake of a heated flat plate. *Proc. Int. Heat Transfer Conf., 5th, Tokyo*, pp. 99–103

Kovasznay, L. S. G., Kibens, V., Blackwelder, R. F. 1970. Large-scale motion in the intermittent region of a turbulent boundary layer. *J. Fluid Mech*. 41:283–325

Kuo, A. Y-S., Corrsin, S. 1971. Experiments on internal intermittency and fine-structure distribution functions in fully turbulent fluid. *J. Fluid Mech*. 50:285–320

Kuo, A. Y-S., Corrsin, S. 1972. Experiment on the geometry of the fine-structure regions in fully turbulent fluid. *J. Fluid. Mech*. 56:447–79

LaRue, J. C. 1974. Detection of the turbulent–nonturbulent interface in slightly heated turbulent shear flows. *Phys. Fluids* 17:1513–17

LaRue, J. C., Libby, P. A. 1974a. Temperature and intermittency in the turbulent wake of a heated cylinder. *Phys. Fluids* 17:873–78

LaRue, J. C., Libby, P. A. 1974b. Temperature fluctuations in the plane turbulent wake. *Phys. Fluids* 17:1956–67

LaRue, J. C., Libby, P. A. 1976. Statistical properties of the interface in the turbulent wake of a heated cylinder. *Phys. Fluids* 19:1864–75

LaRue, J. C., Libby, P. A. 1978. Detailed similarity in the turbulent wake of a heated cylinder. *Phys. Fluids* 21:891–97

Lau, J. C., Fisher, M. J. 1975. The vortex-street structure of turbulent jets. Part I. *J. Fluid Mech*. 67:299–337

Laufer, J. 1975. New trends in experimental turbulence research. *Ann. Rev. Fluid Mech*. 7:307–26

Libby, P. A. 1975. On the prediction of intermittent turbulent flows. *J. Fluid Mech*. 68:273–95

Libby, P. A. 1976. Prediction of the intermittent turbulent wake of a heated cylinder. *Phys. Fluids* 19:494–501

Lu, S. S., Willmarth, W. W. 1973. Measurements of the structure of the Reynolds stress in a turbulent boundary layer. *J. Fluid Mech*. 60:481–512

Moum, J. N., Kawall, J. G., Keffer, J. F. 1979. Structural features of the plane turbulent jet. *Phys. Fluids* 22:1240–44

Mulej, D. J., Goldschmidt, V. W. 1975. Velocity and foldover of the turbulent non-turbulent interface in a plane jet. *Rep. HL 75-32*. Purdue Univ., Lafayette, Ind.

Murlis, J. 1975. *The structure of a turbulent boundary layer at low Reynolds number*. PhD thesis. Imperial College, London

Nakagawa, H., Nezu, I. 1977. Prediction of the contributions to the Reynolds stress from bursting events in open-channel flows. *J. Fluid Mech*. 80:99–128

Nychas, S. G., Hershey, H. C., Brodkey, R. S. 1973. A visual study of turbulent shear flow. *J. Fluid Mech*. 61:513–40

Offen, G. R., Kline, S. J. 1973. A comparison and analysis of detection methods for the measurement of production in a boundary layer. See Kaplan 1973, pp. 289–318

Offen, G. R., Kline, S. J. 1974. Combined dye-streak and hydrogen bubble visual observations of a turbulent boundary layer. *J. Fluid Mech*. 62:223–39

Oswald, L. J., Kibens, V. 1971. Turbulent flow in the wake of a disk. *Tech. Rep. 002820*. Univ. Mich., Ann Arbor, Mich.

Paizis, S. T., Schwarz, W. H. 1974. An investigation of the topography and motion of the turbulent interface. *J. Fluid Mech*. 63:315–43

Paizis, S. T., Schwarz, W. H. 1975. Entrainment rates in turbulent shear flows. *J. Fluid Mech*. 68:297–308

Perry, A. E., Hoffman, P. H. 1976. An experimental study of turbulent convective heat transfer from a flat plate. *J. Fluid Mech*. 77:355–68

Phong-anant, D., Antonia, R. A., Chambers, A. J., Rajagopalan, S. 1980. Features of the organized motion in the atmospheric surface layer. *J. Geophys. Res.* 85:424–32

Rajagopalan, S., Antonia, R. A. 1979a. Interaction between large and small scale motions in a two-dimensional turbulent duct flow. *Rep. TN FM 35.* Dept. Mech. Engrg., Univ. Newcastle, N.S.W., Australia

Rajagopalan, S., Antonia, R. A. 1979b. Properties of the large structure in a slightly heated turbulent mixing layer. *Rep. TN FM 44.* Dept. Mech. Engrg., Univ. Newcastle, N.S.W., Australia

Rao, K. N., Narasimha, R., Badri Narayanan, M. A. 1971. The bursting phenomenon in a turbulent boundary layer. *J. Fluid Mech.* 48:339–52

Roshko, A. 1976. Structure of turbulent shear flows: a new look. *AIAA Pap. 76-78*

Sabot, J., Comte-Bellot, G. 1976. Intermittency of coherent structures in the core region of fully developed turbulent pipe flow. *J. Fluid Mech.* 74:767–96

Schon, J. P., Charnay, G. 1975. Conditional sampling in measurement of unsteady fluid dynamic phenomena. *von Kármán Inst. Fluid Dynamics Lect. Ser. 73*

Sreenivasan, K. R., Antonia, R. A. 1979. Joint probability densities and quadrant contributions in a heated turbulent round jet. *AIAA J.* 16:867–68

Sreenivasan, K. R., Antonia, R. A., Britz, D. 1979. Local isotropy and large structures in a heated turbulent jet. *J. Fluid Mech.* 94:745–75

Sreenivasan, K. R., Antonia, R. A., Stephenson, S. E. 1978. Conditional measurements in a heated axisymmetric turbulent mixing layer. *AIAA J.* 16:869–70

Subramanian, C. S., Antonia, R. A. 1978. Reynolds number dependence of a slightly heated turbulent boundary layer. *Rep. TN FM 21.* Dept. Mech. Engrg., Univ. Newcastle, N.S.W., Australia

Subramanian, C. S., Antonia, R. A. 1979. Some properties of the large structure in a slightly heated turbulent boundary layer. See Andreopoulos 1979, pp. 4.18–4.21

Takeuchi, K., Leavitt, E., Chao, S. P. 1977. Effects of water waves on the structure of turbulent shear flow. *J. Fluid Mech.* 80:535–59

Taylor, R. J. 1958. Thermal structures in the lowest layers of the atmosphere. *Aust. J. Phys.* 11:168–76

Thomas, A. S. W. 1977. *Organized structures in the turbulent boundary layer.* PhD. thesis. Univ. Adelaide, S. A., Australia

Thomas, A. S. W., Brown, G. L. 1977. Large structure in a turbulent boundary layer. See Antonia & Sreenivasan 1977, pp. 407–10

Thomas, R. M. 1973. Conditional sampling and other measurements in a plane turbulent wake. *J. Fluid Mech.* 57:549–82

Townsend, A. A. 1949. The fully developed turbulent wake of a circular cylinder. *Aust. J. Sci. Res.* 2:451–68

Townsend, A. A. 1979. Flow patterns of large eddies in a wake and in a boundary layer. *J. Fluid Mech.* 95:515–37

Ueda, H., Mizushima, T. 1977. Turbulence structure in the inner part of the wall region in a fully developed turbulent tube flow. See Kaplan 1973, pp. 357–66

Van Atta, C. W. 1974. Sampling techniques in turbulence measurements. *Ann. Rev. Fluid Mech.* 6:75–91

Van Atta, C. W. 1980. Conditional sampling techniques. *Handbook of Turbulence Vol. 2.* London/New York:Plenum. In press

Wallace, J. M., Brodkey, R. S. 1977. Reynolds stress and joint probability density distributions in the u–v plane of a turbulent channel flow. *Phys. Fluids.* 20:351–55

Wallace, J. M., Brodkey, R. S., Eckelmann, H. 1977. Pattern recognized structures in bounded turbulent shear flows. *J. Fluid Mech.* 83:673–93

Wallace, J. M., Eckelmann, H., Brodkey, R. S. 1972. The wall region in turbulent shear flow. *J. Fluid Mech.* 54:39–48

Weir, A. D., Bradshaw, P. 1977. The interaction of two parallel free shear layers. See Andreopoulos 1979, pp. 1.9–1.15

Willmarth, W. W. 1975. Structure of turbulence in boundary layers. In *Adv. Applied Mechanics, Vol. 15,* pp. 159–254. New York:Academic

Willmarth, W. W., Lu, S. S. 1972. Structure of the Reynolds stress near the wall. *J. Fluid Mech.* 55:65–92

Winant, C. D., Browand, F. K. 1974. Vortex pairing: the mechanism of turbulent mixing-layer growth at moderate Reynolds number. *J. Fluid Mech.* 63:237–55

Wygnanski, I. 1978. On the possible relationship between the transition process and the large coherent structures in turbulent boundary layers. In *Coherent Structures of Turbulent Boundary Layers,* ed. C. R. Smith, D. E. Abbott, pp. 168–93

*Ann. Rev. Fluid Mech. 1981. 13:157–87*


# POWER FROM WATER WAVES

*8173

*D. V. Evans*
School of Mathematics, University of Bristol, Bristol, England

## 1 INTRODUCTION

"I have a great many reasons why I think wave power can be put to practical and profitable use, but I don't wish to occupy too much of your valuable time, or, perhaps, there are not many of you particularly interested in the subject."

The last part of this statement was probably true when Mr. Stodder made it nearly ninety years ago in the discussion to a long paper by Stahl (1892) to the American Society of Mechanical Engineers on "The Utilization of the Power of Ocean Waves." Few people nowadays would confess an indifference to the possibility of replacing our dwindling fossil fuels by an inexhaustible clean energy source. On the contrary, extensive research programs are underway in Japan, Norway, and, in particular, the U.K. where a number of different wave-energy devices are receiving Government funding together with supporting work on wave data, mooring, transmission, and environmental problems. A full first report on the U.K. program up to 1978 can be found in Quarrell (1978) although many of the cost estimates have since been revised downwards considerably.

The renaissance for wave power as a potential large-scale energy source came in 1974 with the publication in *Nature* of an article by Salter (1974) in which he described experiments carried out on a rocking cam-shaped cylindrical section in a narrow wave tank in the University of Edinburgh. The cylinder, which was mounted with its axis spanning the wave tank, had a circular rear section so that no waves were generated behind it due to the rocking motion induced by an incident sinusoidal wave. The front face was shaped to accommodate the circular motion of neighboring water particles in the wave field. The motion of the duck, as it has become known, was resisted by applying a velocity-proportional opposing torque and the product of torque and



0066-4189/81/0115-0157$01.00

angular velocity, averaged over a cycle, was compared to the incident-wave power. Efficiencies in excess of 80% were reported. Coinciding with the dramatic increase in oil prices announced by the OPEC countries, Salter's paper undoubtedly prompted Government action and within two years the U.K. wave-energy program was in full swing.

Prior to 1974 the use of wave energy on a small scale had already been exploited commercially in the form of a self-powered light-buoy invented by Y. Masuda in Japan. The idea behind the device is to resonate a column of water within the buoy to the predominant wave frequency. The water drives air through a contraction containing an air turbine linked to an electrical generator. A theoretical treatment of the Masuda buoy together with references to previous wave-power systems has been given by McCormick (1974). The oscillating-water-column concept has since been taken up on a larger scale by a number of device teams including the Vickers group, the National Engineering Laboratory group in Glasgow, and the Queen's University Belfast group. In contrast to these devices, usually consisting of large single units with a vertical axis of symmetry, there are so called "terminator" devices, which include the Salter duck whereby a string of ducks oscillate about a long central spine. Also included in this category is the "clam" device invented at Lanchester Polytechnic, Coventry, U.K., the floating articulated raft invented by Sir Christopher Cockerell, and the Bristol submerged-cylinder idea invented by the author. These latter two, while not involving a continuous line of devices, but rather an array of spaced devices, share the common aim of terminator devices which is to absorb as much energy as possible in a single "bite." Contrasting again with these are "attenuator" devices, being elongated articulated or flexible structures whose long axis is perpendicular to the incident-wave crests. Perhaps the best known of these is the flexible-bag idea by Professor M. J. French of the University of Lancaster, U.K., in which the energy is progressively absorbed along the bag by the deformation of flexible rubber sections which in turn drive air through a turbine housed in the center of the structure.

It is not intended in this review to provide an encyclopaedic coverage of all proposed wave-energy devices and further details of most of those mentioned and others are described in Quarrell (1978).

The review that follows will focus primarily on the hydrodynamic properties of various idealized devices for extracting power from waves. The idealization arises out of necessity because of the difficulty of solving the full equations of motion of a (possibly) articulated structure in irregular waves. Thus all theories seeking to explain or predict the behavior of energy-absorbing structures in waves have assumed that wave heights and steepnesses are sufficiently small for classical linearized water-wave theory to be applied. In addition, body motions are assumed small enough for linear body dynamics to be used, so that, for instance, body boundary conditions may be applied on the equilibrium position of the body, rather than on the instantaneous position. An outline of the resulting linearized equations is given in Section 2. In Sections 3 and 4 a review of theoretical results in two and three dimensions is given, while in Section 5 a number of additional topics are discussed including a theoretical treatment of a nonlinear power take-off system and also some experimental fluid dynamics arising from wave-power studies.

## 2 GOVERNING EQUATIONS

Most wave-power machines involve the motion of a large structure similar in size to a small ship and it is not surprising, therefore, that the fluid dynamics involved is similar to that required in ship hydrodynamic theory. The basic differences are that the wave-power machine has zero forward speed, is often an articulated structure, and, of course, has the ability to absorb energy from the neighboring wave field. Nevertheless, the theory for the motion of floating bodies as described by Wehausen (1971) or Newman (1977, Chapter 6) provides an invaluable starting point for the study of wave-energy absorbers. Only a brief résumé of the governing equation will be given here.

The fluid is assumed to be incompressible, irrotational, and inviscid so that a harmonic velocity potential $\Phi(x, y, z, t)$ exists where Cartesian coordinates $x, y, z$ are chosen with $z=0$ the plane of the undisturbed free surface. For motions of period $2\pi/\omega$ the velocity potential can be written

$$\Phi(x, y, z, t)=\mathrm{Re}\{\phi(x, y, z)\exp(i\omega t)\}. \tag{2.1}$$

The linearized free-surface condition is

$$k\phi+\partial\phi/\partial z=0 \text{ on } z=0, k=\omega^2/g. \tag{2.2}$$

In deep water it is assumed that $\phi$, $\nabla\phi\to 0$ as $z\to\infty$. The modification to finite depth is straightforward, usually involving a multiplicative hyperbolic factor. A harmonic potential describing a wave of amplitude $A$ traveling in a direction making an angle $\beta$ with the positive $x$ axis can be written

$$\phi_0(x, y, z)=gA\omega^{-1}\exp\{ikx\cos\beta+iky\sin\beta-kz\}. \tag{2.3}$$

Simple harmonic motions of a wave-energy structure may occur in any or all of its six degrees of freedom plus additional degrees of freedom if the structure is articulated. In general, the resulting potential is expressed as

$$\phi=\phi_s+\Sigma U_n\phi_n$$

where $\phi_s$ describes the effect of $\phi_0$ on the *fixed* structure while $\phi_n$ is the radiation potential appropriate to the *forced* motion with unit velocity in one of the possible degrees of freedom. Each of the harmonic potentials $\phi_n$ and $\phi_s$ satisfies a boundary-value problem involving conditions on the body and at large distances, as well as Equation (2.2), which must first be solved before the total hydrodynamic force on the structure can be determined from $\phi$ in terms of the unknown complex velocity amplitudes $U_n$. The dynamic equations of motion of the structure with known forcing terms, hydrodynamic and dynamic, can then be solved for $U_n$ and the mean absorbed power determined.

## *Power of Water Waves*

For the regular plane wave given by Equation (2.3) the mean power per unit crest length is derived by computing the mean energy flux crossing a vertical plane parallel to the wave crests. This turns out to be

$$P_w=(\rho g A^2/2)\times(g/2\omega)=\rho g^2A^2/4\omega,$$

the product of the energy density $E$ and the group velocity $c_g$ for deep water. For long devices aligned in a direction parallel to the incoming wave crests, and for devices spanning the entire width of a wave tank, it is straightforward to define an efficiency of power absorption $\eta$ as the ratio of the total mean power absorbed per unit length of the device to the incident mean power per unit crest length, $P_w$. For isolated three-dimensional devices the appropriate quantity is a capture width or absorption length $l$, defined as the ratio of the total mean power absorbed, $P$, to $P_w$. For a line of three-dimensional absorbers lying along the $y$ axis and occupying a length $2L$, say, it is possible to define the efficiency $\eta$ as

$$\eta=P/P_w 2L\cos\beta=l/2L\cos\beta$$

where the denominator is the component of the mean energy flux in the incident wave that crosses the line occupied by the devices, whilst the numerator is the total mean power absorbed by all the devices. This definition is not entirely satisfactory as it leads to curious conclusions for waves approaching along the $y$ axis where $\cos\beta=0$ and where $P$ may still be greater than zero. For groups of devices, the capture width is more suitable.

## *Irregular Waves*

Observation of the sea surface makes it clear that a mixture of waves of different amplitudes, periods, wavelengths, and directions coexist at a given time. A mathematical model of the sea surface assumes it to be an infinite superposition of wave trains of different amplitudes, frequencies, and directions with randomly distributed phases. Records of the water-surface level at a given point can be converted, using Fast Fourier Transform techniques, into a one-dimensional energy spectrum $S(\omega)$ where $\sqrt{2S(\omega)\delta\omega}$ is the wave amplitude corresponding to frequency $\omega$, so that the total energy density is $E=\rho g\int_0^\infty S(\omega)d\omega$ and the total mean power per unit crest length is

$$P_{\mathrm{w}}=\tfrac{1}{2}\rho g^2\int_0^\infty \omega^{-1}S(\omega)\,d\omega. \tag{2.4}$$

A more realistic measure of the wave-field is given by the two-dimensional spectrum $S(\omega,\theta)$ which is often modelled by applying a "spreading function" to the one-dimensional spectrum $S(\omega)$. It is then possible to extend (2.4) to be the mean power incident at a point, by which is meant the mean power per unit crest length summed over all frequencies and directions:

$$P_{\mathrm{w}}=\tfrac{1}{2}\rho g^2\int_0^\infty\int_{-\pi}^{\pi} \omega^{-1}S(\omega,\theta)\,d\theta\,d\omega.$$

The total mean power absorbed by an isolated device is then

$$P=\int_0^\infty\int_{-\pi}^{\pi} \omega^{-1}S(\omega,\theta)l(\omega,\theta)\,d\theta\,d\omega$$

where $l(\omega,\theta)$ is the capture width for the device. For a line of devices occupying a length $2L$ along the $y$ axis, say, it is possible to define the mean power crossing that line so that a sea efficiency would be

$$\eta=\int_0^\infty\int_{\pi}^{\pi} \omega^{-1}S(\omega,\theta)l(\omega,\theta)\,d\theta\,d\omega\Big/\int_0^\infty\int_{-\pi}^{\pi} \omega^{-1}S(\omega,\theta)2L\cos\theta\,d\theta\,d\omega.$$

Implicit in the above derivations is the assumption of linear superposition which allows one to infer performance in irregular seas from knowledge of efficiency and capture width in a monochromatic regular wave from a given direction.

The most commonly used empirical spectrum function is that of Pierson & Moskowitz (1964) appropriate to fully developed ocean waves, modified by a cosine-law spreading function, although good directional wave data are scarce and confirmation of the appropriateness of the model spectrum used is usually based on wave-climate synthesis (Hogben 1978). Regardless of the accuracy of the mathematical representations, there is no doubt that an appreciable energy source is available. Recent estimates suggest a mean annual incident power density of about 50 Kw per meter a few kilometers off the west coast of Scotland.

# 3 TWO-DIMENSIONAL THEORY

Although wave-power machines will operate in a three-dimensional world there has been considerable interest in developing a theory for the oscillation of two-dimensional energy-absorbing cylindrical sections, partly due to the simplicity of these restricted motions, and partly to provide a theoretical basis to the original narrow-wave-tank tests carried out by Salter (1974). Such an approach is commonplace in ship hydrodynamic theory and forms the basis of the "strip" method for estimating hydrodynamic characteristics of fully three-dimensional ship hulls.

Thus Mei (1976), Evans (1976), and Newman (1976) independently arrived at expressions for the efficiency of power absorption of an arbitrarily shaped cylindrical section oscillating in a single degree of freedom in a narrow wave tank and absorbing energy from a regular plane incident wave, solely in terms of the wave-making properties of that particular section in the absence of the incident wave.

It is remarkable that the *maximum* efficiency of power absorption for such a section can be predicted without reference to the precise coupling between body and fluid motion provided only that the section responds harmonically with the same frequency as the incident wave.

This was proved by Evans (1976) and Newman (1976) using the far-field behavior of the waves only. Thus Evans (1976) defines the efficiency $\eta$ in terms of mean energy flux so that $1-\eta$ is the ratio of mean energy flux radiated away from the oscillating section to the incident energy flux.

Then

$$\eta=\gamma-\gamma^{-1}|A_+\alpha^*-\gamma|^2 \tag{3.1}$$

where

$$\gamma^{-1}=1+|A_-/A_+|^2$$
$$\alpha=\omega U_0/gA,$$

and * denotes complex conjugate. Here $A_+$ and $A_-$ are the complex potential amplitudes of the waves radiated upstream—in the opposite direction to that of the incident waves—and downstream respectively, due to the forced oscillatory motion of the section with unit amplitude velocity in its energy-absorbing mode in the absence of the incident wave; $U_0$ is the complex amplitude of the as yet unknown velocity of the section in its induced motion, which will subsequently be determined through the equations coupling the fluid and body motions. The expression (3.1) is taken from Evans (1976, Equation 3.7) with a slight change of notation. Newman (1976, Equation 62) uses Kochin functions to achieve what can be shown to be the same result.

Crucial to the derivation of (3.1) is the Newman relation (Newman 1975)

$$RA_+^* + TA_-^* + A_+ = 0 \tag{3.2}$$

which relates the $A_+, A_-$ from the forced motion to $R, T$ the usual complex reflection and transmission coefficients for the *fixed* section in the incident waves.

It follows from (3.1) that if the velocity of the section and hence $\alpha$ can be chosen so that $A_+\alpha^* = \gamma$ then

$$\eta_{\max} = \gamma \tag{3.3}$$

is the maximum efficiency possible for that particular section oscillating in that degree of freedom. But $\gamma$ depends only upon the wave-making properties of the section in the absence of the incident waves. This result explains the high efficiencies achieved by the Salter "duck" whose circular rear section ensures that $|A_-/A_+| \ll 1$ and $\gamma \simeq 1$. Equation (3.3) also shows that for a section oscillating about an axis of symmetry, so that $|A_+| = |A_-|$, $\eta_{\max} = \frac{1}{2}$. To achieve $\eta_{\max} = 1$ it is necessary to have $A_- = 0$, $A_+ \neq 0$ from (3.1) and it follows from (3.2) that $|R| = 1$ and hence $T = 0$. Thus, a necessary condition for a section oscillating in one degree of freedom to be 100% efficient is that an incident wave be totally reflected by the section at that frequency. This can be understood physically as follows. The complete motion of the absorbing section at the particular frequency can be written as a combination of the totally reflected incident wave from the (fixed) section plus a wave of amplitude $CA_+$ in the same direction due to the section's induced motion, where $C$ is a complex constant depending on the coupling between section and fluid. It is clearly possible to eliminate waves traveling away from the section by ensuring that $CA_+ + R = 0$ whence the section is 100% efficient.

This has been confirmed theoretically by Cocklin (1977; also Martin 1977, private communication) who has shown that a half-immersed elliptical section that can roll about a point on its major axis other than its center can be 100% efficient at certain wave frequencies.

A similar result has been noted by Mynett et al (1979) who have computed $A_+, A_-, R, T$ for various "duck" sections. They found that for a "duck" whose rear circular section touched rather than intersected the mean free surface and whose front section intersected the free surface at an angle of 45°, $|R|=1$, $T=0$, and $A_-=0$ at a certain frequency. The peculiar shape of this "duck" section may explain why $T=0$ at some frequency but it is less easy to accept the implication that $T$ can vanish for the symmetric elliptical section studied by Cocklin, as it is generally accepted that for partly submerged sections of ship-hull shape, $R$ and $T$ vary monotonically with frequency (Mynett et al 1979, p. 16).[1]

Some further useful relations to emerge from this far-field approach to the efficiency of two-dimensional sections are

$$|R_1|=1-\gamma, \ |T_1|=\gamma^{1/2}(1-\gamma)^{1/2} \tag{3.4}$$

at maximum efficiency. Here $R_1, T_1$ are the complex reflection and transmission coefficients when the section is absorbing energy from the incident wave.

The choice of an efficient section shape can now be made by forced-motion tests. Such a shape will have $|A_-/A_+|\ll 1$ over the frequencies of interest. In addition, when the section is absorbing energy through an appropriate mechanism, measurement of $|R_1|$ and $|T_1|$ and comparison with (3.4) will show how close the motion is to optimum efficiency.

The result (3.3) for the maximum efficiency was derived purely from the far-field behavior of the waves. It is also possible to derive the result from a consideration of the wave-induced pressure forces on the section, still without knowing the details of the section-fluid coupling. This approach also permits an easier generalisation to three dimensions and was first described by Ambli et al (1977) for three-dimensional problems and has since been used by Evans (1980a, b) in both two and three dimensions. Mei (1976) and Evans (1976) also achieve the same result, but only after first prescribing the dynamics of the cylinder constraints, although their prescription can be shown to be equivalent to the sole assumption needed, which is simple-harmonic response of the section.

The result for the total mean power absorbed by the section is

$$P=\frac{1}{8}|X_s|^2/B-\frac{1}{2}|U_0-\tfrac{1}{2}X_s/B|^2B. \tag{3.5}$$

Here $X_s$ is the (complex) amplitude of the exciting force on the (fixed) section due to the incident wave of amplitude $A$, in the direction of motion of the section when it is free to absorb energy. The term $B$ is the

[1] Closer scrutiny of Cocklin's work reveals that although very high efficiencies can be achieved by a rolling elliptical section the absorption is never 100% at some frequency so that it does not necessarily follow that $T=0$ at that frequency (J. Martin 1980, private communication).

damping or radiation coefficient and is a measure of the mean rate at which work must be done on the section to maintain simple harmonic oscillations in the energy-absorbing mode with complex velocity amplitude $U_0$. In fact, this quantity is just $\frac{1}{2}B|U_0|^2$. Both $X_s$ and $B$ are functions of frequency and have been computed for various ship-hull sections as they play an important role in ship hydrodynamics theory.

The connection between (3.1) and (3.5) becomes clear when use is made of the Haskind relation in two dimensions (Newman 1976),

$$X_s = \rho g A A_+ L, \tag{3.6}$$

and the result, derived using Green's theorem,

$$B = \tfrac{1}{2}\rho\omega(|A_+|^2 + |A_-|^2)L \tag{3.7}$$

where $L$ is the width of the wave tank and the section.

Equation (3.6) relates the exciting force to the wave-making ability of the device in the direction opposite to that of the incident wave, while (3.7) relates the damping coefficient to the energy radiated away from the section by its forced motion. These relations show that the condition $A_+\alpha^* = \gamma$ is entirely equivalent to the relation

$$U_0 = \tfrac{1}{2}X_s/B \tag{3.8}$$

which gives a maximum mean power

$$P_{\text{max}} = \frac{1}{8}|X_s|^2/B. \tag{3.9}$$

Furthermore, from (3.6) and (3.7)

$$|X_s|^2/B = 8P_w L\gamma$$

where $P_w = \frac{1}{4}\rho g^2 A^2/\omega$ is the mean power in a plane sinusoidal traveling wave per unit crest length.

It follows that

$$\eta_{\text{max}} = P_{\text{max}}/P_w L = \gamma$$

as before.

One immediate conclusion from (3.8) is that the body velocity must be in phase with the exciting force (not the total hydrodynamic force) for maximum efficiency. This result has been exploited directly by Falnes & Budal (1978) to improve the performance of an oscillating-buoy wave-power device. Their idea is to force the phase of the velocity of the buoy to follow the phase of the exciting force by arresting its motion during the cycle and letting go at the appropriate time. In this way greatly amplified buoy motions and correspondingly large power levels have been achieved. One difficulty with the method is how to separate the exciting force from the total force, a problem that would presumably be compounded in irregular waves. Also it is not clear how the resulting square-wave rather than sinusoidal velocity-time curves can be used to meet a criterion derived under the assumptions of simple harmonic motion. Nevertheless, this so-called "phase-locking" technique is attracting considerable interest and is being explored further as part of the U.K. wave-energy program.

It is interesting that the same idea is behind a wave-energy float invented by Mr. Dedger Jones in the United States in the early seventies (W. S. Pope, private communication).

## *Equations of Motion*

Up until now no mention of the actual body dynamics has been made, it being assumed that the optimal velocities for maximum power were achievable. With few exceptions authors have assumed linear body dynamics for the power take-off mechanism. For motion in a single degree of freedom with coordinate $\zeta(t)$, it is assumed that the section, of mass $m$ per unit width, moves relative to some fixed reference axis and that its movement is opposed by forces proportional to displacement, with spring constant $k$, due to buoyancy-restoring forces, mechanical constraints, or both, and proportional to velocity, with damping constant $d$. Then

$$m\ddot{\zeta}(t)+d\dot{\zeta}(t)+k\zeta(t)=F(t) \tag{3.10}$$

where $F(t)$ is the hydrodynamical pressure force in the direction of motion, per unit width of section.

Rewriting $F(t)$ as

$$\begin{aligned} F(t)&=F_s(t)+F_R(t)\\ &=\mathrm{Re}\left(X_s e^{i\omega t}\right)-M\ddot{\zeta}(t)-B\dot{\zeta}(t) \end{aligned}$$

and making use of (3.6) and (3.7) gives

$$\eta(\omega)=\frac{4\omega^2 dB\gamma}{\left\{k-(m+M)\omega^2\right\}^2+\omega^2(d+B)^2}. \tag{3.11}$$

Here $F_s(t)$ is the exciting force on the fixed section and $F_R(t)$ the force induced by the motion of the section, expressed in terms of frequency-dependent added mass $(M)$ and damping $(B)$ coefficients, the former being the apparent increase in inertia of the section due to the presence of the fluid.

It follows from (3.11) that the section is "tuned" to an incident wave of frequency $\omega_0/2\pi$ when

$$k=(m+M(\omega_0))\omega_0^2 \quad \text{and} \quad d=B(\omega_0). \tag{3.12}$$

The first condition implies that the section is at resonance while the second condition implies that the energy extraction rate $\frac{1}{2}d|U_0|^2$ equals the rate of radiation damping $\frac{1}{2}B|U_0|^2$.

With (3.12) satisfied at $\omega=\omega_0$,

$$\eta_{max}=\gamma$$

in agreement with (3.3).

Thus estimates of $\eta(\omega)$ for particular sections require knowledge of the hydrodynamic quantities $B$, $M$, $A_+$, $A_-$ for that section. Although these parameters are known for symmetric ship-hull-type sections, very few calculations exist for unsymmetric sections which can be efficient absorbers.

One exception that permits an explicit analytic solution is the clam wave-power device, which is a development from the duck idea and is another terminator device. The long central cylindrical spine on which the ducks rotate is replaced by a long vertical rectangular section. The "ducks" are replaced by vertical plates hinged to the back section along their bottom edges. Incident waves cause the front plates to roll while the back plate remains relatively stationary. The movement of each front plate compresses a flexible bag containing air. The flow of air from all the bags drives a turbine housed inside the structure.

A simple model of the "clam" is obtained by replacing the back section by a thin vertical plate. Although the model "clam" is an articulated structure, it can still be treated by the methods described above and the maximum efficiency of a two-dimensional clam operating in a narrow wave tank is still given by $\eta_{max}=\{1+|A_-/A_+|^2\}^{-1}$. Because of the simple geometry of the model device $A_-$, $A_+$ can be determined explicitly as the sum of the solution of two distinct problems. The first is the rolling of a single vertical plate about its lower edge, whose far-field solution is given by Ursell (1948), and the second is the rolling of a plate about its lower edge when no flow is allowed under the bottom edge. This second problem is particularly simple and can be solved using Havelock (1929) wavemaker theory. Thus, it is found (Evans 1980b) that

$$A_\pm=S\pm A$$

where

$$A/a^2=\pi i\{\pi I_1(ka)-iK_1(ka)\}^{-1}\left\{\frac{1}{2}-\frac{(1-ka)}{ka}[I_1(ka)+L_1(ka)]\right\},$$

$$S/a^2=2i(e^{-ka}-1-ka)/(ka)^2. \tag{3.13}$$

Here $k=\omega^2/g$, $a$ is the depth of submergence of the plates, $I_1$, $K_1$ are modified Bessel functions, and $L_1$ a modified Struve function.

Hurdle (1979) has considered the other hydrodynamic properties of the clam including $B$ and $M$ and provides curves of efficiency versus wavenumber for varying values of the tuned frequency, as well as the envelope of maximum efficiency computed from (3.3) and (3.13).

Values of $M, B$ for a single rolling plate given by Kotik (1963) enabled Evans (1976) to compute $\eta$ for an energy-absorbing rolling vertical plate where of course the maximum efficiency is 50%. He shows further curves for a heaving or swaying half-immersed circular cylinder based on results for $B$, $M$ given by Frank (1967).

Computations for unsymmetric sections including "duck" sections have been made by Kan (1979). He has compared the maximum efficiency of a number of different sections absorbing power in either heave, or roll, including a "duck" section, and a set of sections having horizontal and vertical sides completed by either a straight side to form a triangle, or an arc of a circle to form a convex or concave section. The method used was the source method of Frank (1967). As might be expected on physical grounds, for both the clam and these unsymmetric sections $\eta_{max} \to \frac{1}{2}$ for $ka \to 0$, $\eta_{max} \to 1$ for $ka \to \infty$. In none of the cases considered by Kan does $T$ vanish identically so that $\eta_{max} < 1$ for all wavelengths. For rolling motions Kan considers sections formed from a combination of quadrants and triangles or concave quadrants and compares their maximum efficiency with a "duck" section. In each case the "duck" is found to be superior regardless of whether the radius of the quadrant or the square root of the sectional area is chosen as the reference length.

Count (1978a) has also considered unsymmetric sections by extending the multipole method first used by Ursell (1949) in solving for the hydrodynamic characteristics of a half-immersed circular section. The idea is to construct the solution from a set of singular potentials whose singularities lie inside the section, each of which satisfies all the conditions of the problem except the boundary condition on the section. This condition provides an infinite set of equations for the unknown constants multiplying the potentials, which can be solved by truncation. Count applied the method to a "duck" section and obtained curves of $M$ and $B$ and $\eta$ against normalised frequency $\Omega = \omega/\omega_0$ where $\omega_0$ is the tuned frequency. The efficiency curve agreed well with experimental values provided by Salter et al (1976). Similar computations were carried out on a raft consisting of a pontoon pair, the complete structure being semi-elliptical with a major:minor axis ratio of 10:1, where one section rotates relative to an equal-length fixed section. Count shows that for comparably sized "duck" and raft sections the efficiency curves are remarkably similar. This follows from (3.11) where it is seen that close to resonance, provided $\gamma \simeq 1$, the efficiency is approximately

$4dB/(d+B)^2$ and the damping coefficient $B$ for both "duck" and raft turn out to be very similar. Count argues that $B(\omega)$ is an important parameter in determining the bandwidth of a wave-power structure. The damping coefficient for a given section generally has a maximum at one value of frequency, say, $\omega_0$. Choosing the predominant wave frequency to coincide with $\omega_0$ fixes the size of the section since $B$ is a function of the dimensionless parameter $\omega^2 a/g$ where $a$ is a typical diameter of the section. The external damping constant $d$ should if possible be chosen such that $d=B(\omega_0)$ to ensure the least drop in efficiency for values of $\omega$ close to but different from $\omega_0$.

Now

$$\eta(\omega) \simeq 1 - \left( \frac{d-B(\omega_0)}{d+B(\omega_0)} \right)^2 + O(\omega-\omega_0)^2$$

since $B'(\omega_0)=0$. This shows that if $d \neq B(\omega_0)$ then it is better to choose a higher rather than a lower value of $d$ since for $d/B>1$ the efficiency $\eta(\omega)$ falls less rapidly from its maximum value. If $d=B(\omega_0)$ then further expansion gives

$$\eta(\omega) \simeq 1 - (\omega-\omega_0)^4 \left\{ \frac{B''(\omega_0)}{4B(\omega_0)} \right\}^2$$

giving the requirement that the damping coefficient be both large and smooth at its maximum to maintain a broad-band efficiency curve.

For multi-degree-of-freedom systems the scalar equation (3.10) needs to be replaced by a vector equation for the unknown displacements. The coefficients $M, B$ are then matrices whose diagonal terms represent the usual added mass and damping coefficients for each mode of motion, and whose off-diagonal terms describe the effect on the hydrodynamic radiation force of the interaction between the different motions. The power can again be calculated as the sum of the powers from each energy-absorbing mode, although it is not clear how to maximise the power absorbed in the general case.

Although primarily concerned with single-degree-of-freedom sections, Count goes some way towards solving the general problem of optimal power absorption from multi-mode or articulated devices. In particular, he solves for the efficiency of a "duck" section free to move in heave, sway, and roll of the entire section, as well as of its energy-absorbing mode, which is roll about its central section. The efficiency curve shows a drastic drop in efficiency over the frequency range considered, showing that the extra induced modes of motion have the effect of creating waves which radiate away, with consequent drop in power absorption. No details of the calculations are given by Count who emphasizes that no attempt at optimization has been made for this four-degree-of-freedom case, the same parameters being used as for a single degree of freedom. Subsequent experiments by Salter (1978) have confirmed that a serious drop in efficiency can occur but he has also shown that, by varying the amount of heave and sway permitted, an actual improvement in efficiency can be achieved. This is due to favorable interaction of the wave trains generated by the heave and sway modes with the reflected and transmitted waves. A similar degradation of performance of the two-pontoon raft system is found by Count when it is allowed to move in four degrees of freedom.

The lesson to be learnt from Count's work is that narrow-tank tests on sections that operate in only a single degree of freedom can produce deceptively high efficiencies and that, generally speaking, once realistic motions including additional degrees of freedom without a power-absorbing mechanism are allowed, the efficiency is reduced. A full optimisation study of a "duck" section oscillating in three degrees of freedom with power take-off in one mode has been tackled by Jeffery (1980) using control-theory methods. He replaces the infinite-dimensional fluid-structure system by a finite system of ordinary differential equations and uses the Pontryagin Maximum Principle to show that angular-velocity measurements alone can give enough information for optimal power extraction and heave and surge velocities need not be measured.

A computer model that goes some way towards solving the complete problem has been produced at the U.K. National Maritime Institute (Standing 1978). The program, called NM1 WAVE, has been applied to a string of ducks and has confirmed the drop in efficiency when the spine is allowed to move, but to date has not been extended to confirm the possible increase in efficiency predicted by Salter's experiments. A similar numerical approach has been described by Katory (1977) and Katory et al (1979). In both cases, the hydrodynamic coefficients are determined from the velocity potential in the fluid and on the structure, using the source method described in detail by Frank (1967). In this method the potential on the structure is expressed as a distribution of fundamental wave sources and the resulting integral equation derived by applying the body boundary condition is solved by replacing it by an equivalent algebraic system. The method works well for the moderate wave frequencies needed for wave-power work but for surface bodies it is plagued by the occurrence of the so-called irregular frequencies that are connected with a complementary interior boundary-value problem. See, for example, Ogilvie & Shin (1978).

Haren & Mei (1979) have developed a method applicable to the Cockerell raft system in two dimensions for any number of linked rafts using a long-wave approximation valid in shallow water. This assumption is checked in one special case using the full linearized equations and a hybrid-element method described by Bai & Yeung (1974). This method uses a finite-element approximation near the body and analytical representation elsewhere. The matching conditions involve two natural boundary conditions thrown up by a localized variational principle. Haren & Mei (1979) go further than other authors in their optimization study by building into their model economic constraints such as a price penalty on increasing raft length.

An earlier paper, already mentioned, by Mynett et al (1979) applies the hybrid-element method to the duck in more than one degree of freedom. Again, reduced efficiency is predicted when the duck is allowed to sway and heave as well as roll.

The difficulty with all these theoretical studies lies both in accurately modeling the necessarily complicated wave-power machine and also in predicting an optimal strategy. For instance, it is envisaged that as many as twenty-five "ducks" will be attached to a rigid central cylindrical spine, each capable of responding to the irregular wave field in its vicinity. Again a full-scale raft system is necessarily three dimensional with spacing between adjacent rafts. Only extensive small-scale tank tests can provide confidence in full-scale predictions for such complicated structures, although the necessarily limited theoretical studies described above may provide general guidelines for reducing the number of tests required. It is unlikely, however, that simple powerful design criteria such as those embodied in (3.3) and (3.8) for the single-degree-of-freedom theory will emerge from multi-degree-of-freedom models.

An exception to this last statement is the case of a wave-power machine comprising, for simplicity, just two independently oscillating sections, each with one degree of freedom, or two independent degrees of freedom of one section, where power absorption can take place at each section or in each degree of freedom. Then it can be shown that the theoretical maximum efficiency is 100%. The argument, which is based on the idea of wave cancellation, is most easily seen by considering two identical sections with vertical planes of symmetry, capable of absorbing energy from, say, vertical oscillations. The in-phase motion of both sections produces identical wave amplitudes at either infinity, whilst the anti-phase motion produces equal-amplitude waves in antiphase at either infinity. An appropriate combination of these two motions will reinforce the wave amplitude at one infinity while annihilating the waves at the other. Reversal of the time coordinate now shows that a solution exists in which an incident wave is totally absorbed by a particular motion of the section. This idea has been developed by Srokosz & Evans (1979) who consider in detail two independently rolling vertical plates each having a power take-off facility. They use an approximate method to determine the generalized added mass and damping coefficients required, based on a wider-spacing approximation used with considerable success by Ohkusu (1974) in related problems. The results show that the efficiency peaks at 100% as anticipated, but that it is highly sensitive to spacing and frequency although this may be exaggerated by the approximation used. It is of interest that the two-plate system is being considered as a practical wave-energy device by Q Corporation in the US (R. O. Wilke, private communication), whilst a tri-plate system has been suggested by Farley et al (1978).

The wave-cancellation argument also holds for a single symmetric section oscillating in two independent degrees of freedom, one symmetric, one antisymmetric, and this idea forms the basis for the Bristol submerged-cylinder wave-energy device. The idea exploits the result proved by Ogilvie (1963) that a long completely submerged horizontal circular cylinder that makes small circular motions about its axis generates waves at the surface that travel away from the cylinder in one direction only, that is in the direction of motion of the cylinder at the top of its orbit. It is clear that this is just a special case of the previous argument, since the circular motion is equivalent to equal-amplitude horizontal and vertical motions 90° out of phase. Theory and experiment confirming the high efficiency of the submerged cylinder when working in reverse as an *absorber* are given in Evans et al (1979a). One advantage of the cylinder idea is that when optimally tuned at a given frequency it loses energy at other frequencies due to transmission, never reflection. Evans et al (1979b) show how this can be exploited by using additional cylinders to pick up the lost transmitted energy.

## *Oscillating-Water-Column Devices*

These devices work on the idea of tuning a trapped mass of water to the incident-wave frequency in contrast to the cases considered so far in which part of the structure is made to resonate to the incoming waves.

The simplest example is that of an immersed open-ended vertical tube held fixed in an incident wave-train. The mass of water inside the tube is oscillated by the pressure fluctuation at the bottom end, of frequency $\omega$, and its motion is opposed by the hydrostatic restoring force. A simple calculation shows that, if the mass of water inside the tube is assumed constant, it will resonate if $\omega^2 = g/l$ where $l$ is the length of the water column. Experiments in wave tanks confirm that very large-amplitude column motions can be achieved. This idea forms the basis of a number of wave-energy devices including the Masuda (Masuda & Miyazaki 1978) self-powered light-buoy which has been operating successfully for many years. The same principle is exploited by the U.K. National Engineering Laboratory device (Meir 1978) and the Queen's University, Belfast, buoy device (Long & Whittaker 1978). In each case the water column forces a trapped volume of air at high speed through a turbine.

Another device that makes use of the resonant water column is being developed by Vickers Ltd. (Chester-Browne 1978). The design of the device is undergoing modifications but an early version consisted of a totally axisymmetric submerged sea-bed structure with a centrally placed opening at the top. This duct led into an annular duct of the same cross-sectional area and an annular air-water interface. The water "column" length was governed by the compression of the enclosed air volume. A long paper devoted to two-dimensional aspects of this device was published by Lighthill (1979). He models the device by a pair of parallel, long vertical plates extending downwards from a given distance beneath the free surface. Arguing from a simple harmonic-oscillator model he suggests (correctly) that the maximum power is given by $|F|^2/8D$ where $|F|$ is the forcing-effect amplitude and $D$ the energy-wastage coefficient. The similarity to (3.9) is clear.

He develops a sequence of mathematical models culminating in the full solution to the linearized problem of the interaction of an incident wave-train with the two plates. His main aim is to estimate $K$, the modification factor, defined as the ratio of the forcing-pressure amplitude in the depths of the duct (in excess of hydrostatic) to the pressure amplitude at the level of the mouth if the duct were absent. He shows that, surprisingly, $K$ can exceed 1 for certain combinations of plate spacings and wave frequencies. This does not affect the maximum efficiency of energy absorption since $D_r = (Ke^{-2\pi h/\lambda})^2$ where $h$ is the depth of the duct's mouth and $\lambda$ is the incident wavelength, this being a special case of Equation (3.7), which relates the response of a two-dimensional system to incident waves and the generation of waves by symmetrical oscillation of that system. Nevertheless, the increased $K$ and hence $D_r$ ensure a broad bandwidth to the efficiency curves and provide useful design criteria. Also $D_r \gg D_f$ ($D_f$ is the friction-damping coefficient), ensuring that the efficiency is close to its maximum of 50% for this case. It is of interest that for the problem of two parallel vertical plates intersecting the free surface and having the duct mouth facing *downwards*, $K < 1$ always, showing that although the maximum efficiency is still the same, the fall-off of efficiency with wave frequency either side of the tuned frequency will be more rapid than for the case where the mouth faces upwards.

A simple approximate solution to the mouth-downwards case, readily applicable to the mouth-upwards case also, has been given by Evans

(1978). Under the assumption that the spacing of the plates is small compared with other length scales, he uses a matched-asymptotic-expansion approach, matching the far field (the solution for a single plate with a line source of unknown strength at its end) with the mean field (the potential flow between two semi-infinite plates). The region of fluid between the plates terminates in an oscillating piston which is attached to a spring and damper which absorbs the wave energy. Again, because of the symmetric motion of the symmetric plates, the maximum efficiency is 50%. The method is identical to that used by Newman (1974) for the same problem in the absence of the power-absorbing piston.

Computer models of surface and seabed-mounted unsymmetric oscillating-water-column devices have been developed by the National Engineering Laboratory Wave Energy Group in East Kilbride, Scotland (Meir 1978). An early difficulty with this device was the large structure required to ensure a stable reference for the oscillating water. Later designs were chosen such that any induced motions of the structure produced waves to the rear which cancelled with the transmitted wave and did not affect performance.

An extension of the theory to axisymmetric vertical submerged ducts in three dimensions has been made independently by Simon (1980, submitted for publication) and Thomas (1980, submitted for publication; see also Lighthill & Simon 1980) while Evans' (1978) approximate method also deals with this case. Simon (1980) uses an accurate variational approach to determine the pressure amplification factor $K$ whilst Thomas (1980) uses eigenfunction expansions to set up an integral equation, which he then solves numerically to obtain the damping coefficient and the added length of the column. Although they use different methods of solution, their results are in good agreement and also confirm that the "point absorber" result, Equation (4.5), derived in the next section, for three-dimensional absorbers, is applicable to oscillating-water-column devices.

## 4 THREE-DIMENSIONAL WAVE-ENERGY ABSORBERS

Just as for two-dimensional sections, the energy-absorbing ability of a three-dimensional body depends upon its hydrodynamical characteristics such as added mass and damping coefficients, but now the situation is complicated by the additional degrees of freedom possible. Nevertheless, existing computer programs such as NMI WAVE (Standing 1978) can be used to predict the performance of simple three-dimensional absorbers.

By far the greatest emphasis in the literature has been on single buoys having a vertical axis of symmetry, capable of absorbing power from either vertical or horizontal translational motions. Thus, it may be shown that the expression

$$P_{max} = |X_s(\beta)|^2/8B \tag{4.1}$$

still holds, where $|X_s(\beta)|$ is the amplitude of the exciting force on the fixed body, in its subsequent direction of motion, due to a regular plane wave whose direction makes an angle $\beta$ to a given direction. Again, $X_s(\beta)$ is related to the damping coefficient $B(\omega)$, this time by the relation

$$\int_0^{2\pi} |X_s(\theta)|^2 d\theta = 8B\lambda P_w \tag{4.2}$$

where $P_w$ is, as before, the mean power per unit crest-length in the incident wave and $\lambda$ is the wavelength. Here the integral is over all possible angles of incidence of the incoming plane wave.

It follows that a power-absorption length or capture width $l$ can be defined where $l = P/P_w$, and from (4.1) and (4.2),

$$\begin{aligned} l_{max} &= P_{max}/P_w \\ &= \lambda |X_s(\beta)|^2 \Big/ \int_0^{2\pi} |X_s(\theta)|^2 \, d\theta. \end{aligned} \tag{4.3}$$

This in turn can be written in terms of the far-field angular variation $f(\theta)$ of the wave motion produced by forcing the body to oscillate in the energy-absorbing mode, in the absence of the incident wave. Thus,

$$l_{max} = \lambda |f(\pi - \beta)|^2 \Big/ \int_0^{\pi} |f(\theta)|^2 \, d\theta. \tag{4.4}$$

Equation (4.4), which, like (4.3), can be found in Evans (1980a,b) and, in effect, in Newman (1979) shows that, as for two dimensions, the ability of a body to concentrate wave energy, by its motion in the absence of the incident wave, in a direction *opposite* to that of the incident wave, is characteristic of an efficient energy absorber. For a buoy with a vertical axis of symmetry, so that $f(\theta)$, $X_s(\theta)$ are independent of $\theta$ for vertical motions and proportional to $\cos\theta$ for horizontal motions,

$$l_{max} = \varepsilon\lambda/2\pi \tag{4.5}$$

where

$$\varepsilon = \begin{cases} 1 & \text{for heave} \\ 2 & \text{for surge or sway.} \end{cases}$$

This remarkable result, discovered independently and simultaneously by Budal & Falnes (1975), Evans (1976), and Newman (1976), shows that the maximum capture width for a single buoy is independent of its size. It may thus be possible for such a body to absorb *more* energy than is contained in an incident wave-crest of length equal to the body diameter.

Although the body dimension does not appear in (4.5) explicitly, it does occur in the following sense. The condition for maximum absorption is, as for two-dimensional sections, that the velocity satisfy

$$U=\tfrac{1}{2}X_s/B,$$

and this becomes large for small bodies, thus violating the linearized theory on which the result is based. The conditions for tuning the body to resonate with a given wave frequency are identical to those for the two-dimensional case, namely that the inertia and stiffness terms cancel and the applied damping equal the radiation damping (Equation 3.12).

Experimental confirmation of (4.5) for a single-point absorber has been provided by B. M. Count (private communication) who has achieved capture widths of 70% of the theoretical maximum. Most experiments have been carried out in narrow wave tanks where the wall effect is important and a direct comparison with (4.5) is not possible. For these cases, it can be shown that, provided the wavelength is greater than the tank width, the two-dimensional results apply and it is possible for a symmetric buoy to absorb a maximum of one half the total power in the incident wave-train. This result has been confirmed experimentally by Budal et al (1980). A general theoretical treatment of three-dimensional absorbers in a wave tank, including the case when the wavelength is *less* than the tank width, is given by Srokosz (1980).

Curves showing the variation of capture width with wave frequency for the case of a half-immersed sphere absorbing energy in heave and tuned to a given frequency are given by Evans (1976). Attempts to tune to longer wavelengths are hampered by the built-in stiffness due to the hydrostatic restoring force on the sphere. To avoid this problem and to provide a sheltered environment for the device, Srokosz (1979) has considered a submerged sphere attached to three cables, which connect symmetrically to pumps and springs at the seabed. The use of three cables enables the sphere to absorb energy from its horizontal as well as its vertical motion with a maximum capture width approaching $3\lambda/2\pi$ from (4.5). This cannot, in fact, be achieved in this case since the same spring and damper constants are chosen for each cable to ensure an omni-directional performance for the sphere. Curves are presented showing the variation of capture width with incident-wave frequency for different tuned frequencies together with the corresponding sphere motions. As expected, the increased capture width in long waves implies large amplifications of the sphere motion. Horizontal motions of the sphere are roughly twice vertical motions, reflecting the fact that the damping coefficient in heave is roughly twice that in surge.

For three-dimensional bodies that are not symmetrical about a vertical axis, no simple relation between $X_s(\beta)$ and $B$ exists. It is possible, however, to get some feel for how the elongation of a body affects its capture width, by making a thin-ship approximation. Thus, Evans (1980a, b) has considered a thin wedge-shaped symmetric body of either parabolic or rectangular plan that absorbs energy through vertical (heave) motions. In this case, the vertical exciting force can be derived explicitly, since the contribution from the scattered wave field cancels by symmetry. Now, from (4.3),

$$l_{\max}(\beta)/l_{\max}(0) = |X_s(\beta)|^2/|X_s(0)|^2, \tag{4.6}$$

showing how the capture width varies with direction of incidence $\beta$ of the incoming plane wave measured from a line perpendicular to the elongated axis of the body. For the parabolic absorber the right-hand side of (4.6) is just $\{2J_1(\nu\sin\beta)/\nu\sin\beta\}^2$ while

$$l_{\max}(0) = q_1\lambda/2\pi$$

where

$$q_1^{-1} = \frac{1}{2\pi\nu^2}\int_0^{2\pi}\left\{\frac{2J_1(\nu\sin\theta)}{\nu\sin\theta}\right\}^2 d\theta.$$

For the rectangular body the terms in curly brackets are replaced by $\sin(\nu\sin\theta)/\nu\sin\theta$. Here $\nu = 2\pi L/\lambda$ where $2L$ is the body length. It is clear that $q_1$ reflects the modification to the point-absorber result due to the elongation of the body. Computation of $q_1$ shows that $l_{\max}(0)/2L > 0.4$ for all wavelengths.

The extension to any number of isolated buoys, each absorbing energy, was first considered by Budal (1977). He assumed that the amplitude of each buoy was the same in its optimal motion and then proceeded to optimize this amplitude and the relative phases of the buoys. While clearly correct for either two or an infinite line of buoys parallel to the crests of an incoming wave-train, the assumption of equal amplitude is not valid otherwise. An optimization of both phase *and* amplitude of the buoy motions has been made independently by Evans (1980a, b) and Falnes (1978). It turns out that the mean power absorbed by the system of buoys may be written

$$P = \tfrac{1}{8}X_s^* B^{-1}X_s - \tfrac{1}{2}\left(U_0 - \tfrac{1}{2}B^{-1}X_s\right)^* B\left(U_0 - \tfrac{1}{2}B^{-1}X_s\right) \tag{4.7}$$

which has a maximum when $U_0 = \frac{1}{2}B^{-1}X_s$, giving for the maximum mean power that can be absorbed by the system,

$$P_{\max}=\tfrac{1}{8}X_s^*B^{-1}X_s. \tag{4.8}$$

Here, $B$ is an $N\times N$ matrix whose elements are $B_{mn}$, where $B_{mn}U_{0n}$ is that part of the force on the $m$th body that is due to the (complex) velocity amplitude $U_n$ of the $n$th body and is in phase with the velocity of the $n$th body. Also, $X_s$ is an $N$-vector whose $m$th component is the (complex) amplitude of the exciting force on the $m$th body in the presence of the others. Note that * denotes complex conjugate transpose.

It can be seen that (4.8) is an elegant generalisation of the result for a single body (4.1) and is also applicable to bodies oscillating in more than one mode, provided power can be absorbed from all the modes. Again, $X_s(\beta)$ and $B$ are related, this time by the expression (Srokosz 1979)

$$B_{mn}=\frac{1}{8\lambda P_w}\int_0^{2\pi}X_{sm}(\theta)\,\overline{X_{sn}(\theta)}\,d\theta \tag{4.9}$$

which shows, incidentally, that the matrix $B$ is positive definite and hence invertible, provided only that the exciting force vector $X_s(\beta)$ for the system does not vanish identically.

In the absence of information about $X_s(\beta)$ for a general system of buoys, it is necessary to make assumptions about the interaction between the buoys if further progress is to be made. Thus, for a line of $N$ identical buoys each having a vertical axis of symmetry small enough not to distort the far field created by any one of them in forced motion, it can be shown (Evans 1980a, b) that

$$l_{\max}(\beta)=P_{\max}/P_w=\frac{\lambda}{2\pi}Nq(\mu,\beta)$$

where $\mu=2\pi d/\lambda$ and $d$ is the spacing between buoys. Here

$$q(\mu,\beta)=\frac{1}{N}L^*J^{-1}L \tag{4.10}$$

where

$$J=\{J_{mn}\}=\left\{\frac{1}{2\pi}\int_0^{2\pi}L_m(\theta)\,\overline{L_n(\theta)}\,d\theta\right\}$$

and

$$L=\{L_m\}=\{\exp(i\mu m\sin\beta)\}$$

so that

$$J_{mn}=J_0\{\mu(m-n)\}.$$

Clearly $q(\mu, \beta)$ is a gain factor, being a measure of the interference between the bodies, since it is just the maximum mean power per buoy divided by the maximum mean power from a single isolated buoy.

Curves of the variation of $q(\mu, 0)$ with $\mu$ for two equally spaced buoys show a gain of greater than 60% for a spacing of about $d = .6\lambda$. B. M. Count (private communication) has recently confirmed these results experimentally using a wide wave tank at the University of Edinburgh. He considered lines of two, five, and ten equally spaced buoys.

The theory for arrays of point absorbers and the thin-ship theory for elongated bodies can be combined to assess the capture width for arrays of elongated bodies. For instance, for $N$ identical thin bodies each of length $2L$ a distance $d$ apart, with their axes in line, the gain $q(\mu, \nu, \beta)$ is still given by (4.10) but now for thin wedge-shaped bodies of rectangular plan making vertical oscillations:

$$L_m = \exp(i\mu m \sin\beta)\sin(\nu \sin\beta)/\nu \sin\beta.$$

The method can be extended to any situation in which each body or each articulated section of the same body has its own energy-absorbing mechanism, provided the diffracted wave due to the interaction of the incident wave and the fixed body or section of body can be neglected. For then $X_s(\theta)$, the exciting-force vector, can be derived by integration of the incident pressure field over the body, and with $B_{mn}$ determined from (4.9), $P_{\max}$ can be obtained from (4.8).

The estimation of the exciting-force vector $X_s(\theta)$ is a difficult problem in general and approximations such as those made above for the system of point absorbers have to be made. Nevertheless, analytical progress has been made for the case of two half-immersed heaving spheres by Greenhow (1980). He uses an iterative procedure to partly take into account the interaction between the local wave fields of the two spheres. He is able to derive more accurate expressions for the $B_{mn}$ coefficients and the q-factor for this configuration and obtains qualitative agreement with the Budal/Evans case of two point absorbers. A similar iterative technique has been used by Ohkusu (1973) for arrays of vertical cylinders in three dimensions.

Newman (1979) (see also Mei & Newman 1980) also derives the result (4.3) but his approach is via Kochin functions and far-field behavior rather than the hydrodynamic forces on the bodies. He applies the results to elongated slender devices using a slender-body approximation similar to the assumptions used in deriving (4.10). By considering motions of the device generated by specific modal shapes, Newman determines a range of capture widths all of which asymptote to $\lambda/2\pi$ as

$\lambda \to \infty$. A particular application to a slender articulated raft having three elements leads him to conclude that such a configuration is near optimal since the combination of even and odd modes has a capture width close to the continuous modes considered previously. An increase in raft sections would not increase the capture width appreciably, whilst a reduction to two sections with a single hinge effectively reduces the capture width by one half. This conclusion differs from those of Count (1978a) who, admittedly on the basis of two-dimensional results, appears to favor a single hinge with a rear section twice the length of the front section. Haren & Mei (1979) also favor no more than two or three sections.

Newman's work realistically builds into the calculations an estimate of the effects of body constraints on capture width. By arbitrarily restricting the motion of the slender device to a practical range, the capture-width curves, instead of increasing with $\lambda$, have a maximum and in fact tend to zero as $\lambda \to \infty$. In this way, Newman estimates that the maximum capture width is of the order of the length of the elongated device, a result that is encouraging to the advocates of alternator devices such as the French flexible-bag device. This conclusion, however, does depend fairly critically upon how much motion of the body is allowable.

It is possible to take the optimal power results a little further and include global constraints on the body motion in the general case. If, for instance, the components of the velocity of each body or each body section are constrained by $|U_m| \leqslant \beta_m$, say, then

$$U^*U = \sum_{m=1}^{N} |U_m|^2 \leqslant \sum_{m=1}^{N} \beta_m^2 = \beta^2, \text{ say,} \tag{4.11}$$

and the problem is to optimize (4.7) subject to (4.11). The function $P(U)$ regarded geometrically as the closed surfaces $P(U) = \text{const.}$ centered at $\frac{1}{2}B^{-1}X_s$ in the $2N$-dimensional $U$-space has a maximum at $U = \frac{1}{2}B^{-1}X_s$ given by $\frac{1}{8}X_s^*B^{-1}X_s$, provided $U$ is unconstrained. The appropriate maximum with $U$ satisfying (4.11) occurs when the surfaces $P(U) = \text{const.}$ just touch the "sphere" $U^*U = \beta^2$. Since $B$ is not the identity matrix, $U$ is not in the direction of $\frac{1}{2}B^{-1}X_s$ but the problem can easily be solved using Lagrange multipliers. It is found that (D. V. Evans, unpublished work)

$$P_{\max} = \tfrac{1}{8}X^*(B+\mu I)^{-1}X + \tfrac{1}{2}\mu\beta^2 \left(= \tfrac{1}{2}U^*(B+2\mu I)U\right.$$

where the scalar $\mu$ satisfies $\|(B+I)^{-1}X\| = 2\beta$ and $\|\ \|$ denotes the Euclidean norm. The optimal $U$ is given by

$$U = \tfrac{1}{2}(B+I)^{-1}X_s$$

provided $\|B^{-1}X\| \geqslant 2\beta$.

## 5 MISCELLANEOUS TOPICS

### *Nonlinear Wave Forces*

A perturbation expansion up to second order in wave slope shows that, in addition to the first- and second-order harmonic-oscillatory force on a body in waves, there also exists a mean horizontal drift force which Longuet-Higgins (1977) has shown to be

$$\tfrac{1}{4}\rho g A^2(1+|R_1|^2-|T_1|^2)=\tfrac{1}{4}\rho g A^2(\eta+2|R_1|^2)$$

where $\eta$ is the efficiency of absorption of a two-dimensional cylindrical section. He has confirmed these results experimentally for the Cockerell raft system and Salter (1976) has done the same for his duck in low-amplitude waves. Salter has also shown that the mean horizontal drift force can actually be reversed for waves traveling over the top of a fixed submerged circular cylindrical section. Longuet-Higgins (1977) explains this in terms of wave-breaking over the cylinder. In relatively long waves breaking occurs close to point of minimum depth and continues well beyond it. The resulting loss in horizontal momentum flux is balanced by a corresponding "set-up" or increase in mean level, creating a reverse horizontal force. These results suggest that horizontal drift forces are unlikely to be a serious problem for submerged wave-energy devices.

### *Nonlinear Power Take-Off*

Little attention in the literature has been given to modeling realistic power take-off systems. All work described thus far has assumed velocity-proportional damping forces and displacement-proportional spring forces or equivalently have assumed all motions to be simple harmonic with single frequency $\omega/2\pi$. One exception is McCormick (1974) in his treatment of the Masuda buoy, where a fully nonlinear formulation of the air flow inside the buoy is given. He makes a Froude-Krylov assumption and ignores the effect of the buoy motions on the incoming waves and also assumes a constant value independent of frequency for the buoy added mass. He then gives arguments for neglecting the nonlinear terms in the force due to internal air flow so that the resulting equations are, after all, linear in the internal surface displacement.

Count (1978b) considers the problem of the coupling of a linear wave field with wave-power machines having nonlinear energy-extraction characteristics. In particular, he considers the Salter duck where the damping is provided by a constant resistive torque that reverses sign with velocity. The equation of motion (3.10) is modified to become

$$(m+M_\infty)\ddot{\zeta}(t)+d\,\mathrm{sgn}(\dot{\zeta})+k\zeta+\int_0^t G(t-\tau)\dot{\zeta}(\tau)\,d\tau=F_s(t) \tag{5.1}$$

where $\int_0^\infty G(t)\exp(i\omega t)\,dt = B(\omega) - i\omega\{M(\omega) - M_\infty\}$ and $M_\infty$ is the added mass of the section at infinite frequency. Here $d\,\mathrm{sgn}(\dot{\zeta})$ describes the Coulomb damping. As before $F_s(t) = \mathrm{Re}\, X_s \exp(i\omega t)$ is the exciting force on the fixed section. With velocity-proportional damping replacing the second term in (5.1), a direct correspondence with (3.10) follows after transforming to the frequency domain. The method used by Count to solve (5.1) is a numerical finite-difference scheme whose convergence for all $d > 0$ is attributed to C. Elliott (private communication). Computations of the roll velocity show that it is strongly frequency-dependent and that for small frequency increasing periods of no motion occur as $\theta = d/|X_s|$ increases towards unity. For $\theta > 1$ no motion occurs at all, as the wave force is insufficient to overcome the pump resistive force. Comparison of efficiency with the linear optimum shows that it is possible, by careful choice of the value of $\theta$, to come close to the linear value but the efficiency too is strongly dependent upon wave frequency. An analytic solution for small $\theta$ using Fourier series and assuming periods of rest confirms the numerical finite-difference solution. Count also considers the case of velocity-proportional damping with an imposed torque limit and makes a comparison of efficiency between this and the Coulomb damping.

Parks & Tondl (1978), using Fourier-series methods, have also tackled the Coulomb damping case for a wave-power device consisting, for simplicity, of a single vertical plate. They achieve results similar to those of Count (1978b). Again no attempt is made to predict zero motions by this method. The problem is simplified by the assumption of exponential decay of the motion down the plate thus making this and the more complicated tri-plate system (Farley et al 1978) amenable to solution.

## *Experimental Fluid Mechanics*

While not specifically concerned with wave-energy devices, it seems appropriate to mention a series of papers that developed from attempts to verify theoretical results in the wave-power field. These begin with experiments carried out by Knott & Flower (1979) in which a fully submerged parallel-plate enclosure is subjected to regular incident waves. The intention is to check out the theoretical prediction (Lighthill 1979) of a pressure amplification in the duct compared to the pressure in the absence of the duct. This they are able to confirm and in a further paper (Knott & Flower 1980a) they extend the work to consider circular ducts in both a narrow tank and a wide tank to obtain reasonable agreement with the theoretical prediction of Lighthill & Simon (1980). Both theory and experiment are aimed at an understanding of the Vickers submerged wave-energy device, and the importance of energy losses in oscillating flows through pipes of various exit shapes led to further papers (Knott & Mackley 1980, Knott & Flower 1980b). In these, the dependence of eddy losses on the radius of the lip mouth, the radius of the tube, the oscillation amplitude, and frequency are all considered. In both papers, impressive flow-visualization techniques are used. In the latter paper it is noted that eddy losses only begin to act once a certain amplitude of motion is exceeded and that the critical amplitude of motion is strongly dependent on the radius of the lip.

## *Elongated Flexible Devices*

Although early devices involved rigid structures with few degrees of freedom, interest has shifted recently to elongated flexible structures which operate with their axes heading into the waves and which absorb energy progressively down their length through continuous deformation. The most well known is due to French (1977), but a more recent alternator device has been suggested by Farley (submitted for publication) consisting of a long slender floating flexible beam, tensioned to encourage buckling and to enhance wave-induced vertical oscillations.

The theory outlined in Section 4 for the capture width of $N$ independently oscillating bodies goes some way towards providing a theory for flexible devices as does the work of Newman (1979). An extension to continuously deformable elongated devices suggests that for such a body, of length $2L$ in head seas,

$$l_{\max}=\frac{\lambda}{2\pi}\int_0^{2L}u_0^*(x)\exp(ikx)\,dx \tag{5.2}$$

where $u_0(x)$ satisfies

$$\int_0^{2L}u_0(x)J_0(k(x-y))dx=\exp(iky),\qquad 0\leqslant y\leqslant 2L. \tag{5.3}$$

Here the body has been represented by a distribution of sources of strength $u_0(x)$. A simple approximation to $l_{\max}$ is obtained by combining (5.2) and (5.3) so that

$$\frac{2\pi}{\lambda}l_{\max}=\left(\int_0^{2L}u_0^*(x)e^{ikx}dx\right)^2\Big/\int_0^{2L}\int_0^{2L}u_0(x)u_0^*(y)J_0[k(x-y)]\,dxdy$$

and using the approximating function $u_0(x)=c\exp(ikx)$ in this stationary form for $l_{\max}$. It follows that

$$l_{\max}=\frac{\lambda}{2\pi}q$$

where $q^{-1}=\pi^{-1}\nu^{-2}\int_0^{\pi}\sin^2\{\nu(1+\cos\theta)\}/(1+\cos\theta)^2\,d\theta$ and $q$ measures the improvement over the point absorber result. Further work along these lines, incorporating motion constraints, and arrays of flexible devices, is in progress (D. V. Evans, unpublished work).

## 6 CONCLUDING REMARKS

Theoretical fluid dynamics applied to wave-power problems is a relatively new field exploiting relatively old ideas. The tools of the theoretician used for many years in the field of ship hydrodynamics have proved invaluable and the added complication of an energy-absorbing device has produced some novel results that have provided a useful and immediate input to the overall wave-energy program.

The general results described in the previous sections inevitably reflect this reviewer's own interests and this has meant that individual papers have gone unmentioned because of lack of space. These include, for example, numerous reports on experimental results produced by various device teams, a veritable mountain of papers produced by the Technical Advisory Groups set up under the U.K. Wave Energy program on problems varying from types of mooring chains to environmental effects, and the essential wave-data information collected and processed by the Institute of Oceanographical Sciences in the U.K. The ingenuity of inventors in the wave-power field shows little sign of dwindling and schemes such as wave focusing described by both Mehlum & Stamnes (1980), and Isaacs (1980), the suggestion for using wave-powered systems for desalination by Pleass (1978) and the Kaimei project (Masuda & Miyazaki 1978) all deserve a mention.

As far as future fluid mechanics is concerned, nonlinear body and wave motions loom large as an area needing further study, although sensible interpretation and extension of empirical expressions such as the Morison equation for the prediction of wave loading on structures have already proved successful (Dixon et al 1980). The work of Count (1978b) on nonlinear pump characteristics is another valuable contribution to this difficult area and this needs to be extended to multi-degree-of-freedom systems. Another area of need is in the prediction of hydrodynamic characteristics of arrays of possibly articulated or flexible devices so that the importance of spacing can be assessed. Existing computer programs become increasingly expensive as the number of devices increases and further approximate methods need to be established and validated.

Finally, work on sea efficiencies has been inhibited by lack of information on empirical directional energy spectra. Indeed none of the papers described in this review has included a calculation on sea efficiencies using Equation (2.5) from knowledge of $l(\omega, \theta)$. As confidence in directional buoy measurements and wave-climate-prediction methods grows, it is anticipated that an estimate of performance in realistic sea conditions will become of increasing importance in future theoretical models.

Whilst recognizing that the viability of wave-power machines can only be confirmed by extensive model tests followed ultimately by full-scale prototype tests, it is clear that in many areas hydrodynamic theory can continue to play a useful complementary role for some time to come.

*Literature Cited*


Ambli, N., Budal, K., Falnes, J., Sørenssen, A. 1977. *Proc. World Energy Conf., 10th, Instanbul*, Pap. 4.5–2

Bai, K. J., Yeung, R. 1974. *Proc. Symp. Naval Hydrodyn., 10th*, pp. 609–41

Budal, K. 1977. Theory for absorption of wave-power by a system of interacting bodies. *J. Ship Res.* 21:248–53

Budal, K., Falnes, J. 1975. A resonant point absorber of ocean-wave power. *Nature* 256:478–9, corrigendum 257:626

Budal, K., Falnes, J., Kyllingstad, Å., Oltedal, G. 1980. Experiments with point absorbers. See Jansson et al 1980, pp. 253–82

Chester-Browne, C. V. 1978. The Vickers device. See Quarrell 1978, pp. 55–58

Cocklin, M. L. 1977. *The efficiency of an elliptical cylinder as a wave-energy absorber*. MS. thesis. Univ. Edinburgh. 43 pp.

Count, B. M. 1978a. On the dynamics of wave-power devices. *Proc. R. Soc. London, Ser. A* 363:559–79

Count, B. M. 1978b. The theoretical analysis of wave power devices with non-linear mechanical conditioning. *C.E.G.B. Marchwood Rep. No. R/M/N1008*

Count, B. M., ed. 1980. *Power from Sea Waves—Proc. I.M.A. Conf., Edinburgh, 1979*. London/NY: Academic. In press

Dixon, A. G., Greated, C. A., Salter, S. H. 1980. Wave forces on partially submerged cylinders. *J. Waterway, Port, Coastal & Ocean Div.* 105:421–38

Evans, D. V. 1976. A theory for wave-power absorption by oscillating bodies. *J. Fluid Mech.* 77:1–25

Evans, D. V. 1978. The oscillating water column wave-energy device. *J. Inst. Math. Its Appl.* 22:423–33

Evans, D. V. 1980a. Some theoretical aspects of three-dimensional wave-energy absorbers. See Jansson et al 1980, pp. 77–113

Evans, D. V. 1980b. Some analytic results for two and three dimensional wave energy absorbers. See Count 1980

Evans, D. V., Jeffrey, D. C., Salter, S. H., Taylor, J. R. M. 1979a. Submerged cylinder wave-energy device: theory and experiment. *Appl. Ocean Res.* 1:3–12

Evans, D. V., Davis, J. P., Srokosz, M. A. 1979b. A system of efficient submerged wave-energy absorbers. *Mechanics of Wave Induced Forces on Cylinders*, ed. T. L. Shaw, pp. 673–83. London:Pitman

Falnes, J. 1978. Radiation impedance matrix and optimum power absorption for interacting oscillators in surface waves. Norges Tekniske Høgskole, Internal Rep. May 1978, reissued Sept. 1979

Falnes, J., Budal, K. 1978. Wave-power conversion by point absorbers. *Norw. Marit. Res.* 6:2–11

Farley, F. J. M., Parks, P. C., Altmann, H. 1978. A wave power machine using freely floating vertical plates. See Stephens & Stapleton 1978, Vol. 1, Pap. B2, pp. 35–46

Frank, W. 1967. Oscillations of cylinders on or below the free surface of deep fluids. *Naval Ship Research and Development Center, Washington, DC, Rep. No. 2375*. vi+40 pp.

French, M. J. 1977. Hydrodynamic basis of wave-energy converters of channel form. *J. Mech. Eng. Sci.* 19:90–92

Greenhow, M. 1980. Hydrodynamic interactions of spherical wave power devices in surface waves. See Count 1980

Haren, P., Mei, C. C. 1979. Wave power

extraction by a train of rafts: hydrodynamic theory and optimum design. *Appl. Ocean Res.* 1:147–57

Havelock, T. H. 1929. Forces surface waves on water. *Philos. Mag.* 8:569–76

Hogben, N. 1978. Wave climate synthesis for engineering purposes. *Nat. Marit. Inst. Rep. No. NMI R45*, pp. 1–37

Hurdle, D. P. 1979. *A description of the clam wave-power absorber*. MS. thesis. Univ. Bristol. 68 pp.

Isaacs, J. D. 1980. Ideas and some developments of wave-power conversion, dynamic wave absorption, and deep-sea mooring. See Jansson et al 1980, pp. 204–21

Jansson. K. -G., Lunde, J. K., Rindby, T., eds. 1980. *Proc. 1st Symp. Wave Energy Utilitzation, Gothenburg, 1979*. Gothenburg: Chalmers Tech. Univ., 1980

Jeffrey, E. R. 1980. Device characterisation. See Count 1980

Kan, M., 1979. Wave-power absorption by asymmetric bodies. *Pap. Ship Res. Inst., Tokyo, Jpn. No. 54*, pp. 1–18

Katory, M. 1977. On the motion analysis of interlinked articulated bodies floating among sea waves. *The Naval Architect* 1977:28–29

Katory, M., Lacey, A. A., Tam, P. K. Y. 1979. Application of theoretical hydrodynamics to the design of arbitrarily shaped and multibodied marine structures. *Trans. R. Inst. Nav. Arch.* 121:1–11

Knott, G. F., Flower, J. O. 1979. Wave tank experiments on an immersed parallel-plate duct. *J. Fluid Mech.* 90:327–36

Knott, G. F., Flower, J. O. 1980a. Wave tank experiments on an immersed vertical circular duct. *J. Fluid Mech.* In press

Knott, G. F., Flower, J. O. 1980b. Measurement of energy losses in oscillatory flow through a pipe exit. *Appl. Ocean Res.* In press

Knott, G. F., Mackley, H. R. 1980. On eddy motions near plates and ducts, induced by water waves and periodic flow. *Philos. Trans. R. Soc. Ser. A* 294:599–628

Kotik, J. 1963. Damping and inertia coefficients for a rolling or swaying vertical strip. *J. Ship Res.* 7:19–23

Lighthill, M. J. 1979. Two dimensional analyses related to wave energy extraction by submerged resonant ducts. *J. Fluid Mech.* 91:253–318

Lighthill, M. J., Simon, M. 1980. Mathematical analysis related to the Vickers' project. See Count 1980

Long, A. E., Whittaker, T. J. T. 1978. The Belfast device. See Quarrell 1978, pp. 61–64

Longuet-Higgins, M. S. 1977. The mean forces exerted by waves on floating or submerged bodies with applications to sand-bars and wave-power machines. *Proc. R. Soc. London Ser. A* 352:463–80

Masuda, Y., Miyazaki, T. 1978. Wave power electric generation study in Japan. See Stephens & Stapleton 1978, Vol. 1, Pap. B6, pp. 95–92

McCormick, M. E. 1974. Analysis of a wave energy conversion buoy. *J. Hydronaut.* 8:77–82

Mehlum, E., Stamnes, J. 1980. Power production based on focusing of ocean swells. See Jansson et al 1980, pp. 29–35

Mei, C. C. 1976. Power extraction from water waves. *J. Ship Res.* 20:63–66

Mei, C. C., Newman, J. N. 1980. Wave power extraction by floating bodies. See Jansson et al 1980, pp. 1–28

Meir, R. 1978. The development of the oscillating water column. See Quarrell 1978, pp. 35–44

Mynett, A. E., Serman, D. D., Mei, C. C., 1979. Characteristics of Salter's cam for extracting energy from ocean waves. *Appl. Ocean Res.* 1:13–20

Newman, J. N. 1974. Interaction of water waves with two closely-spaced vertical obstacles. *J. Fluid Mech.* 66:97–106

Newman, J. N. 1975. Interaction of waves with two-dimensional obstacles: a relation between the radiation and scattering problems. *J. Fluid Mech.* 71:273–82

Newman, J. N. 1976. The interaction of stationary vessels with regular waves. *Proc. Symp. Naval Hydrodyn., 11th, London*, pp. 491–501

Newman, J. N. 1977. *Marine Hydrodynamics*. Cambridge:MIT Press

Newman, J. N. 1979. Absorption of wave energy by elongated bodies. *Appl. Ocean Res.* 1:189–96

Ogilvie, T. F. 1963. First- and second-order forces on a cylinder submerged under the free surface. *J. Fluid Mech.* 16:451–72

Ogilvie, T. F., Shin, Y. S. 1978. Integral-equation solutions for time-dependent free-surface problems. *J. Soc. Nav. Arch. Jpn.* 143:41–51

Ohkusu, M. 1973. Wave action on groups of vertical circular cylinders. *Selected papers from J. Soc. Nav. Arch. Jpn.* 11: 37–50

Ohkusu, M. 1974. Hydrodynamic forces on multiple cylinders in waves. *Proc. Int. Symp. Dynam. Mar. Vehicles and Struc-*

*tures in Waves*, Pap. 12, pp. 107–12. Mech. Engrg. Publ. London

Parks, P. C., Tondl, A. 1978. Non-linear oscillations in wave power machines. *Proc. Int. Conf. Non-linear Oscillations, 8th, Prague*

Pierson, W. S., Moskowitz, L. 1964. A proposed spectral form for fully developed wind seas based on the similarity theory of S. A. Kitaigorodski. *J. Geophys. Res.* 69:5181–90

Pleass, C. M. 1978. The use of wave powered systems for desalination: a new opportunity. See Stephens & Stapleton 1978, Vol. 1, Pap. D1, pp. 1–10

Quarrell, P., ed. 1978. *Proc. Wave Energy Conf. London-Heathrow*, London:Publ. H.M.S.O.

Salter, S. H. 1974. Wave power. *Nature* 249:720–24

Salter, S. H. 1976. Edinburgh Wave Power Project. Second Year Rep.

Salter, S. H. 1978. The development of the duck concept. See Quarrell 1978, pp. 17–26

Salter, S. H., Jeffrey, D. C., Taylor, J. R. M. 1976. The architecture of nodding duck wave power generators. *The Naval Architect* 1976:21–24

Srokosz, M. A. 1979. *Some theoretical aspects of wave power absorption*. PhD thesis. Univ. Bristol. 170 pp.

Srokosz, M. A. 1979. The submerged sphere as an absorber of wave power. *J. Fluid Mech.* 95:717–41

Srokosz, M. A. 1980. Some relations for bodies in a canal, with an application to wave power absorption. *J. Fluid Mech.* 99:145–62

Srokosz, M. A., Evans, D. V. 1979. A theory for wave-power absorption by two independently oscillating bodies. *J. Fluid Mech.* 90:337–62

Stahl, A. W. 1892. The utilization of the power of ocean waves. *Trans. A.S.M.E.* 13:438–506

Standing, R. G. 1978. Applications of wave diffraction theory. *Int. J. Numer. Meth. Enrgr.* 13:49–72

Stephens, H. S., Stapleton, C. A., eds. 1978. *Proc. Int. Symp. Wave & Tidal Energy*, BNRA Fluid Engrg., Cranfield, Bedford, England. Vol. 1

Ursell, F. 1948. On the waves due to the rolling of a ship. *Q. J. Mech. Appl. Math.* 1:246–52

Ursell, F. 1949. On the heaving motion of a circular cylinder on the surface of a fluid. *Q. J. Mech. Appl. Math.* 2:218–31

Wehausen, J. V. 1971. The motion of floating bodies. *Ann. Rev. Fluid Mech.* 3:237–68

*Ann. Rev. Fluid Mech. 1981. 13:189–215*


# MENISCUS STABILITY

*8174

*D. H. Michael*
Department of Mathematics, University College, London, Gower Street, London WC1E 6BT, England

## 1 INTRODUCTION

The study of the mechanics of the meniscus, beginning with the formulation of the equilibrium equation by Young (1805) and Laplace (1805), has given rise to a very large literature with applications in many contexts. The equation in its most general form poses a challenge to mathematicians to prove properties of the solutions under prescribed constraints, as, for example, in the recent work of Concus & Finn (1974a, b) and Finn (1974) on the form of the solutions in a wedge-shaped domain. Otherwise, a large body of the literature is devoted to the calculation and classification of solutions, in most cases when plane or axial symmetry is assumed. Such calculations whilst formerly of some difficulty have now become relatively easy to perform with modern computing aids. We distinguish between several different contexts of equilibrium studies, depending on the nature of the force field in which the meniscus is formed, as follows:

(*a*) The meniscus supporting a constant or zero pressure difference as, for example, with soap bubbles and interfaces between neutrally buoyant fluids.

(*b*) The meniscus as an interface between two fluids of different density in an external gravitational field, this being characteristic of drops or bubbles formed in immiscible fluids of different densities under ordinary terrestrial conditions.

(*c*) Equilibrium studies describing the meniscus formed at the surface of rotating bodies of fluid. Such studies are for the most part restricted to rigid-body rotations in which the problem is reduced to a static one by use of d'Alembert's principle to replace the rotation by a radial force field. A recent example of such studies is that of Brown & Scriven (1980a, b).



0066-4189/81/0115-0189$01.00

(*d*) Equilibria of the meniscus under the action of electrostatic fields in which surface electrical stresses enter into the stress balance at the meniscus surface. Such problems in general add an extra difficulty to the solution by the need to know the electric field in an irregular domain for which the meniscus surface forms part of the boundary. Some nontrivial exact solutions have been obtained by numerical computation such as those of Sozou & Hewson-Browne (1976) and Gifford et al (1980). Otherwise, in this category effective approximations to the meniscus equilibrium have been made by Taylor (1964, 1968).

Our reference in each of these categories is in the first place to the study of equilibrium. However, since the existence in a steady state of any meniscus in the physical sense must imply mathematically not only equilibrium but stability, it is clear that in all categories the study of stability of the equilibrium solutions is of paramount importance. Much recent work has been devoted to this and in this article we restrict ourselves to a discussion of some of the developments in the study of stability. We cannot in the space of the article survey problems in all the categories mentioned. In particular, the substantial literature on the stability of rotating systems (*c*) and the stability of the electrified meniscus (*d*) is not discussed here.

In one sense, however, this is no restriction because the patterns of transition in stability repeat themselves in all these cases. These patterns are now well recognised in a wide variety of contexts. The most common forms of transition have been described, for example, in texts by Lyttleton (1953) and Thompson & Hunt (1973). A more complete discussion of the structure of transition in the seven forms of catastrophe associated with four independent loading parameters is given by Thompson & Hunt (1975). In meniscus theory we are able to describe the equilibrium and its stability in terms of a potential energy and we note here the transitions which commonly occur. Most common is the fold or limit point occurring when a controlled loading parameter, in this case usually the volume or the internal pressure, reaches a maximum or minimum position in its loading path. This is illustrated in Figure 1*a*. At the maximum position the energy function incurs third-order changes for the small displacement from one equilibrium configuration to a neighboring one on the loading path which can occur without change in the loading parameter at the maximum of the curve. The instability thus introduced is associated with a change of profile that preserves the symmetry. Instabilities that introduce changes in the symmetry are usually the stable and unstable symmetric perfect cusps illustrated in Figures 1*b* and *c*. The stable symmetric cusp gives rise to a

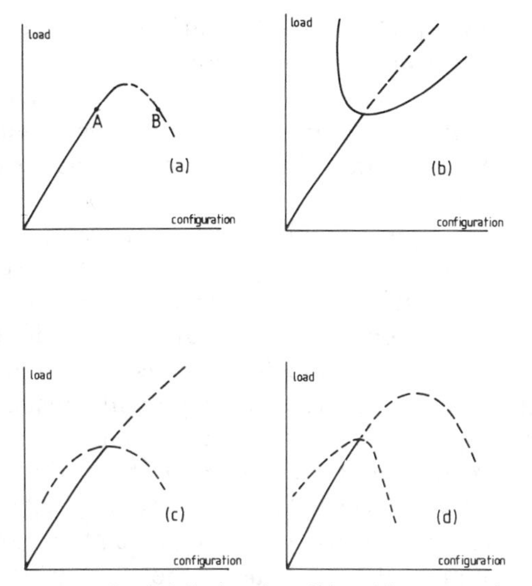


*Figure 1* Loading patterns for (*a*) limit point; (*b*) stable symmetric cusp; (*c*) unstable symmetric cusp; (*d*) limit point and unstable symmetric cusp; _______ stable paths; - - - - unstable paths.

new stable equilibrium path with a change of symmetry. The unstable cusp represents a disintegration of the profile with the same change in symmetry. Calculations of fourth-order changes in energy are necessary to discriminate mathematically between these cases, but the analysis of meniscus stability has not been taken to this stage. Nevertheless, some cases of unstable cusp transitions are easily recognisable from the experimental instability observed.

Figure 1*d* illustrates the bifurcation pattern that applies to a number of meniscus-stability problems as will be seen in this discussion. Here a rising loading path comes to a limit-point transition at the load maximum, and an unstable symmetric cusp bifurcation also occurs on this path. The position of the cusp bifurcation varies depending upon an additional parameter, which in this context is a geometrical parameter such as an orifice radius. Interest centers on whether the cusp occurs before or after the limit-point transition. This bifurcation pattern is familiar in other contexts; see for example Michael, Norbury & O'Neill (1974, 1975) and Thompson & Hunt (1977). When the cusp bifurcation occurs at the limit point and an additional imperfection parameter is present the compound bifurcation becomes a hyperbolic umbilic catastrophe.

All these forms of transition are manifestations of a general principle of bifurcation that is the unifying theme of all this work. According to it, a change in stability occurs when an alternative "equilibrium" profile becomes possible by a small perturbation satisfying the boundary conditions and occurring without change of the control parameter.

The mathematical formulation of problems in meniscus theory must depend upon some modeling of the edge of the meniscus at a contact line. Statical and dynamical studies of the three-point contact were the subject of a recent review in this series by Dussan V. (1979), which showed the difficulties in the way of a full understanding of the mechanics of the meniscus near such contact limits. We shall not enter into any further discussion of spreading at a three-point contact line or of the related problem of the stability of three-phase systems such as the sessile lens. However, there are some important situations in which an effective mathematical edge condition is evident. For example, a film supported on a wire rim is well modeled by taking the rim to be mathematically a line, which provides a prescribed edge to the meniscus, and from which the meniscus can emanate at any angle of inclination. For a membrane, this leads us to the Plateau problem of minimizing the surface area enclosed by the prescribed rim. In another context, drops emanating from an orifice with sharp edges can be well modeled in the same way with the meniscus emanating from the edge, again without restriction on the angle of inclination. Good agreement with experiment can be achieved in this case. The situation is less clear when the meniscus attaches to a geometrically smooth surface, as for example when a pendent drop hangs from a horizontal plane. In this case the contact angle is subject to both static and dynamic hysteresis effects and is not uniquely defined. However, there seems no doubt that the mathematical step of studying the form of equilibria for prescribed values of the contact angle is worthwhile. One can then hope to model the effects of hysteresis by a study of the variation of the equilibria with contact angle. This process can lead to an effective interpretation of some effects, as was shown by Taylor & Michael (1973).

We conclude this introduction with a few general remarks about the mathematical methods employed in the study of meniscus stability. In terms of the potential-energy function, the Young-Laplace equation for equilibria may be derived using an Euler-Lagrange equation as a necessary condition for the energy to be stationary to small variations. The equilibrium is then seen to be stable if all allowable small perturbations give an increase in the energy, whilst it is an unstable equilibrium when there exists any one perturbation that reduces the energy. A proof of instability is clearly easier to obtain than a proof of stability since it requires us to produce just one displacement that lowers the energy, whereas to prove stability we must consider all the allowable displacements. The development of the calculus of variations has shown the importance of specifying the class of displacements within which one seeks to prove stability, and many classical examples exist in which a functional may be shown to be minimal within one class but not within a more general class of displacements. In our context all displacement functions must by their nature be continuous. But we note three features of a displacement system that need to be considered in a proof of stability. The first is the degree of continuity in the displacement system. Most investigations of the second variation of the energy in a small displacement assume the displacement to be perfectly smooth, in which derivatives of all orders exist. The second feature is the distinction usually drawn between strong and weak displacements, the former being displacements of small amplitude but not small in their gradients, whilst weak displacements are such that the displacement function and all its derivatives are small quantities. The third relevant feature is the global stability. For this, we recognise that an equilibrium that is stable to small-amplitude displacement may not continue to be stable when disturbed by larger-amplitude displacements.

In the context of meniscus stability, however, these considerations do not appear to lead to any significant modifications in the criteria for change of stability from those provided by the classical theory of the weak smooth variation. This is borne out by many experimental studies that confirm the criterion for change of stability so obtained. Confirmation also comes in the form of mathematical results. In the matter of the continuity of the displacement we find by rounding arguments that, where a given profile minimises the energy function in the class of displacements with continuous first derivatives, it must also do so in the class that allows displacements with only piecewise continuous first derivatives. Concerning the relations between strong and weak displacements, it is intuitively clear that small displacements having large gradients result in larger increases in surface area, and hence of surface energy, than weak displacements. This is reflected in the application of the Weierstrass Sufficiency Theorem [see, for example, Pars (1962)] to these problems, which ensures stability within a class of strong displacements. These results point to the conclusion that the stability of the meniscus is most vulnerable for small displacements to weak smooth variations within the theory of the second variation. The Sufficiency Theorem also has a bearing on the global stability since a minimizing property of the equilibrium profile can be ensured not only for small perturbations, but also for a displacement contained within a nonsingular field of extremals. However, this does not mean that the meniscus will not be unstable to sufficiently disruptive disturbances. The pendent drop, for example, is seen to lose its stability in axisymmetric form by the limit-point instability illustrated in Figure 1*a* in which the load parameter is, say, the volume of the drop. A large statical displacement, which preserves the volume, from the stable configuration at $A$ to a suitable configuration near to the unstable configuration at $B$ will mean that the drop thereafter detaches from its support and does not return to the configuration at $A$. Joseph (1976) and Dussan V. (1975) have established disturbance conditions sufficient to ensure that the system cannot return to an initial configuration that may be stable to small disturbances.

## 2 MENISCUS STABILITY IN THE ABSENCE OF AN EXTERNAL FORCE FIELD

We turn now to discuss the stability of the meniscus in case (*a*) above, in which it supports a constant pressure difference. The axisymmetric profiles obtained in this case are well known and are described in detail in the treatise of Plateau (1873). These figures are the plane, sphere, circular cylinder, catenoid, unduloid, and nodoid. The plane and the cylinder are the only two-dimensional figures. In Plateau's experiments these figures were produced by soap films and also by suitably suspending quantities of oil in a mixture of alcohol and water adjusted to produce neutrally buoyant conditions. To Plateau must also be credited the first major contributions to the study of stability, as we find in Chapters IX and X of his treatise a detailed discussion of these figures in which the principles governing changes in stability are set out.

The sphere, which is the only closed figure in this group, is evidently stable when it encloses an incompressible fluid since the sphere gives the absolute minimum of the surface area enclosing a given volume. We use this example to illustrate the significance of the control parameter. Plateau's experiments were conducted typically with oil filling the interior of the figure, so that instability when it arises must develop without change of the interior volume. This we describe as volume control. We may, however, consider alternative controls, in particular, the internal pressure. In this case, it is seen that the sphere is unstable since an increase in radius results in a decrease of surface-tension stress whilst the difference in pressure across the surface is unchanged. This result is modified in the case of the spherical cap bubble or soap film grown on a circular base of fixed radius $c$. This was discussed by Searle (1934). Here it is seen that if the bubble is grown from zero volume the radius initially decreases from infinity to $c$ when the bubble is hemispherical, and thereafter increases again. In the equilibrium at radius $r$ the internal excess pressure $p^*$ is $2T/r$ where $T$ is the effective surface tension, so that $p^*$ rises to a maximum of $2T/c$ and then falls as successive equilibria are traced out. The volume continues to increase monotonically and when the bubble is volume-controlled the equilibrium is stable throughout for the same reasons as those given for the complete sphere. However, when the bubble is pressure-controlled, the maximum of $p^*$ is associated with a change of stability. Starting from zero volume, the equilibria are stable until the pressure maximum is reached, after which they become unstable. This is the first example in this account of the limit-point instability referred to earlier. It also illustrates the principle of bifurcation since a small first-order perturbation from one equilibrium to another one occurs without change of the boundary conditions or the control parameter $p^*$, at the maximum of $p^*$. Similar remarks apply in the corresponding two-dimensional case where the plane section is a part circle. Even so, the stability in this case is different because the cylindrical forms are sensitive to perturbations periodic along their length. These perturbations were examined by Majumdar & Michael (1976) for two-dimensional pendent drops, and the analysis applies to this case also. If a sinusoidal perturbation with real axial wave number $k$ is introduced, equilibria of this form begin to occur at the threshold value $k=0$ where the pressure is stationary. This is an unstable cusp bifurcation, which occurs in both cases of volume and pressure control. In the case of volume control, it produces a change of stability at the pressure maximum when the meniscus is otherwise stable to two-dimensional disturbances. When it is pressure-controlled, we get a coincidence of this bifurcation with the limit-point instability remarked on earlier.

The two cases of volume and pressure control are directly analogous with the two forms of control known as rigid and dead loading respectively in the theory of structural stability. In that context, the stabilities have been described respectively as internal and external stability by Thompson (1979). These terms are also appropriate here and will be used in this account.

The axisymmetric figures supporting zero pressure difference are those with zero mean curvature, namely the plane and the catenoid. Plateau produced a catenoid in stable equilibrium by suspending oil on two circular rings and adjusting the quantity of oil so that the interface across the rings was a plane. He found that the catenoid so formed was at the limit of its stability when the ratio of the distance apart of the rings $h$ to the radius $a$ reached a value approximately 0.663. He recognised also that for $(h/a)<0.663$ there is an alternative catenoid solution not producible in the experiments, and that the limit of stability is reached when the two solutions coalesce. Taylor & Michael (1973) photographed an open-ended catenoid soap film near the stability limit, and showed mathematically the instability of the catenoid solution standing nearer the axis of symmetry. The form a soap film may take for a given geometry of support is of course not unique. Taylor & Michael observed an alternative stable equilibrium for a soap film suspended from two rings consisting of catenoid sections connected by a center plane. The stability of this configuration was calculated by them and confirmed by observations. With open-ended arrangements the results given refer to external stability when perturbations to the film can develop without differences in mean pressure arising. Internal-stability criteria for volume control are different. This case arises in liquid bridges between circular discs under neutrally bouyant conditions and also for catenoid soap films between circular discs enclosing air, which for this purpose behaves incompressibly. Erle et al (1970) gave an experimental and theoretical discussion of this following earlier experiments of Terquem (1881) and theoretical work using conjugate-point analysis by Howe (1887) and Hormann (1887). Erle et al showed that the catenoid is now stable or unstable according as the ratio of the end-plate separation to neck diameter is less than or greater than 2.239. This implies that, if the ratio of end-plate separation to diameter is greater than 0.47, both catenoid solutions noted by Plateau become stable. It is interesting to note that the pressure minimum observed in these experiments, which would have been a point of instability under pressure control, is not significant in this case, and the equilibrium solution remains stable until a position of stationary volume is reached on the equilibrium loci. Erle et al point out that this result refutes the conjecture made by Maxwell (1875) that all the catenoid solutions would be stable in this case.

Perhaps the most well known of Plateau's studies of stability is his experimental and theoretical investigation of the stability of the circular cylinder. It was found that a cylinder of oil could be formed between two circular ends in stable equilibrium, provided the ratio of the length to the diameter did not exceed a critical value. The critical value of this ratio, $\pi$, was given first by Beer (1855). It was shown by Plateau to give the length at which an axisymmetric bifurcation from the cylinder begins to appear when the volume is the control parameter. This may be seen by considering the Young-Laplace equation for small axisymmetric displacements from the cylinder. If $a$ is the cylinder radius, and $r$ and $z$ the radial and axial coordinates, the equation in dimensionless form is

$$\frac{r''}{(1+r'^2)^{3/2}} - \frac{1}{r(1+r'^2)^{1/2}} + \mu = 0, \qquad \left(r' = \frac{dr}{dz}, r'' = \frac{d^2r}{dz^2}\right),$$

where $\mu = p^*a/T$ is the dimensionless form of the internal excess pressure $p^*$, and lengths are made dimensionless with respect to $a$. We write $r = 1 + \zeta$ and linearise the equation in $\zeta$. Thus

$$\zeta'' + \zeta + \nu = 0,$$

where $\nu$ is a small perturbation in the internal pressure.

Plateau assumed that $\nu = 0$ so that $\zeta = A \sin z$ for a solution in which $\zeta = 0$ at $z = 0$. Values of the length, $l$, at which $\zeta = 0$ again are $l = \pi, 2\pi, \ldots$ However, the constancy of volume requires $\int_0^l \zeta \, dz = 0$, so that the first bifurcation is at $l = 2\pi$, when the volume is controlled. It is easily verified that the first bifurcation mode in which $\nu \neq 0$ occurs where $\tan(l/2) = l/2$, the first solution of which has $l > 2\pi$. This result is particular to the case of volume control. The experimental verification by Plateau has recently been reaffirmed in more refined experiments by Mason (1970). If the internal pressure is controlled, the solution requires $\nu = 0$, but there is now no volume constraint so that the first bifurcation is at $l = \pi$. This result was recognised by Maxwell (1875) and Searle (1934). The bifurcation from the cylinder in the incompressible case was recognised by Plateau to be the manifestation of another locus of equilibria, for varying volume and gap width, of unduloidal form. But further study of these bifurcations is necessary to establish the structure of the bifurcation. Haynes (1970), in an interesting comment, has compared the surface area of unduloidal and spherical-cap equilibria to the surface area of the cylinder of equal volume for cylindrical lengths below the critical. It is found that equilibrium in the form of two spherical caps of the same radii (one on each circular end) is attainable from the cylinder in the sense that the surface energy is lower than that of the cylinder. The intervening unduloidal equilibrium, however, has a higher energy, and it is suggested that this energy barrier prevents the cylinder from changing into spherical caps until the length reaches $2\pi$, at which point the energy barrier is removed. It was also noted that when this happens the spherical-cap form with equal radii is not attained from the cylinder because its energy is lower and that, in consequence, axial movement of the fluid must take place, which results in spherical-cap equilibrium with caps of different radii.

Plateau went on to discuss the significance of this instability in breaking up a long cylinder into a succession of drops and droplets in a periodic array. He observed that the equivalent length of cylinder associated with each period was greater than $2\pi a$, and further that the greater the resistance in the fluid by way of viscosity the longer the equivalent length appeared. These matters were taken up by Rayleigh who recognised this mechanism of instability as being responsible for the breakup of a fine jet of fluid, the analogy with Plateau's result being exact when the jet has a uniform velocity profile and when dynamical effects exterior to the jet are negligible. Perhaps for this reason the discovery of this instability has been widely credited to Rayleigh in the fluid-dynamics literature. Rayleigh (1879a) gave an analysis of the axisymmetric normal modes of vibration of the cylinder. He thus showed that when inertia is taken into account the exponential growth rate for axisymmetric modes is greatest for a wavelength $4.508 \times 2a$, which is significantly greater than the lowest unstable wavelength $2\pi a$. The normal-mode analysis was extended by Rayleigh (1879b) to include nonaxisymmetric perturbations. Plateau's experimental observation of the instability to axisymmetric perturbations is confirmed by Rayleigh's finding that all asymmetric modes are stable. Plateau's observations on the effects of increasing viscosity on the varicosity of the cylinder were also clarified by Rayleigh (1892a), who showed, by a further normal-mode analysis, that when viscous rather than inertia effects are dominant the mode of maximum growth rate has zero wave number. This illustrates the observations that viscous threads give way to attenuation "at few and distant places," rather than to a varicose structure. As a final remark on the cylinder we note another paper by Rayleigh (1892b) which gives the wavelength of maximum growth rate, $6.48 \times 2a$, for a cylindrical cavity formed in a fluid, which is assumed inviscid, as when a jet of air is directed through water.

Returning to Plateau's treatise we come to his account of the stability of the unduloid and nodoid. Plateau described the transition to instability that occurs in an unduloid section produced by oil suspended between equal circular ends in which the central radius is a maximum as shown in Figure $2a$. The unduloid loses stability when the length of the section extends beyond the interval between consecutive minimum cross sections. The bifurcation of the equilibrium path at this point is evident since the profile can be displaced a small amount along the axis without violating the end conditions and without change of the volume or the internal pressure. It is thus a point of instability for both volume and pressure control, although the former was the only case considered by

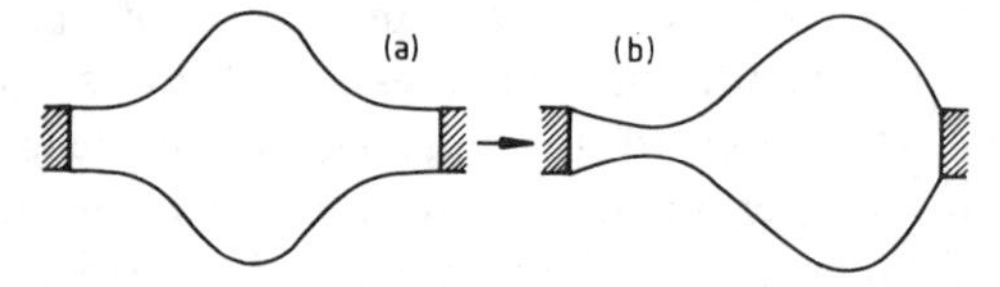


*Figure 2* Sketch figures of the instability of the unduloid. (*a*) The limit of stability; (*b*) the figure after transition.

Plateau. The bifurcation gives a new locus of equilibria whose profiles are no longer symmetric about the center plane between the supports, and the instability develops spontaneously by the central bulge in the profile moving to one end. Instability of this form appears in a number of other contexts and it may be conjectured by analogy with other problems and from the experimental behavior that this is an unstable symmetric cusp bifurcation.

With regard to the nodoid, Plateau remarked that for a section of a nodoid convex towards the axis of symmetry and set up between two equal circular discs two solutions are possible for given volume, radii, and spacing of the discs as sketched in Figure 3*a*. The outer solution is stable but loses its stability when it coincides with the inner solution, in which case the fluid separates axisymmetrically on to the discs. Plateau conjectured from his observations of the figure formed on a single circular-wire support that the stable branch of the nodoid may be continued to enclose almost a complete node, and his observations suggest that the stability of this is ultimately broken by an asymmetric movement of the fluid around the ring, resulting in its collecting at one position on it. Further observations on the stability of the nodoid formed between circular discs, and concave towards the axis, are described by Plateau. In this case, as the discs are brought together the figure remains stable until the profile becames tangential to the discs at the point of support as sketched in Figure 3*b*. At this point instability causes the oil between the discs to collect at one position on the rim. Plateau further noted that in this case a stable asymmetric figure may continue to be followed as the spacing of the discs is further reduced. Mathematical analysis of the stability in this case remains to be done, but we may conjecture from these observations that this instability is a stable symmetric cusp. Plateau observed that this instability together with that of the node formed on a circular wire are the only cases in this category in which the figure is observed to give way to a nonaxisymmetric form. Gillette & Dyson (1971) have given a mathematical analysis of

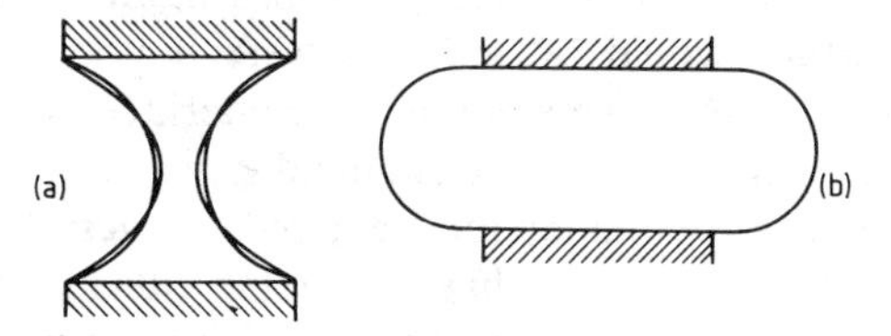


*Figure 3* Sketch figures of the nodoid. (*a*) Solutions for the nodoid convex to the axis; (*b*) concave nodoid at the limit of stability.

the stability of a liquid bridge between two circular discs to axisymmetric disturbances using conjugate-point analysis. Their calculations confirm the existence of a minimum volume of the bridge, this being a consequence of the limit-point instability. The maximum volume is, however, limited by the asymmetric instability described by Plateau. The problem is also of interest in connection with the growth of crystals in liquid zones. Further study in this context is given by Carruthers & Grasso (1972a, b).

We conclude this discussion of axisymmetric menisci in category (*a*) with the remark that, although in many of these cases we can make a strong conjecture as to the structure of the bifurcations involved by analogy with other problems and from descriptions of experimental behavior, considerable further mathematical work is necessary to categorize these instabilities conclusively.

## 3 MENISCUS STABILITY IN A UNIFORM GRAVITATIONAL FIELD

We consider finally the stability of equilibria in case (*b*) of our introduction in which an external gravitational field has an influence. In such problems variations of gravitational pressure are comparable with the stress produced by surface tension and the length scale of the problem is the capillary length $(T/\rho g)^{\frac{1}{2}}$, where $\rho$ is a typical density difference across the meniscus and $g$ the gravitational acceleration. All lengths referred to in dimensionless form in this section are in units of the capillary length.

We refer first to the studies of Duprez (1851, 1854) on the effect of surface tension in stabilizing a horizontal-plane interface against an inversion of density. In Duprez's experiments a vessel containing olive oil was placed with its mouth vertically downwards in a vessel containing alcohol and water, this mixture being initially denser than the olive oil. The surface of separation was initially horizontal and stable. Alcohol was then added until the mixture became lighter than the olive oil, so that the interface would be unstable were it not for the effect of surface tension in maintaining stability. The interface ultimately became unstable when a sufficiently large unstable density difference was established, in which case the lighter fluid entered the vessel in one part of the orifice and drove out the olive oil in the other. Criteria for this instability were given by Plateau (1873) and by Maxwell (1875). The criterion depends on the size and shape of the vessel opening and the assumed conditions at the edges. Let us consider the orifice to be a rectangle of sides $h, k$, given by $x=0, h, y=0, k$. If the interface, taken at $z=0$, is displaced vertically upwards by a small amount $z(x, y)$, assuming the mean pressure to be unaltered the small additional upward surface stress to the first power in $z$ is $T(\partial^2 z/\partial x^2+\partial^2 z/\partial y^2)+\rho g z$. When this quantity has the same sign as $z$ the displacement is accentuated and the equilibrium is unstable. When it is, in all displacements, of the opposite sign, the equilibrium is stable. The marginal case in which this quantity is zero gives the condition for an alternative equilibrium of the surface to be formed, and hence a bifurcation of equilibrium paths. Assuming that the interface adheres to the edge of the vessel, we require examination of displacements for which $z=0$ at $x=0, h, y=0, k$. Such displacements are described in terms of the complete set of Fourier components of the form $z=c\sin px\sin qy$, where $ph=n\pi$, $qk=m\pi$, and $m$ and $n$ are integers. A further constraint in Duprez's experiments is that the interior volume is unchanged, and for this one of $m$ and $n$ must be an even integer. The condition for stability in each component is $\pi^2 T\{(n/h)^2+(m/k)^2\}>\rho g$. Thus if $h>k$ all components are stable when $\pi^2 T(4/h^2+1/k^2)>\rho g$. For a two-dimensional orifice where $h\to\infty$ we have the condition $k<\pi$ in dimensionless form for stability. For a circular orifice a similar calculation shows that the limiting value of the radius $r$ for stability is given by $r=\xi$, where $\xi=3.831\ldots$ is the first zero of the Bessel function $J_1(r)$. In each case the volume constraint imposes a sinuous structure of a whole wavelength on the critical displacement at bifurcation, in the rectangular case by the dependence on $\sin n\pi x/h$, and in the circular case by dependence on the azimuth angle $\theta$ like sin or cos $\theta$. It may be seen that these calculations of the lowest critical dimensions for transition are not altered by the addition of a small change in mean pressure in the bifurcation. The critical displacement thus occurs without change of volume or mean pressure. The bifurcation therefore occurs under both volume and pressure control. In the case of pressure control, however, the stability limits are different because the volume constraint is removed. For the rectangle the condition is then $1/h^2+1/k^2>1/\pi^2$ and for the circle the limiting radius is given by $\xi=2.404\ldots$ the first zero of $J_0(r)$ as was seen by Pitts (1974) and Michael & Williams (1976). These criteria apply when the edge of the interface is held in position. When it is free to move, the interface is unstable as soon as the density inversion takes place. In Duprez's experiments the interface was in fact formed in the interior of the inverted cylinder and it joined the vertical walls at its edge. The observations suggest a statical resistance to the movement of the contact line that is sufficiently large to hold the edge in place until the stability criteria apply. Stability results for the related problem of the two-dimensional meniscus meeting vertical walls using the alternative-edge condition of a prescribed contact angle were given by Concus (1963, 1964), for the case of pressure control. Using conjugate-point analysis he showed that such profiles are unstable when the curvature changes sign, and gave a calculation of critical Bond number as a function of contact angle above which profiles with curvature of one sign are unstable. The results do not directly relate to the Duprez experiments which require the constant-volume constraint. However, the internal stability was considered by Tyuptsov (1966) who also gave calculations of the critical Bond number as a function of contact angle for an axisymmetric meniscus in a circular cylinder. Progress in understanding the relevance of fixed-end-point conditions at a wall was made by Karasalo (1979), who considered the stability of a liquid-vapor annular interface at zero contact angle. It was shown that stability results obtained by using fixed-end-point conditions are appropriate provided the analytic continuation of the meniscus interface does not penetrate the wall. It was also shown that when this condition is not satisfied the meniscus is unstable.

The stability of the plane interface just described has a bearing on the more general problem of the stability of pendent drops, or sessile bubbles, in the plane and axisymmetric cases because the planes represent limiting cases of drops. They show that, at least in these limits, the internal stability is broken by modes that are not of the simplest symmetry, that is, modes that are not plane modes in the two-dimensional case, and modes that are not axisymmetric for the axisymmetric pendent drop. These considerations lead us on to discuss the stability of the pendent drop.

The two-dimensional pendent drop is of interest as representing the two-dimensional form in which fluid will hang from a horizontal rail or edge. Profiles of this drop were studied by Neumann (1894) for zero contact angles and more recently by Pitts (1973) for a series of prescribed contact angles. These studies show that the suspended weight, as measured by the cross-sectional area $A$, grows to a maximum value and then declines as illustrated in Figure 4*a*. Pitts made a study of the second-order changes in energy of the drop in a two-dimensional, weak, smooth, small displacement, symmetric about the center line of symmetry, which preserves the value of $A$ unchanged. He thus gave a direct proof of the instability of the profiles after passing the position of maximum suspended weight, when the volume is the control parameter. The instability is clearly a limit point or fold. Corresponding results for the drop suspended from a horizontal orifice or rail were given by Majumdar & Michael (1976). Calculations here show again a maximum

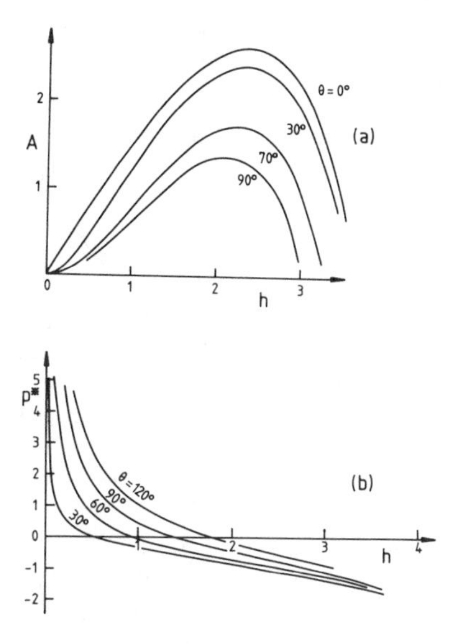


*Figure 4* Plane-pendent-drop data for prescribed contact angles. (*a*) The variation of cross section area $A$ with height $h$. Pitts (1973); (*b*) the variation of excess pressure $p^*$ with height. Majumdar & Michael (1980).

of suspended weight beyond which the profile is similarly two-dimensionally internally unstable. Majumdar & Michael also calculated the internal excess pressure $p^*$ which also shows a maximum value in this case. The pressure maximum occurs before the area maximum in the growth of the drop and it gives a limit-point change of external stability where the pressure is the control parameter. Figures 5*a* and *b* show the typical variation of $A$ and $p^*$ with the growth of the drop from a plane. The pressure maximum here has a greater significance than is indicated by the limit-point instability alone. The analysis of Duprez's experiments indicated an instability sinuous in the direction along the drop generators and this form of instability must therefore be significant for the drop that is near plane. Majumdar & Michael examined this profile for the remaining forms of displacement, in particular those of sinuous variation along the length of the drop, which are here described as three dimensional. It was shown that three-dimensional bifurcations begin to appear in the limiting case of zero wave number when the pressure maximum is reached. This bifurcation, which is a symmetric cusp, takes place without any first-order change in the volume or pressure of the drop, and will therefore give a change of both internal and external stability. Figure 5*a* shows that in the case of internal stability this cusp instability preempts the limit-point instability at the maximum of $A$. For external stability there is a coincidence of transition of the limit point and the three-dimensional instability at the pressure maximum. Figure 5*a* illustrates that the pressure maximum occurs early

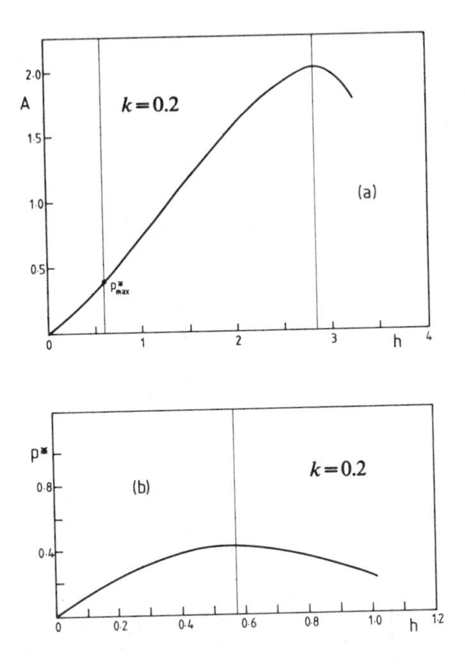


*Figure 5* Plane-pendent-drop data for prescribed orifice width $k=2.0$. ($a$) Variation of cross-sectional area $A$ with depth $h$; ($b$) excess pressure $p^*$. Majumdar & Michael (1976).

in the growth of the two-dimensional drop, and this explains why fluid hanging from a horizontal rail is not usually seen to take up a two-dimensional form.

A fuller understanding of the equilibrium chosen requires a further examination of the three-dimensional bifurcation to ascertain whether it is a stable or unstable symmetric cusp. These results enable us to put the analysis of the two-dimensional plane slit in perspective. As the orifice width $k$ is increased the pressure maximum in Figure 5$b$ approaches the origin and reaches it when $k=\pi$, the critical width seen earlier for the slit. This is consistent with the earlier analysis of the perturbations from the plane-rectangular interface, which showed an alternative equilibrium without change of mean pressure and symmetric about the center line in the limit as $h\rightarrow\infty$. Majumdar & Michael (1980) extended the study of Pitts (1973) to give the variation of pressure in the plane pendent drop with prescribed contact angle. Typical calculations are shown in Figure 4$b$. The pressure $p^*\rightarrow\infty$ as the depth approaches 0, and falls monotonically with increasing $h$. A decline in $p^*$ with increase in $A$ may be expected to correspond with the behavior of the drop past the pressure maximum in the previous case. This was confirmed, and as a consequence the drop is externally unstable in the plane-symmetric mode and unstable in the three-dimensional sense both externally and internally. The drop is thus unstable from its initial stages of growth even though it is internally stable to plane-symmetric perturbations up to the volume maximum.

We turn now to the axisymmetric pendent drop. Early work on this drop is summarised by Bakker (1928), and a detailed mathematical study of the shape of a pendent drop has recently been given by Concus & Finn (1979). It has been seen by a number of authors, for example, by Lohnstein (1906a, b, c, 1907), Pitts (1974), and Boucher & Evans (1975) that, for a drop suspended from a horizontal plane assuming a prescribed contact angle, or for a drop suspended from the end of a vertical tube, the volume reaches a maximum value. A mathematical study of stability was given by Pitts. He considered the second-order variation in the energy for small, axisymmetric, smooth, weak perturbations. Within this class Pitts (1974) showed (*a*) that the pendent drop suspended at constant contact angle becomes internally unstable at the volume maximum and (*b*) that for a drop suspended from a tube the point of maximum pressure is a transition point for external stability. Pitts (1976a) showed further that in the latter case the drop is internally unstable after the volume has reached its maximum. These transitions are evidently further examples of the limit point, and have been recognised by Padday & Pitt (1973), Boucher et al (1976), Shoukry et al (1975). The results give support to the view advanced in this review that small, weak, and smooth perturbations provide the most sensitive test of meniscus stability since many experiments on the internal stability of the drop suspended from a tube have been made, such as those of Kovitz (1975), which confirm that the drop is, in general, stable up to the point of the volume maximum, and becomes unstable at this point in the axisymmetric mode. Levin et al (1976) suggested the measurement of maximum height of a pendent drop in stable equilibrium as a method of measuring surface tension for liquid-liquid interfaces. However, it is necessary to examine the asymmetric modes to complete the picture. This was done by Michael & Williams (1976) for the drop suspended from a tube. They formulated the eigenvalue problem for the appearance of asymmetric bifurcations with angular variations like $\cos m\theta$ ($m=1,2,3,\ldots$) about the axis of symmetry of the drop. Bifurcation in the mode $m=1$ is shown to arise when the profile becomes horizontal at the point of support. Figure 6*a*, giving the variation of suspended volume with depth for given orifice radius $a$, shows the effect of this mode on the internal stability. The limit-point axisymmetric instability arises at the volume maximum whilst the $m=1$ mode becomes unstable at the point of contact with the envelope. If the drop is grown from zero volume the limit point is reached first provided $a<3.219$. For $a>3.219$

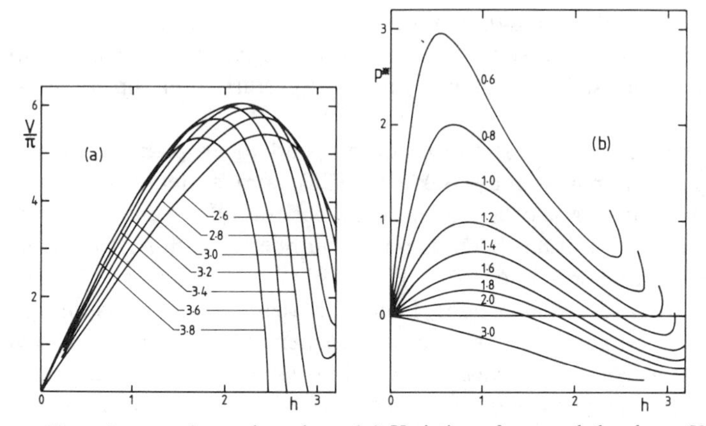


*Figure 6* The axisymmetric pendent drop. (*a*) Variation of suspended volume $V$ with depth $h$; (*b*) variation of excess pressure $p^*$. Michael & Williams (1976).

the asymmetric mode comes first, though the range of values of $a$ for this case is small since we have seen that a pendent drop cannot be formed for $a > 3.831$. At this value of $a$ the point of contact with the envelope in Figure 6*a* reaches the origin and we recover the condition of instability for the plane interface covering the orifice. Again we see that the asymmetric modes occur under conditions of both volume and pressure control. However, in the latter case Michael & Williams showed (see Figure 6*b*) that the axisymmetric instability at the pressure maximum is always reached before the asymmetric instability. The pressure maximum reaches the origin in Figure 6*b* when $a = 2.405\ldots$, the first zero of $J_0(a)$, as noted earlier. Michael & Williams showed also that in this problem the modes $m = 2, 3, 4$, are not significant in the sense that none of them arises before the $m = 1$ mode. This change in the mode of internal instability has been demonstrated by R. Collins using a set of tubes of fine gradations in their radii, for drops of water and liquid paraffin in air. However, detailed experiments to confirm this change of mode have not yet been performed to the author's knowledge. In further studies Shoukry et al (1975) made calculations and experimental observations of the axisymmetric limit-point detachment of pendent drops which wet the supporting surface for the cases of spherically and conically shaped supports.

The stability of the sessile drop or pendent bubble may be considered according to the principles already outlined. Both plane and axisymmetric sessile drops have been considered, but we here restrict our account to the axisymmetric case. Stability to axisymmetric perturbations was discussed by Padday & Pitt (1973) and a further discussion was given by

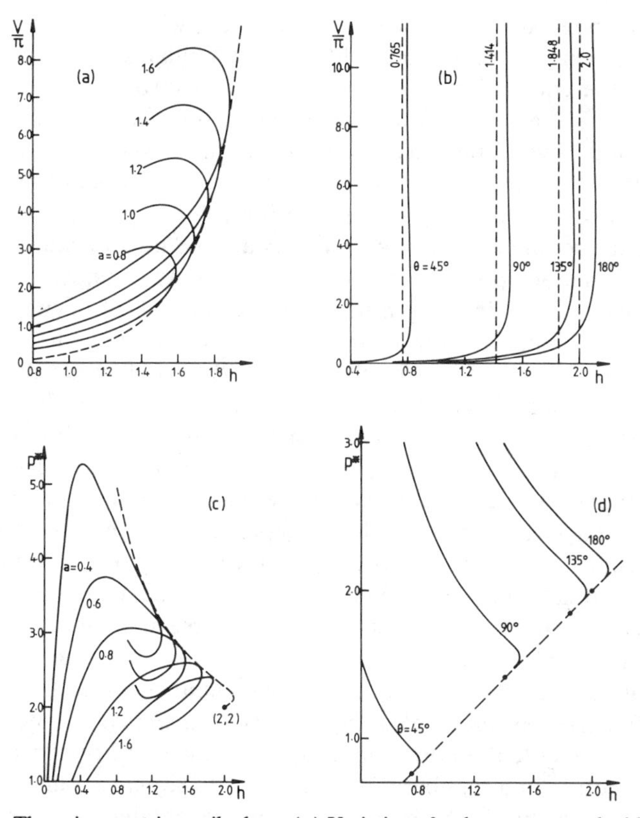


*Figure 7* The axisymmetric sessile drop. ($a$) Variation of volume supported with depth $h$ for given base radius; ($b$) volume supported at given contact angle; ($c$) pressure variation for given base radius; ($d$) pressure variation for given contact angle. Michael & Williams (1977).

Michael & Williams (1977). Equilibrium paths are given in Figure 7*a*, *b* *c* and *d*. For the drop supported on a base of given radius Figure 7*a*, *c* applies. For internal stability the volume maximum in Figure 7*a* signifies an axisymmetric limit-point instability. However, when the drop is grown from zero volume the contact with the envelope is reached first. This is again a point of bifurcation for the first asymmetric mode $m=1$ irrespective of the base radius. For external stability Figure 7*c* shows that the pressure maximum is reached before contact with the envelope so that the instability first occurring is the axisymmetric limit point at the pressure maximum. For prescribed contact angle Figure 7*b* shows that the volume no longer has a maximum and the drop remains internally stable. When the drop is pressure-controlled, the absence of a stationary point in Figure 7*d* shows that there will be no transition of stability in axisymmetric modes. Michael & Williams conjectured from this that the equilibrium was stable, but further study shows that the curves should be regarded as equivalent in the pendent-drop case to pressure loci beyond the maxima, and therefore unstable paths. Padday & Pitt drew the same conclusion in this case since the equilibrium paths represent declining pressure with increasing volume.

Other interesting axisymmetric profiles are the liquid bridges and the rod in free-surface profiles. Studies of the stability of liquid bridges, or liquid zones formed in the gap between circular ends of aligned vertical tubes, are of importance in connection with crystal-growth techniques and have been studied by Coriell et al (1977), Coriell & Cordes (1977), and Surek & Coriell (1977). Coriell et al considered the stability of the liquid bridge and gave calculations of the stability limit for equal end radii for a bridge of given volume. The variation of critical length with Bond number was calculated and close agreement was obtained with experimental measurements. The asymptotic limit for large radii was also discussed, and a study was made of the stability of horizontal zones for small Bond numbers. Calculations of the maximum length of a molten zone with zero contact angle at the freezing end are given by Coriell & Cordes taking account of both axisymmetric limit-point instability and asymmetric modes for wave number $m=1$. The study by Surek & Coriell examines also the stability of the crystal growth in this context. Padday & Pitt (1973) have discussed the stability of rod-in-free-surface profiles in a number of cases. Pitts (1976b) gave a mathematical study of axisymmetric stability of the meniscus obtained when a vertical circular rod is drawn out of a horizontal fluid surface when the meniscus is assumed to attach to the circular rim of the tube. This gives another limit-point transition in which the height of the tube above the free surface may be regarded as the load parameter. Below the critical height there are two profiles, one stable and one unstable, which coalesce at the point of transition at the maximum-height position. Asymmetric instability in these cases remains to be studied. The inverted profile of the rod in free surface is the profile of a hole in a horizontal sheet of fluid. A study of the behavior of holes in sheets of fluid was made by Taylor & Michael (1973) and we end this account with some remarks on that work.

The stability results described here are based upon the assumption that contact angles at a plane may be prescribed and can receive only small changes in critical displacements. It is well known, however, that the contact angle exhibits both statical and dynamical forms of hysteresis, and on this account the behavior can change significantly from that predicted by stability theory based on these assumptions. Some insight into this was obtained by Taylor & Michael (1973) in their study of the behavior of holes in thin fluid sheets under the action of surface tension. Observations of the disintegration of thin fluid sheets in free fall, such as those of Dombrowski & Fraser (1954), suggest that sufficiently large holes in a thin sheet quickly expand into larger holes. On the other hand very small holes appear to close up. For example, experiments in shooting small ball bearings through a free-falling sheet of water failed to produce a hole that persisted, the hole closing over after the passage of the ball bearing. Taylor & Michael conjectured that there exists a critical hole size above which the hole will open out and below which it closes up. Some simple models of holes bear this out. A hole in the form of a right circular cylinder of radius $a$ in a plane sheet of thickness $t$ suffers a decrease in surface area if the hole closes provided $a<t$. But if $a>t$ the surface area is reduced when the hole opens out. The catenoid soap film suspended between two rings is another simple model. The stability characteristics given by Plateau were discussed earlier. The relation between the depth $d$ and the minimum radius $r_0$ are shown in Figure 8$a$. The lower branch of the curve up to the maximum in $d/a$ represents the unstable-catenoid solution. Taylor & Michael suggested that the unstable equilibrium models the critical hole size. For example, taking the family of all catenoids attached to the support rings and rotating them about the axis of the rings, we see that the energy of the unstable-equilibrium catenoid is a maximum in that family. If disturbed inwards it would then collapse to the axis, closing the hole, whilst a displacement outward moves the film towards the stable position thus

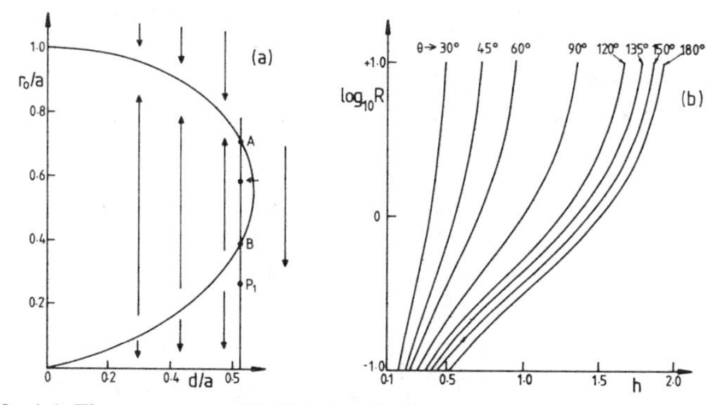


*Figure 8* ($a$) The open catenoid. Relation between minimum radius $r_0$ and depth $d$, scaled with the radius $a$ of supporting rings. ($b$) Holes in a horizontal sheet of depth $h$. Relation between hole radius $R$ and depth for unstable equilibria. Taylor & Michael (1973).

enlarging the hole. Unfortunately the model is not exact for a hole in a plane sheet because the catenoid does not have any asymptotes perpendicular to the axis and cannot be fitted into the sheet. However, with gravity acting vertically, profiles with one horizontal asymptote become available, and this then gives a model for critical hole sizes in a plane layer standing on a horizontal base. Assuming the angle of contact to be prescribed, the relation between hole radius $R$ and height $h$ of the layer is shown in Figure 8*b*. These curves, being analogous to the lower branch in Figure 8*a*, suggest that the equilibrium is unstable and this was proved by Taylor & Michael. This suggests that in cases where the contact angle is well defined the appropriate curve in Figure 8*b* gives the critical hole radius as a function of depth. Taylor & Michael describe two series of experiments designed to test this hypothesis. In the first, holes were made in a horizontal layer of mercury, beneath water and standing on a glass disc. In this experiment, all holes either opened or closed. The results of the experiment show, in Figure 9, a fairly well-defined division between opening and closing holes which follows the loci of the unstable equilibria. The observations suggested that the effect of hysteresis on the contact angle was small in this case and that the behavior could be interpreted in terms of a single contact angle in the neighborhood of 150°. In the second series of experiments, holes were blown by an air jet in a sheet of water standing on a disc coated with paraffin wax. Here contact-angle hysteresis was strongly in evidence. In particular it was found quite easy to produce holes that stood still and could be photographed. This could not happen if the contact angle were uniquely defined, but it can be interpreted in terms of contact-angle hysteresis. If $\theta_a$, the contact angle that applies when the meniscus is on the point of advancing, is greater than the value $\theta_r$, when it is about to retreat, the two loci of unstable equilibria for these angles provide a band of hole radii for a given depth within which the hole is trapped, it being repelled from each end of the range of radii. An

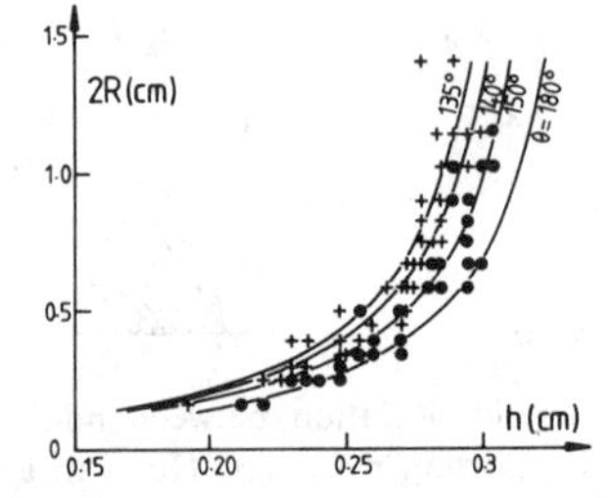


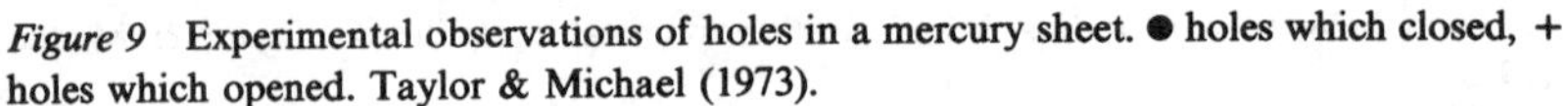

*Figure 9* Experimental observations of holes in a mercury sheet. ● holes which closed, + holes which opened. Taylor & Michael (1973).

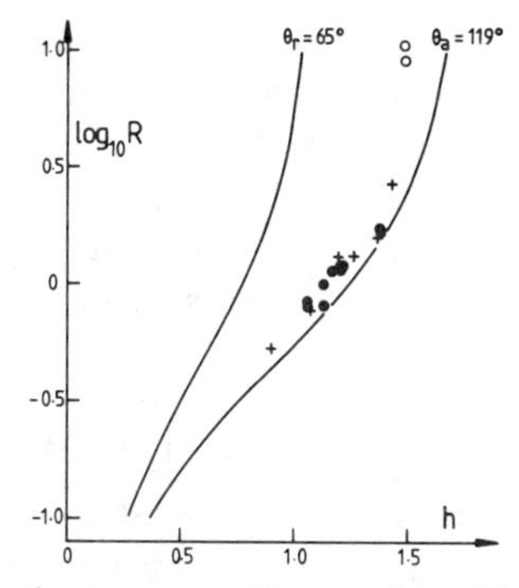


*Figure 10* Holes in water sheets on paraffin wax base. Two sets of observations, in dimensionless form, of the minimum sizes for holes which remained open. $\theta_a = 119°$, $\theta_r = 65°$ were measured values of contact angle for the meniscus on the point of advancing and retreating. Taylor & Michael (1973).

independent measurement of contact angle in these experiments suggested the values $\theta_a = 119°$ and $\theta_r = 65°$. Equilibrium curves for these values are given in Figure 10. Experimental points in this figure indicate minimum sizes for holes that remained open, which tends to confirm the lower limit. Reliable observations on the largest holes that do not expand were not obtainable with this apparatus because the size of the water sheet was not large enough to be regarded as infinite for the larger holes. Contact-angle hysteresis may usefully be interpreted by analogy with solid friction in elementary mechanics. Thus, the behavior of holes described here may be closely modelled by a chain hanging in the form of a catenary over two pegs or pulleys at the same height. When the pegs are very smooth there are two equilibrium positions, the lower one being an unstable one. If the chain is released from a position just above the unstable equilibrium it will rise, whilst it will fall if released from just below this equilibrium. This simulates the opening and closing of holes in a mercury sheet. If friction is introduced at the pegs, it becomes possible for the chain to stand in equilibrium in a range of positions enclosing the previously unstable-equilibrium position, this being the counterpart of stationary holes in the water sheet.

## 4 SUMMARY

Experimental and theoretical studies of the mechanics of the meniscus have given rise to a large literature predominantly concerned with studies of equilibrium. Here we have attempted to describe some of the significant developments in the understanding of the stability of the meniscus. The account is restricted to the study of two main groups of problems, the first in which no external field of force is present, and the second when the meniscus is formed in a uniform external gravitational field. The review is limited almost entirely to statical considerations including some discussion of the effects of statical hysteresis. But some important problems such as the stability of three-phase systems with three-point contact lines such as the sessile lens remain to be studied.

In retrospect, the grouping of the problems we have used will be seen to be unimportant because changes in stability conform to basic laws irrespective of the geometry or the force field. We have seen that, except insofar as an instability may be suppressed by strong enough frictional effects, the meniscus changes stability at a point of bifurcation that can be recognized as an equilibrium position in which a small perturbation of the configuration that satisfies the boundary conditions gives rise to an alternative neighboring equilibrium configuration with the same value of the control or loading parameter. The stability depends on the identity of this control parameter. The two cases considered here are those of volume control or internal stability, and pressure control or external stability. In some cases the structure of the stability transition is not clear, but for most of the problems discussed here the transition can be described in terms of the limit-point or fold instability and the Riemann-Hugoniot cusp, which may be of stable or unstable symmetric form. The relation between criteria for internal and external stability illustrates some features that have been pointed out in a generalized form, for a conservative system, by Katz (1978) and Thompson (1979). We find, for example, in the pendent-drop profiles that where the system is both internally and externally stable, the pressure and volume increase together, and that when the system is externally stable it is always internally stable. The sessile drop illustrates the case in which the drop is externally unstable and internally stable, in which case the pressure decreases with increasing volume.

Stability changes associated with the limit points have the most geometrical symmetry, such as axial symmetry for axisymmetric menisci. The cusp, on the other hand, represents a change of symmetry into asymmetric modes. A situation familiar in many instances is where both these forms of bifurcation occur at different points on a loading path. This occurs in the internal stability of a pendent drop, and the form of the instability first occurring depends on the value of another imposed parameter such as the orifice radius. Another interesting case is the external stability of the two-dimensional drop profile in which the limit-point and cusp instabilities coincide at the pressure maximum. It is typical of all the cusp transitions that they occur in both internal and external stability. This is an illustration of the general result that when the system becomes externally unstable in two different modes together it must also change its internal stability.

The effect of friction on the statics of the meniscus is illustrated in an account of the study of the behavior of holes in a thin sheet of fluid as a plane horizontal support. It is seen there that an unstable equilibrium in the absence of friction determines a critical hole size below which the hole will close, and above which it will open out. The effect of friction acting through the statical hysteresis of the contact angle is to create a band of hole sizes for each of which the hole is maintained in a steady state by the effects of statical hysteresis.

## ACKNOWLEDGMENT

The author wishes to acknowledge with thanks the support of the National Research Council of Canada during the preparation of this work.

*Literature Cited*

Bakker, G. 1928. Kapillarität und Oberflächenspannung. *Handbuch der Experimentalphysik*. Vol. VI, ed. W. Wien, F. Harms. Leipzig:Akademische. 458 pp.

Beer, A. 1855. Ueber die Oberflächen rotirender Flüssigkeiten im Allgemeinen, insbesondere über den Plateau'schen Rotationsversuch. *Ann. d. Phys. u. Chem.* 96:1–18, 210–35

Boucher, E. A., Evans, M. J. B. 1975. Pendent drop profiles and related capillary phenomena. *Proc. R. Soc. London Ser. A* 346:349–74

Boucher, E. A., Evans, M. J. B., Kent, H. J. 1976. Capillary phenomena, II. Equilibrium and stability of rotationally symmetric fluid bodies. *Proc. R. Soc. London Ser. A* 349:81–100

Brown, R. A., Scriven, L. E. 1980a. The shape and stability of rotating captive drops. *Philos. Trans. R. Soc. London Ser. A* In press

Brown, R. A., Scriven, L. E. 1980b. The shapes of rotating axisymmetric fluid interfaces. *Proc. R. Soc. London Ser. A*. In press

Carruthers, J. R., Grasso, M. 1972a. Studies of floating liquid zones in simulated zero gravity. *J. Appl. Phys.* 43:436–45

Carruthers, J. R., Grasso, M. 1972b. The stabilities of floating liquid zones in simulated zero gravity. *J. Crystal Growth* 13:611–14

Concus, P. 1963. Capillary stability in an inverted rectangular tank. *Adv. Astronaut. Sci.* 14:21–37

Concus, P. 1964. Capillary stability in an inverted rectangular channel for free surfaces with curvature of changing sign. *AIAA J.* 2:2228–30.

Concus, P., Finn, R. 1974a. On capillary free surfaces in the absence of gravity. *Acta Math.* 132:177–98

Concus, P., Finn, R. 1974b. On capillary free surfaces in a gravitational field. *Acta Math.* 132:207–23

Concus, P., Finn, R. 1979. The shape of a pendent liquid drop: *Philos. Trans. R. Soc. London Ser. A* 292:307–40

Coriell, S. R., Cordes, M. R. 1977. Theory of molten zone shape and stability. *J. Crystal Growth* 42:466–72

Coriell, S. R., Hardy, S. C., Cordes, M. R. 1977. Stability of liquid zones. *J. Colloid Interface. Sci.* 60:126–36

Dombrowski, N., Fraser, R. P. 1954. A photographic investigation into the disintegration of liquid sheets. *Philos. Trans. R. Soc. London Ser. A* 247:101–30

Duprez, F. 1851, 1854. Mémoire sur un cas particulier de l'équilibre des liquides. I, II. *Mém. Acad. R. Sci. Lett. Beaux-Arts Belgique* 26:42 pp. (1 plate), 28:34 pp. (1 plate)

Dussan V., E. B. 1975. Hydrodynamic stability and instability of fluid systems with interfaces. *Arch. Ration. Mech. Anal.* 57:363–79

Dussan V., E. B. 1979. On the spreading of liquids on solid surfaces: Static and dynamic contact lines. *Ann. Rev. Fluid Mech. II*. 11:371–400

Erle, M. A., Gillette, R. D., Dyson, D. C. 1970. Stability of interfaces of revolution with constant surface tension-the case of the catenoid. *Chem. Engrg. J.* 1:97–109

Finn, R. 1974. A note on the capillary problem. *Acta Math.* 132:199–205

Gifford W. A., Brown, R. A., Scriven, L. E. 1980. The shape and stability of rotating charged or gravitating drops. *Proc. R. Soc. London Ser. A*. In press

Gillette, R. D., Dyson, D. C. 1971. Stability of fluid interfaces of revolution between equal solid circular plates. *Chem. Engrg. J.* 2:44–54

Haynes, J. M. 1970. Stability of a fluid cylinder. *J. Colloid Interface Sci.* 32:652–54

Hormann, G. 1887. Untersuchungen über die Grenzen, zwischen welchen Unduloide und Nodoide, die von zwei festen Parallelkreisflächen begrenztsind, bei gegebenen Volumen ein Minimum besitzen. Dissertation. Univ. Göttingen

Howe, W. 1887. Die Rotations-Flächen welche bei vorgeschriebener Flächengrösse ein möglichst grosses oder kleines Volumen enthalten. Dissertation. Univ. Berlin

Joseph, D. 1976. Interfacial stability. In *Stability of Fluid Motions II*, Chap. XIV, 241–61. *Springer Tracts Nat. Philos.*, Vol. 28. 274 pp.

Karasalo, I. 1979. Stability of axisymmetric annular fluid interfaces at zero contact angle. *Arch. Ration. Mech. Anal.* 71:257–70

Katz, J. 1978. On the number of unstable modes of an equilibrium. *Mon. Not. R. Astron. Soc. London* 183:765–69

Kovitz, A. A. 1975. Static fluid interfaces external to a right circular cylinder—Experiment and theory. *J. Colloid Interface Sci.* 50:125–42

Laplace, P. S. de. 1805. *Traité de Mécanique Céleste*. Suppl. au Xe Livre, Paris, Coureier. Transl. and annotated by N. Bowditch (1839). Vol. 4, pp. 685–1018. Boston:Little, Brown Reprinted by Chelsea Publ. Co., NY (1966)

Levin, P., Pitts, E., Terry, G. C. 1976. New method for measuring surface tension from the height of a pendent drop. *J. Chem. Soc. Faraday Trans. I* 72:1519–25

Lohnstein, T. 1906a. Zur Theorie des Abtropfens mit besonderer Rücksicht auf die Bestimmung der Kapillaritätskonstanten durch Tropfuersuche. *Ann. Physik* (4)20:237–68

Lohnstein, T. 1906b. Zur Theorie des Abtropfens. Nachtrag und weitere Belege. *Ann. Physik* (4)20:606–18

Lohnstein, T. 1906c. Zur Theorie des Abtropfens. Zweiter Nachtrag. *Ann. Physik* (4)21:1030–48

Lohnstein, T. 1907. Weiteres zur Theorie der fallenden Tropfen, nebst einen Rückblick auf ältere theoretische Versuche. Ann. Physik (4)22:767–81

Lyttleton, R. A. 1953. Stability of statical systems. *The Stability of Rotating Liquid Masses*, Chap. II, pp. 6–15. Cambridge Univ. Press. 147 pp.

Majumdar, S. R., Michael, D. H. 1976. Equilibrium and stability of two-dimensional pendent drops. *Proc. R. Soc. London Ser. A* 351:89–115

Majumdar, S. R., Michael, D. H. 1980. The instability of plane pendent drops. *J. Colloid Interface Sci.* 73:186–92

Mason, G. 1970. An experimental determination of the stable length of cylindrical liquid bubbles. *J. Colloid Interface Sci.* 32:172–76

Maxwell, J. C. 1875. Capillary Action. *Encycl. Brit. 9th Ed. Vol. 5, pp. 56–71. Sci. Pap.* vol. 2, pp. 541–91

Michael, D. H., Norbury, J., O'Neill, M. E. 1974. Electrohydrostatic instability in electrically stressed dielectric fluids. Part 1. *J. Fluid Mech.* 66:289–308

Michael, D. H., Norbury, J., O'Neill, M. E. 1975. Electrohydrostatic instability in electrically stressed dielectric fluids. Part 2. *J. Fluid Mech.* 72:95–112

Michael, D. H., Williams, P. G. 1976. The equilibrium and stability of axisymmetric pendent drops. *Proc. R. Soc. London Ser. A* 351:117–27

Michael, D. H., Williams, P. G. 1977. The equilibrium and stability of sessile drops. *Proc. R. Soc. London Ser. A* 354:127–36

Neumann, F. E. 1894. *Vorlesungen über die Theorie der Capillarität*, Chap. 5. Leipzig:Teubner. x+234 pp.

Padday, J. F., Pitt, A. R. 1973. The stability of axisymmetric menisci. *Philos. Trans. R. Soc. London Ser. A* 275:489–528

Pars, L. A. 1962. *An Introduction to the Calculus of Variations*. London: Heinemann Educational Books Ltd. 345 pp.

Pitts, E. 1973. The stability of pendent liquid drops. Part I. Drops formed in a narrow gap. *J. Fluid Mech.* 59:753–67

Pitts, E. 1974. The stability of pendent liquid drops. Part 2. Axial symmetry. *J. Fluid Mech.* 63:487–508

Pitts, E. 1976a. The stability of a drop

hanging from a tube. *J. Inst. Math. Its Appl.* 17:387–97

Pitts, E. 1976b. The stability of a meniscus joining a vertical rod to a bath of liquid. *J. Fluid Mech.* 76:641–51

Plateau, J. 1873. *Statique Expérimentale et Théorique des liquides Soumis aux Seules Forces Moléculaires.* Paris:Gauthier-Villars 495 pp.

Rayleigh, Lord. 1879a. On the instability of jets. *Proc. London Math. Soc.* 10:4–13. *Sci. Pap.*, Vol. I, pp. 361–71

Rayleigh, Lord. 1879b. On the capillary phenomena of jets. *Proc. R. Soc. London* 29:71–97. *Sci. Pap.*, Vol. I, pp. 377–401

Rayleigh, Lord. 1892a. On the instability of a cylinder of viscous Liquid under capillary force. *Philos. Mag.* 34:145–54. *Sci. Pap.*, Vol. III, pp. 585–93

Rayleigh, Lord. 1892b. On the instability of cylindrical fluid surfaces. *Philos. Mag.* 34:177–80. *Sci. Pap.*, Vol. III, pp. 594–96

Searle, G. F. C. 1934. Mathematical discussions of problems in surface tension. In *Experimental Physics*. pp. 128–63. Cambridge Univ. Press. 363 pp.

Shoukry, E., Hafez, M., Hartland, S. 1975. Separation of drops from wetted surfaces. *J. Colloid Interface Sci.* 53:261–70

Sozou, C., Hewson-Browne, R. C. 1976. On two-dimensional electrohydrostatic stability. *Proc. R. Soc. London Ser. A* 349:231–43

Surek, T., Coriell, S. R. 1977. Shape stability in float zoning of silicon crystals. *J. Crystal Growth* 37:253–71

Taylor, G. I. 1964. Disintegration of water drops in an electric field, *Proc. R. Soc. London Ser. A* 280:383–97

Taylor, G. I. 1968. The coalescence of closely spaced drops when they are at different electric potentials. *Proc. R. Soc. London Ser. A* 306:423–34

Taylor, G. I., Michael, D. H. 1973. On making holes in a sheet of fluid. *J. Fluid Mech.* 58:625–39

Terquem, M. A. 1881. Sur les surfaces de révolution limitant les liquides dénués de pesanteur. *C. R. Acad. Sci. Paris* 92: 407–9

Thompson, J. M. T., Hunt, G. W. 1973. *A General Theory of Elastic Stability.* New York:Wiley. 322 pp.

Thompson, J. M. T., Hunt, G. W. 1975. Towards a unified bifurcation theory. *Z. Angew. Math. Phys.* 26:581–604

Thompson, J. M. T., Hunt, G. W. 1977. The instability of evolving systems. *Interdisciplinary Sci. Rev.* 2:240–61

Thompson, J. M. T. 1979. Stability predictions through a succession of folds. *Philos. Trans. R. Soc. London Ser. A* 292:1–23

Tyuptsov, A. D. 1966. Stability of equilibrium forms of a liquid surface. *Mekh. Zhidk. Gaza* 1:78–85. Transl. *Fluid Dynamics* 1:51–55

Young, T. 1805. An essay on the cohesion of fluids. *Philos. Trans. R. Soc. London Ser. A* 95:65–87

*Ann. Rev. Fluid Mech. 1981. 13:217-29*


# SOME ASPECTS OF THREE-DIMENSIONAL LAMINAR BOUNDARY LAYERS

*8175

*Harry A. Dwyer*
Department of Mechanical Engineering, University of California–Davis, Davis, California 95616

## INTRODUCTION

The purpose of this article is to discuss and present some aspects of three-dimensional boundary-layer flows as they apply to numerical solution techniques. The primary numerical method to be discussed will be finite-difference techniques, and other approaches will not be mentioned. The major difficulties that one faces in an attempt to solve three-dimensional boundary-layer flows are essentially the same for all numerical methods, and arguments on the relative merits of finite-difference and finite-element techniques are not central to the problem. What is central to the problem, however, is the many peculiarities of three-dimensional boundary layers such as initial conditions, interaction with the inviscid flow, flow reversal, and separation. In particular, the phenomena of flow reversal and separation in three dimensions are much richer in the number of physical possibilities than in two dimensions.

In this article the most modern and promising techniques, rather than the older work in the field, are reviewed. The general emphasis here is on the incompressible flow equations; however, some important new work in the area of supersonic three-dimensional flows is briefly treated because of the significant progress that has been made in this area.

## BASIC EQUATIONS

The starting point for a derivation of the incompressible, three-dimensional boundary-layer equations is the Navier-Stokes equations;



0066-4189/81/0115-0217$01.00

the boundary-layer equations can be obtained by taking the simultaneous limit of Reynolds number approaching infinity and $n$, the coordinate perpendicular to a no-slip surface, approaching zero (for example, see Rosenhead 1963 and Stewartson 1964). The results for unsteady laminar flow are

$$\frac{\partial u}{\partial x}+\frac{\partial v}{\partial n}+\frac{\partial w}{\partial z}=0, \tag{1}$$

$$\frac{\partial u}{\partial t}+u\frac{\partial u}{\partial x}+v\frac{\partial u}{\partial n}+w\frac{\partial u}{\partial z}=-\frac{1}{\rho}\frac{\partial p_e}{\partial x}+\nu\frac{\partial^2 u}{\partial n^2}, \tag{2}$$

$$\frac{\partial w}{\partial t}+u\frac{\partial w}{\partial x}+v\frac{\partial w}{\partial n}+w\frac{\partial w}{\partial z}=-\frac{1}{\rho}\frac{\partial p_e}{\partial z}+\nu\frac{\partial^2 w}{\partial n^2}, \tag{3}$$

where $x$ and $z$ are two Cartesian coordinates locally parallel to a body or wall surface, $u$, $v$, and $w$ the velocity components in the $x$, $n$, and $z$ direction respectively, $p_e$ pressure, $\rho$ density, and $\nu$ the kinematic viscosity of the fluid. Also implied along with Equations (1)–(3) is the neglect of the momentum equation in the $n$-direction due to the small relative magnitude of $v$.

The mathematical nature of the equation set (1)–(3) is parabolic with boundary conditions needed for $u$ and $w$ along $n$ and initial conditions associated with $t$, $x$, and $z$. The continuity equation (1) determines $v$ from a local initial-value problem associated with the no-slip surface, and the pressure must be determined from other sources (either experimental results or inviscid-flow results). A more physical way of describing this mathematical nature would be to say that convection occurs only in the $x$ and $z$ coordinates while the combined processes of diffusion and convection occur along the $n$ coordinate. Convection by itself defines an initial-value problem while diffusion defines a boundary-value problem, and, of course, time defines an initial-value problem.

The combined processes of convection and diffusion define a very interesting region of influence that is unique to the three-dimensional boundary layer, and which must be considered carefully in any numerical procedure that is employed (Der & Raetz 1962, Wang 1971, Kitchens et al 1975). The region of influence in a three-dimensional boundary-layer problem is defined by the fluid particles that communicate and interact by the convection and diffusion processes. The interaction along the $n$ direction is dominated by the viscous-diffusion term, which causes all fluid particles along a given $n$ coordinate to interact simultaneously at all times. For the $x$, $z$, and $t$ coordinates the interaction causing the region of influence is more complicated and is defined by the sub-characteristics of Equations (2) and (3). These sub-characteristics are determined by the convection or time-like operator,

$$\frac{\partial}{\partial \xi}=\frac{\partial}{\partial t}+u\frac{\partial}{\partial x}+w\frac{\partial}{\partial z}, \tag{4}$$

which reduces to the convection terms in the steady state.

At each point in the boundary layer a characteristic direction in the $t$, $x$, and $z$ space is defined; however, due to the variation of $u$ and $w$ across an $n$-coordinate line the total region defined by the characteristics can be quite complex. For example, for a spinning axisymmetric body with $x$ in the plane through the axis of symmetry and $z$ perpendicular to it, the region of influence is defined by a plane perpendicular to the body axis and all the volume downstream of this plane. This case of surface crossflow $w$ causes the most comprehensive region of influence that can be treated by boundary-layer methods, as is pointed out later in this paper (Dwyer 1974). We also show that the numerical method used to solve the flow is strongly influenced by the region of influence.

## INITIAL AND BOUNDARY CONDITIONS

Because of the parabolic nature of the three-dimensional boundary-layer equations, initial conditions play a significant role in the flow development and the history of most flows. In general, the two types of initial conditions for the spatial coordinates $x$ and $z$ correspond to either a sharp leading edge or a stagnation point, while for time $t$ a wide variety of conditions appear in practice (the impulsively started flow being of general interest). From the numerical viewpoint these various initial conditions present serious scaling problems, and there is a great advantage to be gained with the use of coordinate transformations. Coordinate transformations can remove leading-edge and time-dependent singularities, as well as generally smooth the variation of all variables in the transformed plane; and the small amount of time spent in transforming the equations is usually well worth the effort.

Recently a transformation has been developed (Dwyer & Sherman 1980) that incorporates many features of other investigators. This transformation is

$$\xi=x, \qquad \eta=n\left(\frac{1+U_e t/x}{\nu t}\right)^{1/2}, \qquad \zeta=z, \qquad \tau=t. \tag{4}$$

(It is assumed for the discussion of initial conditions that the leading edge or stagnation point is located at $x=0$ and $U_e$ is the inviscid-flow velocity.) In terms of these new independent variables, Equations (1)–(3)

become

$$\xi\frac{\partial f'}{\partial \xi}+\xi\frac{\partial f'}{\partial \eta}\frac{\partial \eta}{\partial x}+\frac{\partial v'}{\partial \eta}+\beta_x f'+\frac{\xi}{U_e}\frac{\partial w}{\partial z}=0, \tag{5}$$

$$\frac{\xi}{U_e}\frac{\partial f'}{\partial t}+\xi f'\frac{\partial f'}{\partial \xi}+\left[\frac{\xi}{U_e}\frac{\partial \eta}{\partial t}+\xi f'\frac{\partial \eta}{\partial x}+v'\right]\frac{\partial f'}{\partial \eta}$$
$$=(1-f')\beta_\tau+(1-f'^2)\beta_x+\varepsilon\frac{\partial^2 f'}{\partial \eta^2}, \tag{6}$$

$$\frac{\xi}{U_e}\frac{\partial \psi}{\partial t}+\xi f'\frac{\partial \psi}{\partial \xi}+\left[\frac{\xi}{U_e}\frac{\partial \eta}{\partial t}+\xi f'\frac{\partial \eta}{\partial x}+v'\right]\frac{\partial \psi}{\partial \eta}$$
$$=(1-\psi)\psi_t+(1-f'\psi)\psi_x+\xi\frac{\partial w_e/\partial z}{U_e}(1-\psi^2)+\varepsilon\frac{\partial^2 \psi}{\partial \eta^2}, \tag{7}$$

where the following definitions have been employed

$$f'=\frac{U}{U_e}, \qquad \psi=\frac{\partial w/\partial z}{\partial w_e/\partial z}, \qquad \beta_x=\frac{x}{U_e}\frac{\partial U_e}{\partial x}, \qquad \beta_t=\frac{x}{U_e^2}\frac{\partial U_e}{\partial t},$$

$$\varepsilon=\frac{\xi+U_e t}{U_e t}, \qquad \psi_x=\frac{x}{\partial w_e/\partial z}\frac{\partial^2 w_e}{\partial x\,\partial z}, \qquad \psi_t=\frac{x}{U_e\partial w_e/\partial z}\frac{\partial^2 w_e}{\partial z\,\partial t},$$

$$v'=\frac{v\xi}{U_e}\left(\frac{1+U_e t/x}{\nu t}\right)^{1/2}.$$

It should be pointed out that Equation (7) represents the $z$ derivative of Equation (3), this form having been chosen to obtain stagnation-point solutions. An equation for $w$ is useful at a sharp leading edge, and has a form very similar to Equation (6) (Dwyer 1968).

The use of this transformation for the stagnation point and sharp leading edge will now be illustrated. The stagnation-point solution is obtained from the conditions

$$\frac{\partial U_e}{\partial x}=A,\ \frac{\partial W_e}{\partial z}=B,\ C=A/B \qquad \text{as} \quad \begin{matrix}\xi\to 0\\ U_e\to 0\\ W_e\to 0\end{matrix}$$

and Equations (5)–(7) become

$$\frac{\eta}{2}\frac{\partial f'}{\partial \eta}+\frac{\partial v'}{\partial \eta}+f'+\frac{\partial w/\partial z}{A}=0, \tag{8}$$

$$\frac{1}{A}\frac{\partial f'}{\partial t}+\left[-\frac{f'\eta}{2}+v'\right]\frac{\partial f}{\partial \eta}=(1-f'^2)+\frac{\partial^2 f'}{\partial \eta^2}, \tag{9}$$

$$\frac{1}{A}\frac{\partial \psi}{\partial t}+\left[-\frac{f'\eta}{2}+v'\right]\frac{\partial \psi}{\partial \eta}=\frac{\partial w_e/\partial z}{A}(1-\psi^2)+\frac{\partial^2 \psi}{\partial \eta^2}. \tag{10}$$

The above system of equations must be solved simultaneously with Equations (1)–(3) for time-dependent problems or once at the start of a calculation for a steady-state problem.

For a sharp leading edge where $U_e$ and $W_e$ are both nonzero it can be shown that Equations (5)–(7) reduce to the well-known Blasius equation (Dwyer 1968), and therefore tabulated values of $u$, $w$, $v$ can be utilized to obtain initial conditions in the transformed plane. However, it should be pointed out that a singularity exists in the physical-plane solution, and that it is "impossible" to calculate the flow accurately near $x=0$ unless the singularity is removed by a coordinate transformation.

A singularity of a different type also exists near $t=0$ for any three-dimensional boundary layer that is impulsively started. This singularity is again removed by transformation (4) and it can be shown (Dwyer & Sherman 1980) that

$$\frac{\partial \eta}{\partial x} \to 0, \quad \frac{\partial \eta}{\partial t} \to -\frac{\eta}{2t}, \quad \text{at } x \neq 0,$$

and the equations for $u$ and $w$ have the form

$$\frac{\eta}{2}\frac{\partial f'}{\partial \eta} + \frac{\partial^2 f}{\partial \eta^2} = 0. \tag{11}$$

Again, tabulated values can be found in standard references (Rosenhead 1963). These same solutions are also useful along lines of symmetry and it can be shown that Equations (8)–(10) reduce to the form of Equation (11).

As mentioned previously boundary conditions are needed at the effective boundaries of the $n$ or $\eta$ coordinate for the variables $u$ and $w$. For the wall location, $n=0$, the conditions are the conventional no-slip ones unless there is surface mass transfer, which will not be considered in this article. At the outer edge of the boundary layer, or the inner edge of the inviscid flow, both the locations of the boundary-layer edge and the values of $U_e$ and $W_e$ must be determined. For laminar boundary layers without heat or mass transfer the effective edge of the boundary-layer flow is usually located below $\eta=6.0$. (The edge of the boundary layer is defined to be the point where the variables $u$ and $w$ have returned to 99% of their inviscid-flow values.) This value is essentially the same as that for two-dimensional flow, and the three-dimensional nature of the problem does not seem to influence to a great degree the location of the boundary-layer edge.

The primary reason that an *a priori* estimate of the boundary-layer edge can be made is because of the use of coordinate transformations. In the physical plane it is very difficult to make any *a priori* estimates, and the boundary-layer edge will vary considerably from one part of the flow to another. Even with coordinate transformations the edge can vary from $\eta=3$ for favorable conditions to $\eta=10$ for unfavorable ones. Favorable conditions are defined as positive $\partial u/\partial x$ and $\partial w/\partial z$, while unfavorable ones are defined as negative values of these derivatives. The consideration of two derivatives, rather than one, for two-dimensional flow, is one of the primary differences in three-dimensional flow, and negative values of $\partial w/\partial z$ can be responsible for boundary-layer "separation" without primary-flow pressure gradients, for example in supersonic flow over a right circular cone at angle of attack (Dwyer 1971).

The actual values of $u$ and $w$ and the pressure gradients in the boundary layer do not come from boundary-layer theory itself. In most cases these values are obtained from inviscid-flow analysis or experimental measurements. Although the boundary conditions very rarely cause difficulties with numerical simulations, they can easily be said to be the Achilles heel of boundary-layer theory. This weakness is due to the influence of downstream separation, or boundary-layer breakdown, on the inviscid solution upstream of separation. For most non-streamline bodies the influence of the downstream separated-flow region has a major effect on the inviscid flow over the boundary layer, and inviscid analyses that do not take account of this fact lead to large errors. Unfortunately for boundary-layer theory, the only remedy is a complete solution of the Navier-Stokes equations, which effectively negates the need for boundary-layer solutions.

## NUMERICAL SOLUTION OF THE EQUATIONS

With the discussion of the initial and boundary conditions completed we can proceed to develop some techniques of numerical solution. In order to restrict the arguments to a tractable size the numerical-solution procedures will be applied only to the steady-state boundary-layer equations. The first thing that must be tackled is the coordinate system to be used over the body for a finite-difference method. A modern approach is to transform from the Cartesian surface coordinates $x$ and $z$ to generalized nonorthogonal coordinates on the surface (Viviand 1974 and Vinokur 1974). For most applications it is generally recommended to have the surface coordinate system close to orthogonal, but the added freedom of having generalized nonorthogonal coordinates is very valuable (for example, it allows the addition of coordinate lines near regions of high gradient without the need for solving complex orthogonal coordinate generators). Also, it should be pointed out that the coordinate system must originate at the stagnation point or leading edge, if the proper initial-value nature of the problem is to be taken into account.

In order to develop the equations in these coordinates we first rewrite equations (1)–(3) in a steady vector form,

$$\frac{\partial E}{\partial x}+\frac{\partial}{\partial n}(F+F_v)+\frac{\partial G}{\partial z}=0, \tag{12}$$

where

$$E=\begin{bmatrix} u \\ u^2+P/\rho \\ uw \end{bmatrix},\qquad G=\begin{bmatrix} w \\ uw \\ w^2+P/\rho \end{bmatrix},\qquad F=\begin{bmatrix} v \\ uv \\ wv \end{bmatrix},$$

$$F_v=\begin{bmatrix} 0 \\ -\tau_{yx} \\ -\tau_{yz} \end{bmatrix},\qquad \begin{matrix}\tau_{yx}=\nu\partial u/\partial y \\ \tau_{yz}=\nu\partial w/\partial y\end{matrix}.$$

By transforming to generalized nonorthogonal coordinates $\xi$, $\eta$, and $\zeta$ we obtain

$$\frac{\partial}{\partial \xi}(\hat{E}+\hat{E}_v)+\frac{\partial}{\partial \zeta}(\hat{G}+\hat{G}_v)+\frac{\partial}{\partial \eta}(\hat{F}+\hat{F}_v)=0 \tag{13}$$

where

$$\hat{E}=J^{-1}\begin{bmatrix} U \\ uU+\xi_x P/\rho \\ wU+\xi_z P/\rho \end{bmatrix},\qquad \hat{G}=J^{-1}\begin{bmatrix} W \\ uW+\zeta_x P/\rho \\ wW+\zeta_z P/\rho \end{bmatrix},\qquad \hat{F}=J^{-1}\begin{bmatrix} V \\ uV \\ wV \end{bmatrix},$$

$$\hat{F}_v=J^{-1}\begin{bmatrix} 0 \\ -\eta_n\tau_{yz} \\ -\eta_n\tau_{yz} \end{bmatrix},\qquad \hat{E}_v=J^{-1}\begin{bmatrix} 0 \\ -\xi_n\tau_{yx} \\ -\xi_n\tau_{yz} \end{bmatrix},\qquad \hat{G}_v=J^{-1}\begin{bmatrix} 0 \\ -\zeta_n\tau_{yz} \\ -\zeta_n\tau_{yz} \end{bmatrix},$$

$$U=\xi_x u+\xi_n v+\xi_z w,\qquad V=\eta_x u+\eta_y v+\eta_z w,\qquad W=\zeta_x u+\zeta_n v+\zeta_z w,$$

$$J^{-1}=x_\xi y_\eta z_\zeta+x_\zeta y_\xi z_\eta+x_\eta y_\zeta z_\xi-x_\xi y_\zeta z_\eta-x_\eta y_\xi z_\zeta-x_\zeta y_\eta z_\xi.$$

In most problems it is difficult to imagine a reason for not having $\eta$ perpendicular to the surface coordinates, and if $\eta$ is perpendicular to the surface then $\xi_n$ and $\zeta_n$ will be zero. For this case $\hat{E}_v$ and $\hat{G}_v$ will be zero and other terms will also simplify.

It will now be assumed that the $\xi$ coordinate line will be approximately aligned with the primary flow direction (always a somewhat arbitrary direction in the three-dimensional-flow world), and the $\zeta$ coordinate will be called the crossflow direction. The flow along the $\zeta$ coordinate will be allowed to be both positive and negative, while that along $\xi$ will be only positive. If both $u$ and $w$ become negative then it is impossible for the time-like operator

$$\frac{\partial}{\partial \xi} + \frac{\partial}{\partial \zeta}$$

to be positive, and one is effectively marching into the history of the flow. When one marches into the history of a parabolic problem it is generally ill posed due to exponential positive growth for negative effective time.

With both positive and negative crossflow allowed, the region of influence can be complex and it is best to be conservative and allow for total communication in the crossflow plane. With total fluid-particle communication, the crossflow variable $\zeta$ then takes on a semi-elliptic nature in a finite-difference method. The simplest finite-difference scheme applied to Equation (13) that will yield second-order accuracy and allow for only one plane of data storage is a central-difference approximation along the parabolic variable $\xi$. A central-difference approximation to the $\xi$ derivative yields

$$\frac{\hat{E}^{i+1}-\hat{E}^{i}}{\Delta\xi} + \frac{\partial}{\partial\eta}\left(\frac{\hat{F}^{i+1}+\hat{F}^{i}}{2} + \frac{\hat{F}_v^{i+1}+\hat{F}_v^{i}}{2}\right) + \frac{\partial}{\partial\zeta}\left(\frac{\hat{G}^{i+1}+\hat{G}^{i}}{2}\right) = 0 \quad (14)$$

where the index $i$ has been used to identify a plane of constant $\xi$. Since the $i+1$ terms are nonlinear, they should be linearized in general for an efficient solution to be obtained. The most robust strategy for linearization is the Newton procedure and, when applied to Equation (14), yields

$$\hat{E}^{i+1} = \hat{E}^{i} + \left.\frac{\partial\hat{E}}{\partial\hat{q}}\right|_i \Delta\hat{q} = \hat{E}^{i} + \bar{\bar{A}}\Delta\hat{q}, \qquad \Delta\hat{q} = \hat{q}^{i+1} - \hat{q}^{i}, \qquad \hat{q} = \begin{pmatrix} v \\ u \\ w \end{pmatrix},$$

$$\hat{F}^{i+1} = \hat{F}^{i} + \left.\frac{\partial\hat{F}}{\partial\hat{q}}\right|_i \Delta\hat{q} = \hat{F}_i + \bar{\bar{B}}\Delta\hat{q},$$

$$\hat{F}_v^{i+1} = \hat{F}_v^{i} + \left.\frac{\hat{F}_v}{\partial\hat{q}}\right|_i \Delta\hat{q} = \hat{F}_v^{i} + \bar{\bar{C}}\Delta\hat{q},$$

$$\hat{G}^{i+1} = \hat{G}^{i} + \left.\frac{\partial\hat{G}}{\partial\hat{q}}\right|_i \Delta\hat{q} = \hat{G}_i + \bar{\bar{D}}\Delta\hat{q},$$

and Equation (14) becomes

$$\left(2\frac{\bar{\bar{A}}}{\Delta\xi} + \frac{\partial}{\partial\eta}\left[\bar{\bar{B}} + \bar{\bar{C}}\right] + \frac{\partial}{\partial\zeta}\bar{\bar{D}}\right)\frac{\Delta\hat{q}}{2} = \frac{-\partial}{\partial\eta}(\hat{F}^{i} + \hat{F}_v^{i}) - \frac{\partial}{\partial\zeta}\hat{G}^{i}. \quad (15)$$

(The quantities $\bar{\bar{A}}$, $\bar{\bar{B}}$, $\bar{\bar{C}}$, and $\bar{\bar{D}}$ are Jacobian matrices.) The above form of Newton linearization is noniterative and this has been recommended by other investigators (Beam & Warming 1976, Briley & McDonald 1973). However, very little research has been carried out on the relative merits of linearization methods for the equations of fluid dynamics in multi-dimensions, and much is left to be done. Also, it should be remembered that the convergence of the nonlinear equations is an independent problem from the stable solution of the resulting linear system, and many investigators have ignored this point.

The final linear system to be solved is obtained by forming finite-difference approximations to the $\eta$ and $\zeta$ derivatives that appear in Equation (15). If fourth-order central differences are used, then one obtains a rather dense linear system of equations, which in general does not have an efficient solution. A more efficient method is to form a factored or ADI scheme that solves Equation (15) along the $\eta$ and $\zeta$ coordinates successively, and thus limits the size of the linear system to be solved. A good discussion of the details of such a procedure can be found in the work of Steger (1978) and Schiff & Steger (1979).

In general, it is a good idea to use as high an order as possible for the accuracy of a derivative. However, when solving large systems of linear equations there are significant economies to be gained by having the linear systems placed into certain forms. A form to be sought after is a block tri-diagonal form, which can be solved very efficiently (Hindmarsh 1977). In order to obtain this system with the factored or ADI forms of Equation (15) only compact three-point differences can be employed (Warming & Beam 1977) and both fourth- and second-order forms exist. The fourth-order form can be unwieldy when variable coefficients are encountered in the equations, and second-order methods become a good second choice. The use of higher-order methods for the $\xi$ direction generally cause storage problems; however, it could be worth the effort in some particular problem.

There are a few terms in Equation (15) that require special treatment, and they will now be discussed. The first is the $\eta$ derivative of $V$ which appears in the continuity equation and is the primary equation for determining $V$. If a central second-order approximation is employed for $\partial V/\partial \eta$, then the variable $V$ does not appear along the diagonal of the matrix. This situation can cause practical convergence problems with the resulting linear system and should be avoided. The solution to this problem is to form a special finite-difference cell that is located at the $i+\frac{1}{2}$ point (Blottner 1973) and weighed towards the wall. With this type of differencing the linear system then has a diagonal term for $V$, and the solution procedure is much less sensitive to error buildup.

Another potential difficulty is caused by the fact that the problem is posed along the $\zeta$ direction as an initial-value problem, while the finite-difference procedures recommended require boundary conditions. For external flow over a body of finite extent, this problem is solved by employing periodic boundary conditions, and no difficulties are encountered. For a field of infinite extent the problem can be handled by employing a backward one-sided difference for the last downstream node point. In this way, a consistent problem is easily posed.

A detailed discussion of the stability characteristics of the above recommended finite-difference scheme is beyond the scope of the present article. However, it should be mentioned that the linearized implicit system posed is unconditionally stable. For the details of stability analysis applied to the type of system discussed, one should consult the articles of Schiff & Steger (1979) and Warming & Beam (1977).

## SEPARATION AND FUTURE DIRECTIONS

It can be said in general that the most serious problem faced when numerically solving the equations of fluid dynamics is scaling. Scaling problems caused by geometry and high-gradient boundary layers generate the need for clustering of grid points in order to accurately resolve the physical phenomena. Usually, the location of the high-gradient region is not known *a priori*, nor can it be resolved with a simple coordinate system, and the modeler is left with high-gradient regions inadequately resolved. For the three-dimensional laminar boundary layer the scaling problems are not as severe. The reason for this is that the boundary layer is always small compared with a characteristic body dimension, and the viscous phenomena are confined to the coordinate perpendicular to the wall. Also, by the use of coordinate transformations, which are well understood, most of the scaling problems can be eliminated. The one most serious problem faced in three-dimensional boundary-layer flows is outside the scope of the theory itself, and consists of the connected phenomena of flow reversal, separation, and inviscid interaction with the boundary layer.

Whenever a boundary layer is retarded by an adverse pressure gradient there is a tendency for the boundary layer to thicken due to an increase in the normal velocity $v$ and the resulting ejection of mass from the boundary layer. In the three-dimensional boundary layer this tendency to thicken can be less or more than its two-dimensional counterpart. It is more when both $u$ and $w$ are decreasing and less when one is decreasing and the other increasing. This added degree of freedom in the three-dimensional boundary layer is responsible for the rich variety of "separation" patterns that have been observed (Rosenhead 1963 and Wang 1972). In the present article we discuss only the influence of these phenomena on the numerical-solution technique.

One of the more serious problems is flow reversal and the inadequacy of present numerical methods to march into the history of the flow. This problem can be eliminated by using an adaptive flow-orientated convective derivative; however, the numerical methods then have a semi-elliptic nature along all coordinates. For this situation the numerical-solution effort is equivalent to the Navier-Stokes equations, and the advantages of a boundary-layer approach have been effectively lost (Dwyer & Sherman 1980). Also, problems still exist even if this approach is taken, associated with the classic problem of a singularity existing in the boundary-layer solution at the "separation" line.

The singularity is usually the result of unlimited growth of the boundary layer, and the lack of interaction with the inviscid flow. At the present time there are no efficient methods for interacting the boundary layer and the inviscid flow when separation has occurred. For streamline bodies without separation, approximate methods are useful in practice (Sanders & Dwyer 1976), but they are usually incomplete. The only alternative at the present time for the general problem is to solve the Navier-Stokes equations.

This present situation is frustrating because many terms in the Navier-Stokes equations are small at high Reynolds number, and simplifying assumptions should be extremely useful (Davis 1970). The major difficulty is the pressure field, which is fully elliptic, and causes significant changes in the inviscid flow when separation exists. At high Reynolds number the viscous term is important only perpendicular to the streamlines, and reversed convection can usually be treated with adaptive convective differences. The challenge of the future is to find ways of interacting the pressure field with the viscous flow without giving up the initial value or parabolic nature of a boundary-layer-type approach for the viscous flow. If this challenge is met, there will be considerable improvement in our ability to calculate three-dimensional viscous flows with reasonable amounts of computer time.

Finally, it should be mentioned that some progress has been made in the above-mentioned direction for supersonic flow with partial separation (Schiff & Steger 1979). The calculations take advantage of the fact that many supersonic flows exhibit only a crossflow reversal and separation, while the primary-flow direction does not reverse. Also, for supersonic flow the inviscid flow is incapable of transmitting upstream the effects of boundary-layer growth, and the pressure interaction is only exerted downstream. These two factors allow for an effective parabolization of the Navier-Stokes equations, and a major reduction in computational effort required.

However, even for this rather specialized case there are still some serious and thus far unsolved problems. The problems are associated with approximations made on the pressure field in the viscous subsonic portion of the flow (where potential pressure interactions could be transmitted upstream), and disquieting behavior of the methods when the numerical step-size is decreased. At the present time the methods only "work" (agree with to some undefined accuracy) for a step size larger than some minimum value. If the step size in the $\xi$ direction is decreased below this minimum value the method goes unstable. Since this behavior goes against the basic convergence requirements of a finite-difference method, it must be said that the parabolized Navier-Stokes methods are still theoretically inadequate. However, it should be stated again that this type of approach must be classified as a very promising one in the future.

## CONCLUSIONS

It has been shown in the above article that our knowledge about numerical-solution techniques for solving three-dimensional boundary layers is very mature in many areas. It is at present possible to employ generalized surface coordinates and scaling-law coordinate transformations together with efficient and stable finite-difference schemes to solve complex flow problems. The central problems that remain, such as separation, are somewhat outside the scope of boundary-layer theory itself, and usually involve the solution of fully elliptic flow problems. However, there does seem to be hope that simplifying assumptions can be made to the full Navier-Stokes equations that do have the boundary-layer spirit. Also, there are large economies to be gained at high Reynolds number if these new potential methods can be used to solve complex flow problems with flow separation occurring in the solution.

### ACKNOWLEDGMENT

At this moment I can think of many scientists who have contributed much to my knowledge on boundary-layer theory and numerical fluid dynamics in general. To all those friends and colleagues I express a hearty thanks. However, one of these friends must be recognized for his very large contribution, and that friend is Fred Blottner.

*Literature Cited*

Beam, R., Warming, R. F. 1976. An implicit finite-difference algorithm for hyperbolic systems in conservative-law-form. *J. Comput. Phys.* 22:87–110

Blottner, F. G. 1973. Three-dimensional, incompressible boundary layer on blunt bodies. *Rep. SLA-73-0366.* Sandia Labs., Albuquerque

Briley, W. F., McDonald, E. 1973. An implicit numerical method for the multidimensional compressible Navier-Stokes equations. *Rep. M911363-6.* United Aircraft Res. Labs.

Davis, R. T. 1970. Numerical solution of

the hypersonic viscous shock-layer equations. *AIAA J.* 8:843–51

Der, J. Jr., Raetz, G. S. 1962. Solution of general three-dimensional laminar boundry layer problems by an exact numerical method. *IAS Pap. 62-70*. Jan.

Dwyer, H. A. 1968. Solution of a three-dimensional boundary-layer flow with separation. *AIAA J.* 6(7):1336

Dwyer, H. A. 1971. Boundary layer on a hypersonic sharp cone at small angle of attack. *AIAA J.* 9:277

Dwyer, H. A. 1974. A physically optimum difference scheme for three-dimensional boundary layers. *Proc. 4th Int. Conf. Numerical Methods in Fluid Mechanics*. New York: Springer

Dwyer, H. A., Sherman, F. R. 1980. Some characteristics of unsteady two- and three-dimensional reversed boundary layer flows. *AIAA J.* In press

Hindmarsh, A. C. 1977. Solution of block-tridiagonal systems of linear algebraic equations. *Lawrence Livermore Lab. Rep. UCID-30150*. Livermore, Calif.

Kitchens, C. W. Jr., Gerber, N., Sedney, R. 1975. Computational implications of the zone of dependence concept for three-dimensional boundary layers on a spinning body. *USA Ballistic Res. Labs. Rep. No. 1774*. Aberdeen Proving Grounds, Maryland

Rosenhead, L. 1963. *Laminar Boundary Layers*. Oxford Univ. Press

Sanders, B. R., Dwyer, H. A. 1976. Magnus forces on spinning supersonic cones. *AIAA J.* 14(5):576–82

Schiff, L. B., Steger, J. L. 1979. Numerical simulation of steady supersonic viscous flow. *17th Aerosp. Sci. Meet., Pap. 79-0130*, New Orleans

Steger, J. L. 1978. Implicit finite-difference simulation of flow about arbitrary two-dimensional geometries. *AIAA J.* 16: 679–86

Stewartson, K. 1964. *The Theory of Laminar Boundary Layers in Compressible Fluids*. Oxford Univ. Press

Vinokur, M. 1974. Conservation equation of gas-dynamics in curvilinear coordinate systems. *J. Comput. Phys.* 14:105–25

Viviand, H. 1974. Conservative forms of gas dynamic equations. *Rech. Aérosp.* 1:65–68

Wang, K. C. 1971. On the determination of the zones of influence and dependence for three-dimensional boundary layer equations. *J. Fluid Mech.* 48(2):397–404

Wang, K. C. 1972. Separation patterns of boundary layer over an inclined body of revolution. *AIAA J.* 10(8):1040–50

Warming, R. F., Beam, R. 1977. On the construction and application of implicit factored schemes for conservation laws. *Proc. Symp. Comput. Fluid Mech., SIAM-AMS Proc.*, Vol. II, New York

*Ann. Rev. Fluid Mech. 1981. 13:231–52*


# STABILITY OF SURFACES THAT ARE DISSOLVING OR BEING FORMED BY CONVECTIVE DIFFUSION

*8176

*Thomas J. Hanratty*
Department of Chemical Engineering, University of Illinois, Urbana, Illinois 61801

## 1 INTRODUCTION

If a turbulent fluid flows over a plane soluble surface, material need not be removed uniformly from the surface. Recent papers by Ashton (1972), Ashton & Kennedy (1972), Thorsness & Hanratty (1979b), and a PhD thesis by Thorsness (1975) explore the stability of a dissolving surface. It is shown that a growing two-dimensional wave pattern can emerge if the function describing the variation of the convective mass-transfer rate along a small-amplitude sinusoidal waveform has a maximum somewhere in the trough region. Depending on whether the maximum is on the downstream side of the trough, the upstream side of the trough, or right at the bottom of the trough, the wave will propagate in the direction of flow, will propagate in a direction opposite to the flow, or will remain stationary.

It is clear from these results that the determination of the stability of a flat surface in a flowing stream that is being dissolved or that is being formed by a diffusional process requires an understanding of how the concentration field is modulated by the presence of waviness on the surface. For laminar flows the wave-induced variation of the velocity normal to the surface gives rise to variations of the concentration field and of the mass-transfer rate. For turbulent flows, wave-induced variation of the turbulence properties is also an important mechanism. At present, a completely satisfactory theory for predicting the wave-induced variation of turbulence properties and its influence on turbulent transport is not available.



0066-4189/81/0115-0231$01.00

Two recent studies of mass and heat transfer to small-amplitude waves are particularly useful in guiding theoretical work. Electrochemical techniques have been used to study the variation of the mass transfer along a sinusoidally shaped wave with a length, $\lambda$, of 5.09 cm and an amplitude (1/2 of the distance between the crest and trough) of 0.03175 cm (Thorsness 1975, Thorsness & Hanratty 1979a). The flow was turbulent and fully developed. The Schmidt number characterizing the process was 729. A portion of a wavy wall over which a liquid was flowing was used as one of the electrodes of an electrolysis cell. The cell was operated under conditions so that the current flowing was controlled by the rate of mass transfer to the wavy electrode. The variation of the mass-transfer rate along the surface was determined by measuring the current flowing to a series of wires embedded flush with the electrode surface, but insulated from it. Kennedy and his students (Ashton & Kennedy 1972, Hsu 1973, Hsu, Locher & Kennedy 1979) studied the melting of ice by a flowing water stream. In the initial stages, long-crested ripples of approximately uniform wave length appeared on the surface. Local convective heat-transfer rates were determined by carefully measuring the changes with time of the profiles of these ripples. The Prandtl number characterizing the convective heat transfer was 13.7.

This review summarizes the results of these studies of stability and convective transport to wavy surfaces. In particular, it describes work on the stability of a plane surface that is dissolving or that is being formed in a flowing stream. A disturbance at the interface of the following form has been examined:

$$h(t)=\bar{h}(t)+a(t)\cos(\alpha x-\alpha ct). \tag{1}$$

This represents a wave of wave number $\alpha=2\pi/\lambda$ propagating with velocity $c$. The term $\bar{h}(t)$ is the average height of the surface at any time. The term $a(t)$ is the amplitude of the disturbance wave. The choice of a two-dimensional disturbance is clearly a limitation of presently available work because many observed dissolution patterns are three dimensional.

## 2 OBSERVATIONS OF WAVELIKE PATTERNS

The chief motivation for the stability analyses is the appearance on soluble solid surfaces of wavelike patterns that are caused by diffusional processes. The goal has been to explain the origin of these patterns and to relate their wave length and velocity to the flow conditions that prevailed during their formation.

Studies of the wavelike patterns on limestone formations that are or have been exposed to flowing water have been reported in literature dating back to 1915. The patterns usually appear as closely packed hollows in the surface that have approximately the same dimension in the direction of flow as in the lateral direction, and are called "scallops" by geologists. However, they also occasionally appear as two-dimensional periodic structures characterized by sharp, parallel crests that lie transverse to the flow direction. These have been identified as "flutes" by Curl (1966). Comprehensive reviews of observations and some excellent photographs of flutes and scallops on limestone surfaces have been given by Curl (1966), Goodchild & Ford (1971), and Blumberg & Curl (1974). Goodchild & Ford give the following description: "Scallops are found in open channels, on the walls, ceilings, and floors of cave conduits, and on boulders and cobbles in stream courses. Although on any one surface there is little variation in scallop dimensions, these range widely between scattered locations. To quote extremes measured by the authors, scallop lengths of 2 mm are typical of steeply sloping conduits developed in Cambrian marbloid rocks in the Nakimu Caves high in the Selkirk Mountains of British Columbia. They are commonly 2 m long in the great horizontal trunk aquifers of Mammoth Cave in Central Kentucky Karst, developed in Mississippian limestones."

It is now generally accepted that solution of limestone is responsible for the formation of scallops and flutes on rock surfaces. This involves the diffusion of $CO_2$ to the surface and diffusion of bicarbonate back to the main stream. Arguments to support this explanation are presented by Curl (1966).

Similar patterns have been observed on snow or ice that is ablating into an air stream. Goodchild & Ford (1971) present a review of literature on this subject and argue that these patterns are due to convective heat transfer. They give the following description of the process: "Snow scallops are generated when the dominant mode of removal is by warm moving air, rather than by direct solar heating. They are common on high alpine snowfields in late summer (where we have measured mean lengths ranging from 10 to 30 cm), and often occur during mild winter spells in temperate latitudes." Photographs of ablation patterns in snow fields and in ice caves are presented by Goodchild & Ford (1971) and by Blumberg & Curl (1974).

Curl (1966) examined examples of flute patterns in limestone-water systems and in ice-air systems where the mainstream velocity $U_B$ could be estimated and noted that the Reynolds numbers based on the wavelength were roughly 23,000 for the two cases, despite an order-of-magnitude difference in the kinematic viscosity. More recently Blumberg & Curl (1974) and Curl (1974) have argued that the friction velocity, $v^*$, should be used rather than the mainstream velocity and that flutes are characterized by $\lambda v^*/\nu = 2200$.

Several studies of the dissolution of plaster of paris have been carried out under carefully controlled laboratory conditions (Rudnicki 1960, Allen 1971, Goodchild & Ford 1971, Blumberg & Curl 1974). Only scallops appeared spontaneously from an initially flat surface. In order to generate flutes it was necessary to inscribe transverse grooves at chosen intervals. To investigate the stability conditions for flutes, Blumberg & Curl carried out experiments at two different flow rates using grooves spaced at distances of 2.54 cm and 5.08 cm. They imposed a pattern characterized by a Reynolds number based on the wavelength and the maximum velocity, $Re_m$. At $Re_m \cong 25{,}000$ the pattern that formed appeared stable both for the 2.54-cm and the 5.08-cm markings. An instability that tended to increase the average wavelength of the pattern was noted at $Re_m \cong 12{,}000$ for markings spaced at 2.54 cm. An instability that decreased the average wavelength was noted for 5.08 cm markings at $Re_m \cong 51{,}000$. The experiments of Blumberg & Curl showed that the $Re_m$ characterizing stable flutes is between 12,000 and 51,000 and that the value of 23,000 suggested by Curl (1966) does maintain flute wavelength. Blumberg & Curl (1974) and Curl (1974) chose to characterize the average length of the scallops that appear spontaneously from a flat surface as a Sauter mean length, $L_{32} = \Sigma l_i^3 / \Sigma l_i^2$, where $l_i$ is the largest longitudinal dimension of the $i$th scallop. In two experiments in which $U_m/\nu$ differed by a factor of nearly four, it was found that $L_{32}U_m/\nu$ had values of 24,300 and 26,200.

Wave forms similar to flutes have been observed on the underside of ice covers on rivers and on canals (Carey 1966, 1967, Larsen 1969, Ashton & Kennedy 1972). These have been described as regular, long-crested, downstream-migrating waves. Larsen describes tests in which an ice floe with a smooth upper side was turned and left in this position for 18 days. When turned again a small-amplitude long-crested wave pattern was observed. During this test, the air temperature was low with night temperatures of $-20°C$ and day temperatures from $-7$ to $-10°C$. Larson points out that heat transfer from the ice to the air would tend to smooth out the surface and that the patterns must be associated with interaction between the water and the ice.

Ashton & Kennedy (1972) identify three different principal stages in the evolution of the geometry of the ice-water interface: "an initial plane-interface stage which accompanies very cold weather and rapid thickening of the ice cover; a growth stage in which smooth undulations develop on the underside of the ice cover, generally when the ice is thickening less rapidly or is melting from below; and a final stage in which the ripples become so steep that the flow separates downstream from each ripple crest leading to the development of asymmetric sharp-crested wave forms." This description was confirmed (Ashton & Kennedy 1972) by flowing water over a 12-meter-long ice slab frozen to the bottom of a recirculating flume that was housed in a cold room of the Iowa Institute of Hydraulic Research. The flow was heated and its temperature was controlled to approximately $\pm 0.01\,^{\circ}$C. As the initially plane slab melted, wave forms developed at the interface and amplified with time.

Ashton & Kennedy (1972) measured the distance between adjacent crests and found that its average value varied inversely with the liquid velocity. A more recent paper from the University of Iowa research group (Hsu, Locher & Kennedy 1979) gives $U_B\lambda/\nu = 5.9\times10^4$ and $\lambda v^*/\nu = 3{,}180$. Good agreement was noted for lengths determined from zero crossings and from the wave number of the peak of the spectral function describing the surface profile.

Records of well-documented deposition patterns arising because of diffusional processes in a flowing stream are difficult to find. Blumberg & Curl (1974) cite wavy roughness in the Ecker Valley pipeline (Wiederhold 1949, Seiferth & Krüger 1950) "caused, ostensibly, by deposition of colloidal aluminum hydroxide." Blumberg & Curl estimate a value of $\lambda v^*/\nu$ for this pipeline in the range of $116-662$. They point out that this is much smaller than what is observed for dissolution patterns.

## 3 STABILITY CRITERION

A criterion for stability can be established by considering the mass balance for a surface that is dissolving or depositing solely because of convective transfer. For a small-amplitude wave,

$$-\rho_s \frac{dh}{dt} = N, \tag{2}$$

where $\rho_s$ is the density of the solid and $N$ is the local rate of convective transfer per unit area from the solid to the fluid. The surface flux can be decomposed into a mean and a fluctuating component defined in the following way:

$$N = \overline{N} \pm a|\hat{n}|\cos[(\alpha x + \theta) - \alpha c t], \tag{3}$$

where the minus sign is to be used when convective transfer is occurring from the fluid to the solid. The term $\overline{N}$ is the mean flux from the solid to the fluid and is a constant. The term $a(t)$ is the wave amplitude defined by (1). For a linear process the amplitude of the mass-transfer variation is linearly dependent on $a$. This dependence is expressed explicitly in (3) by designating the mass-transfer amplitude as $a|\hat{n}|$. By comparing (1) and (3) it is seen that at a given time the spatial variation in the mass-transfer rate is allowed to be out of phase with the wave profile. The phase angle $\theta$ is the number of degrees by which the maximum in the mass-transfer rate precedes the maximum in the wave profile.

By using the trigonometric relation for the cosine of the sum of two angles, (3) can be written in the alternate form

$$N=\bar{N}\pm a|\hat{n}|\cos\theta\cos(\alpha x-\alpha ct)\mp a|\hat{n}|\sin\theta\sin(\alpha x-\alpha ct). \tag{4}$$

If (4) and (1) are substituted into (2) the following relations are obtained for a dissolving surface:

$$-\rho_s\frac{d\bar{h}}{dt}=\bar{N}, \tag{5}$$

$$-\rho_s\frac{da}{dt}=a|\hat{n}|\cos\theta, \tag{6}$$

$$c=\frac{|\hat{n}|\sin\theta}{\alpha\rho_s}. \tag{7}$$

Equation (6) can be solved to get

$$a=a_0e^{\beta t}, \tag{8}$$

with

$$\beta=-|\hat{n}|\frac{\cos\theta}{\rho_s}. \tag{9}$$

If $\beta$ is positive, the surface waves will grow and if $\beta$ is negative, they will decay. For a dissolution process, positive $\beta$ requires $\cos\theta$ to be negative so that $\pi/2<\theta<3\pi/2$. This implies that for dissolution waves to grow the maximum in the mass-transfer rate must occur somewhere in the trough. If it is in the downstream half of the trough, the dissolution wave will propagate in the downstream direction and if it is in the upstream half, it will propagate upstream.

For processes in which a surface is being deposited by convective diffusion

$$c=-\frac{|\hat{n}|\sin\theta}{\alpha\rho_s},\ \beta=+|n|\frac{\cos\theta}{\rho_s}, \tag{10}$$

and the stability criterion is just opposite that for dissolution, i.e. the maximum mass-transfer rate must occur along the wave crest for waves to grow.

It is clear from the above considerations that the stability of surfaces that are dissolving by convective diffusion is determined by the phase of the wave-induced variation of the surface flux. Ashton & Kennedy (1972) have argued that for turbulent flows the local transfer rate is proportional to the mean squared value of the turbulent velocity fluctuations near the boundary. They cited previous work that shows that flow convergence (positive pressure gradient) results in an attenuation of the turbulence intensity while flow divergence (negative pressure gradient) enhances it. For the case of a sinusoidal boundary the flow close to the boundary is exposed alternatively to positive and negative pressure gradients. Because the streamlines are squeezed together at the crest and spread apart at the trough one would expect that the flow velocity would roughly be a maximum at the crest and a minimum at the trough. From the Bernoulli equation the pressure gradient would have a minimum at $\theta = +\pi/2$ and a maximum at $\theta = -\pi/2$.

Ashton & Kennedy argued that, if the local turbulence intensity adjusted instantaneously to the value that it would have at equilibrium conditions under the local pressure gradient, the maximum in the local turbulence would be shifted downstream of the maximum in the wave height by $\gamma$ radians, where $\gamma \cong \pi/2$. However, since the flow does not adjust instantaneously, they introduce an additional lag distance, $\Delta$, by which the local turbulence lags the boundary displacement. The final expression for $\theta$ suggested by Ashton & Kennedy is

$$\theta = -\left(\gamma + \frac{2\pi\Delta}{\lambda}\right). \tag{11}$$

No method was presented by Ashton & Kennedy to predict the parameters $\gamma$ and $\Delta$.

Thorsness & Hanratty (1979a) have calculated values of $\theta$ and $|\hat{n}|$ by solving the mass-balance equation for the concentration field. The results obtained by them will now be discussed.

# 4 CALCULATION OF THE CONCENTRATION FIELD

## *Equations for the Wave-Induced Concentration Variations*

Thorsness & Hanratty formulated the mass-balance equation in a boundary-layer coordinate system in which the $x$-direction is parallel to the wave surface and the $y$-direction perpendicular to it. The time-averaged velocity components in the $x$- and $y$-directions are represented by $U$ and $V$, the time-averaged concentration by $S$, and the Reynolds transport in the $y$-direction by $M$. For small-amplitude waves a linear response can be assumed. Then $U$, $V$, $S$, and $M$ at a fixed value of $y$ are given as the sum of the average over a wavelength and a wave-induced component having the same wave number as the profile of the solid surface:

$$U = \overline{U}(y) + a\hat{u}(y)e^{i\alpha x}, \tag{12}$$

$$V = \overline{V}(y) + a\hat{v}(y)e^{i\alpha x}, \tag{13}$$

$$S = \overline{S}(y) + a\hat{s}(y)e^{i\alpha x}, \tag{14}$$

and

$$M = \overline{M}(y) + a\hat{m}(y)e^{i\alpha x}, \tag{15}$$

where $a$ and $\alpha$ are real and $\hat{u}$, $\hat{v}$, $\hat{s}$, $\hat{m}$ can be complex. Since for a linear response the amplitudes of the wave-induced changes vary linearly with amplitude this dependency is shown explicitly. Thus $a\hat{u}(y)$ is the complex amplitude of the wave-induced variation of $U$, and $\hat{u}(y)$ is independent of $a$. The above equations imply that the changes of $U$, $V$, $S$, and $M$ are small over one wavelength, and, consequently, that the thickness of the velocity and concentration boundary layers does not change appreciably over one wavelength.

Thorsness & Hanratty formulated the mass-balance equation in a curvilinear coordinate system embedded in the wave surface. They used a boundary-layer assumption whereby the influence of turbulent and molecular diffusion terms in the $x$-direction is ignored. Equations (12), (13), (14), (15) are substituted into this mass-balance equation and terms of order $a^2$ are neglected. The following equation is derived for $\hat{s}(y)$, where primes denote differentiation with respect to $y$ and $D$ is the diffusion coefficient:

$$e^{i\alpha x}\left[i\alpha\overline{U}a\hat{s} + a\hat{v}\overline{S}' + a\hat{u}\frac{\partial\overline{S}}{\partial x} + \overline{V}a\hat{s}'\right] + \overline{U}\frac{\partial\overline{S}}{\partial x} + \overline{V}\overline{S}'$$

$$= D\overline{S}'' - \overline{M}' + e^{i\alpha x}\left[aD\hat{s}'' - a\hat{m}' + a\alpha^2\left(y\overline{U}\frac{\partial\overline{S}}{\partial x} - \overline{M} + D\overline{S}'\right)\right]. \tag{16}$$

The last three terms on the right side of (16), which arise because of the curvature of the coordinate system, can be neglected. In addition, if interest is focused on fully developed flows in a channel or on boundary layers that are sufficiently developed, then $\hat{u}\partial\overline{S}/\partial x$ and $\overline{V}\hat{s}'$ can be neglected. The following differential equations are obtained for $\overline{S}$ and $\hat{s}$:

$$\overline{U}\frac{\partial\overline{S}}{\partial x} + \overline{V}\overline{S}' = D\overline{S}'' - \overline{M}', \tag{17}$$

$$i\alpha\overline{U}\hat{s} + \hat{v}\overline{S}' = D\hat{s}'' - \hat{m}'. \tag{18}$$

These are to be solved using the boundary conditions

$$\bar{S}(0)=0, \qquad \bar{S}(\infty)=\bar{S}_\infty, \tag{19}$$

$$\hat{s}(0)=0, \qquad \hat{s}(\infty)=0. \tag{20}$$

The rate of mass transfer from the surface to the fluid can also be represented by an equation of the same form as (12)–(15):

$$N=\bar{N}+a\hat{n}e^{i\alpha x}. \tag{21}$$

The complex amplitude $\hat{n}$ can be evaluated once a solution to (18) is obtained since

$$-D\hat{s}'(0)=\hat{n}. \tag{22}$$

The amplitude and phase angle appearing in (3) are then calculated as follows:

$$|\hat{n}|^2=\hat{n}\hat{n}^{\neq}, \tag{23}$$

$$\tan\theta=\frac{\hat{n}_I}{\hat{n}_R}, \tag{24}$$

where $\hat{n}^{\neq}$ is the complex conjugate of $\hat{n}$.

From (18) it is seen that wave-induced variations in the concentration field occur because of wave-induced variations in the velocity normal to the surface, through the $\hat{v}\bar{S}'$ term, and because of wave-induced variations of the turbulence, through the $\hat{m}'$ term.

## *Solution for Laminar Flow*

For laminar flows $\hat{m}'=0$. Thorsness & Hanratty present a solution to (15) for the case of high Schmidt number. Under these circumstances, the scalar boundary layer is much thinner than the velocity boundary layer so that the velocity field may be taken as $\bar{U}=\bar{\tau}_w y/\mu, \hat{v}(y)= -i\alpha\hat{\tau}_w y^2/2$, where $\bar{\tau}_w$ and $a\hat{\tau}_w$ are the average wall shear stress and the wave-induced amplitude of the wall shear stress. The variation of the mean concentration was assumed to have the form

$$\bar{S}=-\left[2\left(\frac{y}{\delta_c}\right)-2\left(\frac{y}{\delta_c}\right)^3+\left(\frac{y}{\delta_c}\right)^4\right]\frac{\bar{N}\delta_c}{2D}. \tag{25}$$

By using the method of variation of parameters to solve (18) the following result is obtained:

$$\frac{a\hat{n}}{\bar{N}}=i\frac{a\hat{\tau}_w}{\bar{\tau}_w}\frac{1}{2Ai(0)}\int_0^{\delta_c^*}\mathrm{Ai}(y^*i^{1/3})y^{*2}\left[1-3\left(\frac{y}{\delta_c}\right)^2+2\left(\frac{y}{\delta_c}\right)^3\right]dy^*, \tag{26}$$

where $y^*=(\alpha^+Z)^{1/3}y^+, \delta_c^*=(\alpha^+Z)^{1/3}\delta_c^+$ and Ai designates the Airy function. The terms $\alpha^+, y^+$, and $\delta_c^+$ are made dimensionless using the wall parameters $\nu$ and $v^* = (\bar{\tau}_w/\rho)^{1/2}$. The term $Z = \nu/D$ is the Schmidt number.

Thorsness & Hanratty present a numerical evaluation of (26). This shows that for large $\alpha$ the amplitude $a\hat{n}$ has the same phase as $a\hat{\tau}_w$ and that $a\hat{n}/\bar{N}$ is approximately equal to 0.5 $a\hat{\tau}_w/\bar{\tau}_w$. With decreasing $\alpha$ the ratio $\hat{n}\bar{\tau}_w/\bar{N}\hat{\tau}_w$ decreases and there is an increase in the relative phase angle so that the maximum mass-transfer rate precedes the maximum shear stress.

Calculations for a laminar boundary layer on a flat plate for $Z = 1000$ show that, for large $\alpha$, $a|\hat{n}|$ can be quite large, being equal to 0.1 to 0.2 of the mean flux for $a^+ = av^*/\nu = 1$. The phase angle, $\theta$, which is largely governed by the phase angle of the wall shear-stress variation, has a value between 30–70 degrees.

Since a phase angle greater than 90° is needed for instability of a dissolving surface, this calculation indicates that dissolution instabilities are not likely to occur for laminar flows.

## *Solution for Turbulent Flow*

Thorsness & Hanratty (1979a) also give solutions of (18) for turbulent flows that are sufficiently developed that the variation of the mean scalar field is given as a solution of

$$\bar{N} = -D\bar{S}' + \bar{M}. \tag{27}$$

If an eddy diffusivity is defined as

$$\bar{M} = -\bar{\varepsilon}\bar{S}' \tag{28}$$

the mean scalar field is obtained by integrating (27),

$$\bar{S} = \int_0^y \frac{-\bar{N}\,dy}{D + \bar{\varepsilon}}\,. \tag{29}$$

The eddy-diffusion coefficient, $\bar{\varepsilon}$, was calculated with the Van Driest relation

$$\frac{\bar{\varepsilon}}{\nu} = \frac{\bar{\nu}_T}{\nu} = (Ky^{+2})\left[1 - \exp\left(-\frac{y^+}{A}\right)\right]^2 \bar{U}', \tag{30}$$

with $K = 0.41$ and $A = 25$. The term $\bar{\nu}_T$ is the eddy viscosity so that for a constant-stress boundary layer

$$\bar{U}^+ = \frac{\bar{U}}{v^*} = \int_0^{y^+} \frac{2\,dy^+}{1 + \sqrt{1 + K^2 y^{+2}\left[1 - \exp\left(-\frac{y^+}{A}\right)\right]^2}}\,. \tag{31}$$

The wave-induced variation of the turbulent transport $\hat{m}$ is assumed to be given by the equation

$$-M=(\bar{\varepsilon}+a\hat{\varepsilon}e^{i\alpha x})(\bar{S}'+a\hat{s}'e^{i\alpha x}). \tag{32}$$

If terms of order $a^2$ are neglected

$$-\hat{m}=\hat{\varepsilon}\bar{S}'+\bar{\varepsilon}\hat{s}'. \tag{33}$$

The term $a\hat{\varepsilon}e^{i\alpha x}$ is the wave-induced variation in the eddy diffusivity and arises because of the wave-induced variation of the turbulence properties. Thorsness & Hanratty estimated $\hat{\varepsilon}$ by using the analogy

$$\hat{\varepsilon}=\hat{\nu}_T. \tag{34}$$

The wave-induced variation of the velocity was obtained from the solution of the linearized momentum equations given by Thorsness et al (1978) for the boundary-layer coordinate system. This solution requires the specification of the variation of eddy viscosity. Thorsness et al (1978) have compared a number of models for $\nu_T$ with respect to their ability to predict the variation of the wall shear stress along the wave surface and found that a modified version of the Van Driest relation presented by Loyd et al (1970) did a satisfactory job. According to this model

$$\nu_T=L^2(2e_{xy}) \tag{35}$$

and

$$L=Ky\left[1-\exp\left(-\frac{y^+}{A}\right)\left(\frac{\tau_w}{\bar{\tau}_w}\right)^{1/2}\right], \tag{36}$$

where $e_{xy}$ is a component of the rate-of-strain tensor, and the coefficient, $A$, which is a measure of the thickness of the viscous wall region, is a function of the pressure gradient. For equilibrium boundary layers where the pressure gradient is not varying, $A$ increases in favorable pressure gradients and decreases in unfavorable pressure gradients.

For small enough pressure gradients,

$$A=A_0\left(1+k_1\frac{dp}{dx}\right), \tag{37}$$

where $k_1$ is a negative number. For nonequilibrium flows, Loyd et al (1970) have argued that an effective pressure gradient, $(dp/dx)_{\text{eff}}$, should be used in (37). Thus,

$$\frac{d\left(\frac{dp}{dx}\right)_{\text{eff}}}{dx}=\frac{\left(\frac{dp}{dx}\right)-\left(\frac{dp}{dx}\right)_{\text{eff}}}{k_L}, \tag{38}$$

where $k_L$ is a relaxation constant. According to this model, wave-induced variation in turbulent transport of heat, mass, or momentum occurs because of variation of $S', e_{xy}, dp/dx$, and $\tau_w$. The turbulent transport terms are made to respond instantaneously to changes in $S'$, $e_{xy}$, and $\tau_w$, but are relaxed with respect to changes in $dp/dx$. This is identified as Model D in the paper by Thorsness et al (1978). Loyd et al (1970) suggested that $K=0.41, \bar{A}=25, k_1=-30$, and $k_L=3000$. Thorsness et al (1978) found that the use of these values gave approximate agreement between the measured and predicted variation of the wall shear stress along the wave surface. However, much closer agreement could be obtained if values of $k_1=-60, k_L=3000$ or $k_1=-30$, $k_L=1500$ were used.

Thorsness & Hanratty (1979a) gave solutions to (18) where $\bar{U}$ is given by (31), $\bar{S}$ by (29), $\hat{m}$ by (33) and (34), and $\hat{v}$ by solution of linear momentum equations presented by Thorsness et al (1978). The eddy-diffusion coefficient $\bar{\varepsilon}$ is given by (30) and $\hat{\nu}_T$ by (35), (36), (37), and (38) with $k_1=-60, k_L=3000$. Calculated values of $\theta$ and $|\hat{n}|\nu/|\bar{N}|v^*$ are plotted in Figure 1 as a function of $\alpha^+=\alpha\nu/v^*$ for $Z=729$ and for $Z=13.7$.

The results are found not to be strongly dependent on $Z$. For large $\alpha^+$ the concentration boundary layer is thin enough that the $\hat{m}$ term is negligibly small and that $\bar{S}=y|\bar{N}|/D$. The solution is (26) with $\delta_c^*=\infty$. The relation of $\hat{n}$ to $\hat{\tau}_w$ for turbulent flows at large $\alpha$ is, therefore, the same as what has been found for a laminar boundary layer. That is, the relative phase of $\hat{n}$ and $\hat{\tau}_w$ is zero and $n/\bar{N}=0.5\ \hat{\tau}_w/\bar{\tau}_w$. Thorsness &

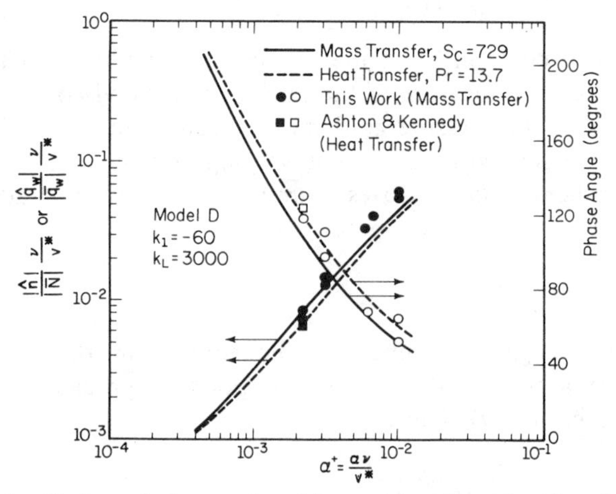


*Figure 1* **Amplitudes and phase angles of the mass-transfer or heat-transfer variation.**

Hanratty (1979a) show that the same result also holds for $\alpha^+ \to 0$. For intermediate $\alpha^+$, however, the phase angle $\theta$ can be larger than the phase angle of $\hat{\tau}_w$. It is of particular interest that there is a range of $\alpha^+$ for which it is possible to have an instability in a dissolving surface; i.e. $90° < \theta < 270°$. From Figure 1 it is seen that this requires $\alpha^+ < 3 \times 10^{-3}$ for $Z = 729$ and $\alpha^+ < 4 \times 10^{-3}$ for $Z = 13.7$. The calculated results shown in Figure 1 were not extended to low enough $\alpha^+$ to determine the lower limit of the range of unstable $\alpha^+$. However, other calculated results presented in the paper by Thorsness & Hanratty (1979a) for $k_1 = -30$ and $k_L = 1500$ reveal $3.4 \times 10^{-4} < \alpha^+ < 3 \times 10^{-3}$. These two calculations indicate that the range of instability for dissolving surfaces is approximately $3.0 \times 10^{-4} < \alpha^+ < 3.0 \times 10^{-3}$.

# 5 MEASUREMENTS OF CONVECTIVE TRANSFER TO WAVY SURFACES

## *Mass Transfer at Z = 729*

A typical mass-transfer profile measured by Thorsness & Hanratty (1979a) along a wavy surface is shown in Figure 2. Measurements of the wall shear-stress profile obtained by Thorsness et al (1978) under similar conditions are shown in Figure 3. The maximum mass-transfer rate is upstream of the wave crest and of the maximum in the wall shear-stress profile.

The curve shown in Figure 2 is the best fit of a sinusoidal relation to the variation of the mass-transfer rate. Values of $|\hat{n}|$ and $\theta$ were then

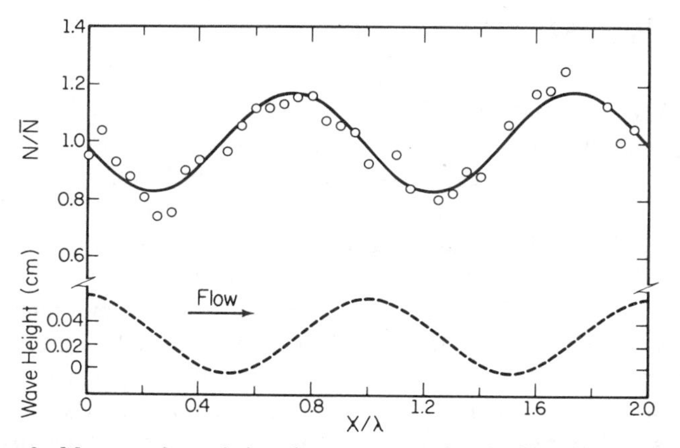


*Figure 2* Mass-transfer variation along a wavy surface for Re = 20,900, $Z = 729$.

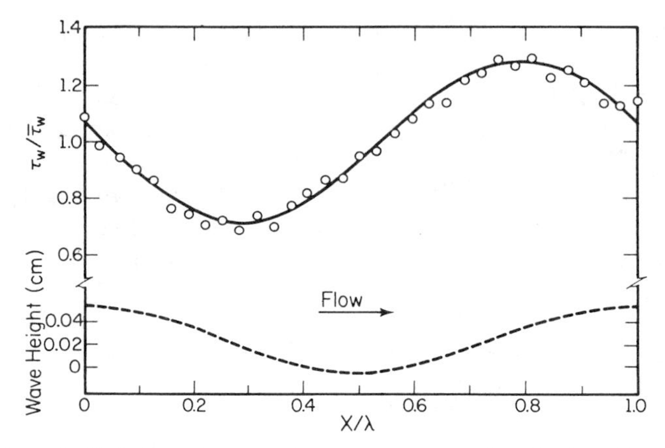


*Figure 3* Wall shear-stress variation along a wavy surface for Re=25,000.

obtained from this best-fit curve. The dimensionless wave number $\alpha^+$ was varied by changing the liquid-flow rate. A comparison of measured values of $|\hat{n}|$ and $\theta$ with calculations is given in Figure 1. Good agreement is noted.

## *Heat Transfer at* Z=*13.7*

The studies by Kennedy and his students of the change of the height of ice ripples with time enable the calculation of the variation of the convective heat-transfer rate along a wavy surface. Ashton & Kennedy present the following heat-balance equation for melting ice:

$$-\rho_s \Delta H_f \frac{dh}{dt} = q_i(x,t) - q_w(x,t), \tag{39}$$

where $\Delta H_f$ is the latent heat of fusion of ice, $q_w$ is the convective heat flux from the wavy side of the ice to the flowing fluid, and $q_i$ is the heat flux to the wavy surface due to heat conduction through ice. The term $+ q_w/\Delta H_f$ then has the same meaning as $N$ in (2). The heat flux was calculated from (39) using measurements of $dh/dt$. The term $q_i$ was calculated by solving the heat-conduction equation.

Values of $|\bar{q}_w|\nu/|\bar{q}_w|v^*$ and of $\theta$ obtained from the measurements of Aston & Kennedy (1972) for a Prandtl number $Z$ of 13.7 are presented in Figure 1. It is noted that these are in close agreement with the mass-transfer measurements of Thorsness & Hanratty, when compared at similar values of $\alpha^+$, even though the experiments were carried out at quite different values of $Z$.

# 6 INTERPRETATION OF PATTERNS ASSOCIATED WITH CONVECTIVE TRANSFER FROM THE SOLID-FLUID INTERFACE

## *Formulation of the Dimensionless Groups*

The analysis of mass-transfer rates to wavy surfaces presented in the previous section suggests that the growth factor, given by (9), and the wave velocity, given by (7), be expressed in the following dimensionless forms:

$$\beta^* = \frac{\beta \nu \rho_s}{\bar{N} v^*} = -\left(\frac{|\hat{n}|\nu}{|\bar{N}|v^*}\right)\cos\theta, \tag{40}$$

$$c^* = \frac{c\rho_s}{\bar{N}} = \left(\frac{|\hat{n}|\nu}{|\bar{N}|v^*}\right)\frac{\sin\theta}{\alpha^+}. \tag{41}$$

Since $|\hat{n}|\nu/|\bar{N}|v^*$ and $\theta$ are strong functions of $\alpha^+$ and only weakly dependent on $Z$, it is concluded that $\beta^*$ and $c^*$ should be primarily dependent on $\alpha^+$.

From (40) it is seen that the maximum in $\beta^*$ is determined by two competing factors. Large values of $|\hat{n}|\nu/|\bar{N}|v^*$ are favored by large values of $\alpha^+$. However, at large $\alpha^+$ the phase angle decreases to zero. For waves to grow $-\cos\theta$ must be positive, $90° < \theta < 270°$, and the most favorable value is 180°. Consequently, it would be expected that the maximum in $\beta^*$ will occur at large values of $\alpha^+$, but not so large

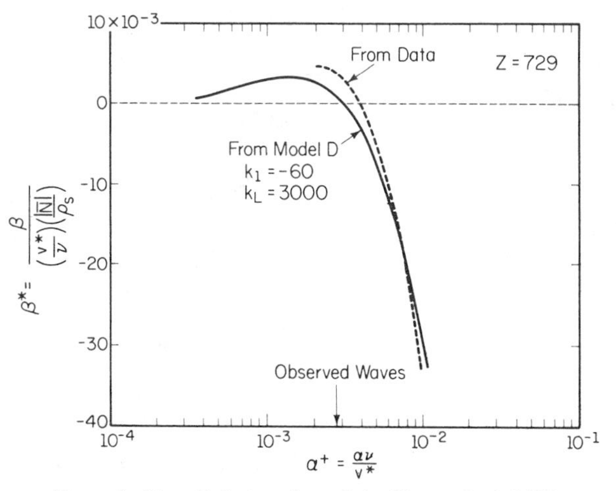


*Figure 4* Growth factors for a Schmidt number of 729.

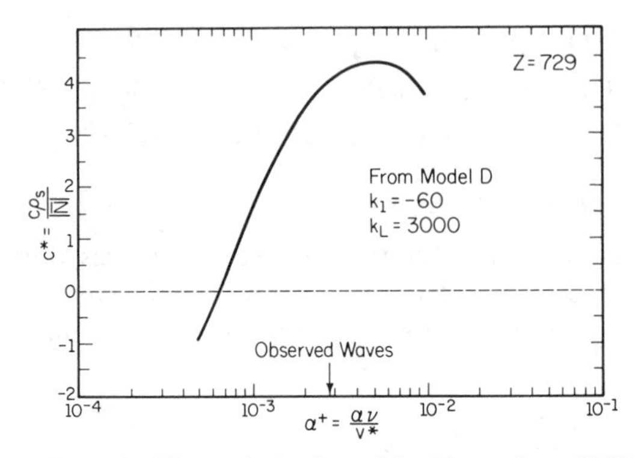


*Figure 5* **Wave velocity for a Schmidt number of 729.**

that the wave-induced variation in the mass-transfer rate will have an unfavorable phase.

Values of $\beta^*$ and $c^*$, calculated by Thorsness & Hanratty (1979b) from the results presented in Figure 1, are shown in Figures 4 and 5 for $Z=729$ and in Figures 6 and 7 for $Z=13.7$. For $Z=729$ the calculated wave-induced mass-transfer variation indicates that dissolution waves are possible for $\alpha^+ < 0.003$ and that a maximum in $\beta^*$ occurs at $\alpha^+ = 0.001-0.002$. The lower limit for the range of $\alpha^+$ for which $\beta^*$ is positive is not shown because the calculations in Figure 1 do not extend to low enough $\alpha^+$. However, as indicated in the previous section, this should be at $\alpha^+ \cong 3.4 \times 10^{-4}$.

Values of $\beta^*$ determined from the mass-transfer measurements of Thorsness & Hanratty are also shown in Figure 4. The measurements

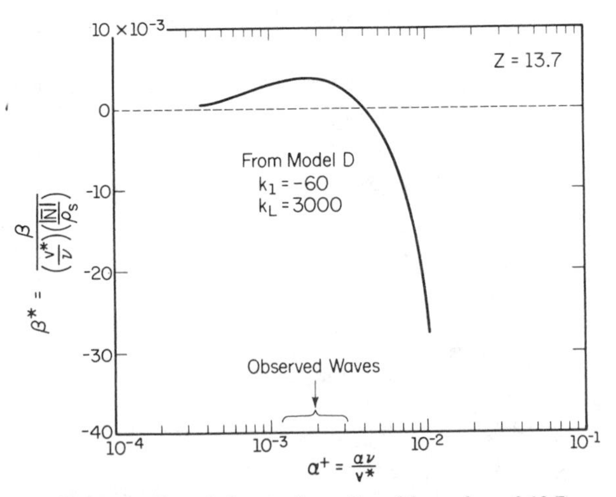


*Figure 6* **Growth factors for a Prandtl number of 13.7.**

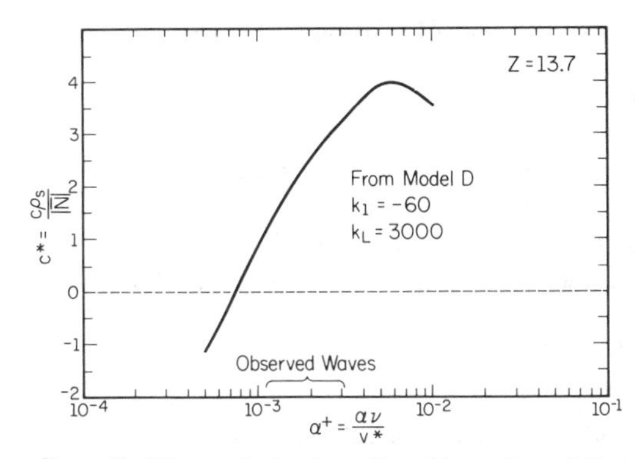


*Figure 7* **Wave velocity for a Prandtl number of 13.7.**

were not carried out over a wide enough range of $\alpha^+$ to establish the maximum in $\beta^*$. However, they do suggest the calculated $\alpha^+$ for maximum growth could be somewhat low because of errors in modeling the wave-induced variation of turbulent transport.

Calculated values of $\beta^*$ for $Z=13.7$ shown in Figure 6 indicate instability for $\alpha^+ < 4\times10^{-3}$ and a maximum in $\beta^*$ at $\alpha^+ = 0.002$.

## *Instability of Ice Surfaces*

The formation of waves on a soluble flat surface can be considered as arising from minute bumps or hollows on the solid surface or from chance spatial variations in the mass-transfer rate arising from disturbances in the flowing fluid. The spatial variation in the height of the surface can be considered as the sum of Fourier components. If the initial surface disturbance is small enough, the rates of change of individual components can be considered independent of one another so that subsequent change of the surface structure can be calculated from solutions of the mass-balance equation for single harmonics. Thorsness & Hanratty (1979b) assumed that the wave that ultimately appears on the surface is rather regular and is the harmonic whose amplitude is increasing most rapidly. This assumption can be checked by examining experimental results on melting ice. According to the above assumption, the wavelength of the observed ice waves should be given by $\alpha^+ =$ constant and $c^* =$ constant, provided $q_i$ is small.

Ashton & Kennedy and later Hsu et al (1979) measured the wavelength and the velocity of small-amplitude ice waves. These covered a range of fluid velocities from 9.1 cm/sec to 83.2 cm/sec. Wavelengths determined by measuring the distance between zero crossings were found to decrease with increasing velocity from 106.4 cm to 13.1 cm.

These results are represented quite well by the relation

$$\frac{\lambda v^*}{\nu} = 3180. \tag{43}$$

This corresponds to a value of $\alpha^+ = 0.002$ in good agreement with predicted $\alpha^+$ for maximum growth.

Measurements of wave velocity shown in Figure 11 of the paper by Ashton & Kennedy (1972) are fitted quite well by the relation

$$c^* = \frac{c\rho_s \Delta H_f}{|q_w|} = 2.2. \tag{44}$$

This is in good agreement with the value of $c^* = 2.5$ predicted for $\alpha^+ = 0.002$ by the theoretical calculation (see Figure 7).

Hsu, Locher & Kennedy (1979) have shown that heat conduction through the ice to the interface is given by

$$q_i = -\frac{k_s \Delta T_i}{\bar{h}} + \frac{k_s \Delta T_i}{\bar{h}} \frac{\alpha a}{\tanh(\alpha \bar{h})} \cos(\alpha x - \alpha c t) \tag{45}$$

where $k_s$ is the thermal conductivity of the ice and $\Delta T_i = T_w - T_0$ is the difference in temperature between the ice-water interface and the bottom of the ice. If this is substituted into the heat-balance equation (39) and an analysis similar to that outlined in Section 3 is carried out, it is found that

$$c = \frac{|\hat{q}_w| \sin\theta}{\alpha \rho_s \Delta H_f}, \tag{46}$$

$$\beta = -\frac{|\hat{q}_w| \cos\theta}{\rho_s \Delta H_f} + \frac{k_s \Delta T_i}{\rho_s \Delta H_f \bar{h}} \frac{\alpha}{\tanh(\alpha \bar{h})}. \tag{47}$$

Thus, $q_i$ has no effect on the wave velocity $c$, but it decreases the time constant for the growth of unstable waves by the amount given by the second term on the right-hand side of the above equation.

Thorsness (1975) used (47) to estimate the influence of conduction of heat away from the surface on the measurements of Ashton & Kennedy. For the experiments, the ratio of the fluctuating component of heat transfer through the ice to that to the water was estimated to be 0.11 or less. Calculations of $\beta$ showed negligible influence of heat transfer through the ice on the wave number for maximum growth. However, consistent with the observation of Larsen (1969) that heat transfer through the ice should have a stabilizing influence, the calculated range of unstable wavelengths is decreased. The comparison of the results of Ashton & Kennedy with theoretical stability calculations for $q_i = 0$ therefore appears justified.

In order to consider the general case of the stability of melting or freezing of ice it is useful to write (47) in dimensionless terms. Define $\beta^* = \beta \nu \rho_s \Delta H_f / |\bar{q}_w| v^*$. Then, for the case where heat is being transferred from the water to the ice

$$\beta^* = -\frac{|\hat{q}_w|\nu}{|\bar{q}_w| v^*}\cos\theta + \frac{\bar{q}_i}{|\bar{q}_w|}\frac{\alpha^+}{\tanh(\alpha h)}, \tag{48}$$

where $\bar{q}_i = -|\bar{q}_i|$ if heat is being conducted away from the water-ice interface. Since

$$\rho_s \Delta H_f \frac{d\bar{h}}{dt} = +\bar{q}_w - \bar{q}_i, \tag{49}$$

it follows that the ice sheet will be thickening for $|\bar{q}_i|/|\bar{q}_w| > 1$.

Recently Gilpin, Hirata & Cheng (1979) have carried out an extensive investigation of the stability of ice surfaces under conditions where conduction of heat away from the ice-water interface is important. They characterize the influence of heat conduction by using a growth parameter $G = |\bar{q}_i|/|\bar{q}_w|$. For increasing $G$ the range of unstable wave numbers and the wave number of maximum growth decreases. They find that for $G > 2.3$ heat conduction is completely stabilizing and that for a given value of $G$ there is a minimum value of $\bar{h}^+ = \bar{h}v^*/\nu$ below which the surface is stable. According to the above calculation, waves can appear while the ice sheet is thickening if $1 < G < 2.3$ and $\bar{h}^+$ is large enough. Thus, there does not have to be a net dissolution of the ice to have an instability. Gilpin, Hirata & Cheng (1979) also find that $G$ does not have a strong influence on the wave number of maximum growth. The range is $0.00105 < \alpha_M^+ < 0.00205$. For the interesting case of an ice cover that has reached a steady-state condition ($G = 1$), they give $\alpha_M^+ = 0.00175$.

## *Instabilities of Dissolving Limestone*

The final wave pattern that develops in a dissolving limestone surface is governed by nonlinear effects, such as flow separation, so that the linearized equations used in stability theory are no longer applicable. Arguments have been presented by Blumberg & Curl (1974) that the final wavelength of the pattern is determined by the distance between the upstream crest, i.e. the separation point, and the average point of flow reattachment.

However, another explanation is that the final wavelength is governed by linear, rather than nonlinear, processes. According to this explanation, small-amplitude ripples occur on the surface because of the instability of the dissolving surface. The final wave form that appears develops from these ripples and has the same wavelength as the ripples but, possibly, a different wave velocity.

Blumberg & Curl characterized two-dimensional dissolution patterns on limestone with $\lambda^+=2200$. This is equivalent to $\alpha^+=2.9\times10^{-3}$. This is in approximate agreement with the calculated value of $\alpha^+=0.001$–$0.002$ for $Z=729$ and in somewhat better agreement with the $\alpha^+$ for maximum growth inferred from the mass-transfer measurements of Thorsness & Hanratty.

The agreement between the calculated and observed wavelengths indicates that a stability mechanism is a possible explanation for the dissolution patterns observed on limestone surfaces.

However, most often these dissolution patterns are three dimensional. Therefore, the comparison between theory and observation would be more convincing if the stability calculations were done for three-dimensional rather than two-dimensional waves. This would be an interesting extension of the analysis of Thorsness & Hanratty.

## 7 INSTABILITY OF SURFACES DUE TO CONVECTIVE TRANSFER BETWEEN A SOLID AND A FLUID

The criterion for deposition patterns, Equation (10), shows that growth is possible provided $-\pi/2<\theta<\pi/2$ and that the most favorable phase angle for growth is $\theta=0°$. From Figure 4 it is found that for large $Z$ deposition patterns are possible provided $\alpha^+>0.003$. The results in Figure 1, as well as the analysis presented by Thorsness & Hanratty (1979a) for large $\alpha^+$, show that $|\hat{n}|$ becomes arbitrarily large and $\theta\rightarrow0°$ for $\alpha^+\rightarrow\infty$. This implies that deposition waves will be of much smaller length than dissolution waves, possibly so small that they will not be observed.

This result seems to be consistent with the observations made by Blumberg & Curl about the dissolution patterns in the Ecker Valley pipeline.

It is of interest to note that deposition instabilities are possible for both laminar and turbulent flows. This is in contrast to dissolution instabilities, which seem to require turbulent flows in order to obtain a large enough phase difference between the mass-transfer variation and the surface profile.

## 8 LIMITATIONS IN THE STABILITY ANALYSIS

In addition to the limitation because of the assumption of a two-dimensional disturbance, the stability analysis for turbulent flows also is open for improvement because of possible inaccuracies in the method for calculating mass transfer to wavy surfaces.

The turbulence model (Model D) for the wave-induced variation of the Reynolds stress uses the notion that the variation of the pressure along the surface gives rise to a variation of the thickness of the viscous wall region. It recognizes that the turbulence does not adjust immediately to a change in pressure gradient. However, the method for taking into account this lag in response is not very sophisticated. Clearly a more rigorous method for modeling the turbulence field is needed. The model used in the calculations summarized in this paper was tested by Thorsness et al (1978) by comparing calculated and measured wall shear-stress variations over a range of wave numbers which is of primary importance in analyzing dissolution patterns, $0.002 < \alpha^+ < 0.01$. Consequently, it should do a reasonably good job in predicting the wave-induced variations in the velocity field. The chief difficulty is in using the analogy between momentum and mass transfer to approximate the Reynolds transport term. As discussed by Thorsness & Hanratty (1979a) this assumption has a weak theoretical basis, so work needs to be done to model the wave-induced variation of the Reynolds transport terms in a more direct way.

## ACKNOWLEDGMENT

This work is being supported by the National Science Foundation under Grant NSF ENG 76-22969.

*Literature Cited*

Allen, J. R. L. 1971. Bed forms due to mass transfer in turbulent flows: a kaleidoscope of phenomena. *J. Fluid Mech.* 49:49–63 (4 plates)

Ashton, G. D. 1972. Turbulent heat transfer to wavy boundaries. *Proc. 1972 Heat Transfer Fluid Mech. Inst.*, pp. 200–13. Stanford University Press

Ashton, G. D., Kennedy J. F. 1972. Ripples on underside of river ice covers. *J. Hydraul. Div. Proc. A.S.C.E.* 98:1603–24

Blumberg, P. N., Curl, R. L. 1974. Experimental and theoretical studies of dissolution roughness. *J. Fluid Mech.* 65:735–51(4 plates)

Carey, K. L. 1966. Observed configuration and computed roughness of the underside of river ice. *US Geol. Surv. Prof. Pap. 550*-B, pp. B192-98

Carey, K. L. 1967. The underside of river ice, St. Croix River, Wisconsin. *US Geol. Surv. Prof. Pap. 575-C*, pp. C195-99

Curl, R. L. 1966. Scallops and flutes. *Trans. Cave Res. Group, Gt. Britain* 7:121–60

Curl, R. L. 1974. Deducing flow velocity in cave conduits from scallops. *Bull. Natl. Speleol. Soc.* 36:1–5.

Gilpin, R. R., Hirata, T., Cheng, K. C. 1979. *Wave Formation and Heat Transfer at an Ice-water Interface in Presence of a Turbulent Flow*, Rep. Dept. Mech. Eng., Univ. of Alberta, Edmonton. *J. Fluid Mech.* In press

Goodchild, M. F., Ford, D. C. 1971. Analysis of scallop patterns by simulation under controlled conditions. *J. Geol.* 79:52–62

Hsu, K. S. 1973. *Spectral evolution of ice ripples*. PhD thesis. Univ. Iowa, Iowa City, Iowa

Hsu, K. S., Locher, F. A., Kennedy, J. F. 1979. *Forced-convection heat transfer from irregular melting wavy boundaries*, Rep.

Iowa Inst. Hydraul. Res., Univ. Iowa, Iowa City, Iowa

Larsen, P. A. 1969. Head losses caused by an ice cover in open channels. *J. Boston Soc. Civil Engrs*. 56:45–67

Loyd, R. J., Moffat, R. J., Kays, W. M. 1970. *Rep. No. MHMT-13*. Thermosci. Div., Dept. Mech. Engrg., Stanford Univ.

Rudnicki, J. 1960. Experimental work on flute development. *Speleologia* (*Warsaw*) 2:17–30

Seiferth, R., Krüger, W. 1950. Überraschend hohe Reibungsziffer einer Fernwasserleitung. *VDI-Z*. 92:189–91

Thorsness, C. B. 1975. *Transport phenomena associated with flow over a solid wavy surface*. PhD thesis. Univ. Ill., Urbana, Ill.

Thorsness, C. B., Morrisroe, P. E., Hanratty, T. J. 1978. A comparison of linear theory with measurements of the variation of shear stress along a solid wave. *Chem. Engrg. Sci*. 33:579–92

Thorsness, C. B., Hanratty, T. J. 1979a. Mass transfer between a flowing fluid and a solid wavy surface. *AIChE J* 25:686–97

Thorsness, C. B., Hanratty, T. J. 1979b. Stability of dissolving or depositing surfaces. *AIChE J* 25:697–701

Wiederhold, W. 1949. Über den Einfluss von Rohrablagerungen auf den hydraulischen Druckabfall. *GWF* (*Gas- u. Wasserfach*) 90:634–41

*Ann. Rev. Fluid Mech. 1981. 13:253–72*


# PROGRESS IN THE MODELING OF PLANETARY BOUNDARY LAYERS

*8177

*O. Zeman*
Sibley School of Mechanical and Aerospace Engineering, Cornell University, Ithaca, New York 14853

## 1 INTRODUCTION

Planetary boundary layers are flows of water or air in the vicinity of planetary surfaces. The planetary boundary layers are known to be mostly turbulent and stratified. Because of large Reynolds numbers associated with planetary motions, boundary-layer turbulence is fully developed and three dimensional. A classic example of a planetary boundary layer is the atmospheric boundary layer over land or water, which has been studied in great detail both theoretically and experimentally. We shall concern ourselves primarily with the modeling problems related to the atmospheric boundary layers, although these problems are equally valid for other planetary boundary layers, such as upper-ocean mixed layers.

Turbulent-flow modeling techniques deal with approximations to the full Navier-Stokes or Reynolds equations. Of considerable and practical interest for planetary boundary layers are the large-eddy simulation and second-order closure techniques and to a lesser degree the integral-method (slab models) and first-order closure techniques (eddy-viscosity models.)

As a matter of historical interest, the modeling of turbulent boundary layers dates back to Reynolds (1895), who established the notion of turbulent (Reynolds) stresses. Reynolds reduced the problem of solving the full, time-dependent Navier-Stokes equations to a seemingly easier task: a closure assumption concerning the Reynolds stresses in terms of mean field quantities and prescribed eddy viscosities. We have learned that this first-order closure often fails in planetary boundary layers, and, therefore, the closure problem has been relegated to a higher order by



0066-4189/81/0115-0253$01.00

formulating conservation equations for Reynolds stresses. The second-order closure technique consists, then, in approximating the third-order unknown terms in the second-order equations, and thus obtaining a more accurate approximation to the second-order quantities (Reynolds stresses). The third-order term approximations, however, present some difficulties especially near the surface, in the presence of large convective structures, or in strong stable stratification.

In the large-eddy simulation technique, large-scale energetic turbulence is resolved exactly on a three-dimensional finite-difference grid and the closure consists in approximating the effect of subgrid scales on the large scales. Large-eddy simulations can be fairly accurate, but prohibitively expensive and, at present, the second-order closure technique remains as the only practical method for simulating a general planetary boundary layer.

In the following section we look at alternative closures in the equations of motion governing the flows in planetary bounday layers. Section 3 is entirely devoted to the review of developments in the second-order modeling of planetary boundary layers.

## 2 GOVERNING EQUATIONS AND THE CLOSURE PROBLEM

The Boussinesq-approximated equations of motion applicable to planetary boundary layers are (Lumley & Panofsky 1964).

$$\dot{U}_i + U_{i,j} + U_j + 2\varepsilon_{ijk}\Omega_j U_k = -\frac{1}{\rho_0} P_{,i} + \beta_i \Theta + \nu U_{i,jj},$$

$$\dot{\Theta} + \Theta_{,j} U_j = \kappa \Theta_{,jj}, \qquad U_{j,j} = 0. \tag{1}$$

Here the $x_3$-axis is in the vertical direction upwards and, unless otherwise noted, $x_1$ and $x_2$ are in the direction West-East and South-North in the Northern hemisphere. $U_i$ is velocity, $\Theta$ potential temperature, $P$ the deviation from hydrostatic pressure, $\Omega_i$, the Earth's angular rotation vector, and $\nu$, $\kappa$ are respectively kinematic viscosity and thermal diffusivity. The buoyancy-acceleration vector is defined as $\beta_i = (0, 0, \alpha g)$, where $\alpha$ is the expansion coefficient.

In a humid noncondensing atmospheric boundary layer, Equation (1) is extended by an equation for water-vapor content and, if important, a radiative-flux divergence has to be included in the $\Theta$ equation. A possibility of condensation and cloud formation in atmospheric boundary layers poses a need for additional equations to describe cloud dynamics (see, for example, Yamada & Mellor 1979 and Deardorff

1980). In the upper ocean (1) may be supplemented by a conservation equation for salinity. In the following, we restrict our discussions to closure problems associated with solving Equations (1) where for convenience the density $\rho_0$ is set at unity.

## *First-Order Closure and Integral Methods*

We substitute in (1)

$$U_i = \overline{U}_i + u_i,\ \Theta = \overline{\Theta} + \theta,\ P = \overline{P} + p,$$

where the overbar designates an average and lower-case letters stand for fluctuating quantities that have, by definition, zero average. Then the averaged (Reynolds) equations are (Lumley & Panofsky 1964)

$$\dot{\overline{U}}_i + \overline{U}_{i,j}\overline{U}_j + 2\varepsilon_{ijk}\Omega_j U_k = -\overline{P}_{,i} + \beta_i\overline{\Theta} + \nu\overline{U}_{i,jj} - \left(\overline{u_i u_j}\right)_{,j},$$

$$\dot{\overline{\Theta}} + \overline{\Theta}_{,j}\overline{U}_j = \kappa\overline{\Theta}_{,jj} - \left(\overline{\theta u_j}\right)_{,j},$$

$$U_{j,j} = 0. \tag{2}$$

Now, the problem of closure consists in modeling the unknown quantities $\overline{u_i u_j}$, $\overline{\theta u_i}$, also called the Reynolds stresses and fluxes.

EDDY-VISCOSITY MODELS The first attempts at modeling the stresses were to assume an analogy between molecular and turbulent transport and make the following closure:

$$\overline{u_i u_j} - \frac{1}{3}\overline{u_k u_k}\,\delta_{ij} = -K_m\left(\overline{U}_{i,j} + \overline{U}_{j,i}\right),$$

$$\overline{\theta u_i} = -K_t\overline{\Theta}_{,i}, \tag{3}$$

where $K_m$ and $K_t$ are the so-called eddy viscosity and diffusivity. The first-order closure techniques (eddy-viscosity models) deal with modeling $K_m$ and $K_t$ in terms of the known averaged quantities.

The simplest eddy-viscosity models assume $K_m = K_t = \text{const}$; more sophisticated ones pattern these after the mixing-length theory with corrections for buoyancy and Coriolis effects (see, for example, Businger & Arya 1974).

INTEGRAL METHODS Integral techniques circumvent entirely the problem of specifying eddy viscosities by integrating (2) over the boundary-layer depth $h$. The closure problem reduces to the assumptions concerning mean profiles, surface-flux parameterizations, and the rate of change of $h$. This technique is particularly useful in convective entraining planetary boundary layers (so-called mixed layers) where mean temperature is known to be uniformly distributed in the core and where a fairly sharp change $\Delta\Theta$ is known to exist at the inversion base $x_3=h$ (Wyngaard et al 1978). The crucial closure problem in this technique is parameterization of the downward heat flux $(\overline{\theta u_3})_i$ at the inversion base; this flux is related to $\Delta\Theta$ and the entrainment rate $\dot{h}$ through the following relationship (Lilly 1968):

$$\dot{h}\Delta\Theta = -(\overline{\theta u_3})_i. \tag{4}$$

Simpler models assume $-(\overline{\theta u_3})_i$ to be a constant fraction of the known surface flux $(\overline{\theta u_3})_0$ (see, for example, Tennekes 1973); more complex ones derive the closure from the turbulence-energy budget at the inversion base (e.g. Zeman & Tennekes 1977). Deardorff (1979) succeeded in closing (4) by making an assumption concerning the temperature-profile shape at the inversion base (see also Wyngaard & LeMone 1980). Integral methods enjoy popularity in modeling mixed layers in the upper ocean (see the review paper by Sherman et al 1978). In these mixed layers the negative buoyancy flux $(\overline{\theta u_3})_i$ at the thermocline is usually maintained by wind-induced turbulence.

In neutral or stable planetary boundary layers, turbulence is maintained solely by shear and (4) is not applicable. Zeman (1979) assumed a planetary boundary layer to consist of an outer, uniform-shear layer and a thin, constant-stress layer at the ground. This allowed him to infer the entrainment rate from the mean kinetic-energy budget:

$$\dot{h} = \frac{Kh^{-1}|\Delta\mathbf{U}|^2 - \mathbf{S}_0\cdot\Delta\mathbf{U}}{|\Delta\mathbf{U}|^2}. \tag{5}$$

Here, $K$ is a bulk eddy viscosity dependent on stability parameters, $\mathbf{S}_0$ is the surface-stress vector, and $\Delta\mathbf{U}$ is the velocity-defect vector in the boundary layer. The rate of change of $h$ is determined by the balance between the turbulence-shear production $\propto K|\Delta\mathbf{U}|^2$ in the outer layer and the work $\mathbf{S}_0\cdot\Delta\mathbf{U}$ done on the outer layer by the mean flow. Equation (5) is capable of predicting collapse of an atmospheric boundary layer during evening transition. A simple general model for the entire diurnal cycle of an atmospheric boundary layer can be generated by linearly combining (4) and (5); then

$$\dot{h} = \frac{Kh^{-1}|\Delta\mathbf{U}|^2 - \mathbf{S}_0\cdot\Delta\mathbf{U} - \beta_3(\overline{\theta u_3})_i h}{|\Delta\mathbf{U}|^2(1+R_{i0})}, \tag{6}$$

where $R_{i0}=\beta_3\Theta h|\Delta\mathbf{U}|^{-2}$ is the outer-layer (bulk) Richardson number. Zeman (1980) employed (6) for predicting convective gravity currents resulting from spills of liquid natural gas.

## *Second-Order Closure Techniques*

It is a logical step further from first-order closures to investigate a possibility of closing the conservation equations for $\overline{u_i u_j}$ and $\overline{\theta u_i}$ in (2). Writing the rate equations for the fluctuating quantities $u_i$, $\theta$, and $p$ and forming the appropriate averages we obtain (see, for example, Lumley & Khajeh-Nouri 1974)

$$\frac{D\overline{u_i u_k}}{Dt} + U_{i,j}\overline{u_k u_j} + U_{k,j}\overline{u_i u_j} - \beta_i \overline{\theta u_k} - \beta_k \overline{\theta u_i}$$

$$= -\underbrace{\left(\overline{p_{,i}u_k + p_{,k}u_i}\right)}_{\langle \Pi_{ik} \rangle} - \underbrace{\left(\overline{u_i u_k u_j}\right)_{,j}}_{T_{ijk,j}} - \underbrace{2\nu\overline{u_{i,j}u_{k,j}}}_{D_{ik}}, \tag{7}$$

$$\frac{D\overline{\theta u_i}}{Dt} + \Theta_{,j}\overline{u_i u_j} + U_{i,j}\overline{\theta u_j} - \beta_i \overline{\theta}^2$$

$$= -\underbrace{\overline{p_{,i}\theta}}_{\Pi_i} - \underbrace{\left(\overline{\theta u_i u_j}\right)_{,j}}_{T_{ij,j}} - (\nu+\kappa)\underbrace{\overline{\theta_{,j}u_{i,j}}}_{D_i}, \tag{8}$$

and

$$\frac{D\overline{\theta}^2}{Dt} + \Theta_{,j}\overline{\theta u_j} = -\underbrace{\left(\overline{\theta^2 u_j}\right)_{,j}}_{T_{j,j}} - \underbrace{2\kappa\overline{\theta_{,j}\theta_{,j}}}_{D}. \tag{9}$$

Here, the overbar over mean quantities was dropped and the molecular-transport terms were discarded on account of a high turbulent Reynolds number $R_t = K_m \nu^{-1}$. The third-order terms on the right-hand side of (7)–(9) are unknown and the second-order closure consists in parameterizing those as functionals of second-order and mean quantities.

The pressure terms $\langle \Pi_{ik} \rangle$, $\Pi_i$ contain gradients of fluctuating pressure and their formal expressions involve two-point turbulent moments up to third-order operated on by a Green's function operator $\int |\mathbf{x} - \mathbf{x}'|^{-1} d\mathbf{x}'$. In boundary-layer flows these formal expressions must include the surface contributions which, according to Green's formula (Krall 1973), involve surface integrals of two-point correlations of $u_i(\mathbf{x})$ (or $\theta(\mathbf{x})$) and the surface fluctuating pressure $p_s(x_3 = 0)$.

The turbulent transport terms $T_{ijk}$, $T_{ij}$, and $T_i$ are one-point third-order moments, which are modeled by gradient diffusion, or inferred from the third-moment conservation equations.

It is now well established that the terms $D_{ij}$, $D_i$, and $D$ in (7)–(9) involving the molecular properties of fluid are important in three-dimensional turbulence. Theoretical and experimental evidence indicates that at high Reynolds numbers these terms involve small-scale isotropic fluctuations and it is customary to model these as follows (Tennekes & Lumley 1972, Lumley & Khajeh-Nouri 1974):

$$D_{ik}=\frac{2}{3}\bar{\varepsilon}\delta_{ik}, \qquad D_i=0, \qquad \text{and } D=2\bar{\varepsilon}_\theta. \tag{10}$$

The closure problem is to model the mechanical dissipation rate $\bar{\varepsilon}$ (of turbulence kinetic energy $q^2=\overline{u_iu_i}$) and the thermal dissipation rate $\bar{\varepsilon}_\theta$ (destruction of $\overline{\theta^2}$) in terms of large-scale quantities.

Apart from $\bar{\varepsilon}, \bar{\varepsilon}_\theta$, there is often a need to model the molecular destruction rates of scalar covariances (e.g. humidity and temperature). These quantities, although analogous to $\bar{\varepsilon}_\theta$, require some special care (see Section 3).

## *Large-Eddy Simulation Technique*

As the title suggests, the idea is to calculate a three-dimensional, time-dependent, large-scale turbulence field and parameterize the subgrid-scale unresolved stresses $\langle u_i'u_j'\rangle$. The large-scale variable is formally defined as

$$\langle f(\mathbf{x})\rangle=\int_v f(\mathbf{x}')G(\mathbf{x},\mathbf{x}')\,d\mathbf{x}',$$

where $G(\mathbf{x},\mathbf{x}')$ is a three-dimensional space filter whose volume is of the scale $\Delta x=(\Delta x_1\,\Delta x_2\,\Delta x_3)^{1/3}$; $\Delta x_i$ are grid spacings in the finite-difference computation domain of the flow. After subjecting the momentum equations in (1) to the filtering operation one obtains (neglecting $\Omega_i$ and $\theta$ for convenience)

$$\langle\dot{U}_i\rangle+\langle\langle U_i\rangle\langle U_j\rangle\rangle_{,j}+\langle P\rangle_{,j}=-\underbrace{\langle\langle U_i\rangle u_j'+\langle U_j\rangle u_i'+u_i'u_i'\rangle_{,j}}_{\sigma_{ij,j}}. \tag{11}$$

Standard closure approximations in (11) are (Deardorff 1970)

$$\langle\langle U_i\rangle\langle U_j\rangle\rangle=\langle U_i\rangle\langle U_j\rangle,$$

and

$$\sigma_{ij}-\frac{1}{3}\sigma_{kk}\,\delta_{ij}=-K(\langle U_i\rangle_{,j}+\langle U_j\rangle_{,i}) \tag{12}$$

where it is usually assumed that the first two terms in the "stress" tensor $\sigma_{ij}$ sum up to zero; $K$ is an eddy viscosity related in neutral flows to $\Delta x$ and gradients of $\langle U_i\rangle$. In convective turbulence, Deardorff (1974) found it necessary to replace the eddy-viscosity model in (12) with a second-order model for the subgrid quantities $\langle u_i' u_j' \rangle$, $\langle \theta' u_i' \rangle$, etc. Obviously $K$ is not buoyancy independent unless the grid spacing is small enough so that fluctuating quantities at scales smaller than $\Delta x$ are not directly affected by buoyancy.

In neutral turbulence, the validity of the approximations in (12) and optimal filter functions $G$ have been investigated in some detail by Leonard (1974). Clark et al (1979) were able to verify (12) by numerical experiments. One of their findings was that the presently used closure schemes for $K$ are equally valid. Notable examples of the applications of this technique in planetary boundary layers are simulations of a convective mixed layer (Deardorff 1974) and most recently of a cloud-topped mixed layer (Deardorff 1980).

### *Critique of Modeling Techniques*

Eddy-viscosity models appear to be quite satisfactory in neutral or even stable planetary boundary layers; however, they fail in convective situations. This failure may be traced to the importance of third moments such as $\overline{u_3^3}$, $\overline{\theta u_3^2}$ or, $\overline{\theta^2 u_3}$ in a convective planetary boundary layer. Under such conditions these third moments contribute significantly to the levels of Reynolds stresses and fluxes. Turbulence may become entirely decoupled from the mean gradients and the concept of eddy viscosity is then physically meaningless (Zeman 1975). According to Tennekes & Lumley (1972), the eddy-viscosity concept makes sense only in flows with a single characteristic scale. Planetary boundary layers possess, in general, many major scales of motion (due to buoyancy and Coriolis forces) and therefore the use of eddy-viscosity models in planetary boundary layers is of questionable value.

Slab models are useful and convenient tools for predicting averaged properties of planetary boundary layers. Because in this integral technique the need for detailed knowledge of the eddy viscosity is eliminated, slab models are often more realistic in predicting growth of a daytime convective atmospheric boundary layer than eddy-viscosity models.

The second-order models remedy the shortcomings of the eddy-viscosity models (in convective planetary boundary layers, for example) by relegating the problem of closure to higher-order moments. This, of course, leads to proliferation of equations needed to simulate a given flow. For example, in a quasi-horizontal planetary boundary layer, a full second-order model consists of 16, or more, prognostic equations compared to 4 or 5 equations in an eddy-viscosity model. Multiplicity of scales in planetary boundary layers, large inhomogenity, and anisotropy pose a serious problem in modeling the unknown third-order quantities.

The eddy-viscosity concept with its shortcoming is simply shifted to a higher-order level.

The large-eddy simulation technique is devoid of the problems that plague the first- and second-order techniques, since the large scales are simulated exactly. There is, however, a concern with the accuracy of the simulations near the surface, where some important quantities may reside entirely in subgrid scales. For example, the Deardorff (1974) simulations show that the subgrid pressure-velocity covariance $\langle p'u_3' \rangle$ is larger near the surface than the resolved scale correlation $\langle\langle P \rangle \langle U_3 \rangle\rangle$ (see also Wyngaard 1979). So far, no attempt has been made to simulate a stable atmospheric boundary layer. It is anticipated that such simulations may lead to unrealizability of subgrid-scale turbulence. Simulations of unsteady and horizontally inhomogeneous flows would require forming statistical averages by ensembling many realizations of the flow. This would further increase the already excessive computer costs of this technique.

## 3 RECENT DEVELOPMENTS IN SECOND-ORDER MODELING

The relative success of second-order closure techniques in predicting engineering flows led to proliferation of a variety of second-order models and, at times, to their indiscriminate use in geophysical flows. It is not our intention to review the work on the modeling of planetary boundary layers, but, rather, to focus on a few conceptual ideas, developed over the past few years, which have some importance for the modeling of planetary boundary layers. These are the concepts of invariant modeling and realizability, and buoyant-transport modeling.

### *Pressure Terms*

The pressure terms $\langle \Pi_{ik} \rangle$ in (7) and $\Pi_i$ in (8) involve integrals over the entire space of turbulence, mean-shear, and buoyancy quantities. Formally, $\langle \Pi_{ik} \rangle$ and $\Pi_i$ constitute three distinct contributions due to (*a*) nonlinear, turbulence (self) interactions, (*b*) mean shear-turbulence interactions, and (*c*) buoyancy-turbulence interactions. The term $\langle \Pi_{ik} \rangle$ in the Reynolds-stress equations may be separated into the zero-trace part $\Pi_{ik}$ and the deviatoric part $\langle \Pi_{ik} \rangle - \Pi_{ik}$, which involves derivatives of pressure-velocity correlations and contributes only in inhomogeneous turbulence. As we shall see later, this separation is not without ambiguity. Presently we concern ourselves only with $\Pi_{ik}$ which has to satisfy the incompressibility and symmetry condition $\Pi_{jj}=0$ and $\Pi_{ik}=\Pi_{ki}$.

A standard approach has been to model the contributions ($a$), ($b$), and ($c$) separately. In the case of the so-called rapid contributions ($b$) and ($c$), it was then plausible to form linear expansions around an isotropic state in terms of first-order turbulence quantities $\overline{u_i u_j}$, $\overline{\theta u_i}$, and $\overline{\theta^2}$. Except for one free coefficient (in the mean-shear contribution to $\Pi_{ik}$), these linear functionals were determined exactly by applying symmetry, incompressibility, and integral constraints (see, for example, Zeman 1975). Thus, standard models for $\Pi_{ik}$ and $\Pi_i$ each contain three separate terms, which can be written as follows (see Launder et al 1975, Lumley 1979, and others):

$$\Pi^{\mathrm{a}}_{ik} = \frac{C_1}{\tau}\left(\overline{u_i u_k} - \frac{1}{3}q^2\,\delta_{ik}\right) = C_1 b_{ik}\bar{\varepsilon},$$

$$\Pi^{\mathrm{b}}_{ik} = U_{l,m}F_{lmik}\{q^2, b_{ij}\} = -2q^2\left[\frac{1}{5}S_{ik} + \alpha_1\left(S_{ij}b_{jk} + S_{kj}b_{ji} - \frac{2}{3}Sb\,\delta_{ik}\right) + \alpha_2\left(R_{ij}b_{jk} + R_{kj}b_{ji}\right)\right],$$

$$\Pi^{\mathrm{c}}_{ik} = \beta_l G_{lik} = -\frac{3}{10}\beta_l\left(\overline{\theta u_i}\,\delta_{lk} + \overline{\theta u_k}\,\delta_{li} - \frac{2}{3}\,\overline{\theta u_l}\,\delta_{ik}\right), \tag{15}$$

and

$$\Pi^{\mathrm{a}}_i = C_\theta \overline{\theta u_i}/\tau,$$

$$\Pi^{\mathrm{b}}_i = U_{l,m}H_{lmi}\{\overline{\theta u_i}\} = -\frac{4}{5}U_{l,m}\left(\delta_{li}\overline{\theta u_m} - \frac{1}{4}\delta_{mi}\overline{\theta u_l}\right),$$

$$\Pi^{\mathrm{c}}_i = \beta_l J_{li}\{\overline{\theta^2}\} = \frac{1}{3}\beta_i\overline{\theta^2}. \tag{16}$$

Here, superscripts a, b, and c designate, respectively, the individual contributions ($a$), ($b$), and ($c$). Thus $\Pi^{\mathrm{a}}_{ik}$ and $\Pi^{\mathrm{a}}_i$ are the standard return-to-isotropy terms due to the nonlinear turbulence self-interactions (Rotta 1951), where the free coefficients $C_1$ and $C_\theta$ are typically taken to be constant ($b_{ij} = \overline{u_i u_j} - 1/3q^2\,\delta_{ij}$ is the departure-from-isotropy tensor, and $\tau = q^2/\bar{\varepsilon}$ is the turbulence time scale). The rapid (linear) terms $\Pi^{\mathrm{c}}_{ik}$, $\Pi^{b}_i$, and $\Pi^{\mathrm{c}}_i$ are determinate; the mean-shear rapid term $\Pi^{\mathrm{b}}_{ik}$ was conveniently rewritten to separate the part depending on the mean strain-rate tensor $S_{ij} = 1/2(U_{i,j} + U_{j,i})$ and the rotation (asymmetric) tensor $R_{ij} = 1/2(U_{i,j} - U_{j,i})$ (Zeman & Tennekes 1975, Reynolds 1976). The constants $\alpha_1$ and $\alpha_2$ are mutually dependent and by matching $\Pi^{\mathrm{b}}_{ik}$ to the rapid-distortion calculations (Townsend 1970), Lumley (1975b) determined $\alpha_1 = 3/7$ and $\alpha_2 = 0$.

With the return-to-isotropy constants determined from experiment it would seem that the problem of modeling the pressure term has been closed. Unfortunately, the models in (15) and (16) are inconsistent with the majority of boundary-layer data and various authors took the liberty of altering the invariant constants in (15) and (16) to achieve better agreement with data. For example, in order to match the model predictions to the observed heat-flux relations in a stably stratified surface layer Zeman & Lumley (1979) found it necessary to alter the sign in the rapid term $\Pi_i^b$ in (16). Similarly, Wyngaard (1975) found the buoyancy term $\Pi_i^c$ troublesome in stable atmospheric boundary layers near critical Richardson numbers. Clearly some of the causes of this disparity between the theoretical (rapid) models and the data have to do with large departures from isotropy, inhomogeneity, and buoyancy effects in real boundary layers; the assumptions of weak anisotropy and locality of turbulence are violated and the separation of the various effects in the pressure terms may not be justified. Furthermore, Lumley & Newman (1977) and Lumley (1979) showed that the assumption of constant $C_1$ and $C_\theta$ is inconsistent, within the range of observed anisotropies, with the so-called realizability condition (discussed later). Gibson & Launder (1978) attributed the model-data disagreement to the surface effect on fluctuating pressure and suggested near-surface corrective functions to the pressure terms.

PRESSURE TERMS FOR LARGE ANISOTROPY Lumley's (1979) approach to modeling pressure terms is based on the concept of realizability. This approach removes some of the noted deficiencies of the linear models in (15) and (16). We shall briefly discuss Lumley's realizable model and its applicability in planetary boundary layers.

The common feature of near-wall turbulence or very stable turbulence is the suppression of vertical fluctuations. Typical values of the ratio $\overline{u_1^2}/\overline{u_3^2}$ near a smooth wall is about 5.0 (Launder et al 1975); in the atmospheric surface layer it is about 2.6. The near-surface heat-flux ratio $\overline{\theta u_1}/\overline{\theta u_3}$ is approximately $-3.0$ in a weakly stable atmospheric boundary layer. In the convective atmospheric boundary layer, Panofsky et al (1977) find the near-surface ratio $\overline{u_1^2}/\overline{u_3^2}$ to increase with the depth $h$ of the atmospheric boundary layer, i.e. large convective eddies supply their energy into horizontal components near the surface; $\overline{u_1^2}/\overline{u_3^2}$ may be of order 10 or more depending on the ratio $-h/L$. Apparently we can view near-surface or stable turbulence as approaching the limit of two dimensionality. Assuring proper (realizable) behavior of the second-order equations in this two-dimensional limit is a way of dealing with large anisotropic (but still three-dimensional) turbulence encountered under some of the extreme conditions in planetary boundary layers.

To recount briefly the realizability concept: consider the rate equation for a diagonal element $\overline{u_\alpha u_\alpha}$, in (7). In the absence of any driving mechanisms, we obtain, with the aid of (10) and (15)

$$\dot{\overline{u_\alpha u_\alpha}} = -\bar{\varepsilon}\left(C_1 b_{\alpha\alpha} + \frac{2}{3}\right).$$

The Schumann (1977) realizability condition $\lim_{u_\alpha \to 0} \dot{\overline{u_\alpha u_\alpha}} = 0$ requires that unless $\bar{\varepsilon}=0$, we must have $C_1=2$ (since $b_{\alpha\alpha}=-1/3$). Lumley & Newman (1977) proposed a less ambiguous formulation

$$\bar{\varepsilon}\Phi_{ik} = \Pi^{a}_{ik} + D_{ik} - 2/3\bar{\varepsilon}\,\delta_{ik}$$

requiring that in the limit $u_\alpha = 0$, $\Phi_{\alpha\alpha} = -2/3$ and suggested

$$\Phi_{ik} = \Phi b_{ik}$$

$$\Phi = 2 + F_1(\mathrm{II},\mathrm{III})F_0(\mathrm{II},\mathrm{III}) \approx C_1, \tag{17}$$

where $\Phi$ is a scalar function that is identical with $C_1$ if the approximation $D_{ik} = \frac{2}{3}\bar{\varepsilon}\,\delta_{ik}$ is valid. The empirical function $F_1$ depends on the second and third invariants II and III defined as $\mathrm{II} = \frac{1}{2}(b_{ij}b_{ij})$ and $\mathrm{III} = \frac{1}{3}(b_{ij}b_{jk}b_{ki})$. [Lumley (1979) defines II with an opposite sign.] The function $F_0 = 1/9 + 3\mathrm{III} - \mathrm{II}$ has an interesting property: $F_0$ vanishes in two-dimensional or one-dimensional limit (when one or two velocity components become zero). Thus $F_0$ is an indicator function of turbulence dimensionality.

RETURN-TO-ISOTROPY MODEL IN PLANETARY BOUNDARY LAYERS In the attempt to match the pressure-term model to boundary-layer data, Zeman & Tennekes (1975) noted the decreasing tendency of the best-fit values of $C_1$ with increasing anisotropy (represented by the invariant II). In view of the realizable return-to-isotropy model (17) this tendency is now understood; in fact, the Reynolds-stress equations contain enough information to define uniquely the function $C_1(\mathrm{II})$. This we demonstrate as follows: in the constant-stress approximation, the Reynolds-stress equations in (7) with the pressure terms $\Pi^{a}_{ik}$ and $\Pi^{b}_{ik}$ from (15) may be reduced to three independent equations [see the Zeman-Tennekes Equations (18), (20), and (21)]:

$$\frac{4}{3}\left(1 - 2\alpha_1 - \frac{3}{4}\alpha_2\right) - C_1 b_{11} = 0,$$

$$-\frac{2}{3}(1 + \alpha_1 - 3\alpha_2) - C_1 b_{33} = 0, \tag{18}$$

$$-\mathrm{Sgn}(U_{1,3})\frac{(b_{33}(1-2\alpha_1) - b_{11}(\alpha_1 - \alpha_2) + 2/15)}{|b_{13}|} - C_1 b_{31} = 0.$$

We now observe that the realizability condition $U_{1,3}b_{13}=0$ for $u_3=0$ is satisfied if $\alpha_1=\alpha_2=0.3$ and, substituting in the definition $\mathrm{II}=\frac{1}{2}(b_{11}^2+b_{22}^2+b_{33}^2+2b_{33}^2)$ from (18), we obtain the quadratic equation $C_1^2\mathrm{II}-\frac{2}{15}C_1-\frac{2}{3}(1-2\alpha_1)^2$, which has a unique positive solution (with $\alpha_1=0.3$)

$$C_1=(1/15\mathrm{II})\left[1+(1+24\mathrm{II})^{1/2}\right]. \tag{19a}$$

Similarly, we can infer from the data on axisymmetric convective turbulence (Adrian & Ferreira 1979)

$$C_1=\tfrac{14}{15}(4/3\mathrm{II})^{-1/2}. \tag{19b}$$

As shown in Figure 1, the functions (19a) and (19b) follow the tendency of the boundary-layer data. With a suitable cutoff, say $2<C_1<6$, these functions provide a realistic return-to-isotropy model. Substituting for II and III in the Lumley (1979) updated version of (17) from boundary-layer data, we obtained excessively large values of $\Phi=C_1$. It appears that the Lumley expression is too sensitive to the Reynolds-number correction causing $\Phi$ to be unrealistically large for high Reynolds numbers.

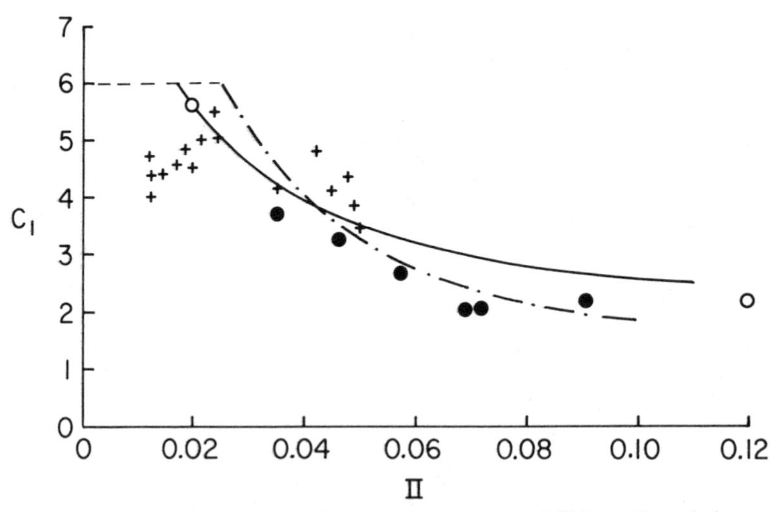


*Figure 1* Relationship between the return-to-isotropy coefficient $C_1$ and the second invariant II; ●: the best fit to boundary-layer data (Zeman & Tennekes 1975); O: the core of a convective mixed layer (Adrian & Ferreira 1979, Willis & Deardorff 1974); the solid and dash-dot lines are, respectively, equations (19a) and (19b).

REALIZABLE PRESSURE TERMS IN THE HEAT-FLUX EQUATIONS Lumley (1979) generalized the realizability concept to scalar-vector correlations. He introduced the tensor

$$D_{ij} = \overline{u_i u_j}/q^2 - \overline{\theta u_i}\,\overline{\theta u_j}/q^2\overline{\theta^2},$$

which has nonnegative eigenvalues that vanish for a perfect correlation in a given component. Utilizing the realizability conditions on $D_{ij}$, he was able to derive nonlinear models to supercede the model in (16) [Lumley's equations (3.26), (4.12), and (4.13)]. The inevitable complexity of these nonlinear models makes them difficult to physically interpret. An interesting aspect of the Lumley return-to-isotropy model $\Pi_i^a$ is the coupling between the return-to-isotropy mechanism in the Reynolds-stress and heat-flux equations. If we neglect the contribution $D_{ij}$ in the Lumley expression (3.26) we obtain (in our notation)

$$C_\theta = \tau/\tau_\theta + C_1/2$$

where $\tau_\theta = \overline{\theta^2}/\bar{\varepsilon}_\theta$ is the thermal time scale. In the constant-stress layer of the atmospheric boundary layer the time-scale ratio $\tau/\tau_\theta$ is about 2, thus $C_\theta = (4 + C_1)/2$. Previous (standard) models use typically $C_\theta = 2C_1$ (Launder et al 1975, Zeman & Lumley 1979).

## *Transport Terms*

The transport terms $T_{ijk}$, $T_{ij}$, and $T_i$ in (7)–(9) are not significant in neutral or stably stratified planetary boundary layers. However, in convective planetary boundary layers these terms often govern the flow dynamics. It has been shown that the transport terms, in fact, drive the entrainment in convective planetary boundary layers (Tennekes 1973, Zeman 1975). The conventional invariant models (Donaldson (1972) were not capable of predicting the significant vertical fluxes $\overline{q^2 u_3}$ observed in convective atmospheric boundary layers (Wyngaard & Coté 1974) or measured in laboratory experiments (Willis & Deardorff 1974). The conservation equations for $T_{ijk}$ revealed buoyancy-related terms that have been identified as important contributors to the flux $\overline{q^2 u_3}$ (Zeman & Lumley 1976, Lumley et al 1978). Next we review the transport models that include these buoyancy contributions.

BUOYANCY-TRANSPORT MODEL Writing the full-rate equation for the third moment $T_{ijk}$, neglecting the mean shear and molecular transport, and expressing the fourth-order moments as

$$\overline{u_i u_j u_k u_l} = (\overline{u_i u_j})(\overline{u_k u_l}) + (\overline{u_i u_k})(\overline{u_j u_l}) + (\overline{u_i u_l})(\overline{u_j u_k}) + C_{ijkl},$$

in terms of their cumulant $C_{ijkl}$, we obtain

$$\frac{DT_{ijk}}{Dt}+\overline{u_iu_l}\left(\overline{u_ju_k}\right)_{,l}+\overline{u_ju_l}\left(\overline{u_iu_k}\right)_{,l}+\overline{u_ku_l}\left(\overline{u_iu_j}\right)_{,l}-\beta_i\overline{\theta u_ju_k}$$

$$-\beta_j\overline{\theta u_iu_k}-\beta_k\overline{\theta u_iu_j}=-\underbrace{\left(\overline{p_{,i}u_ju_k}+\overline{p_{,j}u_iu_k}+\overline{p_{,k}u_iu_j}\right)}_{\Pi_{ijk}}$$

$$-C_{ijkl,l}+\text{molecular terms}. \tag{20}$$

Hanjalic & Launder (1972) closed (20) by making the following approximations:

1. The fourth moments are jointly Gaussian distributed, hence the cumulants $C_{ijkl}=0$.
2. The pressure terms $\Pi_{ijk}$ tend to relax the third moments to Gaussian distribution on the time scale $\tau$, $\Pi_{ijk}=C_3T_{ijk}/\tau$ (the so-called eddy-damping assumption with $C_3$ being an unknown coefficient).
3. The molecular, buoyancy, and temporal terms are neglected.

The above approximations lead to the invariant model (Donaldson 1972)

$$T_{ijk}=-C_3^{-1}\tau\left[\overline{u_iu_l}\left(\overline{u_ju_k}\right)_{,l}+\overline{u_ju_l}\left(\overline{u_iu_k}\right)_{,l}+\overline{u_ku_l}\left(\overline{u_iu_j}\right)_{,l}\right]. \tag{21}$$

In a free convective atmospheric boundary layer (with axisymmetric $\overline{u_iu_j}\propto\delta_{ij}$) (21) gives

$$\overline{u_3^3}=-3\tau C_3{}^{-1}\overline{u_3^2}\left(\overline{u_3^2}\right)_{,3}$$

and

$$\overline{q^2u_3}=-\tau C_3^{-1}\overline{u_3^2}\left(q_{,3}^2+2\left(\overline{u_3^2}\right)_{,3}\right).$$

Comparison with experiments (Willis & Deardorff 1974, Adrian & Ferreira 1979) show that the gradient model above is unrealistic. Buoyancy-driven turbulence is capable of maintaining countergradient transport. In the lower half of the convective atmospheric boundary layer, $\overline{u_3^3}$ and $\overline{q^2u_3}$ are positive, while (21) predicts $\overline{u_3^3}\propto\overline{q^2u_3}\propto-(\overline{u_3^2})_{,3}<0$. Zeman & Lumley (1976) and André et al (1976) removed the shortcomings of the gradient model (21) by including the buoyancy terms on the left-hand side of (20). The complete buoyant transport model in a planetary boundary layer then consists of a set of third-moment equations coupled through buoyancy terms. This, in a general flow, requires solving time-dependent third-moment equations (André et al 1978). In a free convective atmospheric boundary layer the situation is somewhat simplified; only the fluxes

$$F_i = -\{\overline{u_1^2 u_3}, \overline{u_3^3}, \overline{\theta u_3^2}, \overline{\theta^2 u_3}\}$$

are needed and by neglecting the temporal terms and making the standard approximations 1. and 2. one is able to reduce the system of the third-order equations to the tensorial form (Zeman & Lumley 1976)

$$F_i \equiv K_{ij}(G_j)_{,3}, \tag{22}$$

where $K_{ij}$ is an eddy-viscosity tensor and $G_j \equiv \{\overline{u_1^2}, \overline{u_3^2}, \overline{\theta u_3}, \overline{\theta^2}\}$. Thus (22) is still a gradient model; however, now the flux of a given quantity depends on the gradient of another. For example, Zeman & Lumley (1976) give

$$\overline{u_3^3} = -\left[K_m(\overline{u_3^2})_{,3} + \beta_3 \tau K_m(\overline{\theta u_3})_{,3} + 3/4\beta_3^2\tau^2 K_t(\overline{\theta^2})_{,3}\right],$$

where $K_m$ and $K_t$ are the principal eddy coefficients of the tensor $K_{ij}$ having a form

$$K_m \propto \tau(\overline{u_3^2} + \text{const}\,\beta_3 \overline{\theta u_3}\,\tau_\theta),$$

$$K_t \propto K_m\left[1 + \text{const}(N\tau)^2\right]^{-1}.$$

The two buoyancy contributions in $\overline{u_3^3}$ due to gradients $(\overline{\theta u_3})_{,3}$ and $(\overline{\theta^2})_{,3}$ counteract the principal down-gradient term $\propto(\overline{u_3^2})_{,3}$ and maintain, in the lower half of the convective atmospheric boundary layer, countergradient transport. The eddy viscosities $K_m$ and $K_t$ are significantly influenced by the orientation of heat flux and by the Brunt-Väisälä frequency $N = (\beta_3 \Theta_{,3})^{1/2}$. These buoyancy contributions appear to be indispensable when modeling an entraining convective planetary boundary layer.

PRESSURE TRANSPORT Lumley (1975a) and, for example, Launder et al (1975) suggested the respective decompositions of the pressure term $\langle \Pi_{ik} \rangle$ in (7)

$$\langle \Pi_{ik} \rangle = \Pi_{ik} + \tfrac{2}{3}(\overline{pu_j})_{,j}\delta_{ik}, \tag{23}$$

$$\langle \Pi_{ik} \rangle = \Pi_{ik}^* + (\overline{pu_i})_{,k} + (\overline{pu_k})_{,i}. \tag{24}$$

Both $\Pi_{ik}$ and $\Pi_{ik}^*$ have zero trace but the divergence parts differ. In homogeneous turbulence this formal separation is immaterial; however, in inhomogeneous flows the ambiguity is worrisome in the convective atmospheric boundary layer where the pressure fluxes $\overline{pu_i}$ are important.

The invariant form of $\overline{pu_i}$ due to turbulence self-interactions is (Lumley 1975b)

$$\overline{pu_i} = -\tfrac{1}{5}\overline{q^2 u_i}. \tag{25}$$

This model appears to agree, at least in sign, with observations as inferred from data by Wyngaard (1979). Now the decomposition (23) with the model (25) leads in the horizontal-variance equations to

$$\overline{\dot{u_2^2}} = \overline{\dot{u_1^2}} = \tfrac{2}{15}\left(\overline{u_3^3}\right)_{,3} + \dots .$$

while the decomposition (24) gives $-2(\overline{pu_3})_{,3} = \frac{2}{5}\overline{q^2 u_{3,3}}$ entirely in the $\overline{u_3^3}$ equation, and no contribution in the $\overline{u_1^2}$ equation. It is speculated sthat near the surface the horizontal variances $\overline{u_1^2}$ and $\overline{u_2^2}$ derive their energy from the pressure-flux divergence (Wyngaard 1979). This is plausible only with the decomposition (23) since the term $\propto(\overline{u_3^3})_{,3}$ in $\overline{u_1^2}$ is a gain near the surface.

CRITIQUE OF THE TRANSPORT MODEL Apart from the success of the buoyancy transport model in predicting some features of the dynamics of convective turbulence, the validity of assumptions 1. (fourth-cumulant discard) and 2. (eddy damping) is now questioned. Wyngaard (1979) inferred from observations that the pressure term $\overline{p_{,3}u_3^2}$ should be a gain in the $\overline{u_3^3}$ budget evaluated near the surface in a convective atmospheric boundary layer; while by assumption 2. this pressure term is modeled as $\propto\overline{u_3^3}/\tau$ and appears in the $\overline{u_3^3}$ budget as a loss. Deardorff (1978) inferred from exact solutions to passive scalar diffusion that the time-dependent, third-moment equations must contain a diffusive term. By forming equations for moment-generating functions and for cumulants, Lumley (1979) derived a Gaussian, zeroth-order model for third moments that is at variance with the model (21). At the first order, Lumley's model accounts for departure from Gaussian behavior of the fourth moments and at that level the diffusive terms of Deardorff also emerge.

A rather transparent flaw of the model (21) is that it is nonrealizable. This is shown by letting one component in (21) (say $u_i$), approach zero; since $(\overline{u_i u_l})_{,l}$ and $(\overline{u_i u_k})_{,l}$ need not be zero, the condition $u_i = \overline{u_i u_j u_k} = 0$ is violated. It can be verified that the original (exact) equation yields (with $u_i = \dot{u}_i = 0$)

$$\overline{\dot{u_i u_j u_k}} = -\overline{u_i(u_j u_k u_l)_{,l}}$$

which clearly satisfies the realizability.

In summary, there remain some fundamental problems to be resolved in transport modeling; these are not as critical, however, as the problems with modeling the pressure and dissipation terms. We feel that there is not much sense in perfecting the transport models far beyond the present level before the same level is achieved in modeling other terms—mainly the dissipation rates.

## *Mechanical and Thermal Dissipation*

The mechanical and thermal dissipation rates $\bar{\varepsilon}$ and $\bar{\varepsilon}_\theta$ cause most difficulties in second-order models. Apart from setting the rate of energy loss to small dissipative eddies, $\bar{\varepsilon}$ determines the turbulence time and length scales ($\tau$ and $l$); thus $\bar{\varepsilon}$ indirectly affects all the modeled terms where $\tau$ or $l$ appear. In the modeling context, one should consider $\bar{\varepsilon}$ not as a small-scale molecular process, but rather as a spectral flux of turbulent energy at the large-scale end of the inertial subrange (see, for example, Tennekes & Lumley 1972). In that sense we may, at least formally, accept the idea of modeling $\dot{\bar{\varepsilon}}$ ($\dot{\overline{\varepsilon_\theta}}$) by large-scale quantities that are accessible in the second-order model.

According to Lumley (1979) there are, in general, 22 invariants that may qualify for dependent variables in the rate equation for $\bar{\varepsilon}$ and, in applications to planetary boundary layers, the well-behaved $\bar{\varepsilon}$ equations are constructed by trial and error rather than through a rational approach. Many authors (e.g. Brost & Wyngaard 1978) avoid entirely the $\bar{\varepsilon}$ equation, preferring length-scale parameterization and relating $\bar{\varepsilon}$ to $q^2/l$. The length-scale prescriptions cause, of course, fewer problems in modeling, but they often oversimplify the physics by constraining the modeled turbulence to some predetermined or anticipated state. For example, a neutral limit $l \propto u_* f^{-1}$ has to be externally imposed in the length-scale scheme while the most rudimentary model equation for $\bar{\varepsilon}$ recovers this limit implicitly (Wyngaard et al 1974, Zeman & Tennekes 1975).

In nonbuoyant homogeneous turbulence some irreducible, universal forms of the dissipation models have been found and in the $\bar{\varepsilon}_\theta$ equation the number of unknown coefficients has been reduced by utilizing the concept of the equilibrium-time-scale ratio $r_e = (\tau/\tau_\theta)_e$ (Newman et al 1980).

Apart from $\bar{\varepsilon}$ and $\overline{\varepsilon_\theta}$ there is a need in the models of planetary boundary layers for modeling the molecular destruction of crosscorrelations between two scalars. Utilizing the realizability condition on the crosscorrelation coefficient, Lumley (1979) was able to derive a fundamental form of the scalar-covariance decay rate. The Lumley formulation appears to be supported by the wind-tunnel experiments of Warhaft (1980). For details on modeling the molecular terms, the reader is referred to the review papers of Reynolds (1976) and Lumley (1979).

# 4 SUMMARY

Comparison of various turbulence-simulation techniques favors second-order modeling as the practical computational method in applications to planetary boundary layers. Still, the complexity of the dynamics of planetary boundary layers presents a number of challenging problems for second-order modeling. Some of these yet to be resolved are

1. Identification and separation of the surface effects from the effects due to large anisotropy or inhomogeneity on pressure terms (because the near-surface contributions to the pressure terms involve the surface pressure, these contributions could be measured).

2. The effect of stratification on turbulence dynamics near the critical Richardson number. How are the turbulence-energy spectrum and the dissipation rate altered in this stable limit with respect to the neutral conditions?

3. Modeling the effects of large eddy–surface interactions in convective atmospheric boundary layers by exploring models that allow for anisotropy of turbulence scales and of dissipation partitioning among the energy components.

4. Apart from the theoretical (linear) models, the actual buoyancy contributions in the pressure terms are largely unknown. It would be useful to attempt to identify these buoyancy contributions with the aid of large-eddy or direct turbulence simulations.

ACKNOWLEDGMENTS

This work was supported in part by the US National Science Foundation Meteorology Program, under Grant No. ATM77-22903, and in part by the US Office of Naval Research under the following programs: Fluid Dynamics (Code 438), Power (Code 473), and Physical Oceanography (Code 481).

*Literature Cited*

Adrian, R. J., Ferreira, R. T. S. 1979. Higher order moments in turbulent thermal convection. *Proc. 2nd Symp. Turbulent Shear Flows*, July, 1979, London. 12:1–5

André, J. C., DeMoor, G., Lacarrere, P., Du Vachat, R. 1976. Turbulence approximations for inhomogeneous flows Part II. *J. Atmos. Sci.* 33:482–91

André, J. C., De Moor, G., Lacarrere, P., Therry, G., Du Vachat, R. 1978. Modeling the 24-hour evolution of the mean and turbulent structure of the planetary boundary layer. *J. Atmos. Sci.* 35: 1861–83

Brost, R. A., Wyngaard, J. C. 1978. A model study of the stably stratified planetary boundary layer. *J. Atmos. Sci.* 35:1427–40

Businger, J. A., Arya, S. P. S. 1974. Height of the mixed layer in the stably stratified planetary boundary layer. *Adv. Geophys.* 18A:73–92

Clark, R. A., Ferziger, J. H., Reynolds, W. C. 1979. Evaluation of subgrid-scale models using an accurately simulated turbulent flow. *J. Fluid Mech.* 91:1–16

Deardorff, J. W. 1970. A numerical study of three dimensional turbulent flow at large Reynolds numbers. *J. Fluid Mech.*

42:453–61

Deardorff, J. W. 1974. Three-dimensional numerical study of turbulence in an entraining mixed layer. *Boundary-Layer Meteorol.* 7:199–226

Deardorff, J. W. 1978. Closure of second- and third-moment rate equations for diffusion in homogeneous turbulence. *Phys. Fluids* 21:525–30

Deardorff, J. W. 1979. Prediction of convective mixed-layer entrainment for realistic capping inversion structure. *J. Atmos. Sci.* 36:424–36

Deardorff, J. W. 1980. Cloud top entrainment instability. *J. Atmos. Sci.* 37:131–47

Donaldson, C. 1972. Calculation of turbulent shear flows for atmospheric and vortex motions. *AIAA J.* 10:4–12

Gibson, M. M., Launder, B. E. 1978. Ground effects on pressure fluctuations in the atmospheric boundary layer. *J. Fluid Mech.* 86:491–511

Hanjalic, K., Launder, B. E. 1972. A Reynolds-stress model of turbulence and its application to this shear flows. *J. Fluid Mech.* 52:609–38

Krall, A. M. 1973. *Linear Methods of Applied Analysis*. Reading, Mass: Addison-Wesley. 706 pp.

Launder, B. E., Reese, G. J., Rodi, W. 1975. Progress in the development of a Reynolds-stress turbulence closure. *J. Fluid Mech.* 68:537–66

Leonard, A. 1974. Energy cascade in large-eddy simulations of turbulent fluid flows. *Adv. Geophys.* 18A:237–48

Lilly, D. K. 1968. Models of cloud-topped mixed-layers under a strong inversion. *Q. J. R. Meteorol. Soc.* 94:292–309

Lumley, J. L. 1975a. Pressure-strain correlation. *Phys. Fluids* 18:750

Lumley, J. L. 1975b. Prediction methods for turbulent flows. Lecture notes, presented at the von Kármán Institute, Rhode-St. Genese, Belgium, March 3–7

Lumley, J. L. 1979. Computational modeling of turbulent flows. *Adv. Appl. Mech.* 18:123–76

Lumley, J. L., Khajeh-Nouri, B. 1974. Computational modeling of turbulent transport. *Adv. Geophys.* 18A:169–92

Lumley, J. L., Newman, G. R. 1977. The return-to-isotropy of homogeneous turbulence. *J. Fluid Mech.* 82:161–78

Lumley, J. L., Panofsky, H. A. 1964. *The Structure of Atmospheric Turbulence*. New York: Wiley. 239 pp.

Lumley, J. L., Zeman, O., Siess, J. 1978. The influence of buoyancy on turbulent transport. *J. Fluid Mech.* 84:581–97

Newman, G. R., Launder, B., Lumley, J. L. 1980. Modeling the behavior of homogeneous scalar turbulence. *J. Fluid Mech.* In press

Panofsky, H. A., Tennekes, H., Lenschow, D. H., Wyngaard, J. C. 1977. The characteristics of turbulent velocity components in the surface layer under convective conditions. *Boundary-Layer Meteorol.* 11:355–61

Reynolds, O. 1895. On incompressible viscous flow and the determination of the criterion. *Philos. Trans. R. Soc. Ser. A* 186:123.

Reynolds, W. C. 1976. Computation of turbulent flows. *Ann. Rev. Fluid Mech.* 8:183–208

Rotta, J. C., 1951. Statistische Theorie nicht-homogener Turbulenz. *Arch. Phys.* 129:547–72

Schumann, V. 1977. Realizability of Reynolds stress turbulence models. *Phys. Fluids* 20:721–25

Sherman, F. S., Imberger, J., Corcos, G. M. 1978. Turbulence and mixing in stably stratified waters. *Ann. Rev. Fluid Mech.* 10:267–88

Tennekes, H. 1973. A model for the dynamics of the inversion above a convective boundary layer. *J. Atmos. Sci.* 30:558–67

Tennekes, H., Lumley, J. L. 1972. *A First Course in Turbulence*. Cambridge, Mass: MIT Press. 300 pp.

Townsend, A. A. 1970. Entrainment and the structure of turbulent flow. *J. Fluid Mech.* 21:13–46

Warhaft, Z. 1980. Experiments on the dissipation rate of scalar covariance in grid generated turbulence. *J. Fluid Mech.* Submitted

Willis, G. E., Deardorff, J. W. 1974. A laboratory model of the unstable planetary boundary layer. *J. Atmos. Sci.* 31:1297–1307

Wyngaard, J. C. 1975. Modeling the planetary boundary layer—extension to the stable case. *Boundary-Layer Meteorol.* 9:441–60

Wyngaard, J. C. 1979. The atmospheric boundary layer—modeling and measurements. *Proc. 2nd Symp. Turbulent Shear Flows*, London, 13:25–30

Wyngaard, J. C., Coté, O. R. 1974. The evolution of a convective planetary boundary layer—a higher-order closure model study. *Boundary-Layer Meteorol.* 7:289–308

Wyngaard, J. C., Coté, O. R., Rao, K. S. 1974. Modeling the atmospheric boundary layer. *Adv. Geophys.* 18A: 193–211

Wyngaard, J. C., LeMone, M. A. 1980. Behavior of the refractive index structure parameter in the entraining convective boundary layer. *J. Atmos. Sci.* In press

Wyngaard, J. C., Pennell, W. T., Lenschow, D. H., LeMone, M. A. 1978. The temperature-humidity covariance budget in the convective boundary layer. *J. Atmos. Sci.* 35:47–58

Yamada, T., Mellor, G. L. 1979. A numerical simulation of BOMEX data using a turbulence closure model coupled with ensemble cloud relation. *Q. J. R. Meteorol. Soc.* 105:915–44

Zeman, O. 1975. *The Dynamics of Entrainment in the Planetary Boundary Layer: A Study in Turbulent Modeling and Parameterization*. PhD thesis. Penn. State Univ., University Park.

Zeman, O. 1979. Parameterization of the dynamics of stable boundary layers and nocturnal jets. *J. Atmos. Sci.* 36:791–804

Zeman, O. 1980. The dynamics and modeling of heavier-than-air cold gas releases. *J. Fluid Mech.* Submitted

Zeman, O., Lumley, J. L. 1979. Buoyancy effects in entraining turbulent boundary layers: A second order closure study. *Lecture Notes in Physics: Turbulent Shear Flows I*, pp. 295–306. Berlin/New York: Springer

Zeman, O., Lumley, J. L. 1976. Modeling buoyancy driven mixed layers. *J. Atmos. Sci.* 33:1974–88

Zeman, O., Tennekes, H. 1975. A self-contained model for the pressure terms in the turbulent stress equations of a neutral atmospheric boundary layer. *J. Atmos. Sci.* 32:1808–13

Zeman, O., Tennekes, H. 1977. Parameterization of the turbulent energy budget at the top of the daytime atmospheric boundary layer. *J. Atmos. Sci.* 34:111–23

*Ann. Rev. Fluid Mech. 1981. 13:273–328*


# CAVITATION IN FLUID MACHINERY AND HYDRAULIC STRUCTURES



*Roger E. A. Arndt*
St. Anthony Falls Hydraulic Laboratory, University of Minnesota, Minneapolis, Minnesota 55455

## INTRODUCTION

Cavitation is a design consideration for a broad variety of devices handling liquids. Cavitation can affect performance, for example through increased drag of hydronautical vehicles, limitations on the thrust produced by various propulsion systems, decreased power output and efficiency of turbines, and a drop in head and efficiency produced by pumps. The accuracy of flow meters can be affected by cavitation. Noise and vibration occur in many applications, ranging from all forms of turbomachinery to large valves in industrial plants. Associated with the deleterious effects of performance breakdown, noise, and vibration, there is the possibility of erosion. Erosion can range in extent from a relatively minor amount of pitting after many years of operation to disastrous failure of large and expensive structures in a relatively short period of time. An extreme example is shown in Figure 1, which illustrates the damage sustained in the spillway tunnel of the Yellowtail Dam in South Central Montana. Numerous descriptions of cavitation damage are available in the literature (e.g. Vennard et al 1947). Cavitation can also sometimes be a useful byproduct of fluid flow, for example in homogenization of milk and industrial cleaning. A very common application is the use of ultrasonic cavitation for the cleaning of false teeth.

There is a vast quantity of cavitation literature on topics ranging from basic mechanisms to practical application. Several excellent reviews already exist (Eisenberg 1961, Acosta 1974, Acosta & Parkin 1975). There is one textbook in English (Knapp et al 1970), a few more



0066-4189/81/0115-0273$01.00

*Figure 1* Cavitation damage in the spillway tunnel of the Yellowtail Dam in Montana. Courtesy of US Bureau of Reclamation.

textbooks in foreign languages (e.g. Karelin 1963), and a brief monograph (Pearsall 1972). Various aspects of the problem have been considered in previous Annual Reviews (Wu 1972, Acosta 1973, Plesset & Prosperetti 1977). Related phenomena have also been addressed (Rohsenhow 1971, van Wijngaarden 1972, Holt 1977). Several major symposia have been devoted to the topic (Vennard et al 1947, Cavitation in Hydrodynamics 1956; Brown 1963, Numachi 1963, Davies 1964, Holl & Wood 1964, Wood et al 1965, Robertson & Wislicenus 1969, Institute of Mechanical Engineers 1974, IAHR 1976), not to mention the Annual Forum on Cavitation and Polyphase Flow held by the American Society of Mechanical Engineers, the periodic review of the topic in the Proceedings of the International Towing Tank Conference, and other symposia devoted to hydraulic machinery, marine propulsion, naval hydrodynamics, etc. This article attempts to provide a broad overview. In addition, emphasis is placed on topics of current interest that have not been reviewed previously.

Since the topic of cavitation in fluid machinery in hydraulic structures cuts across many disciplines, we cannot provide an in-depth review of the entire topic here. Instead, we review our present state of knowledge of the physical phenomena that are common to all types of applications, particularly the factors leading to the occurrence of cavitation, subsequent performance breakdown, and other deleterious side effects such as vibration, noise, and damage. We attempt to bridge the gap between the description of fundamental mechanisms and practical experience. Cavitation is generally considered here as an unwanted side effect of fluid flow, although, as already mentioned, there are useful applications of the phenomenon.

## *Nature of the Problem*

From an engineering-design point of view there are two basic questions. First, will cavitation occur? Second, if cavitation is unavoidable, then the question is whether a given design can still function properly. Often economic or other operational considerations necessitate operation with some cavitation. Under these conditions it is necessary to understand the effects of cavitation on steady loading as well as any unsteady forces that may be induced.

Practical experience indicates that we must rely heavily on experimentation in the development of fluid machinery and hydraulic structures. In order to reliably predict cavitation inception or performance in cavitating flow in the laboratory, it is essential that the physics of the phenomenon be clearly understood so that proper modeling techniques are applied. In addition, estimates of operational lifetime and maintenance must rely on erosion tests carried out in the laboratory. Many cavitation problems involve liquids other than cold water, yet there is little information on cavitation in liquids such as oil, liquid metals, and cryogenics. Related to this problem is the question of whether cold water can be substituted in the laboratory to simulate prototype conditions involving an exotic and difficult-to-handle liquid such as liquid hydrogen or molten sodium. Obviously further research is required to answer some of these questions.

## *Occurrence of Cavitation*

Cavitation is normally defined as the formation of the vapor phase in a liquid. The term cavitation (originally coined by R. E. Froude) can imply anything from the initial formation of bubbles (inception) to large-scale, attached cavities (supercavitation). The formation of individual bubbles and subsequent development of attached cavities, bubble clouds, etc, is directly related to reductions in pressure to some critical value, which in turn is associated with dynamical effects, either in a flowing liquid or in an acoustical field. Cavitation is distinguished from boiling in the sense that the former is induced by the lowering of hydrodynamic pressure, whereas the latter is induced by the raising of vapor pressure to some value in excess of the hydrodynamic pressure. The two phenomena are related. Cavitation inception and boiling can be compared in terms of the vapor-bubble dynamics of sub-cooled and super-heated liquids (Plesset 1957). Quite often a clear distinction between the two types of phenomena cannot be made. This is especially true for cavitation in liquids other than cold water.

Examples of cavitation are shown in Figures 2 through 5. Limited cavitation in the form of a ring of bubbles surrounding a hemispherical-nosed body is shown in Figure 2. Careful study indicates that the cavitation shown here is actually occurring *downstream* from the minimum pressure point. Figure 3 illustrates cavitation occurring away from the surface of a body in the turbulent shear layer surrounding the wake of a sharp-edged disk. Figure 4 is a classic photograph of cavitation occurring in tip vortices trailing from a propeller. Figure 5, which is another example of cavitation on a marine propeller, is presented to indicate several salient features of the cavitation problem. More than one type of cavitation can occur simultaneously (sheet cavitation on the blade, vortex cavitation at the tips). It is also important to note that in many practical applications cavitation is a nonsteady phenomenon

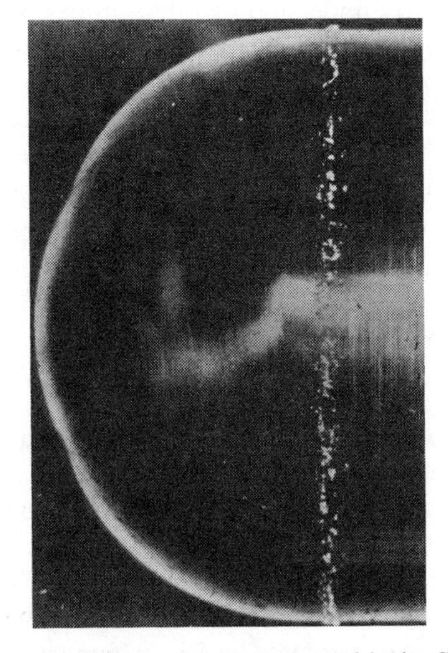

*Figure 2* Cavitation inception on a hemispherical-nosed body. Courtesy of the Applied Research Laboratory.

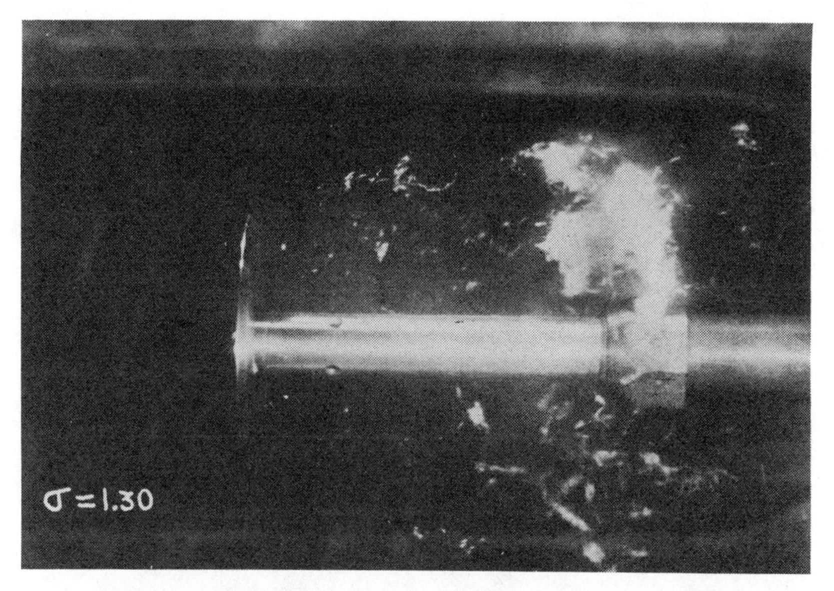


*Figure 3* Cavitation in the wake of a sharp-edged disk. Courtesy of the Applied Research Laboratory.

*Figure 4* Cavitation in the tip vortices of a propeller. Courtesy of the Applied Research Laboratory.

*Figure 5* Vortex and sheet cavitation on a marine propeller. Courtesy of the Netherlands Ship Model Basin.

(note the varying degrees of cavitation on each blade). The nonsteadiness in this case is induced by periodic passage of each propeller blade through the ship's wake. These photographs provide only an inkling of the complexity of the problem.

## PHYSICAL DESCRIPTION OF CAVITATION

### *Cavitation Scaling*

The fundamental parameter in the description of cavitation is the cavitation index

$$\sigma = \frac{p_0 - p_v}{\frac{1}{2}\rho U_0^2} \tag{1}$$

wherein $p_0$ and $U_0$ are a characteristic pressure and velocity, respectively, $\rho$ is the density, and $p_v$ is the vapor pressure of liquid. Various hydrodynamic parameters such as lift and drag coefficient, torque coefficient, and efficiency, are assumed to be unique functions of $\sigma$ when there is correct geometric similitude between the model and prototype. Generally speaking, these parameters are independent of $\sigma$ above a critical value. This critical value is often referred to as the

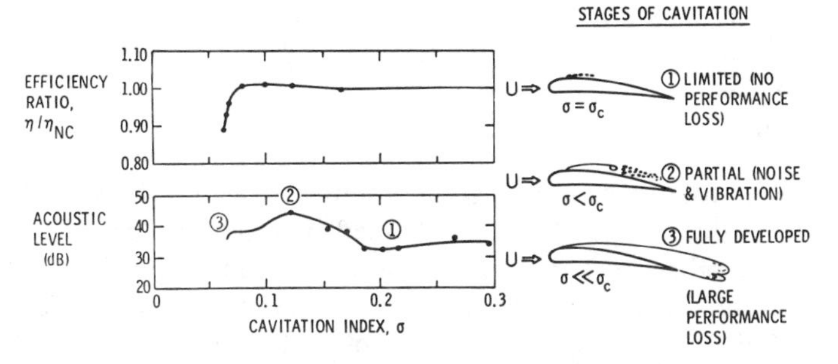


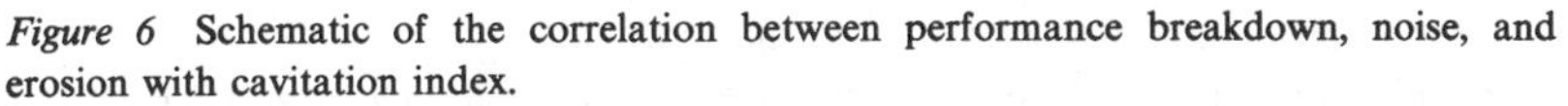
*Figure 6* Schematic of the correlation between performance breakdown, noise, and erosion with cavitation index.

incipient cavitation number. It should be emphasized, however, that the point where there is a measureable difference in the performance is not the same value of $\sigma$ where cavitation can first be detected visually or acoustically. This is illustrated in Figure 6, which is based on some data from Deeprose et al (1974). In this figure we see that a measureable drop in the head produced relative to the head produced under non-cavitating conditions occurs at a value of the cavitation index that is much lower than the value where cavitation can first be detected acoustically.

There are two ways of defining the critical value of $\sigma$. $\sigma_i$ is defined as the incipient cavitation number, which is normally determined in a water tunnel at constant velocity by slowly lowering the test-section pressure until cavitation is first observed. A more repeatable quantity is the desinent cavitation index, $\sigma_d$, which is again usually determined at constant speed in a water tunnel by first lowering the pressure until cavitation occurs and then slowly *raising* the pressure until the cavitation is extinguished. In order to eliminate confusion, the term $\sigma_c$ will be used to imply a critical value throughout the rest of this paper. $\sigma_c$ can be thought of as a performance boundary, such that

$\sigma > \sigma_c$ no cavitation effects,

$\sigma < \sigma_c$ cavitation effects: performance degradation, noise, and vibration.

Another form of the cavitation index is the Thoma number defined as

$$\sigma_T = \frac{H_{sv}}{H} \tag{2}$$

where $H_{sv}$ is the net positive suction head (Wislicenus 1965) and $H$ is the total head under which a given machine is operating. In a given situation, $\sigma$ and $\sigma_T$ are proportional. Another scaling parameter used quite often in pumps is the suction specific speed

$$S=\frac{\omega\sqrt{Q}}{[gH_{sv}]^{3/4}} \tag{3}$$

with $\omega$ being rotational speed, $Q$ the volumetric rate of flow, and $g$ the acceleration due to gravity. $S$, as defined in Equation (3), is a nondimensional number. In the United States, the quantity $g$ is normally set equal to unity, $\omega$ is replaced by $n$ in rpm, $Q$ in gpm, and $H_{sv}$ in feet. The value of $S$ then depends on the system of units used.

The suction specific speed is a natural consequence of considering dynamic similarity in the low-pressure region of a turbomachine (e.g. a pump). The dynamic relations are

$$\frac{gH_{sv}D_e^4}{Q^2}=\text{const}, \qquad \frac{gH_{sv}}{nD_e^2}=\text{const}, \tag{4}$$

where $D_e$ is the eye or throat diameter. These relations hold when the kinematic condition for similarity of inlet flow is satisfied, i.e.

$$\frac{Q}{nD_e^3}=\text{const}. \tag{5}$$

Elimination of $D_e$ in Equation (4) yields the suction specific speed, Equation (3). Thoma's $\sigma_T$, $S$, and specific speed $n_s$ are interrelated,

$$\sigma_T=\left(\frac{n_s}{S}\right)^{4/3}. \tag{6}$$

Hence when the operating conditions are held fixed ($n_s$=const), there is a direct relationship between $\sigma_T$ and $S$. The suction specific speed is used extensively for pumps but has not received the same use in the turbine field.

## *Scaling Cavitation Inception*

In scaling cavitation inception (or desinence) it is assumed that cavitation occurs when the minimum pressure in the flow is equal to the vapor pressure. For steady flow over a streamlined body this implies

$$\sigma_c=-C_{pm} \tag{7}$$

where $C_{pm}$ is the minimum pressure coefficient. In the case of a hydrofoil or a strut, $C_{pm}$ is a function of angle of attack, $\alpha$, as well as the shape of the body. The equivalent parameter for a turbomachine would be the normalized flow rate, $Q/nD^3$. In the case of a propeller, the effective parameter is the advance coefficient, $J=U/nD$. Hence, for fixed geometry

$$\sigma_c = f\begin{cases} \alpha & \text{—hydrofoils, struts} \\ Q/nD^3 & \text{—pumps and turbines} \\ \dfrac{U}{nD} & \text{—propellers} \end{cases} \tag{8}$$

Inception on hydrofoils has been studied by numerous investigators (Kermeen 1956a, b, Daily 1949). Much of this work is summarized in Knapp et al (1970). Typical data for a hydrofoil (Kermeen 1956a) are shown with angle of attack as the independent parameter (in Figure 7). Typical cavitation data for a propeller using $J$ as the independent parameter are shown in Figure 8. It is no accident that the two sets of

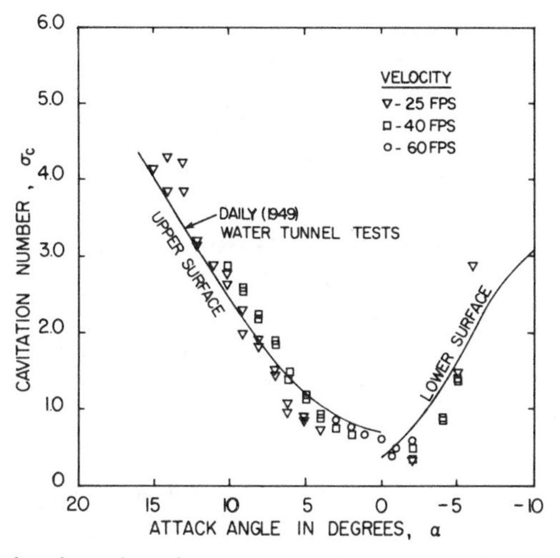


*Figure 7* Cavitation inception characteristics of a NACA 4412 hydrofoil (Kermeen 1956a).

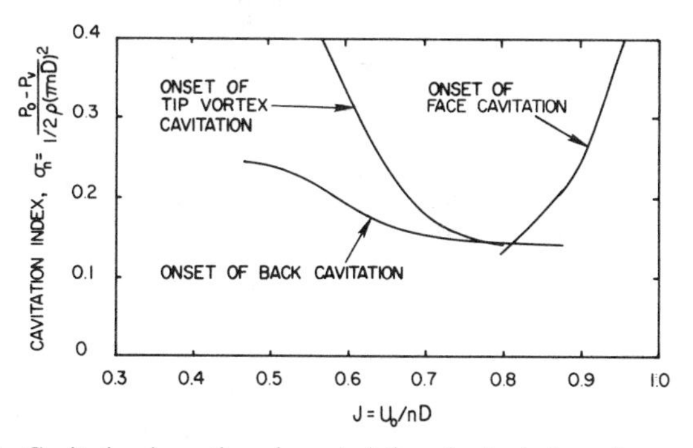


*Figure 8* Cavitation inception characteristics of a typical marine propeller.

data are qualitatively similar. In fact, hydrofoil data of the type shown in Figure 7 can be used to predict the cavitating performance of a propeller (Brockett 1966).

As already pointed out, resistance to cavitation in a turbomachine is normally defined by the value of $\sigma_T$ where an observable reduction in performance is noted as $\sigma_T$ is lowered below this critical value. The same definition can be used for hydrofoil data, as shown in Figure 9. Again the similarity of these data with those presented in Figure 6 is no accident. In some situations, performance breakdown can also be calculated (Acosta & Hollander 1959, Stripling & Acosta 1962, Stripling 1962, Wade 1967). However, the bulk of our information concerning the resistance of a given machine to cavitation has been developed empirically. An example is shown in Figure 10, where the minimum allowable $\sigma_T$ for hydraulic turbines is plotted as a function of specific speed defined in a nondimensional way (standard practice results in different values of $n_s$ depending on the system of units). Superimposed in this diagram are the limits on cavitation performance for pumps [assuming $S$ remains constant, Equation (6)]. It is apparent that the assumption $S$ = constant is not applicable to turbine practice.

Cavitation has an influence not only on steady-state phenomena but on nonsteady phenomena as well. An example is shown in Figure 11, which indicates the dependence of the frequency of vortex shedding behind wedges on the cavitation index (Young & Holl 1966). Here the Strouhal number is normalized with respect to its non-cavitating value $S_{t_\infty}$, while the cavitation number is normalized with respect to its incipient value (denoted as $\sigma_c$).The values of $S_{t_\infty}$ and $\sigma_c$ as a function of wedge angle are shown in the upper portion of the diagram. The

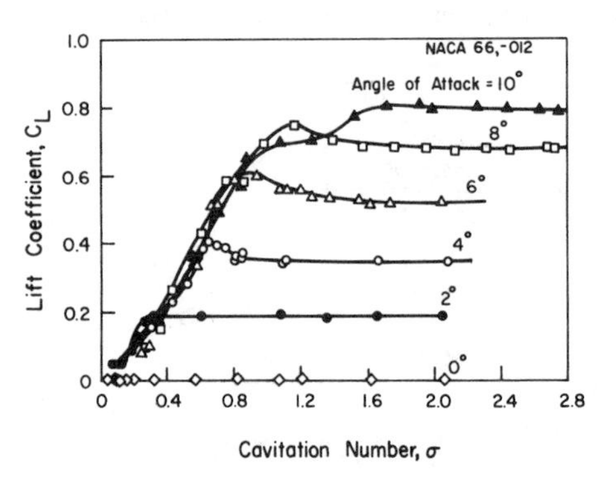


*Figure 9* Variation in lift coefficient with cavitation number (Kermeen 1956b)

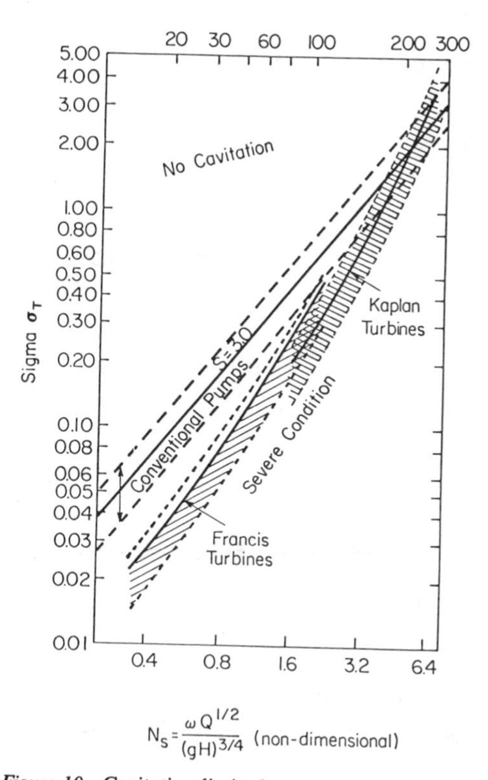


*Figure 10* Cavitation limits for pumps and turbines.

possible influence of cavitation on hydroelastic vibration is evident from these data.

It should also be pointed out at this juncture that developed cavitation is basically a nonsteady phenomenon even though the external flow is steady. Clear, glassy cavitation can be observed attached to bodies, hydrofoils, etc. However, examination of these cavities with high-speed cinematography indicates that, under certain conditions, the process is periodic. Sample data for two different bodies are shown in Figure 12. The frequency of the process is quite sensitive to the headform shape. This question is discussed in detail by several authors (Kermeen 1956a, b, Acosta 1974, Knapp 1955, Stinebring 1976, Shen & Peterson 1978).

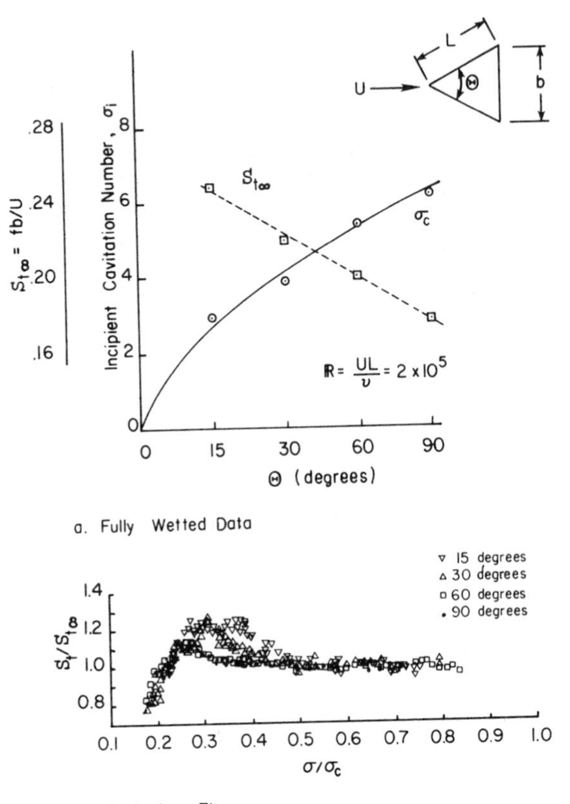


*Figure 11* Influence of cavitation on vortex-shedding frequency (Young & Holl 1966).

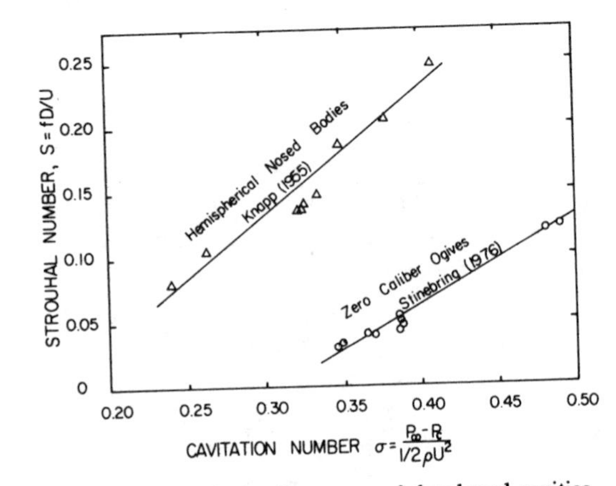


*Figure 12* Oscillation frequency of developed cavities.

## *Limitations of Cavitation Scaling Laws*

The suitability of the cavitation index $\sigma$, and its derivatives $\sigma_T$ and $S$, is dependent on the assumption of an inviscid fluid (i.e. pressure scaling with velocity squared) and the assumption that cavitation occurs when the minimum pressure in a given flow field is equal to the vapor pressure at the bulk temperature. It is further assumed, in the case of developed cavitation, that the cavity pressure is equal to the vapor pressure. With these assumptions in mind, $\sigma$ is the primary scaling parameter for inception and is the primary independent variable for scaling all hydrodynamic parameters involving cavitating flow.

It is imperative to understand the limitations of the basic scaling law since much of the information about the effects of cavitation are deduced experimentally. The question of cavitation scale effects has been considered by many authors in the past (Kermeen et al 1956, Holl & Wislicenus 1961, Arndt 1974a, b). Unfortunately, most of the experimental evidence is based on laboratory data obtained over a limited range of independent variables. There is little information available on the correlation between model and full scale. Since the development of fluid machinery and various hydraulic structures is based on considerable testing, usually in a laboratory, it is necessary to establish accurate modeling or scaling laws in order to be able to extrapolate model data to field or service conditions with confidence.

A few examples of cavitation scale effects are illustrated in the next few figures. In Figure 13 (Parkin & Holl 1954) the critical cavitation index for two streamlined bodies is plotted as a function of Reynolds

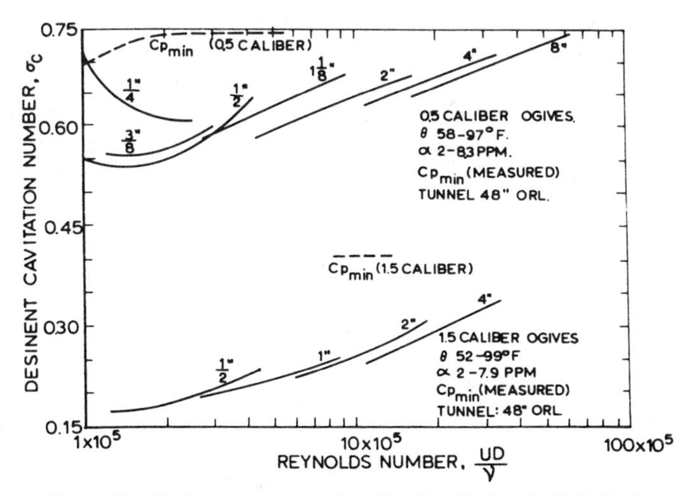


*Figure 13* Cavitation on streamlined bodies (Parkin & Holl 1954).

number. In both cases the data fall below the value predicted by the basic scaling law [Equation (7)]. Reasonable correlation with Reynolds number is achieved, but a definite size effect is still apparent. Holl & Robertson (1956) provided some convincing evidence that Weber number was also a significant parameter. This is in keeping with the unique measurements of Kermeen et al (1956) who found that the flow was in tension in the minimum pressure region of a streamlined body when cavitation was incipient.

Another aspect of the scale-effects problem is illustrated in Figure 14 which contains data for cavitation in the wake of a sharp-edged disk. The Reynolds-number dependence is even stronger and the critical cavitation number is greatly in excess of the predicted value. The results shown are typical of separated flows where the minimum pressure is found in the cores of eddies formed in the shear layer emanating from the separation point. Obviously the minimum pressure measured at a flow boundary is not a good indication of the inception characteristics for flows of this type.

Another type of scale effect was found by Holl (1960a). Two types of cavitation are noted on the same NACA 16012 hydrofoil section. One set of data corresponds to a band type of cavitation that disappears uniformly across the span when $\sigma = \sigma_c$. A second type, consisting of random spots, has a much higher value of cavitation index at desinence. The first type follows the same trend shown for streamlined bodies in Figure 13. The latter type shows a *decrease* in $\sigma_c$ with an increase in velocity, and $\sigma_c$ is higher than the predicted value [Equation (7)]. In his analysis, Holl (1960a) conjectured that the former type was true vaporous cavitation, whereas the latter type of cavitation was dominated by the effects of gaseous diffusion. This type of bubble growth is not unlike the bubble formation that occurs when the pressure in a bottle of champagne is suddenly reduced when the cork is popped. This type of cavitation follows a different scaling law [compare with Equation (2)]:

$$\sigma_c = -C_{pm} + \frac{p_g}{\frac{1}{2}\rho U_0^2} \tag{9}$$

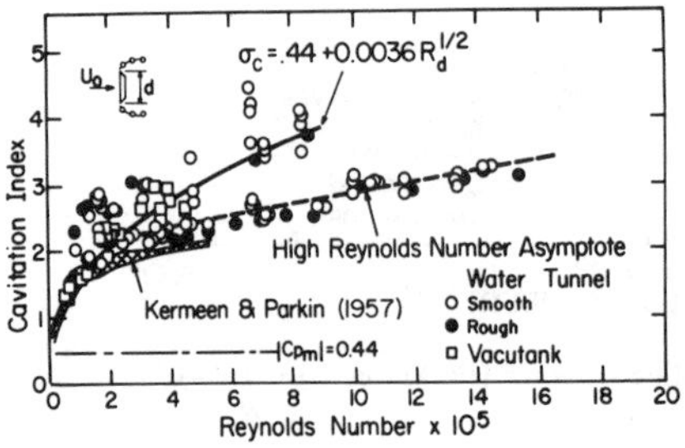

*Figure 14* Cavitation data for a sharp edged disk (Arndt 1978). Data from Chivers' test series I.

where $p_g$ is the saturation pressure at a given gas content.

The foregoing examples illustrate that cavitation inception (or desinence) is not adequately described by a single parameter such as $\sigma$. There are many reasons for the discrepancies noted. Fundamentally these fall into two categories:

1. The pressure field is not adequately described by an ideal-fluid-flow model; the effects of turbulence and other viscous phenomena dominate the problem.
2. The critical pressure at inception is not the vapor pressure and can be either higher or lower than $p_v$ depending on the situation. Ordinary liquids *can* sustain tension and more than one type of cavitation process can occur in the same flow field. Bubble growth can be a result of the formation of the vapor phase, or be due to the release of dissolved gas, or can be some combination thereof. It is not always possible to clearly distinguish between vaporous and so-called gaseous cavitation.

A final example of a cavitation scale effect is shown in Figure 15. Here efficiency ratio for a pump is plotted as a function of cavitation number. The only difference in the operating conditions for the various curves is the temperature of the water. Note that at higher temperature the pump appears to be more resistant to cavitation. A similar effect is noted when comparing a device tested in cold water with the same device tested in another liquid. It will be shown that the effect is most evident when there is a large change in vapor pressure with a small change in temperature ($\Delta p_v/\Delta\theta$ large).

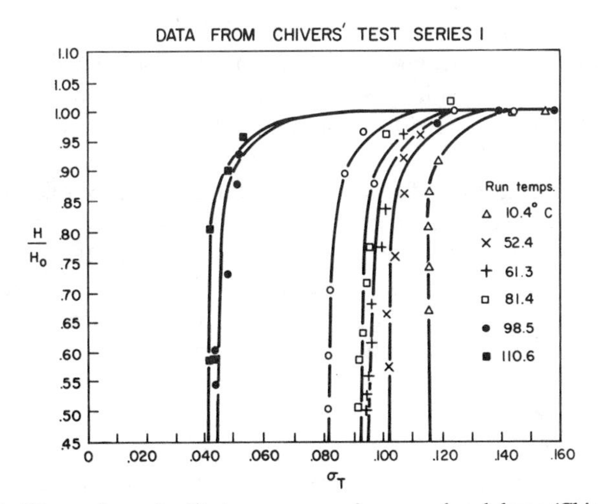


*Figure 15* **Thermodynamic effect on pump performance breakdown (Chivers 1969).**

# CAVITATION AND BUBBLE DYNAMICS

## *Cavitation Nuclei*

It is generally accepted that cavitation inception occurs as a consequence of the rapid or explosive growth of small bubbles or nuclei that have become unstable due to a change in ambient pressure. These nuclei can be either imbedded in the flow or find their origins in small cracks or crevices at the boundary surface of the flow (Acosta & Hamaguchi 1967, Holl & Treaster 1966, Gupta 1969, Peterson 1968, van der Meulen 1972). The details of how these nuclei exist have been considered by many investigators. A coherent summary of this work is offered by Holl (1969, 1970).

Most of the progress in understanding the details of the inception process has been made through the consideration of the dynamic equilibrium of a spherical bubble containing vapor and non-condensible gas. The Rayleigh-Plesset equation describes this equilibrium (Plesset & Prosperetti 1977):

$$R\ddot{R}+\frac{3}{2}\dot{R}^2=\frac{1}{\rho}\left[p_i-p_\infty(t)-\frac{2S}{R}-4\mu\frac{\dot{R}}{R}\right] \tag{10}$$

wherein $R$ is the bubble radius, $p_i$ the internal pressure, $S$ the surface tension, $\mu$ the dynamic viscosity, and $\rho$ the liquid density. Dots denote differentiation with respect to time. Neglecting the dynamical terms in Equation (10) results in

$$p_i-p_\infty=\frac{2S}{R_e}, \tag{11}$$

$R_e$ being an equilibrium radius.

Consider first a nucleus in the form of a small spherical bubble containing incondensible gas in a saturated solution. As shown by Epstein & Plesset (1950), Equation (11) represents an unstable equilibrium. A slight reduction in $R_e$ increases $p_i$ and drives gas into solution resulting in a further reduction in size. Similarly, a slight increase in $R_e$ decreases $p_i$ and the bubble grows by diffusion of gas across the bubble wall. The postulate of Harvey et al (1944a, b, 1947) circumvents this problem, as illustrated in Figure 16. If a solid particle containing a fissure, as shown, is hydrophobic, then the gas-liquid interface is concave and the stability equation is now

$$p_\infty-p_i=\frac{2S}{R_e} \tag{12}$$

which represents a stable equilibrium. The argument against small gas

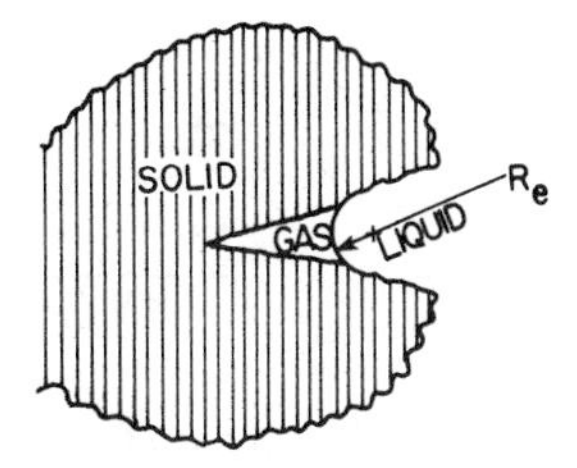


*Figure 16* Schematic of a cavitation nucleus in the form of a solid particle (Harvey et al 1947).

bubbles as nuclei is only valid in a static environment. Ripken & Killen (1962) discuss the source of nuclei and point out that in the turbulent flow of a water tunnel, gas bubbles *can* be a source of nuclei. Presumably the bubbles are stabilized by rectified diffusion due to turbulent pressure fluctuations. Arndt & Keller (1976) measured the nuclei in a large water tunnel and in a depressurized towing tank using a laser scattering technique. They concluded that the only source of nuclei in the towing tank was in the form of suspended solid material, whereas both solid particles and gas bubbles were sources of nuclei in the water tunnel. Gates & Acosta (1978) report considerable differences in the makeup of the nuclei population in two different tunnels (one having essentially small particles, the other essentially micro-bubbles). There is generally some relationship between the number density of nuclei and the level of dissolved gas in a given test facility. This relationship is not unique and depends on several factors including the facility configuration and operating conditions (Ripken & Killen 1962, Arndt & Keller 1976, Keller & Weitendorf 1976).

The general influence of nuclei size on the critical pressure can be deduced from Equation (11), assuming $p_i$ to be made up of vapor and non-condensible gas that is free to expand or contract isothermally

$$p_i = p_g + \frac{Nk\theta}{\frac{4}{3}\pi R^3}, \tag{13}$$

$N$ being the total mass of the gas in the bubble, $k$ the universal gas constant, and $\theta$ the temperature. There is a critical value of $p_v - p_\infty$ below which static equilibrium is not possible (Blake 1949)

$$(p_v - p_\infty)_{cr} = \frac{4S}{3R_*} \tag{14}$$

wherein $R_*$ is the critical radius which is related to the amount of gas in the bubble

$$R_* = \left[\frac{9}{8\pi}\frac{Nk\theta}{S}\right]^{\frac{1}{2}}. \tag{15}$$

Equation 14 implies that the basic cavitation scaling law should be modified to read

$$\sigma_c = -\left[C_{pm} + \frac{\frac{4}{3}S/R_*}{\frac{1}{2}\rho U^2}\right]. \tag{16}$$

Measured nuclei size distributions are shown in Figure 17. The data are for two water tunnels whose nuclei are mostly micro-bubbles and for two other facilities (a water tunnel and a depressurized towing tank) whose nuclei consist mostly of solid particles (Arndt & Keller 1976, Gates & Acosta 1978). The data indicate that nuclei population that are dominated by micro-bubbles are sensitive to the relative saturation level of dissolved gas.

The data in Figure 17 also indicate that a reasonable number of larger bubbles exist in most facilities (corresponding to tensions of less

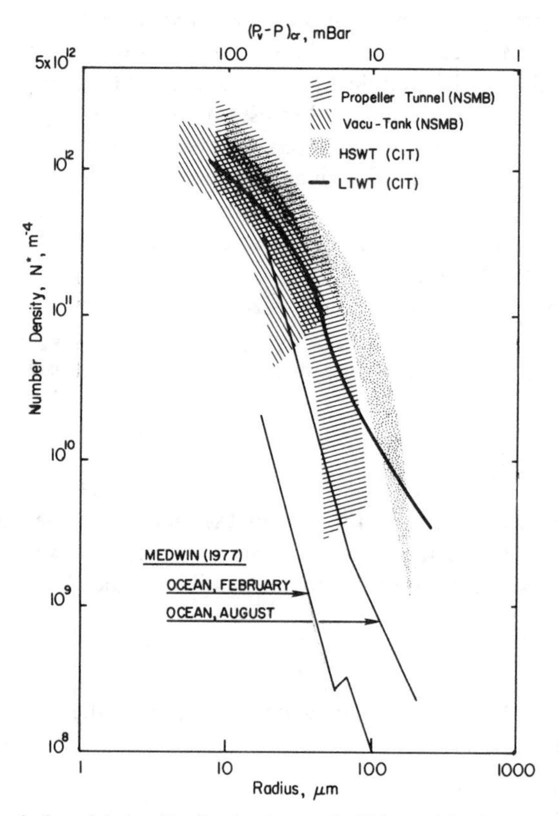


*Figure 17* Typical nuclei size distribution in test facilities and in the ocean (Arndt 1978, Gates & Acosta 1978, Medwin 1977).

than 50 mb). However, there is evidence that the flow field around a given body produces a screening effect such that larger bubbles tend to move out of the critical-pressure region leaving only smaller bubbles with more negative critical pressures to be active in the cavitation process (Johnson & Hsieh 1966). Very little has been done to investigate the question of the influence of nuclei size distribution on cavitation. What information we do have does indicate that this can be an important factor in laboratory experiments (Keller 1972, 1974, Gates & Acosta 1978). Most of the theoretical work on the influence of nuclei on cavitation assumes that the bubbles are always in static equilibrium until a critical size is reached and then explosive growth occurs. If the exposure time to negative pressure is short (less than the natural period of the bubble), then much greater tensions than predicted by an equilibrium model are possible (Kermeen et al 1956, Keller 1974, Arndt 1974a, Arndt & George 1978). A universal relationship is not available, but the amount of tension will vary inversely with a characteristic time scale of the flow

$$(p_v - p_\infty)_{cr} = \frac{4S}{3R_*} f\left\{\frac{d}{U}\left[\frac{S}{\rho R_*^3}\right]^{\frac{1}{2}}\right\} \tag{17}$$

where $f$ is an unknown function which tends to unity for large values of its argument.

The question of an adequate distribution of nuclei in a test facility can be important for modeling developed cavitation as well as determining cavitation-inception limits. This is especially true for unsteady cavitation [e.g. vibration due to cavitation on a propeller operating in a nonuniform inflow (Keller & Weitendorf 1976, Kuiper 1978)].

Techniques for the measurement of cavitation nuclei have been developed over only the past 20 years (Ripken & Killen 1959, 1962, Keller 1972, Peterson et al 1975, Peterson 1972, Feldberg & Shlemenson 1971, Morgan 1972, Gates & Acosta 1978). It is only recently that reliable measurements are possible and it is anticipated that more information on this important topic will be made available shortly.

### *Viscous Effects on Cavitation Inception*

Although Parkin & Kermeen (1953) appear to be the first investigators to consider the influence of boundary-layer development on the inception process, viscous effects have been studied in detail only more recently (Arakeri 1973, Arakeri & Acosta 1973, 1976, Casey 1974, van der Meulen 1976, 1978, Gates & Acosta 1978, Arakeri et al 1978). Related studies of the influence of viscosity on developed cavities are also available (Brennen 1970, Arakeri 1975b).

The most obvious influence of viscosity is the displacement effect on the mean pressure distribution due to boundary-layer growth on a given body. In the case of hydrofoils at an angle of attack, this can be an important factor since the minimum pressure peak occurs close to the leading edge and its magnitude is very sensitive to the geometry of the nose (Casey 1974).

Flow separation and transition to turbulence can have a measurable influence on the cavitation process. A graphic illustration is presented in Figure 18 (van der Meulen 1976, 1978). Cavitation data for a hemispherical-nosed body and a second body which is one of a series developed by Silberman & Schiebe (1974) are displayed. Both bodies have an identical value of $C_{pm}$. The Silberman-Schiebe body has a much lower value of $\sigma_c$, which is fairly insensitive to Reynolds number. As illustrated schematically, traveling bubble cavitation occurs in the minimum pressure zone on this body, whereas cavitation is observed to occur in a stabilized ring *aft* of the minimum pressure region on the hemispherical-nosed body (refer to Figure 2). Careful examination of the boundary-layer characteristics (Arakeri & Acosta 1973) under these conditions shows that cavitation occurs in the reattachment region of a laminar-separation bubble. Since the separation point is independent of Reynolds number, whereas the length of the separation bubble decreases with increasing Reynolds number, it is obvious that $\sigma_c$ should be asymptotic to $-C_{pm}$ as the length of the bubble decreases. Above a certain critical Reynolds number, transition to turbulence occurs upstream of the separation point and the separation bubble is eliminated. Intense pressure fluctuations have been measured in the reattachment region with peak values as high as 25% of the mean-stream dynamic pressure (Arakeri 1975a, Huang & Hannan 1976). Another anomaly observed in cavitation inception data is shown in Figure 19 (Arakeri et

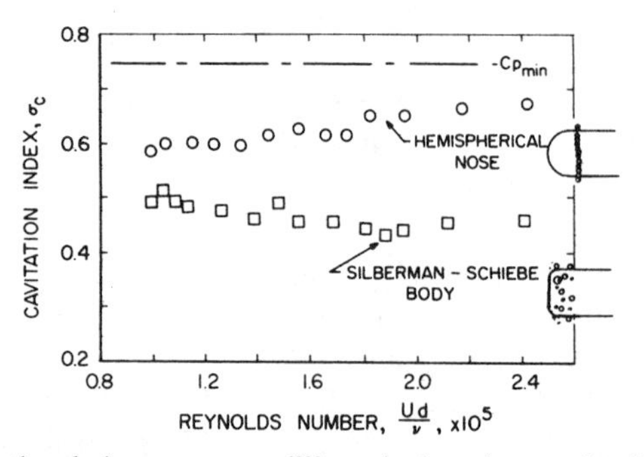


*Figure 18* Cavitation desinence on two differently shaped streamlined bodies of the same size that have the same minimum pressure coefficient.

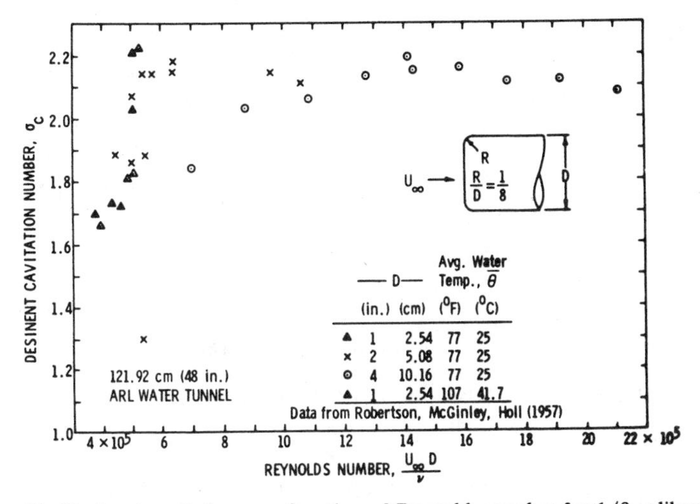


*Figure 19* Desinent cavitation as a function of Reynolds number for 1/8 caliber ogives (Arakeri et al 1978).

al 1978). A sudden jump in $\sigma_c$ with a small increase in Reynolds number for a 1/8 caliber ogive is traced to the sudden bursting of a "short" separation bubble when critical conditions are reached. Cavitation normally occurs in the separated shear layer of a "short" bubble and in the reattachment zone of a "long" bubble. "Bursting" of a leading-edge laminar-separation bubble can also occur on moderately thick hydrofoils (Roberts 1979). The entire question of laminar separation is discussed in detail by Gaster (1966) and Tani (1964).

In the absence of separation, the intense pressure fluctuations associated with natural transition to turbulence can control the position of cavitation (Arakeri & Acosta 1973). As an example, it is found that $\sigma_c$

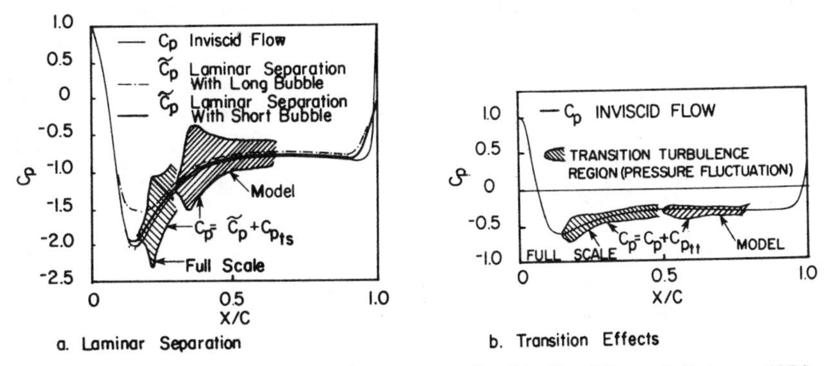


*Figure 20* Viscous effects on cavitation on streamlined bodies (Huang & Peterson 1976).

correlates with the value of $-C_p$ in the predicted region of transition on a 1.5 caliber ogive.

Viscous effects on streamlined bodies can be summarized as follows (refer to Figure 20 based on Huang & Peterson 1976):

1. Long separation bubbles affect the mean static pressure distribution. Long bubbles normally do not occur full scale. In a model

$$\sigma_c < -C_{pm} \{C_{pm} \text{ is potential-flow value}\}. \tag{18}$$

Cavitation can occur at the minimum pressure point where

$$\sigma_c = -\hat{C}_{pm} \{\hat{C}_{pm} \text{ is based on mean pressure}\} \tag{19}$$

or if it occurs at the trailing edge of the separation bubble,

$$\sigma_c = -\hat{C}_{ps} - C_{pts} \tag{20}$$

where $\hat{C}_{ps}$ is the mean pressure coefficient at reattachment and $C_{pts}$ is an *effective* correction to account for the intense fluctuations in pressure.

2. Short bubbles can occur in model and prototype. Little is known except that short bubbles have little influence on the mean pressure distribution. For cavitation occurring at a position $\xi$ we have

$$\sigma_c = -\hat{C}_{p_\xi} - C_{pts}. \tag{21}$$

3. Without separation, transition can determine the position of cavitation. Typically, in a model we have

$$\sigma_c < -C_{pm}, \tag{22}$$

$$\sigma_c = -C_{ptr} - C_{ptt}, \tag{23}$$

where $C_{ptr}$ is the potential-flow pressure coefficient in the region of transition and $C_{ptt}$ is an effective pressure drop due to turbulent-pressure fluctuations. In a prototype situation (see Figure 20*b*)

$$\sigma_c > -C_{pm}, \tag{24}$$

$$\sigma_c = -C_{ptr} - C_{ptt}, \tag{25}$$

where $-C_{ptr}$ is greater in the prototype. Relatively little is known about $C_{ptt}$ and $C_{pts}$. Suggested values are $-0.1$ and $-0.35$, respectively.

The effect of viscosity is to change the magnitude of $\sigma_c$ in model and prototype as well as the position of cavitation. Viscous effects can dominate at model scale and are relatively unimportant in the prototype. Typically, the flow can be in a state of tension in the model (if $\sigma_c < -C_{pm}$), whereas the critical pressure in the prototype is typically $p_v$

or higher. However, it must be borne in mind that cavitation-inception performance of a full-scale body is typically predicted from model tests. This underscores the necessity for understanding the scaling problem in detail.

## *Influence of Surface Roughness*

A knowledge of the effects of surface roughness is of great practical significance since the degree of surface finish can be a controlling factor in the inception process. Strictly speaking, the problem is a subset of viscous effects. The problem can be broken down into consideration of the effects of isolated asperities, distributed roughness, or a patch of roughness (Bechtel 1971). Only the two former cases have been considered in any depth (Arndt et al 1979).

Most of the information is based on experiment; an exception is the theoretical considerations of Holl (1958, 1960b). The critical cavitation index for a roughened body, $\sigma_{cR}$, is

$$\sigma_{cR} = -C_p + (1 - C_p)\sigma_{cfp} \tag{26}$$

or

$$\sigma_{cR} = \sigma_c \tag{27}$$

whichever is greater. $C_p$ denotes the pressure coefficient at the position of the roughness; $\sigma_c$ is the critical cavitation index for a smooth body ($-C_{pm}$ for many practical problems); $\sigma_{cfp}$ denotes the cavitation number for a flat plate of equivalent roughness (zero pressure gradient). In the case of isolated irregularities

$$\sigma_{cfp} = f\left(\frac{h}{\delta}, H, \mathrm{Re}_h\right) \tag{28}$$

wherein $h$ is roughness height, $\delta$ is boundary-layer thickness, $H$ boundary-layer shape factor, $\delta_1/\delta_2$, and $\mathrm{Re}_h$ is Reynolds number based on roughness height. Available data are presented in Figure 21, wherein the following correlation is used:

$$\sigma_{cfp} = C\left(\frac{h}{\delta}\right)^a \left(\frac{U\delta}{\nu}\right)^b \ \{\text{isolated asperity}\}. \tag{29}$$

There is insufficient information for correlation with $H$ and this parameter was necessarily deleted from Equation (29).

The influence of distributed roughness is even more complex than the case of an isolated asperity since there is an interrelationship between the boundary-layer development and the degree of surface roughness. An isolated asperity produces a localized perturbation to the flow with only a small influence on the overall development of the boundary

| SYMBOL | IRREGULARITY | FLOW DIMENSIONS | DATA SOURCE | a | b | C | |
|---|---|---|---|---|---|---|---|
| △ | TRIANGLES | TWO | HOLL 1960 | 0.361 | 0.196 | 0.152 | h |
| ○ | CIRCULAR ARCS | TWO | HOLL 1960 | 0.344 | 0.267 | 0.041 | h h=0.175l |
| ▲ | HEMISPHERES | THREE | BENSON 1966 | 0.439 | 0.298 | 0.0108 | h = 0.5d |
| ● | CONES | THREE | BENSON 1966 | 0.632 | 0.451 | 0.00328 | h=d |
| □ | CYLINDERS | THREE | BENSON 1966 | 0.737 | 0.550 | 0.00117 | h=d |
| ⊔ | SLOTS | TWO | BOHN 1972 | 0.041 | 0.510 | 0.000314 | h |

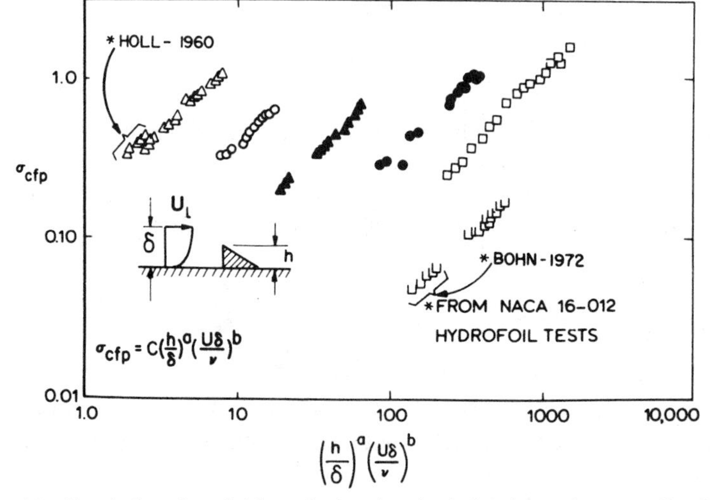


*Figure 21* Correlation of available cavitation data for isolated protuberances (Arndt et al 1979).

layer. Cavitation on a uniformly roughened surface occurs in the cores of large eddies in the boundary layer in much the same way as cavitation occurs in the turbulent boundary layer near a smooth wall (Arndt & Ippen 1968, Daily & Johnson 1956). Turbulence is the dominant mechanism here; the major influence of the roughness is an increased turbulence intensity. The details of the inception process are obviously quite complex. However, a simple relationship between $\sigma_{cfp}$

and the skin friction coefficient, $C_f$, has been determined (Arndt & Ippen 1968):

$$\sigma_{cfp} = 16C_f \{\text{distributed roughness}\}. \tag{30}$$

Thus, the best available method for estimating the effect of surface finish is through the use of Equation (26). $\sigma_{cfp}$ can be determined from experimental data [Equation (29)] or from Equation (30), whichever is appropriate. Holl (1965) presents a semi-empirical method for estimating $\sigma_{cfp}$. An application of this method for predicting the inception characteristics of boundary-layer trip wires on a propeller model is discussed by Arndt (1976a).

Examples of the relative effect of isolated and distributed roughness on a NACA 16-212 hydrofoil are displayed in Figures 22 and 23. The information presented is based on calculations using Equations (26), (29), and (30). The cavitation index is normalized to an assumed value of $\sigma_c = -C_{pm}$ for a smooth hydrofoil. In the first diagram, three salient features are obvious: there is a critical position for an isolated asperity, the effect of isolated protuberances is very sensitive to Reynolds number, and the effect can be very large. The second diagram shows the effect of a uniform sand-grain roughness. There is a relatively minor effect up to about $h_s/c = 10^{-3}$ and even at $h_s/c = 4 \times 10^{-3}$, the increase in cavitation index is only 50%, as compared with increases in the range of 300–600% for an equivalent isolated protuberance. The practical

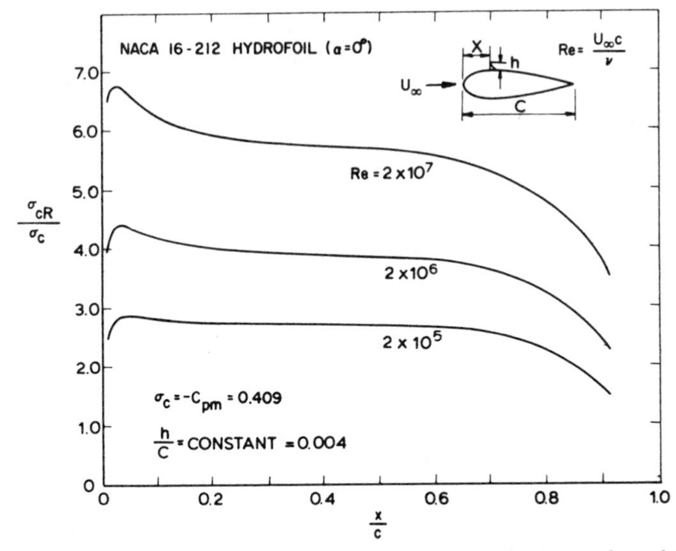


*Figure 22* Effect of an isolated roughness on hydrofoil cavitation (Arndt et al 1979).

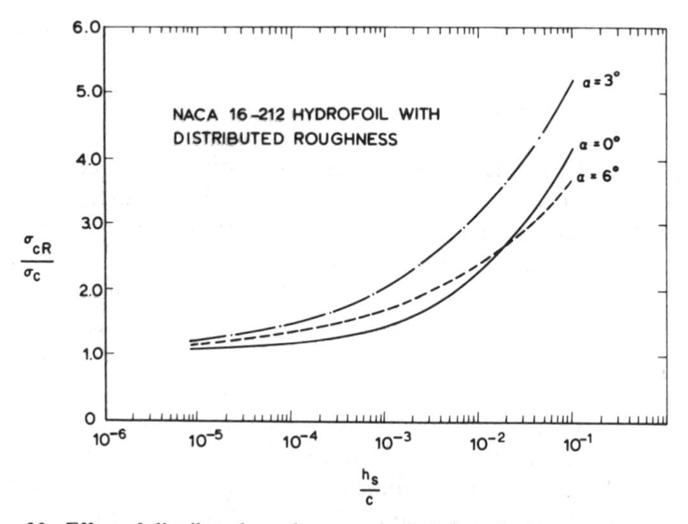


*Figure 23* Effect of distributed roughness on hydrofoil cavitation (Arndt et al 1979).

implications of these results are enormous and the effects of surface finish on service life of various devices has not been seriously considered. (Recall that significant cavitation damage can occur when there is limited cavitation with no detectable change in performance.)

## *Vortex Cavitation*

Vortical motion of all types is common in fluid machinery and hydraulic structures, in tip vortices on propellers, and in the clearance passages of pumps and turbines, hub vortices trailing from propellers and Kaplan turbines, secondary flows in corners, etc. Vortex cavitation is a major consideration for erosion in ducted propellers, surging in draft tubes, propeller-induced hull vibrations, etc. There are two significant aspects to this problem from a cavitation point of view: the minimum pressure in the cores of vortices can be very Reynolds-number dependent and cavitation nuclei can be entrained in the low-pressure core for relatively lengthy periods of time.

The minimum pressure in a vortex can generally be predicted from a Rankine model

$$C_{pm} = -2\left[\frac{\Gamma}{2\pi r_c U}\right]^2 \tag{31}$$

wherein $\Gamma$ is the circulation in the flow and $r_c$ is the core radius. The radius of the core is generally unknown and related to several parameters. In addition, it is often difficult to estimate $\Gamma$, especially for secondary flows, hub vortices, etc. We can expect cavitation to occur when $r_c$ is at its smallest and can normally disregard the effects of laminar and turbulent diffusion of vorticity which act to thicken the core and reduce $C_{pm}$ in the downstream direction. The problem is further complicated by the phenomenon of vortex breakdown which has been considered by several investigators (Hall 1972, Sarpkaya 1971, Bossel 1967, Harvey 1962). However, the effects are imperfectly understood at this time.

The pressure field in a vortex is very sensitive to variations in Reynolds number. McCormick (1954, 1962) has conducted an extensive study of tip-vortex cavitation. He found

$$\sigma_c = KR_e^n \tag{32}$$

where $K$ varied linearly with angle of attack and flow configuration. $\sigma_c$ was found to be almost independent of aspect ratio for a given plan form. The strong Reynolds-number dependence was related to the thickness of the boundary layer on the pressure face of the wing generating the vortex. This concept was verified qualitatively with a hydrofoil having its upper surface artificially roughened. At positive angle of attack, the roughened surface is on the suction side, and the roughness exerts little influence on the cavitation. A different set of circumstances is noted at negative angles of attack. The roughened surface is now on the pressure side and a significant reduction in the cavitation index is noted, presumably due to a thickening of the vortex core. Arndt (1976a) found an entirely different effect on a model propeller. He noted a significant attenuation of the tip vortex with the *suction* side roughened. Presumably, centrifugal effects can significantly change the manner in which the tip vortex develops.

Similar trends are found for hub vortices (Holl et al 1972, 1974). Figure 24 contains the results of pressure measurements in a hub vortex. $\Gamma$ is controlled by the blade setting of a vortex generator and $r_c$ is controlled by varying the Reynolds number. The effects of vortex breakdown and Reynolds number are clearly evident. Wind-tunnel data are presented for various values of circulation and, for comparison purposes, equivalent water-tunnel data are presented. These data indicate that $-C_{pm}$ increases monotonically with Reynolds number until a critical value of Reynolds number is reached, at which point a slight rise in the minimum pressure is noted. This is indicative of vortex breakdown, which is a function of both Reynolds number and a swirl parameter, $\Gamma/Ud$, where $\Gamma$ is the circulation and is proportional to the blade setting. Thus the value of the Reynolds number at the threshold of vortex breakdown varies inversely with the swirl parameter. This

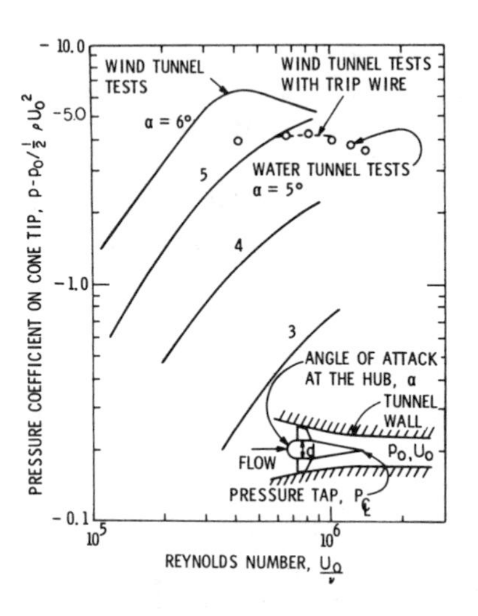


*Figure 24* Minimum pressure in the core of a hub vortex (Holl et al 1972).

conjecture is supported by dye studies. These dye studies also showed that vortex breakdown could be induced prematurely by the presence of a small measuring probe positioned near the vortex core. This implies that more sophisticated flow-measurement techniques, such as the use of the laser-doppler velocimeter, must be used to investigate flows on the verge of vortex breakdown.

An extensive study of the factors involved in determining the circulation in a hub vortex indicates that the vortex strength is very sensitive to the inlet velocity profile near the root of the blades. Upstream disturbances can have a very significant effect. The cavitation number for a given rotor can vary by almost a factor of two, with and without upstream struts (see, for example, Billet 1978).

In addition to the sensitivity of $\Gamma$ and $r_c$ to flow configuration and Reynolds number, the influence of dissolved gas has a major effect because of entrainment of nuclei and exposure to low pressure for long periods of time. An equilibrium theory (Edstrand 1950, Holl 1960a) suggests the following law

$$\sigma_c = -C_{pm} + \frac{K\alpha\beta}{\frac{1}{2}\rho U^2} \tag{33}$$

wherein $K$ is an unknown factor ($0.3 < K < 1$), $\alpha$ is the dissolved gas content, and $\beta$ is Henry's constant.

Figure 25 contains cavitation data for the same configuration shown in Figure 24 obtained at various velocities and with water containing varying amounts of gas. In all cases the cavitation numbers are higher than the absolute value of the pressure coefficient measured at the cone

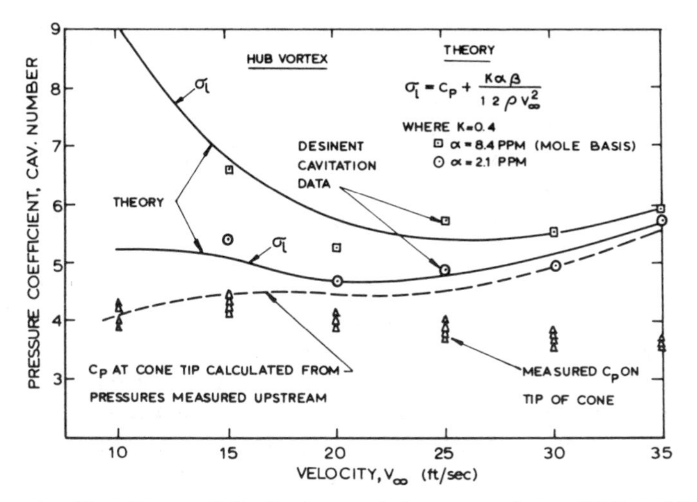


*Figure 25* The influence of dissolved gas on hub vortex cavitation (Holl et al 1972).

tip and in most cases higher than the calculated values based on upstream pressure measurements. The data are well represented by Equation (33), using $K = 0.4$. The theoretical curves are based on calculated values of $C_p$. It is interesting to note that during the entire test program, cavitation was found to occur at critical pressures higher than the vapor pressure.

## *Cavitation in Turbulent Shear Flows*

It has already been shown that cavitation inception on a smooth streamlined body can occur first in the turbulent transition region associated with either the reattachment of laminar separation or with a fully attached boundary layer. However, turbulence can also be a major, if not primary, factor in free turbulent shear flows and fully developed boundary-layer flows (Arndt & Daily 1969, Arndt 1974a, b, Arndt & George 1978). Scant attention has been paid to this problem, even though the topic is of practical significance. For example, the critical cavitation index for different-sized submerged jets is observed to vary from 0.15 to 3.0 over a size range of 0.1 cm to 13 cm (Lienhard & Goss 1971). Available data for cavitation in the wake of a sharp-edged disk show a monotonic increase with Reynolds number and no upper limit on the critical cavitation index can be determined with the available information (Arndt 1976b). At present, laboratory experiments simply do not provide a reasonable estimate of the critical cavitation index that can be encountered under prototype conditions. From a practical point of view the situation is much more critical than the scaling problems associated with streamlined bodies since at present there is no definable upper limit on the cavitation index for free shear flows.

The problem is highly complex, involving the diffusion of nuclei into the minimum pressure region from both the free stream and from cracks and crevices in the body and the intensity and spectrum of turbulent pressure fluctuations. In principle, a knowledge of both the size distribution of nuclei and the spectrum of the turbulent pressure fluctuations is necessary to determine the critical pressure in the inception region. The problem is further complicated by dissolved gas in the flow, which under certain circumstances can markedly change the desinent value of the cavitation index (e.g. Baker et al 1976).

In order to provide a basis for discussion we must distinguish between the time-mean pressure, $p$, sensed at a surface adjacent to the region of cavitation and the pressure actually sensed by cavitating nuclei, $p_{ml}$. Assuming cavitation occurs when $p_{ml}=p_v$, we have

$$\sigma_c = -C_p + \frac{p-p_{ml}}{\frac{1}{2}\rho U_0^2}. \tag{34}$$

Equation (34) is another version of the superposition equation already discussed.

The pressure difference, $(p-p_{ml})$, is quite complex. In fact, the major contribution to this factor is related to the *unsteady* pressure field, since spatial variations in the time-mean static pressure due to turbulence are normally minor. Many practical flows, such as boundary layers, jets, and wakes all fall in the general category of mixing layers. In many cases the variation in time-mean pressure across the layer is minor (Arndt 1974a, b) and is equal to the pressure in the potential flow at the edge of the layer. Perfect examples are submerged jets and boundary layers (Rouse 1953, Daily et al 1959). In the case of a jet, for example, variations in mean pressure are found to be less than three per cent of the dynamic pressure at the nozzle, yet $\sigma_c$ is usually measured to be of the order 0.5 or greater. Thus, temporal fluctuations in pressure must be at the heart of the cavitation problem in mixing layers. The problem is complicated by the fact that scaling laws must be found for both the pressure amplitudes (which may not scale with $U^2$) and the spectral characteristics of the pressure field. In general, the incipient cavitation index can be written in the form:

$$\sigma_c = -C_{pm} + C_1 \frac{p'}{\frac{1}{2}\rho U_0^2} \tag{35}$$

wherein $-C_{pm}$ is usually negligible. The factor $C_1$ takes into account both the statistics of the pressure field and any effects associated with the influence of turbulence on the nuclei size distribution and their response to the unsteady pressure. As already discussed, nuclei size influences the critical pressure. Negative pressure peaks that induce cavitation should have a statistical occurrence that is relatively high, and a duration that is comparable to the natural period of the nuclei for cavitation to occur at a local pressure approaching the vapor pressure. Thus $C_1$ is a factor involving both the probability density and frequency content in the pressure field. If all the energy in the pressure field has a frequency content much higher than the natural frequency of the nuclei, there would be a lowering of $\sigma_c$ associated with a lower value of $p_{cr}$. Hence, one could conclude that if the turbulent intensity remains constant, a reduction in the scale of a turbulent shear flow could reduce $\sigma_c$. Generally, this is unlikely (cf Arndt & Daily 1969), but as already noted $\sigma_c$ does vary with orifice size (Lienhard & Goss 1971). Measured values of $C_1$ range from about 7 in a turbulent boundary layer (Arndt & Ippen 1968) to about 10 in a jet (Rouse 1953).

The theoretical determination of a turbulent pressure field is extremely complex. However, considerable progress has been made over the last five years in the understanding of turbulent pressure fluctuations in free shear flows in an Eulerian frame of reference. Of particular importance is the development of pressure-sensing techniques which under certain circumstances can lead to reliable measurements of pressure fluctuations. The first theoretical arguments on the pressure fluctuations associated with turbulent flow appear to be due to Obukhov and Heisenberg (Batchelor 1953). Heisenberg argued that Kolmogorov scaling should be possible for small-scale pressure fluctuations. Batchelor (1951) was able to calculate the mean-square intensity of the pressure fluctuations as well as the mean-square fluctuating pressure gradient in a homogeneous, isotropic, turbulent flow. This work was extended by Kraichnan (1956) to the physically impossible but conceptually useful case of a shear flow having a constant mean velocity gradient and homogeneous and isotropic turbulence. Apparently there were no attempts made to extend this theoretical work until the 1970s when George (1974) and Beuther et al (1977a, b, George & Beuther 1977) applied the concepts developed by Batchelor and Kraichnan to the calculation of the turbulent pressure spectrum in homogeneous, isotropic, turbulent flows with and without shear. When compared with experimental evidence gathered in turbulent mixing layers, the theory is found to be remarkably accurate.

All this work indicates that relatively higher-intensity pressure fluctuations can be expected in a shear flow than in the idealized case of isotropic turbulence. In general we can say

$$\frac{p'}{\frac{1}{2}\rho U^2} = C\frac{u'}{U^2} \qquad \text{where} \begin{cases} C \cong 1.4 \text{ isotropic flow,} \\ C > 1.4 \text{ shear flow.} \end{cases} \tag{36}$$

Measurements by Barefoot (1972) qualitatively confirm Kraichnan's work (see Figure 26). More recent works (George 1974, Beuther et al 1977a, b, George & Beuther 1977) appear to indicate that the spectral characteristics of the pressure field can also be inferred theoretically. As shown in Figure 27, the experimental data and the theory are remarkably consistent, especially in light of the fact that several different experimental techniques and different flow facilities are involved.

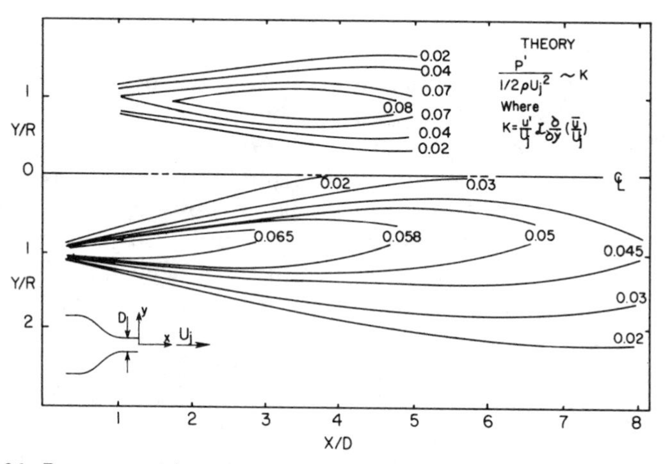


*Figure 26* Rms pressure intensity in the mixing zone of a submerged jet, theory and experiment (Barefoot 1972).

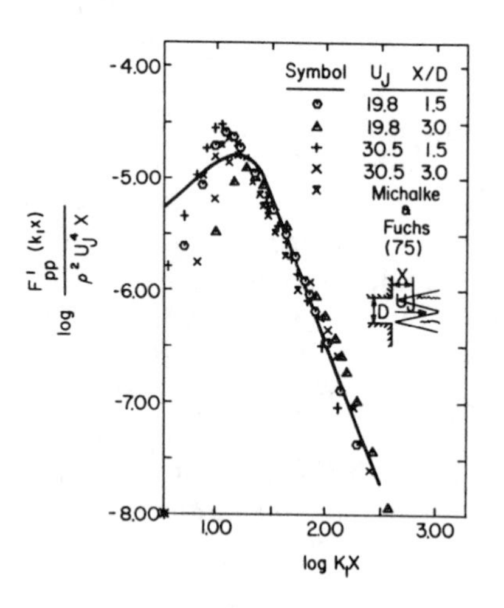


*Figure 27* Measured and theoretical spectra of pressure fluctuations in the mixing zone of a free jet (George & Beuther 1977).

This work is supplemented by the considerable body of evidence that coherent structures play an important role in the determination of at least the large-scale pressure fluctuations (Fuchs & Michalke 1975, Fuchs 1972a, b, and Chan 1974a, b 1977). For example, cavitation in highly turbulent jets is observed to occur in ring-like bursts, smoke rings if you will. These bursts appear to have a Strouhal frequency, $fd/U_J$, of approximately 0.5. This point is underscored by some recent work of Fuchs (1973), which indicates that the pressure field in a jet is highly correlated in the circumferential direction at $fd/U_J = 0.5$.

Because of the relative ease of measurement, there exists a considerable body of experimental data for wall pressure due to turbulent boundary-layer flow. However, in many ways less is known about the turbulent pressure field for boundary layers than for free turbulent shear flows. Not only is the theoretical problem made more difficult (impossible to the present) by the presence of the wall, the experimental problem is considerably complicated by the dynamical significance of the small scales near the wall. Thus, in spite of over two decades of concentrated attention, we cannot say with confidence even what the rms wall pressure level is, although recent evidence points to a value of (Willmarth 1975)

$$p' = C\rho u_*^2, \qquad C = 2 \text{ to } 3, \tag{37}$$

where $u_* = \sqrt{\tau/\rho}$. The basic problem is that the most interesting part of a turbulent boundary layer appears to be the region near the wall where intense dynamical activity apparently gives rise to the overall boundary-layer activity. While the details of the process are debatable, most investigators concur on the importance of the wall region in overall boundary-layer development. Unfortunately, under most experimental conditions, the scales of primary activity are smaller than standard wall pressure probes can resolve. Thus we have virtually no information concerning the contribution of the small scales to the pressure field. Arndt & George (1978) suggest that the small scales *are* significant and can possibly be a dominant factor in very-large-scale flows (e.g. boundary layers on ships, spillways, etc).

Both experimental and theoretical information on the spectral characteristics of turbulent pressure fields are cast in an *Eulerian* frame of reference. The problem of cavitation inception requires pressure information in a *Lagrangian* frame of reference. Arndt & George (1978) point out that Kolomogorov scaling has been successful in an Eulerian frame of reference and suggest that similar scaling will be valid for Lagrangian spectra. Based on this assumption, they derived expressions for the spectrum of a turbulent pressure field. However, there is no experimental evidence available to support their findings.

Whether or not the pressure fluctuations play a role in the cavitation-inception process depends on previously cited criteria:

1. The minimum pressure must fall below a critical level.
2. The minimum pressure must persist below the critical level for a finite length of time.

The first criterion depends greatly on the yet unresolved question of intermittency and its effect on the probability density of the pressure fluctuations. It is possible that the critical cavitation index will increase with Reynolds number because larger excursions from the mean pressure are more likely (Arndt & George 1978).

Some specific comments can be made concerning the question of residence time. In a turbulent boundary layer most of the energy in the pressure spectrum scales with $u_*$ and $\delta$. It is clear that the criteria for bubble growth without appreciable tension reduce to

$$\frac{u_* T_B}{\delta} \leqslant 1$$

where $T_B$ is a characteristic response time of the nuclei. That is to say, a pressure fluctuation must persist for a time that is long in comparison with the response time of a typical nucleus.

Since $\nu/u_*^2$ is the shortest time scale in a smooth-wall boundary layer, *all* of the pressure spectrum is sampled by the nuclei when

$$\frac{u_*^2 T_B}{\nu} < 1.$$

This criterion is especially important in view of the highly intermittent process near the wall.

For rough walls, the last criterion can be expressed in terms of the roughness height $h$ by

$$\frac{u_* T_B}{h} < 1.$$

Since in fully rough flow $(u_* h)/\nu > 1$, it is clear that the small-scale criterion is more easily satisfied with rough-wall experiments.

It is extremely difficult to relate the available data on cavitation inception in turbulent flows to the information available on turbulent pressure fields. Only a rather limited amount of cavitation data has been collected under controlled conditions. The types of flows considered to date include the wake behind a sharp-edged disk, submerged jets from nozzles and orifices, and smooth and rough boundary layers. There is little information relating the observed cavitation inception with the

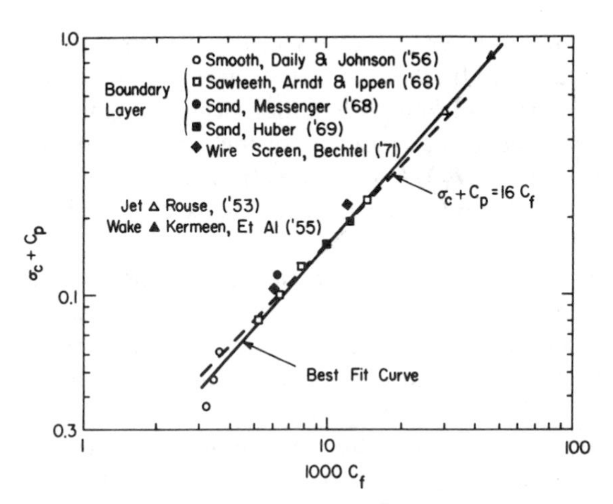


*Figure 28* Cavitation inception in turbulent shear flows.

turbulence parameters. A collation of available data is presented in Figure 28. Here the data are presented in a form suggested by Equations (35), (36), and (37), namely

$$\sigma_c + C_p = f(C_f) \tag{38}$$

$$C_f = \begin{cases} 2\tau_0/\rho U^2 & \text{boundary-layer flow} \\ \overline{(u_1 u_2)}\,/\,U_0^2. & \text{free-shear flows} \end{cases} \tag{39}$$

In this figure, $C_f$ is computed either from the measured wall shear in the case of boundary-layer flows or from turbulence measurements made in air at comparable Reynolds numbers for the case of a free jet and a wake. The measured value of $C_p$ is only significant for the case of the disk wake and the pressure data were determined from the experimental work of Carmody (1964). The available data seem to be well approximated by the relation

$$\sigma_c + C_p = 16 C_f \tag{40}$$

which was originally proposed for boundary-layer flow by Arndt & Ippen (1968). These data might seem to imply that a relatively simple scaling law already exists and then further imply that the previous discussion in this paper on turbulence effects is superfluous. *This is not the case*. Observations of the bubble growth in turbulent boundary layers indicate that the growth rate stabilizes at a constant value during most of the growth phase. Using the measured growth rate, the levels of local tension are computed to be quite small, of the order of 20 to 100 millibar (Arndt & Ippen 1968). Observations of bubble growth in turbulent boundary-layer flow over rough walls indicate the life time for bubble growth is a fraction of the Lagrangian time scale, $\mathfrak{T}=\delta/u'$, actually of the order of $h/u_*$. Cavitation occurs roughly in the center of the boundary layer with a tendency for the zone of maximum cavitation to shift inward as $(u_* T_B)/h_s$ decreases from about 1.5 to approximately 0.1. In the cited boundary-layer experiments $C_p$ is negligible. For this case, $\sigma_c \cong 16C_f$.

The fact that a strong dependence on Reynolds number can be observed even in free-shear flows has already been illustrated in Figure 14. These data were obtained in two water tunnels and a new depressurized tow-tank facility located at the Netherlands Ship Model Basin. The water-tunnel data are for cavitation desinence, whereas the tow-tank data are for cavitation inception determined acoustically. The cross-hatched data were determined in a water tunnel at high velocities by Kermeen & Parkin (1957). All other data were obtained at relatively low velocities (2–10 m/s). There is considerable scatter in these data and this is traceable to gas-content effects which dominate at low velocities. The low-Reynolds-number data appear to be satisfied by a semi-empirical relationship discussed by Arndt (1976b):

$$\sigma_c = 0.44 + 0.0036\left[\frac{Ud}{\nu}\right]^{\frac{1}{2}}. \tag{41}$$

It was found that the tow-tank data agreed with this relationship at relatively high Reynolds numbers. Equation (41) was developed from a model that assumes laminar-boundary-layer flow on the face of the disk. It would be expected that this condition would be satisfied at higher Reynolds number in a tow tank than in a highly turbulent water tunnel. At high Reynolds number (and also high velocity, where gas content effects are negligible), there is a continuous upward trend in the data with increasing Reynolds number. This underscores the need for further work as suggested in the introduction to this section.

To summarize, cavitation inception in turbulent shear flows is the result of a complex interaction between an unsteady pressure field and distribution of free stream nuclei. There is a dearth of data relating cavitation inception and the turbulent pressure field. What little information is available indicates that negative peaks in pressure having a magnitude as high as ten times the root-mean-square pressure can induce cavitation inception. This fact alone indicates that consideration should be given to the details of the turbulent pressure field. It is suggested that two basic factors related to the pressure field enter into the scale effects. First, as the scale of the flow increases, cavitation nuclei are relatively more responsive to a wider range of pressure fluctuations. Second, the available evidence indicates that large deviations from the mean pressure are more probable with increasing Reynolds number. This would explain some of the observed increases in cavitation index with physical scale. In view of the almost total lack of information on the statistics of turbulent pressure fields (aside from some correlation and spectral data) and the potential importance of this knowledge to understanding cavitation, it is strongly recommended that further work be carried out.

Direct application of the pressure-field information to cavitation problems is hampered by the effect of dissolved gas which tends to increase the cavitation index with increasing exposure time. Although not considered in this article, there is also an influence of dissolved gas on erosion and noise.

## DISSOLVED-GAS EFFECTS

The influence of free and dissolved gas on cavitation inception has already been alluded to. However, the fact that dissolved gas can influence hydrodynamic loads in cavitating flows is not given much consideration in the literature. For example, Edstrand (1950) found that measured thrust and torque on a model propeller could vary by a factor of 2 with varying amounts of dissolved gas with $\sigma$ held constant. Measurements on a model hydrofoil differed from the expected value by as much as 20% (Arndt 1979). The discrepancy was traced to the influence of dissolved gas. Wade & Acosta (1966) measured the cavity pressure on a hydrofoil with developed sheet cavitation and found that there were substantial differences between $\sigma$ defined by the measured cavity pressure and that defined in the usual manner using the vapor pressure. This is in agreement with previous measurements by Kermeen (1956a). This could account for a difference of 10% in the measured lift coefficient if completely degassed water were used in place of the saturated water used in the experiments.

A systematic investigation of gas-content effects on cavitation inception in free-shear flow was recently reported by Baker et al (1976). Cavitation inception in confined jets, generated either by an orifice plate or a nozzle, was determined as a function of total gas content in the liquid. When the liquid was undersaturated at test-section pressure, the critical cavitation index was independent of gas content and roughly equal to that observed by Rouse (1953) for an unconfined jet. When the flow is supersaturated, the cavitation index is found to vary linearly with gas content as predicted by the equilibrium theory, Equation (33). This effect occurs even though the Lagrangian time scale is much shorter than typical times for bubble growth by gaseous diffusion. For example, in the cited cavitation data, a typical residence time for a nucleus within a large eddy is roughly 1/15 of a second. At a gas content of 7 ppm and a jet velocity of approximately 10 m/s, inception occurs at a mean pressure equivalent to a relative saturation level of 1.25. Epstein & Plesset (1950) show that for growth by gaseous diffusion alone, 567 seconds is required for a $10^{-3}$ cm nucleus to increase its size by a factor of 10. However, the local pressure within an eddy is much less than the mean pressure and highly supersaturated conditions can occur locally. Arndt & Keller (1976) also reported large gas-content effects in their experiments with disks when the flow was supersaturated (Figure 29). The magnitude of the effect also depends on the number of nuclei in the flow. Gas-content effects on cavitation inception and desinence were noted only in their water-tunnel experiments (where there is a healthy supply of nuclei). No gas-content effects on inception were noted in a depressurized tow tank (where the flow is highly supersaturated but there is a dearth of nuclei). Thus the picture becomes cloudier as the influence of dissolved, non-condensible gas is taken into consideration.

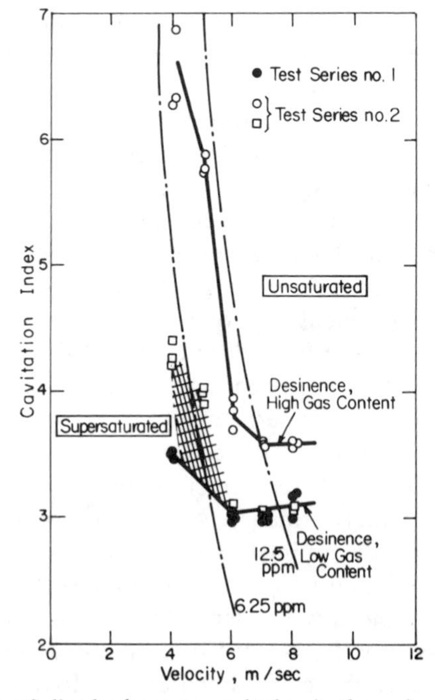


*Figure 29* Influence of dissolved gas on cavitation in the wake of a sharp-edged disk (Arndt & Keller 1976).

Three basic problems have been considered theoretically: the growth or dissolution of spherical bubbles in the absence of convection effects (Epstein & Plesset 1950), the growth of spherical bubbles when convection is important (Parkin & Kermeen 1962, van Wijngaarden 1967), and the turbulent diffusion of gas in a developed cavity (Brennen 1969). There is a natural progression in the degree of complexity in the physical processes involved, beginning with the Epstein-Plesset theory.

In the Epstein-Plesset theory, the growth of a spherical bubble in an unbounded liquid containing a certain amount of gas in solution, is modeled by Fick's second law[1] (convection terms are neglected)

$$\frac{\partial C_g}{\partial t} = D_{g1}\left\{ \frac{\partial^2 C_g}{\partial r^2} + \frac{2}{r}\frac{\partial C_g}{\partial r} \right\} \tag{42}$$

where $C_g$ is the concentration of gas in the liquid and $D_{g1}$ is a diffusion coefficient. The concentration at some distance from the bubble wall is assumed specified, $C_\infty$. The liquid is assumed to be saturated at the bubble wall ($C_g = C_s$). The concentration gradient at the bubble wall, determined from Equation (42) can be substituted into Fick's first law to obtain the flux of gas across the bubble wall

$$\frac{dN}{dt} = 4\pi R^2 D_{g1}(C_\infty - C_s)\left[ \frac{1}{R} + \frac{1}{\sqrt{\pi D_{g1} t}} \right]. \tag{43}$$

Inertial effects are neglected and the difference between the partial pressure of gas and vapor and the external pressure is balanced by surface tension [Equation (11)]. If one uses the perfect-gas law, the mass of gas in the bubble at any instant is

$$N = \frac{4}{3}\pi R^3\left( \frac{p_\infty - p_v + 2S/R}{k\theta} \right). \tag{44}$$

Differentiation of Equation (44) and equating to the right-hand side of Equation (43) yield the bubble wall velocity

$$\frac{dR}{dt} = \frac{D_{g1}(C_\infty - C_s)}{\dfrac{p_\infty - p_v}{k\theta} + \dfrac{4}{3}\dfrac{S/R}{k\theta}}\left[ \frac{1}{R} + \frac{1}{\sqrt{\pi D_{g1} t}} \right]. \tag{45}$$

Equation (45) can be integrated after noting $C_s$ is a function of bubble

[1] In this section $C_g$, $C_s$, and $C_\infty$ are expressed in terms of mass (k moles) per unit volume. Normal water-tunnel practice is to express the concentration in ppm on a mole/mole basis. To avoid confusion, $\alpha$ will symbolize concentration measured on a mole/mole basis.

**Table 1** Time in seconds for growth of air bubbles in water[a]

| $\frac{c_\infty}{c_s}$ | Exact | $R_0=10\mu$ Approximate | | $R_0=100\mu$ Approximate | |
|---|---|---|---|---|---|
| | $S=0$ | $S=0$ | $S$ finite | $S=0$ | $S$ finite |
| 1.25 | 466 | 496 | 567 | 496 | 501 |
| 1.50 | 228 | 248 | 266 | 248 | 249 |
| 1.75 | 149 | 165 | 174 | 165 | 166 |
| 2.00 | 110 | 124 | 129 | 124 | 124 |
| 5.00 | 24.6 | 30.9 | 31.7 | 30.9 | 31.0 |

[a]From $R_0$ to $10R_0$, $\theta=22°$C.

radius

$$C_s = \frac{p_g}{\beta} = \frac{\frac{2S}{R} + p_\infty - p_v}{\beta}. \tag{46}$$

The results are tabulated in Table 1. They indicate that a relatively lengthy period of time is required for bubble growth strictly by gaseous diffusion. This does not mean that gaseous diffusion can be neglected in vaporous cavitation (Oldenziel 1976).

This work was extended to include convection terms in Equation (42) by Parkin & Kermeen (1962) and van Wijngarden (1967). The original impetus for this work was the observation by Kermeen et al (1956) that small bubbles stabilized by pressure-gradient and surface-tension forces at a position upstream of the zone of cavitation on a hemispherical body would slowly grow to a critical size, be lifted up, and swept downstream where they grow explosively to many times their original size. The growth rate in the initial stage exceeded that predicted by the Epstein-Plesset theory. Since the bubble is fixed to the wall, there is a flow velocity relative to the bubble. Hence the convection term in Fick's second law dominates. Under these conditions

$$\frac{dR}{dt} = \frac{1}{\pi R^{1/2}} \frac{(2\pi U D_{gl})^{1/2} k\theta (C_\infty - C_s)}{\left(p_\infty - p_v + \frac{4}{3}\frac{S}{R}\right)}. \tag{47}$$

In comparing the time for a bubble to grow to ten times its original value ($R/R_0 = 10$), the ratio of growth times is

$$\frac{t_g\text{ (without convection)}}{t_g\text{ (with convection)}} = 1.93\sqrt{\mathrm{Pe}} \tag{48}$$

where Pe is the gaseous diffusion Péclet number $\mathrm{Pe} = UR_0/D_{gl}$. In the experiments of Kermeen et al (1956), Pe was $4 \times 10^4$ so that the effects of convection are important. Reasonably good comparison with observed bubble growth is achieved, especially when appropriate corrections are made (Arndt 1979).

The shape of a developed cavity is a function only of $\sigma$ when it is based on cavity pressure

$$\sigma' \equiv \frac{p_0 - p_c}{\frac{1}{2}\rho U_0^2} \tag{49}$$

wherein $p_c$ is the cavity pressure. $p_c$ differs from $p_v$ in two ways. First, there is the partial pressure of non-condensible gas, $p_g$. Second, $p_v$ within the cavity can differ from $p_v$ at bulk temperature because of thermal gradients across the cavity wall. Brennen (1969) considered the balance of heat and mass flux into the cavity behind a sphere as sketched in Figure 30*a*. An estimate of the total mass flow of gas into the cavity can be determined by using the idealization of the physical process as shown in Figure 30*b*. The theory assumes that the cavity wall consists of a turbulent shear layer having constant momentum thickness. (There is no shear at the cavity wall.) Separation occurs at a point $x_s$ from the front of the sphere and transition to turbulent flow occurs a short distance downstream from that point. Because of the turbulent flow, the diffusion processes for momentum and mass are assumed to be identical, implying similarity of velocity and concentration profiles. The result of

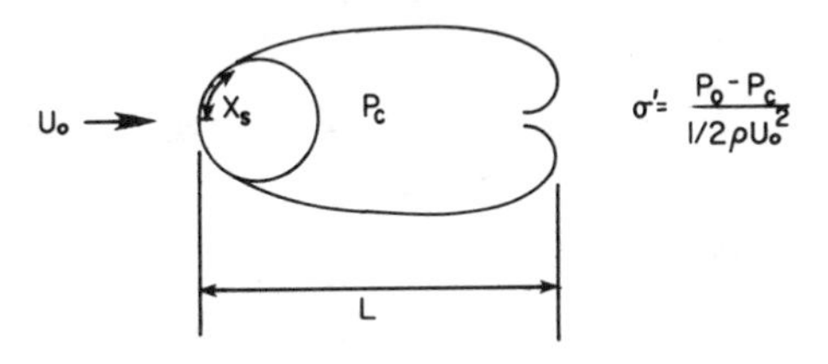


a. Physical Situation in Brennen's Experiments

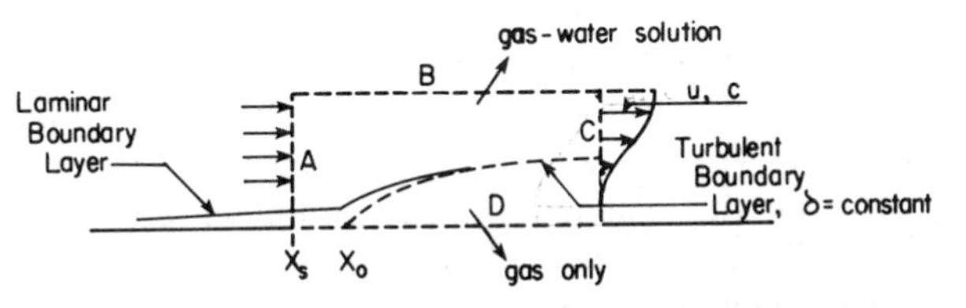

b. Two Dimensional Mathematical Model

*Figure 30* Idealized entrainment model for heat and mass transfer effects (Brennen 1969)

Brennen's calculations is

$$\frac{dN}{dt}=1.298\frac{(C_{\infty}-C_{s})U_{0}d^{2}}{\mathrm{Re}_{d}^{1/4}}\left(\frac{r_{c}}{d}\right)\left(\frac{L-x_{s}}{d}\right)^{1/2}(1+\sigma)^{1/2}, \tag{50}$$

where $r_c$ is cavity radius, $d$ is bubble diameter, and $\mathrm{Re}_d=U_0d/\nu$. By comparison, if the Parkin-Kermeen theory is applied to the same problem, the result is (Arndt 1979)

$$\frac{dN}{dt}=14.18r_{c}(C_{\infty}-C_{s})\left\{D_{gl}(1+\sigma)^{1/2}(L-x_{s})\right\}^{1/2}. \tag{51}$$

The ratio of the two solutions is

$$\mathrm{Ratio}=0.0915\frac{(1+\sigma)^{1/4}\mathrm{Pe}^{1/2}}{\mathrm{Re}_{d}^{1/4}},\qquad \mathrm{Pe}=\frac{U_{0}d}{D_{gl}}. \tag{52}$$

In Brennen's experiments, the ratio is roughly 60, implying that turbulence has increased the diffusion rate by almost two orders of magnitude. The theory implies that whenever a cavity is maintained in a flow that is locally supersaturated ($C_{\infty}-C_{s}>0$) there will be diffusion of gas into the cavity and $p_c>p_v$. This accounts for the observed effects of gas content already cited. For example, Billet & Weir (1975) measured the flow into a ventilated cavity on a zero-caliber ogive. They maintained the cavity pressure at a value equal to saturation pressure for the dissolved gas in the flow and determined the following empirical relationship:

$$\frac{dN}{dt}=0.0042\frac{p_{g}}{k\theta}\left(\frac{L}{d}\right)^{0.69}\mathrm{Re}_{d}^{0.16}\mathrm{Fr}^{0.13}U_{0}d^{2} \tag{53}$$

wherein $\mathrm{Fr}=U_0/\sqrt{gd}$. Noting that $C_s=p_g/\beta$ and that $p_c=p_g+p_v$, Equations (53) and (50) can be combined to yield

$$\sigma=\sigma'+\frac{\beta C_{\infty}}{\frac{1}{2}\rho U_{0}^{2}}\,\frac{K\dfrac{k\theta}{\beta}}{\mathrm{Re}^{0.41}\mathrm{Fr}^{0.13}}(1+\sigma')^{1/2}(\sigma')^{1/2} \tag{54}$$

where differences in geometry between the Brennen theory and the Billet-Weir experiments are absorbed in the constant $K$. Equation (54) implies a new scaling parameter for gas content ($\beta\alpha/\frac{1}{2}\rho U_0^2$). Although the experiments of Wade & Acosta (1966) display the trend indicated by Equation (54), no dependence on the parameter $\beta\alpha/\frac{1}{2}\rho U_0^2$ is noted. This question will have to be explored further.

In summary, cavitation inception can exceed the expected value by an amount $K\alpha\beta/\frac{1}{2}\rho U_0^2$, $0.3<K<1$ in vortical flows, separated flow, and large-scale turbulence, the factor $K$ being unknown beforehand. The same parameter $\alpha\beta/\frac{1}{2}\rho U_0^2$ is an implicit factor in scaling developed cavitation, but little is known about how important this factor really is.

## THERMODYNAMIC EFFECTS

As illustrated in Figure 15, the use of $\sigma$ based on vapor pressure at bulk temperature can lead to a scale effect for developed cavitation. A similar effect may occur for cavitation inception but the issue is cloudier.

In both cases the key factor is the difference between $p_v$ computed from bulk temperature and the actual value of $p_v$ within the bubble or cavity wall which is necessary for the supply of heat to vaporize liquid as required during bubble growth or to make up vapor diffused into the wake of a stable cavity. This whole question is of critical concern when attempting to relate cold-water tests to cavitation inception in liquid metals or to pump performance in hot water or in more exotic fluids such as cryogenics. A question of economics is involved since cold-water tests always give conservative results and a particular device may end up being over-designed for the task at hand.

The B-factor theory was first offered as an explanation of the observed scale effect on performance breakdown (Stahl & Stepanoff 1956, Stepanoff 1961, 1964). The essence of the theory is that there is a necessary temperature depression in the liquid to produce the vapor. Thus the vapor exists in equilibrium at temperatures less than the bulk temperature of the liquid. This results in a correction to $\sigma$ (always positive):

$$\Delta\sigma = \frac{\Delta p_v}{\frac{1}{2}\rho U^2} = B\left(\frac{\rho_v}{\rho}\right)^2 \frac{\lambda^2}{\theta_\infty C_{pl} U^2} \tag{55}$$

wherein $\lambda$ is the latent heat of evaporation, $\theta_\infty$ the bulk temperature, $\rho_v$ the density of the vapor, and $C_{pl}$ the specific heat of the liquid. The factor $B$ is the ratio of the volume of vapor formed relative to the volume of liquid. The value of $B$ for $\Delta\sigma = 0.5$ at $U = 3.5$ m/s is given in Table 2. (The larger the value of $B$ the greater the expected change in performance.)

According to Table 2, propane and hydrogen should be most resistant to cavitation whilst the excessively high value of $B$ for cold water indicates that a turbomachine operating with that liquid would be most sensitive to change in $\sigma$. Thus cavitation tests in cold water are conservative as shown in Figure 15. In other words the thermodynamic

**Table 2** Cavitation susceptibility parameter $B$ for various fluids[a]

| Substance | Temp. | $B$ for $\Delta\sigma = 0.5$ @3.5 m/s |
|---|---|---|
| Water | 21.1°C | 1800. |
| Water | 37.8°C | 353. |
| Water | 82.2°C | 7.87 |
| Water | 100.0°C | 2.08 |
| Water | 121.1°C | 0.66 |
| Water | 148.9°C | 0.16 |
| Butane | 21.1°C | 0.17 |
| Butane | −.5°C | 0.57 |
| Propane | 37.8°C | 0.008 |
| Propane | 6.7°C | 0.92 |
| Hydrogen | 20.0°K | 0.01 |
| Nitrogen | 76.7°K | 0.40 |
| Oxygen | 90.0°K | 0.69 |
| Freon 11 | 23.9°C | 1.63 |
| Freon 22 | 32.8°C | 1.28 |
| Freon 113 | 47.8°C | 1.82 |

[a] Acosta & Hollander (1959).

scale effect is proportional to $1/B$. As pointed out by Acosta & Hollander (1959), the B-factor method is equivalent to an adiabatic problem in thermostatics and does not take into account the dynamics of the fluid flow. Salemann (1959) suggests that performance breakdown in cold water occurs primarily as a large attached cavity, whereas the larger $\Delta p_v$ required of the higher-vapor-pressure liquids results in an extensive region of superheated liquid within a given machine, and a more or less homogeneous distribution of individual bubbles occurs. This is consistent with observations of C. Brennen (1979, private communication) and Sarosdy & Acosta (1961). The B-factor theory has been extended by many authors (Gelder et al 1966, Moore & Ruggeri 1968, Hord et al 1972, Hord 1973a, b). In this work the B-factor is assumed to be functionally dependent on the flow features as well as the physical properties of the liquid, where the dependence on flow velocity, etc, is determined empirically. This work forms the basis for the entrainment theory (Brennen 1969, Billet 1970, Holl et al 1975, Billet et al 1978). It is assumed that a given drop in performance ($\Delta H$, $\Delta\eta$) corresponds to an identical attached cavity on the blades of a turbomachine regardless of the type of liquid. Implicit in this argument is the theory that the geometry of attached cavities is uniquely a function of cavitation index based on cavity pressure (Hutton & Furness 1974):

$$\frac{l_c}{D}=f(\sigma'), \tag{56}$$

$$\sigma' \equiv \frac{p_0-p_c}{\frac{1}{2}\rho U_0^2} \tag{57}$$

wherein $l_c$ is cavity length and $p_c$ is the cavity pressure. It is further assumed that a certain amount of heat must be constantly supplied to a steady cavity in order to vaporize additional liquid to offset the amount of vapor constantly being entrained into the flow at the trailing edge of the cavity (Brennen 1969). The temperature drop across the cavity wall is presumably a function of the thermodynamic properties of the liquid and the flow configuration

$$\Delta\theta = \frac{C_Q}{C_A} \cdot \frac{\mathrm{Pe}}{\mathrm{Nu}} \cdot \frac{\rho_v}{\rho} \cdot \frac{\lambda}{C_{pl}} \tag{58}$$

wherein $C_Q$ is a vapor flow rate coefficient ($C_Q=\dot{N}_v/\rho_v UD^2$), $C_A$ is a cavity surface-area coefficient ($C_A=A_w/D^2$), Pe and Nu are the Péclet number and Nüsselt number for heat transfer across the bubble wall. Equation (58) compares with the B-factor theory

$$\Delta\theta = B\frac{\rho_v}{\rho}\frac{\lambda}{C_{pl}}. \tag{59}$$

Obviously $C_Q$ and $C_A$ depend on the geometry of the cavity and must be determined empirically for a given flow condition. Billet et al (1978) argue that $C_Q$ can be determined from the amount of air necessary to maintain a ventilated cavity at the same value of $\sigma'$ as a vaporous cavity.

The balance between the mass flow of gas and the mass flow of vapor is related to the partial pressure of each within the cavity. Assuming an isothermal, steady-state process, the volume-flow rate into the cavity is

$$Q_g = \frac{k\theta}{p_g}\dot{N}_g. \tag{60}$$

Similarly the total volume-flow rate of vapor and gas is

$$Q_T \sim \frac{\dot{N}_T}{p_c} = \frac{\dot{N}_v + N_g}{p_v + p_g} \tag{61}$$

or

$$p_g = \frac{\dot{N}_g}{\dot{N}_v + \dot{N}_g} p_c, \tag{62}$$

$$p_v = \frac{\dot{N}_v}{\dot{N}_v + \dot{N}_g} p_c \tag{63}$$

where $\dot{N}_v$ and $\dot{N}_g$ are the mass-flow rates of vapor and non-condensible gas (in moles per second). In the case of a ventilated cavity, there are two sources of non-condensible gas: from the ventilation system and via the process of diffusion across the cavity wall. Ideally, the cavity can be ventilated at a partial pressure $p_g$ such that the saturation concentration at the cavity wall is identical with the concentration in the free stream (Billet & Weir 1975):

$$C_s = \frac{p_g}{\beta} \tag{64}$$

and when $C_s = C_\infty$

$$p_g = \beta C_\infty. \tag{65}$$

At this value of partial pressure, $\dot{N}_g$ is solely a function of ventilation. The cavity can be artificially ventilated such that

$$\frac{p_g}{p_v} \gg 1, \quad \frac{\dot{N}_g}{\dot{N}_v} \gg 1. \tag{66}$$

Thus the mass flow of vapor into a cavity under natural conditions can be estimated by simulating the same cavity development (same $\sigma'$) by ventilating and measuring the ventilation flow rate. The assumption is then made that the mass flow of vapor in a natural cavity is equal to the gas-flow rate measured in a ventilating cavity. The unknowns in a given problem are $C_A$, $C_Q$, and Nu. These can be measured and correlated with $l_c/D$, Reynolds number, Froude number, Weber number, and Prandtl number. Substitution of the empirically derived coefficients in Equation (58) yields the temperature drop, $\Delta\theta$. Comparison between theory and experiment is presented in Figure 31 for two different ogives at two different velocities. Good agreement between theory and experiment is achieved. Unfortunately, these are simple flow configurations and the results cannot be directly extrapolated to the case of a pump or turbine. The relationship between $\Delta\sigma$ and $\Delta\theta$ can be determined from the Clausius-Clapeyron equation

$$\Delta\sigma = \frac{2p_v}{\rho} \frac{\lambda}{U^2} \frac{\Delta\theta}{\theta_\infty}. \tag{67}$$

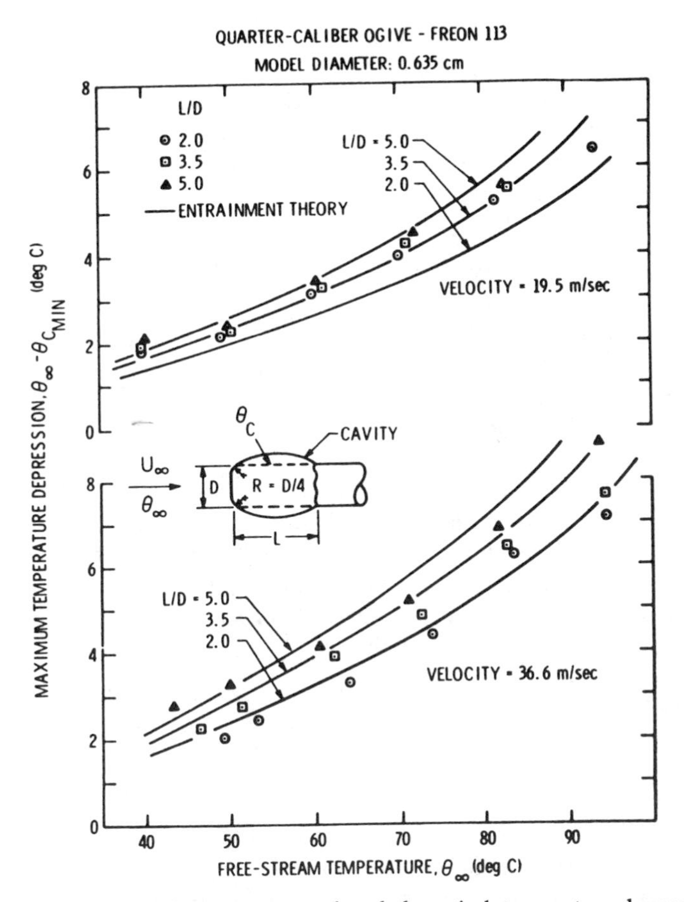


*Figure 31* Comparison between measured and theoretical temperature depression in Freon 113 (Holl et al 1975).

The same thermodynamic effect controls the rate of growth of individual bubbles. In the Rayleigh-Plesset equation [Equation (10)], the internal pressure is made up of the partial pressure of vapor (assumed in equilibrium with the temperature at the bubble wall $\theta_R$) and noncondensible gas

$$p_i = p_v(\theta_R) + P_g. \tag{68}$$

The thermodynamic correction $p_v$ is a function of time. The maximum correction is (Plesset 1957)

$$\Delta p_v = \left[ \frac{R_{max}}{3\sqrt{D_1 t_m}} \frac{\lambda}{\dot{C}_{p1}} \frac{\rho_v}{\rho_1} \right] \frac{dp_v}{d\theta} \tag{69}$$

wherein $t_m$ is the time available for growth and $D_l$ is the thermal diffusivity of the liquid. The bracketed quantity in Equation (69) is the temperature drop across the bubble wall. In cold water ($\theta = 15°C$), for example, $\Delta\theta = 1.3°$ and the thermodynamic effect is negligible.

The growth rate of a bubble has two asymptotic extremes. When inertial effects dominate,

$$\dot{R} = \sqrt{\frac{2}{3}\frac{p_v - p}{\rho}}\ ; \tag{70}$$

and when thermal effects dominate (Plesset & Zwick 1954),

$$\dot{R} = \sqrt{\frac{3}{\pi}}\ \frac{K}{\lambda \rho_v}\ \frac{\theta_\infty - \theta(R)}{\sqrt{D_l t}} \tag{71}$$

wherein $K$ is the thermal conductance. The question whether inertial effects or thermodynamic effects control bubble growth in a given situation depends on both the fluid properties and the flow situation. Bonnin (1972) suggests that when the time available for growth, $t_m$, is small in comparison to a characteristic thermal time scale, then inertial effects dominate, i.e.

$$B' = \frac{t_m}{\dfrac{\rho_v}{\rho_l}}\ \frac{\lambda \dfrac{dp_v}{d\theta}}{K} \begin{cases} \ll 1 & \text{inertial effects dominate,} \\ \gg 1 & \text{thermodynamic effects dominate.} \end{cases}$$

No concrete evidence is available to support this conjecture. Holl & Kornhauser (1970) defined cavitation inception to occur when a given nucleus grows to a macroscopic size in the time available for growth. Their calculations of bubble growth including heat-transfer effects indicate that there should be a strong thermodynamic effect for inception on hemispherical-nosed bodies in hot water and in Freon 113. However, the experimental results were found to agree more closely with isothermal calculations. Presumably other effects, such as non-condensible gas and variations in viscosity with temperature, reverse the predicted decrease in $\sigma_c$ with temperature (Holl et al 1972, 1976). Thus the subject of thermodynamic effects on inception remains open to further investigation.

## CONCLUDING REMARKS

Consideration has been given to the occurrence of cavitation and its impact on the performance of fluid machinery and hydraulic structures. Emphasis has been placed on the mechanics of the inception process, and on thermodynamic and gaseous-diffusion effects on bubble growth and developed cavitation. Many aspects of the problem have been deleted and some facets of the field have been overemphasized because of their omission in previous reviews. For example, considerable advances have been made in instrumentation which, because of the empiricism involved, is a major factor in the rate of growth of knowledge. Space limitations do not permit a coverage of this important topic. The theory of supercavitation as well as the design of supercavitating machinery have been purposely omitted. Reviews of some aspects of two-dimensional and axisymmetric supercavitation are already available (Wu 1972, Tulin 1961, Knapp et al 1970). Recent advances have been made in the design of supercavitating machinery (Yin 1978). There has also been a closer examination of unsteady developed cavitation (Shen & Peterson 1978). It is well known that small amounts of free gas can significantly change the bulk modulus of water, the speed of sound becoming as low as 15 m/s. This has a significant impact on pump performance and stability (Brennen 1973, 1978, Brennen & Acosta 1973, 1976). The exact mechanism for attenuation of cavitation erosion through air injection is not understood in detail but is probably also related to changes in compliance. Certain features of developed cavitation may also be related to shockwave formation in the bubbly flow at the trailing edge of cavities. Because of space limitations, cavitation erosion has not been considered. Some aspects of the problem have already been reviewed (Knapp et al 1970, Thiruvengadam 1971, 1974a, b).

Most of our understanding of cavitation evolves from a great deal of experimentation. Review of the factors leading to inception indicates that careful interpretation of laboratory data is necessary before extrapolation to field conditions can be successful. Separation and transition to turbulence are important factors in model tests. The influence of nuclei, both size and number density, is significant, with recent advances in instrumentation allowing more careful analysis of this problem. Cavitation nuclei in the flow can be in the form of small gas bubbles as well as in the form of solid particles with small quantities of gas stabilized within small cracks and fissures.

Cavitation in vortical motion is a complex topic. Considering its importance as a source of noise and vibration as well as surging in draft tubes, etc, it has received relatively little attention. Cavitation in turbulent shear flows is even less understood. Laboratory investigations indicate the negative pressure peaks with amplitudes as high as ten times the rms pressure incite cavitation. Reliable estimates of critical cavitation indices for large-scale turbulent shear flows are nonexistent. A full understanding of the problem is dependent on information concerning the statistics of turbulent pressure fields as well as an understanding of the diffusion of nuclei into a turbulent flow and their response to pressure fluctuations. Information of this type is almost totally lacking in the literature. Dissolved gas has an impact on the number density of nuclei and is an important factor in bubble growth in vortical flows. Surface roughness is an important consideration, very small protuberances (say 1/20 of a boundary-layer thickness) can have a significant effect on cavitation inception. Although the effects of surface finish on performance and performance breakdown are negligible, the influence of surface roughness as a catalyst for erosion and subsequent reduced life has not been addressed.

Under certain conditions the temperature gradients necessary for the production and supply of vapor during bubble growth or in the maintenance of stable cavities can be significant. This implies that under certain circumstances the results of tests in cold water cannot be extrapolated to conditions in different fluids. In the case of performance breakdown, tests in cold water are generally conservative.

A great deal has been accomplished in the past eighty years. We have a considerably clearer understanding of bubble growth in boiling or cavitation and a grasp of the essentials of growth by gaseous diffusion. We still lack an understanding of cases where inertia, heat transfer, and mass diffusion are of equal importance and further work along these lines is crucial to many aspects of nuclear, marine, petroleum, mechanical, civil, and aerospace engineering. The influence of roughness and viscosity is reasonably well understood in certain classes of flow and not understood at all in other problems. Performance breakdown in turbomachinery involves developed cavitation, but the thermodynamics of the process remain unclear. There is some evidence that the flow patterns in geometrically similar machines can vary with the thermodynamic properties of the liquid in question, yet theoretical methods for correlation of the data neglect this possibility. Although cavitation erosion has not been considered in this review, it should be mentioned that the problem of accurately determining service life in the field on the basis of screening tests in the laboratory is still beyond our grasp. This can be of considerable importance, especially in the field of nuclear power generation.

The continual increase in power and speed as well as the rapidly growing degree of sophistication in applications provides considerable impetus for continual advancements in the field of cavitation. It is expected that very rapid progress will be made in the development of new experimental techniques as well as theoretical approaches in the next quarter century.

ACKNOWLEDGMENTS

This article was prepared under grants from the US Air Force of Scientific Research and the Seed Research Fund of the St. Anthony Falls Hydraulic Laboratory. A special word of thanks should go to Professor Dr. Ir. J. D. van Manen and Dr. Ir. M. W. C. Oosterveld, of the Netherlands Ship Model Basin, for making possible a sabbatical leave that allowed time for a portion of the literature review. Dr. Ir. J. H. J. van der Meulen, also of NSMB, was my gracious host and colleague during that period.

*Literature Cited*

Acosta, A. J. 1973. Hydrofoils and hydrofoil craft, *Ann. Rev. Fluid Mech.* 5:161–84

Acosta, A. J. 1974. Cavitation and fluid machinery. *Cavitation*, pp. 383–96. Conf. held at Heriot-Watt Univ., Edinburgh, Scotland, Sept. 1974. London: Inst. Mech. Engrs.

Acosta, A. J., Hamaguchi, H. 1967. Cavitation inception on the I.T.T.C. standard head form. *Calif. Inst. Technol. Hydrodyn. Lab. Rep. No. E-149*

Acosta, A. J., Hollander, A. 1959. Remarks on caviation in turbo machines. *Calif. Inst. Technol. Eng. Div. Report No. 79.3.*

Acosta, A. J., Parkin, B. R. 1975. Cavitation inception—a selective review, *J. Ship Res.* 19:193–205

Arakeri, V. H. 1973. *Viscous effects in inception and development of cavitation on axi-symmetric bodies.* Ph.D. thesis Calif. Inst. Technol., Pasadena, Calif.

Arakeri, V. H. 1975a. A note on the transition observations on an axisymmetric body and some related fluctuating wall pressure measurements. *J. Fluids Engrg.* 97:82–87

Arakeri, V. H. 1975b. Viscous effects on the position of cavitation separation from smooth bodies, *J. Fluid Mech.* 68:779–99 (3 plates)

Arakeri, V. H., Acosta, A. J. 1973. Viscous effects in the inception of cavitation on axisymmetric bodies, *J. Fluids Engrg.* 95:519–27

Arakeri, V. H., Acosta, A. J. 1976. Cavitation inception observations on axisymmetric bodies at supercritical Reynolds numbers, *J. Ship Res.* 20:40–50

Arakeri, V. H., Carroll, J. A., Holl, J. W. 1978. A note on the effect of short and long laminar separation bubbles on desinent cavitation. *Proc. Meas. in Polyphase Flows*, ASME, New York, pp. 115–20

Arndt, R. E. A. 1974a. Cavitation inception and how it scales: A review of the problem with a summary of recent research, *Proc. Symp. High Powered Propulsion of Large Ships*, Wageningen, The Netherlands, Part 2, pp. XXI, 1–65 (available from the Netherlands Ship Model Basin).

Arndt, R. E. A. 1974b. Pressure fields and cavitation. *Trans. 7th IAHR Symp.*, Vienna, Part 1, pp. IX, 1–20

Arndt, R. E. A. 1976a. Cavitation on model propellers with boundary layer trips. *ASME Cavitation and Polyphase Flow Forum*, pp. 30–32

Arndt, R. E. A. 1976b. Semiempirical analysis of cavitation in the wake of a sharp edged disk, *J. Fluids Engrg.* 98:560–62

Arndt, R. E. A. 1978. Investigation of the effects of dissolved gas and free nuclei on cavitation and noise in the wake of a sharp edged disk. *Proc. Joint IAHR/ASME/ASCE Symp. Design and Operation of Fluid Machinery*, Colo. State Univ., Ft. Collins, Colo. Vol. II, pp. 543–56

Arndt, R. E. A. 1979. The influence of dissolved and free gas content on cavitation. *Boeing Co. Doc. No. D 324-51505-1*

Arndt, R. E. A., Daily, J. W. 1969. Cavitation in turbulent boundary layers. *Cavitation State of Knowledge*, pp. 64–86. ASME

Arndt, R. E. A., George, W. K. 1978. Pressure fields and cavitation in turbulent shear flows. *12th Symp. Naval Hydrodyn.*, Washington, DC, pp. 327–39

Arndt, R. E. A., Ippen, A. T. 1968. Rough surface effects on cavitation inception. *J. Basic Engrg.* 90:249–61

Arndt, R. E. A., Keller, A. P. 1976. Free gas content effects on cavitation inception and noise in a free shear flow. *Proc. IAHR Symp. Two Phase Flow and Cavitation in Power Generation Systems*, Grenoble, France, pp. 3–16

Arndt, R. E. A., Holl, J. W., Bohn, J. C., Bechtel, W. T. 1979. The influence of surface irregularities on cavitation performance. *J. Ship Res.* 23:157–70

Baker, C. B., Holl, J. W., Arndt, R. E. A. 1976. The influence of gas content and polyethelene oxide additive upon confined jet cavitation in water. *ASME Cavitation and Polyphase Flow Forum*, pp. 6–8

Barefoot, G. L. 1972. *Fluctuating pressure characteristics of perturbed and unperturbed round free jets*. M. S. thesis. Penn. State Univ., University Park, Pa.

Batchelor, G. K. 1951. Pressure fluctuations in isotropic turbulence, *Proc. Cambridge Philos. Soc.* 47:359–74

Batchelor, G. K. 1953. *The Theory of Homogeneous Turbulence*. Cambridge: Univ. Press. xi + 197 pp.

Bechtel, W. T. 1971. *The influence of surface irregularities on cavitation: field study and limited cavitation near wire screen roughness*. M. S. thesis. Penn. State Univ., University Park, Pa.

Beuther, P. D., George, W. K., Arndt, R. E. A. 1977a. *Modelling of pressure spectra in a turbulent shear flow*. Presented at 93rd Meet. Acoust. Soc. Am., State College, Pa.

Beuther, P., George, W. K., Arndt, R. E. A. 1977b. Pressure spectra in homogeneous isotropic turbulent flow, *Bull. Am. Phys. Soc.* 22:1285

Billet, M. L. 1970. *Thermodynamic effects on developed cavitation in water and Freon 113*, MS thesis. Dept. Aerospace Engrg., Penn. State Univ., University Park, Pa.

Billet, M. L. 1978. Secondary flow generated vortex cavitation, *12th Symp. Naval Hydrodyn.*, Washington, DC., pp. 340–47

Billet, M. L., Weir, D. S. 1975. The effect of gas diffusion on the flow coefficient for a ventilated cavity. *Cavity Flows*, ASME Fluids Engrg. Conf., Minneapolis, Minn., pp. 95–100

Billet, M. L., Holl, J. W., Weir, D. S. 1978. Correlations of thermodynamic effects for developed cavitation. *Polyphase Flow in Turbomachinery. ASME Spec. Publ.*, pp. 271–89

Blake, F. G., Jr. 1949. The onset of cavitation in liquids, I. *Harvard Acoustics Res. Lab. TM 12*, Sept.

Bonnin, J. 1972. Theoretical and experimental investigations on incipient cavitation in different liquids. *ASME Pap. No. 72WA/FE31*

Bossel, H. H. K. 1967. Inviscid and viscous models of the vortex breakdown phenomenon. PhD thesis. Univ. Calif., Berkeley, *Rep. No. AS-67-14*. viii + 78 + 60 pp.

Brennen, C. 1969. The dynamic balances of dissolved air and heat in natural cavity flows. *J. Fluid Mech.* 37:115–27 (1 plate)

Brennen, C. 1970. Cavity surface wave patterns and general appearance. *J. Fluid Mech.* 44:33–49 (2 plates)

Brennen, C. 1973. The dynamic behavior and compliance of a stream of cavitating bubbles. *J. Fluids Engrg.* 95:533–42

Brennen, C. 1978. The unsteady dynamic characterization of hydraulic systems with emphasis on cavitation and turbomachines. *Proc. Joint IAHR/ASME/ASCE Symp. Design and Operation of Fluid Machinery*, Vol. 1, pp. 97–107, Colo. State Univ., Fort Collins, Colo.

Brennen, C., Acosta, A. J. 1973. Theoretical quasi-static analysis of cavitation compliance in turbopumps. *J. Spacecraft Rockets* 10: pp. 175–80

Brennen, C., Acosta, A. J. 1976. The dynamic transfer function for a cavitating inducer. *J. Fluids Engrg.* 98:182–91

Brockett, T. 1966. Minimum pressure envelopes for modified NACA-66 sections with NACA a = 0.8 camber on Bu Ships type I and type II sections. *David Taylor Model Basin Rep. 1780*. iv + 39 pp.

Brown, F. R. 1963. Cavitation in hydraulic structures: Problems created by cavitation phenomena. *J. Hydraul. Div., Proc. ASCE* 89(HY1):99–115

Carmody, T. 1964. Establishment of the wake behind a disk. *J. Basic Engrg.* 86:869–82

Casey, M. V. 1974. The inception of attached cavitation from laminar separation bubbles on hydrofoils. *Cavitation*, pp. 9–16. Conf. held at Heriot-Watt Univ., Edinburgh, Scotland, Sept. 1974. London: Inst. Mech. Engrs.

*Cavitation in Hydrodynamics*. 1956. *Proc. Symp.*, Natl. Physical Lab., Sept. 1955. London: HMSO

Chan, Y. Y. 1974a. Pressure sources for a wave model of jet noise. *AIAA J.* 12: 241–42

Chan, Y. Y. 1974b. Spatial waves in turbulent jets. *Phys. Fluids* 17:46–53

Chan, Y. Y. 1977. Noise generated wave-like eddies in a turbulent jet. *AIAA J.* 15(7):992–1001
Chivers, T. C. 1969. Cavitation in centrifugal pumps. *Proc. Inst. Mech. Eng.* 184, Pt. 1, No. 2
Daily, J. W. 1949. Cavitation characteristics and infinite-aspect ratio characteristics of a hydrofoil section. *Trans. ASME* 71:269–84
Daily, J. W., Johnson, V. E. 1956. Turbulence and boundary layer effects on cavitation inception from gas nuclei. *Trans. ASME* 78:1695–1706.
Daily, J. W., Lin, J. D., Broughton, R. S. 1959. The distribution of the mean static pressure in turbulent boundary layers in relation to inception of cavitation. *MIT Hydrodyn. Lab., Rep. 34* 54 pp.
Davies, R., ed. 1964. *Cavitation in Real Liquids*. Amsterdam: Elsevier
Deeprose, W. M., King, N. W., McNulty, P. J., Pearsall, I. S. 1974. Cavitation noise, flow noise, and erosion. *Proc. Conf. Cavitation*, Inst. Mech. Engrs. London.
Edstrand, H. 1950. Cavitation tests with model propellers in natural sea water with regard to the gas content of the water and its effect upon cavitation point and propeller characteristics. *SSPPA Rep. No. 5*. Göteborg, Sweden
Eisenberg, P. 1961. Mechanics of cavitation. *Handbook of Fluid Dynamics*, ed. V. L. Streeter, pp. 12.2–12.24. New York: McGraw-Hill
Epstein, P. S., Plesset, M. S. 1950. On the stability of gas bubbles in liquid-gas solutions. *J. Chem. Phys.* 18:1505–9
Feldberg, L. A., Shlemenson, K. T. 1971. The holographic study of cavitation nuclei. *Proc. IUTAM Symp. Non-Steady Flow of Water at High Speeds*, Leningrad, USSR, pp. 239–42 (English version, Moscow 1973, pp. 106–11)
Fuchs, H. V. 1972a. Measurements of pressure fluctuations within subsonic turbulent jets. *J. Sound Vib.* 22:361–78
Fuchs, H. V. 1972b. Space correlations of the fluctuating pressure in subsonic turbulent jets. *J. Sound Vib.* 23:77–99
Fuchs, H. V. 1973. Resolution of turbulent jet pressure into azimuthal components. *Advis. Group Aerosp. Res. & Develop. Conf. Proc. No. 131, Pap. 27*
Fuchs, H. V., Michalke, A. 1975. On turbulence and noise of an axisymmetric shear flow. *J. Fluid Mech.* 70:179–205
Gaster, M. 1966. The structure and behaviour of laminar separation bubbles. *Advisory Group for Aerospace Res. Develop. Conf. Proc.*, No. 4, Part 2, pp. 813–54. London: Technical Editing and Reproduction, Ltd.
Gates, E. M., Acosta, A. J. 1978. Some effects of several free-stream factors on cavitation inception on axisymmetric bodies. *12th Symp. Naval Hydrodyn.*, Washington, DC, pp. 86–108
Gelder, T. F., Ruggeri, R. S., Moore, R. D. 1966. Cavitation similarity considerations based on measured pressure and temperature depressions in cavitated regions of Freon 114. *NASA TN D-3509*
George, W. K. 1974. The equilibrium range of turbulent pressure spectra. *Bull. Am. Phys. Soc.* 19:1158
George, W. K., Beuther, P. D. 1977. Pressure spectra in a turbulent shear flow. *Bull. Am. Phys. Soc.* 22:1285
Gupta, S. K. 1969. *The influence of porosity and contact angle on incipient and desinent cavitation*. MS thesis. Penn. State Univ., University Park, Pa.
Hall, M. G. 1972. Vortex breakdown. *Ann. Rev. Fluid Mech.* 4:195–218
Harvey, E. N., Barnes, D. K., McElroy, W. D., Whiteley, A. H., Pease, D. C., Cooper, K. W. 1944a. Bubble formation in animals, I. Physical Factors. *J. Cell. Comp. Physiol.* 24:1–22
Harvey, E. N., Whiteley, A. H., McElroy, W. D., Pease, D. C., Barnes, D. K. 1944b. Bubble formation in animals, II. Gas Nuclei and Their Distribution in Blood and Tissues, *J. Cell. Comp. Physiol.* 24:23–34
Harvey, E. N., McElroy, W. D., Whiteley, A. H. 1947. On cavity formation in water. *J. Appl. Phys.* 18:162–72
Harvey, J. K. 1962. Some observations of the vortex breakdown phenomenon. *J. Fluid Mech.* 14:585–92 (2 plates)
Holl, J. W. 1958. *The effect of surface irregularities on incipient cavitation*. PhD Thesis. Penn. State Univ., University Park, Pa.
Holl, J. W. 1960a. An effect of air content on the occurrence of cavitation. *J. Basic Engr.* 82:941–46
Holl, J. W. 1960b. The inception of cavitation on isolated surface irregularities. *J. Basic Engr.* 82:169–83
Holl, J. W. 1965. The estimation of the effect of surface irregularities on the inception of cavitation. *Cavitation in Fluid Machinery*, ASME pp. 3–15.
Holl, J. W. 1969. Limited cavitation. *Cavitation State of Knowledge*, pp. 26–63.

ASME

Holl, J. W. 1970. Nuclei and cavitation. *J. Basic Engrg*. 92: 681–88

Holl, J. W., Kornhauser, A. L. 1970. Thermodynamic effects on desinent cavitation on hemispherical nosed bodies in water at temperatures from 80 deg F to 260 deg F. *J. Basic Engrg*. 92:44–58

Holl, J. W., Robertson, J. M. 1956. Discussion of Kermeen et al. (1956). *Trans. ASME* 78:540

Holl, J. W., Treaster, A. L. 1966. Cavitation hysteresis. *J. Basic Engrg*. 88:199–212

Holl, J. W., Wislicenus, G. F. 1961. Scale effects on cavitation. *J. Basic Engr*. 83:385–98

Holl, J. W., Wood, G. M., eds. 1964. *Symp. Cavitation Res. Facilities and Techniques*, ASME

Holl, J. W., Arndt, R. E. A., Billet, M. L. 1972. Limited cavitation and the related scale effects problem. *Proc. 2nd Int. Symp. Fluid Mech. and Fluidics*, pp. 303–14. JSME, Tokyo, September

Holl, J. W., Arndt, R. E. A., Billet, M. L., Baker, C. B. 1974. Cavitation research at the Garfield Thomas water tunnel. *Cavitation* pp. 81–83. Conf. held at Heriot-Watt Univ., Edinburgh, Scotland. Sept. 1974. London: Inst. Mech. Engrs.

Holl, J. W., Billet, M. L., Weir, D. J. 1975. Thermodynamic effect on developed cavitation. *Cavity Flows*, ASME Fluids Engrg. Conf., Minneapolis, Minn. pp. 101–9

Holl, J. W., Billet, M. L., Weir, D. J. 1976. Thermodynamic effects on limited cavitation. *Proc. IAHR Symp. Two Phase Flow and Cavitation in Power Generation Systems*, Grenoble, France, pp. 53–64

Holt, M. 1977. Underwater Explosions. *Ann. Rev. Fluid Mech*. 9:187–21

Hord, J., 1973a. Cavitation in liquid cryogens. II. Hydrofoil. *NASA CR-2156*. 157 pp.

Hord, J. 1973b. Cavitation in liquid cryogens. III. Ogives. *NASA CR-2242*. 235 pp.

Hord, J., Anderson, L. M., Hall, W. J., 1972. Cavitation in liquid cryogens. I. Venturi. *NASA CR-2054*. 83 pp.

Huang, T. T., Hannan, D. E. 1976. Pressure fluctuation in the regions of flow transition. *David W. Taylor Naval Ship Res. and Develop. Cent. Rep. 4723*

Huang, T. T., Peterson, F. B. 1976. Influence of viscous effects on model/full-scale cavitation scaling. *J. Ship. Res*. 20:215–23

Hutton, S. P., Furness, R. A. 1974. Thermodynamic scale effects in cavitating flows and pumps. *Cavitation*, pp. 329–40. Conf. held at Heriot-Watt Univ., Edinburgh, Scotland, Sept. 1974. London: Inst. Mech. Engrs.

IAHR. 1976. *Proc. IAHR Symp. Two Phase Flow and Cavitation in Power Generation Systems*, Grenoble, France

Institute of Mechanical Engineers. 1974. *Cavitation*. Conf. held at Heriot-Watt Univ., Edinburgh, Scotland, Sept. 1974. London: Inst. Mech. Engrs.

Johnson, V. E., Hsieh, T. 1966. The influence of the trajectories of gas nuclei on cavitation inception. *6th Symp. Naval Hydrodyn*., Washington, DC, pp. 163–82

Karelin, V. Ya. 1963. *Kavitatsionnye Yavleniya v Tsentrobezhnykh i Osevykh Nasosakh*. (*Cavitation Phenomena in Centrifugal and Axial Pumps*.) Moscow: Mashgiz. 255 pp. Transl. available through Boston Spa (Engl.) Nat. Lending Lib. for Sci. & Engrg. xiii + 354 pp.

Keller, A. P. 1972. The influence of the cavitation nucleus spectrum on cavitation inception, investigated with a scattered light counting method. *J. Basic Eng*. 94:917–25.

Keller, A. P. 1974. Investigations concerning scale effects of the inception of cavitation. *Cavitation*, pp. 109–17. Conf. held at Heriot-Watt Univ., Edinburgh, Scotland, Sept. 1974. London: Inst. Mech. Engrs.

Keller, A. P., Weitendorf, E. A. 1976. Influence of undissolved air content on cavitation phenomena at the propeller blades and on induced hull pressure amplitudes. *Proc. IAHR Symp. Two Phase Flow and Cavitation in Power Generation Systems*, Grenoble, France, pp. 65–76

Kermeen, R. W. 1956a. Water tunnel tests of NACA 4412 and Walchner profile 7 hydrofoils in noncavitating and cavitating flows. *Calif. Inst. Technol. Hydrodyn. Lab., Rep. 47–5*

Kermeen, R. W. 1956b. Water tunnel tests of NACA 66, −012 hydrofoil in noncavitating and cavitating flows. *Calif. Inst. Technol. Hydrodyn. Lab. Rep. 47–7*. 12 pp.

Kermeen, R. W. McGraw, J. T., Parkin, B. R. 1956. Mechanism of cavitation inception and the related scale-effects problem. *Trans. ASME* 78:533–41

Kermeen, R. W., Parkin, B. R. 1957. Incipient cavitation and wake flow behind sharp-edged disks. *Calif. Inst. Technol. Hydrodyn. Lab. Rep. 85-4* 34 pp.

Knapp, R. T. 1955. Recent investigations of the mechanics of cavitation and cavitation damage. *Trans. ASME* 77:1045-54

Knapp, R. T., Daily, J. W., Hammitt, F. G. 1970. *Cavitation*. New York: McGraw-Hill. xix + 578 pp.

Kraichnan, R. H. 1956. Pressure fluctuations in turbulent flow over a flat plate. *J. Acoust. Soc. Am.* 28:378–90

Kuiper, G. 1978. Scale effects on propeller cavitation inception. *12th Symp. Naval Hydrodyn.* Washington, DC, pp. 400–29

Lienhard, J. H., Goss, C. D. 1971. Influence of size and configuration on cavitation in submerged orifice flows. *ASME Pap. No. 71-FE-39*

McCormick, B. W. 1954. *A study of the minimum pressure in a trailing vortex system*. PhD thesis. Dept. Aerospace Engrg., Penn. State Univ., University Park, Pa.

McCormick, B. W. 1962. On cavitation produced by a vortex trailing from a lifting surface. *J. Basic Engrg.* 84:369–79

Medwin, H. 1977. In situ acoustic measurements of microbubbles at sea. *J. Geophys. Res.* 82:921–76

Moore, R. D., Ruggeri, R. S. 1968. Prediction of thermodynamic effects of developed cavitation based on liquid hydrogen and Freon 114 data in scaled venturis. *NASA TN D-4899*

Morgan, W. B. 1972. Air content and nuclei measurement. *Proc. 13th Int. Towing Tank Conf.*, Berlin/Hamburg, vol. 1, pp. 657–74

Numachi, F., ed. 1963. Cavitation and hydraulic machinery. *Proc. IAHR Symp.*, Inst. High Speed Mech., Tohoku Univ., Sendai, Japan. 570 pp.

Oldenziel, D. M. 1976. Gas transport into a cavitation bubble during the explosion. *Proc. IAHR Symp. Two Phase Flow and Cavitation in Power Generation Systems*, Grenoble, France

Parkin, B. R., Kermeen, R. W. 1953. Incipient cavitation and boundary layer interaction on a streamlined body. *Calif. Inst. Technol. Hydrodyn. Lab. Rep. E-35.2*

Parkin, B. R., Holl, J. W. 1954. Incipient cavitation scaling experiments for hemispherical and 1.5 calibre ogive nosed bodies. Penn. State Univ. Ordnance Res. Lab. *Rep. Nord. 7958-264*, 33 pp.

Parkin, B. R., Kermeen, R. W. 1962. The roles of convective air diffusion and liquid tensile stresses during cavitation inception. *Proc. IAHR Symp.*, Sendai, Japan

Pearsall, I. S. 1972. *Cavitation*. London: Mills & Boon

Peterson, F. B. 1968. Cavitation originating at liquid-solid interfaces. *Naval Ship Res. and Develop. Cent. Rep. 2799*

Peterson, F. B. 1972. Hydrodynamic cavitation and some considerations of the influence of free gas content. *9th Symp. Naval Hydrodyn.*, Paris, vol. 2, pp. 1131–86

Peterson, F. B., Danel, F., Keller, A., Lecoffe, Y. 1975. Determination of bubble and particulate spectra and number density in a water tunnel with three optical techniques. *Proc. 14th Int. Towing Tank Conf.* Ottawa, vol. 2, pp. 27–52

Plesset, M. S., Zwick, S. A. 1954. The growth of vapor bubbles in superheated liquids. *J. Applied Phys.* 25:493–500

Plesset, M. S. 1957. Physical effects in cavitation and boiling. *Proc. 1st Symp. Naval Hydrodyn.*, Washington, DC, 1956, pp. 297–323

Plesset, M. S., Prosperetti, A. 1977. Bubble dynamics and cavitation. *Ann. Rev. Fluid Mech.* 9:145–85

Ripken, J. R., Killen, J. M. 1959. A study of the influence of gas nuclei on scale effects and acoustic noise for incipient cavitation in a water tunnel. *St. Anthony Falls Hydraul. Lab. Tech. Pap. 27-B*. 49 pp.

Ripken, J. R., Killen, J. M. 1962. Gas bubbles: their occurrence, measurement, and influence in cavitation testing. *Proc. IAHR Symp.*, Sendai, Japan, pp. 37–57

Roberts, W. B. 1979. Calculation of laminar separation bubbles and their effect on airfoil performance. *AIAA Pap. No. 79-0285*. 17th Aerosp. Sci. Meet., New Orleans, La.

Robertson, J. M., Wislicenus, G. F., eds. 1969. *Cavitation State of Knowledge*, ASME, New York

Rohsenhow, W. M. 1971. Boiling. *Ann. Rev. Fluid Mech.* 3:211–36

Rouse, H. 1953. Cavitation in the mixing zone of a submerged jet. *Houille Blanche* 8:9–19

Salemann, V. 1959. Cavitation and NPSH requirements of various liquids. *J. Basic Engrg.* 81:167–80

Sarosdy, C. R., Acosta, A. J. 1961. Note on observations of cavitation in different fluids. *J. Basic Engrg.* 3:399–400

Sarpkaya, T. 1971. On stationary and traveling vortex breakdowns. *J. Fluid Mech.* 45:545–59 (9 plates)

Shen, Y. T., Peterson, F. B. 1978. Unsteady cavitation on an oscillating hydrofoil. *12th Symp. Naval Hydrodyn.*, Washington, DC, pp. 362–84

Silberman, E., Schiebe, F. R. 1974. A method for determining the relative cavitation susceptibility of water. *Cavitation*, pp. 101–8. Conf. held at Heriot-Watt Univ., Edinburgh, Scotland, Sept. 1974. London: Inst. Mech. Engrs.

Stahl, H. A., Stepanoff, A. J. 1956. Thermodynamic aspects of cavitation on centrifugal pumps. *Trans. ASME* 78:1691–93

Stepanoff, A. J. 1961. Cavitation in centrifugal pumps with liquids other than water. *J. Engrg. Power* 83:79–90

Stepanoff, A. J. 1964. Cavitation properties of liquids. *J. Engr. for Power* 86:195–200

Stinebring, D. R. 1976. *Scaling of cavitation damage*. MS thesis. Penn. State Univ., University Park, Pa.

Stripling, L. B. 1962. Cavitation in turbopumps. Part II. *J. Basic Engrg.* 84:339–50

Stripling, L. B., Acosta, A. J. 1962. Cavitation in turbopumps. Part I. *J. Basic Engrg.*, 84:326–38

Tani, I. 1964. Low speed flows involving bubble separations. *Prog. Aeron. Sci.* 5:70–103

Thiruvengadam, A. 1971. Cavitation erosion. *Appl. Mech. Rev.* 24:245–53

Thiruvengadam, A. 1974a. Basic and applied aspects of cavitation erosion. *Proc. Symp. High Powered Propulsion of Large Ships, Pt. 2*, Wageningen, The Netherlands, pp. xxiii, 1–31

Thiruvengadam, A. 1974b. *Handbook of cavitation erosion. Tech. Rep. 7301-1*. Hydronautics, Inc.

Tulin, M. P. 1961. Supercavitating flows. *Handbook of Fluid Dynamics*, ed. V. L. Streeter, pp. 12.24–12.46. N.Y.: McGraw-Hill

van der Meulen, J. H. J. 1972. Cavitation on hemispherical nosed teflon bodies. *Int. Shipbuilding Prog.*, 19:333–41

van der Meulen, J. H. J. 1976. *A holographic study of cavitation on axisymmetric bodies and the influence of polymer additives*. PhD thesis. Univ. Twente, The Netherlands

van der Meulen, J. H. J. 1978. A holographic study of the influence of boundary layer and surface characteristics on incipient and developed cavitation on axisymmetric bodies. *12th Symp. Naval Hydrodyn.*, Washington DC, pp. 433–51

van Wijngaarden, L. 1967. On the growth of small cavitation bubbles by convective diffusion. *Int. J. Heat Transfer* 10:127–34

van Wijngaarden, L. 1972. One-dimensional flow of liquids containing small gas bubbles. *Ann. Rev. Fluid Mech.* 4:369–96

Vennard, J. K., Harrold, J. C., Warnock, J. E., Hickox, G. H. 1947. Cavitation in hydraulic structures: a symposium. *Trans. ASCE* 112:2–72; discussion 73–124

Wade, R. B. 1967. Linearized theory of a partially cavitating plano-convex hydrofoil including the effects of camber and thickness. *J. Ship Res.* 11:20–27

Wade, R. B., Acosta, A. J. 1966. Experimental observations on the flow past a plano-convex hydrofoil. *J. Basic Engrg.* 88:273–83.

Willmarth, W. W. 1975. Pressure fluctuations beneath turbulent boundary layers. *Ann. Rev. Fluid Mech.* 7:13–38

Wislicenus, G. F. 1965. *Fluid Mechanics of Turbomachinery, Vol. 1.* N.Y.: Dover 2 vols. 744 pp.

Wood, G. M., Hartz, F. H., Hammitt, F. G., Agostinelli, A. eds. 1965. *Symp. Cavitation in Fluid Machinery*, ASME. 267 pp.

Wu, T. Y. 1972. Cavity and wake flows, *Ann. Rev. Fluid Mech.* 4:243–84

Yin, B. 1978. A preliminary design theory for polyphase impellers in unbounded flow. *Proc. Symp. Polyphase Flow in Turbomachines*, ASME, San Francisco, pp. 253–70

Young, J. O., Holl, J. W. 1966. Effects of cavitation on periodic wakes behind symmetric wedges. *J. Basic Engrg.* 88:163–76

*Ann. Rev. Fluid Mech. 1981. 13:329-50*


# THE FLUID DYNAMICS OF INSECT FLIGHT

*8179

*T. Maxworthy*
Departments of Mechanical and Aerospace Engineering, University of Southern California, Los Angeles, California 90007

## INTRODUCTION

The subject of this review is one that has produced a fruitful interaction between zoologists and aerodynamicists, requiring each to learn something of the specialization of the other. It is written from the point of view of one interested in fluid dynamics and attempting to organize conflicting evidence on the detailed mechanisms by which insects, covering a large range of size and flight speed, are able to remain aloft. In most cases, to date, the basic approach has been to study the wing motions of a variety of insects and then use established aerodynamic principles to calculate the forces resulting from such motions. These forces are then compared, in an averaged sense, to the weight of the insect, and, in a few cases, to the instantaneous forces measured from tethered insects. Almost invariably, steady-state aerodynamics principles are used. That is, the lift-drag polar of an actual wing or wing model is measured in a wind tunnel at a steady Reynolds number equal to a suitably averaged wing Reynolds number ($Uc/\nu$, where $U$ is the mean wind speed relative to the moving wing, $c$ a maximum chord, and $\nu$ the kinematic viscosity of the test fluid) of the prototype. Forces on the actual wing are then calculated from these results based on the measured instantaneous relative velocity between the wing chord and the relative velocity vector. Except in a few cases, unsteady effects are ignored, as are some subtleties of wing motion, e.g. aerodynamically or insect-induced changes in wing profile, that are difficult, if not impossible, to measure with any precision.

It has been considered appropriate to ignore unsteady effects after realizing that the magnitude of the appropriate dimensionless parameters is often of unit order so that the unsteady perturbation from the steady lift is likely to be small.



0066-4189/81/0115-0329$01.00

Two such parameters are

$$N = nL/V$$

(Holst & Kuchemann 1941, 1942) where $n$ is the wing-flapping frequency, $L$ a characteristic wing dimension (semi-span or chord), and $V$ the flight velocity of the insect, or

$$k = 2n\phi r/V$$

(Walker 1925, 1927) where $\phi$ is the angular excursion of the wing and $r$ the distance to the plane of interest along the wing span. Both parameters measure in some sense the ratio of the unsteady velocity of the flapping wing to the flight speed, which under many circumstances have values between 0.1 and 1.0, although they clearly become zero when the insect is gliding and infinite when it is hovering. Calculation of values below unity from measured data has justified the use of the steady-state approach, although recent experimental results suggest that confidence in the detailed results obtained may be misplaced. With less justification, this method has also been used to study hovering flight, resulting in the need for unrealistically large lift coefficients $C_L$ ($C_L$ = lift force/$\frac{1}{2}\rho V^2 \times$ wing area, $\rho$ is the density of air) to balance the insect's weight (Weis-Fogh 1973, Ellington 1975).

In what follows, we discuss each of these topics in turn, concentrating mainly on fluid-dynamical matters and introducing only enough zoology to give the reader an adequate background. We also note several other excellent reviews that are to be highly recommended for their clarity and attention to completeness and zoological detail beyond that attempted here (Weis-Fogh & Jensen 1956, Pringle 1957, Lighthill 1975, 1977, Nachtigall 1974, Weis-Fogh 1975, 1977).

## THE FLUID DYNAMICS OF FLAPPING, FORWARD FLIGHT

### *The Determination of Wing Motions*

Two approaches to the problem of measuring wing motions are common. Numerous investigators have designed and constructed ingenious wind-tunnel facilities in order to measure both wing motions and the average forces applied by the insect to the surrounding fluid; in some cases instantaneous forces have been measured also. These studies include those of Nachtigall (1966, 1974), Weis-Fogh (1956), Jensen (1956), Vogel (1966, 1967a, b), Wood (1970), Bennett (1966), Cloupeau, Devillers & Devezeaux (1979), and Buckholtz (1980). Such wind-tunnel studies have the disadvantage that the insect must be restrained in some way and, therefore, may not fly in a completely natural manner nor at the correct flight velocity, although comments to the contrary are presented by Vogel (1966, 1967a, b) and Jensen (1956), among others. Other workers (e.g. Weis-Fogh 1973, Ellington & Runnells 1976, and Norberg 1975) have avoided this problem by photographing the insect in free flight, in which case forces (except mean values that must balance the insect weight) cannot be measured and usually only hovering or near-hovering can be photographed successfully.

In Figure 1, we show a typical wind-tunnel type of apparatus and its associated force balance and control system (Nachtigall 1966, 1974). The insect was suspended in front of an open return wind tunnel (Figure 1*a*) and the wind speed adjusted automatically until the net average horizontal force on the insect (i.e. drag minus thrust) was zero. No attempt was made to ensure that the net average vertical force (i.e. weight minus lift) was also zero. The insect was then photographed from three orthogonal directions (Figure 1*b*). From these photographs, the

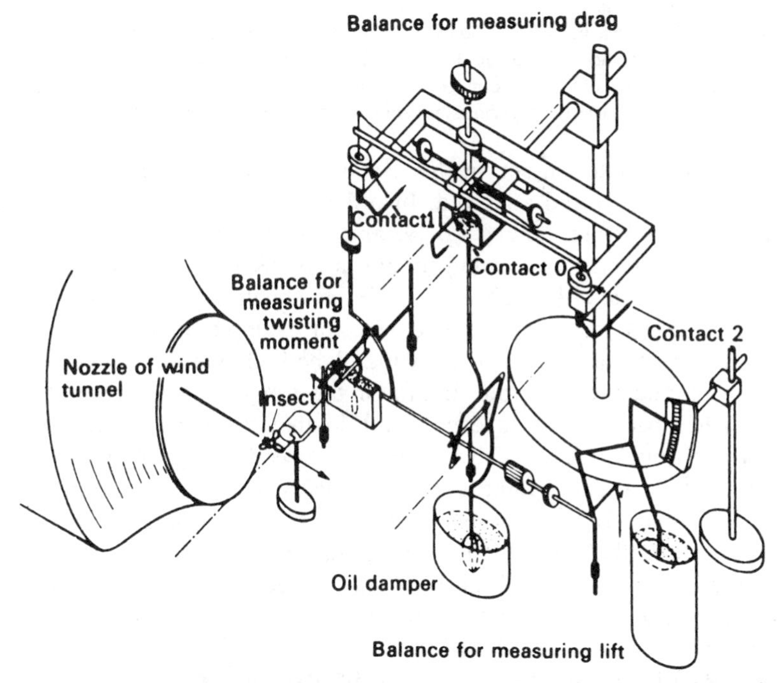


*Figure 1a* Three-component balance used to support a flying insect in front of an open-return wind tunnel (as also shown in Figure 1*b*). Each component of the balance is set on knife edges, is oil damped, and has adjustment weights. The contacts switch a servo motor that in turn controls the tunnel wind speed to a condition at which the thrust generated by the insect exactly balances its drag. Reproduced with permission from Nachtigall (1966, 1974).

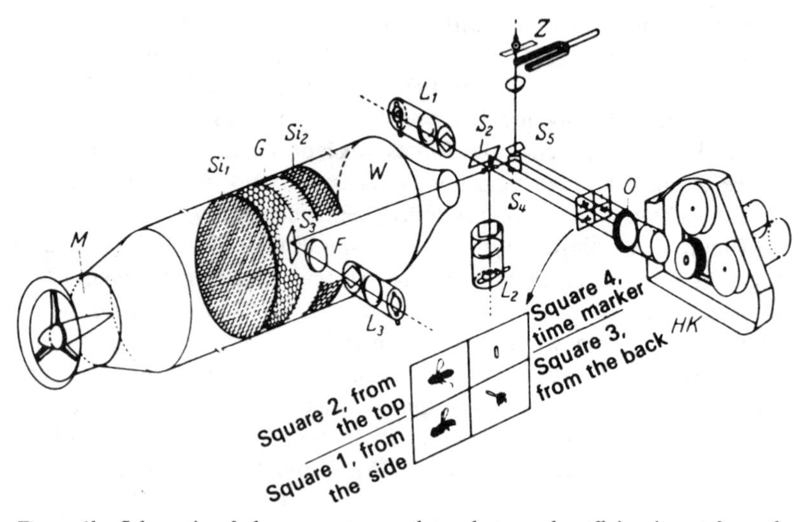


*Figure 1b* Schematic of the apparatus used to photograph a flying insect from three orthogonal directions, using three lamps ($L$) and five mirrors ($S$). The vibrating tuning fork ($Z$) is also filmed to provide a known timing mark on each frame. Wind-tunnel details include the motor ($M$), honeycomb straightening section ($G$), screens ($S_i$), and nozzle ($W$). Reproduced with permission from Nachtigall (1966, 1974).

wing motion of the insect, in this case the fly *Phormia regina*, could be drawn (Figure 2) and analyzed. In order to do so, it was necessary to define at least three angles ($\beta$, $\gamma$, and $\delta$ in Figure 3) between a line drawn from wing root to wing tip and a convenient coordinate system. These angles vary through each cycle, as shown in Figure 4. Further data reduction was possible as shown, for example, in Figure 5 where the motion of one section of the wing plane is shown in a space-related frame of reference, i.e. one in which the translational motion of the insect has been added to each point so that the wing appears to be moving through still air. This figure also shows the angle of attack of the wing section at each point and can be used to calculate the instantaneous velocity of the chosen wing section and associated forces, as discussed in the Introduction. A qualitative description of such a calculation is shown in Figure 6, reproduced from Nachtigall (1974) where the lift and drag on the wing are shown resolved into vertical and horizontal forces, while the wing-velocity vector is shown as the local tangent to the curve of wing motion. In particular, these figures (5 and 6) show that even when the insect wing is moving backwards, the angle of attack is such that a positive lift and thrust are produced (see page 336 of the present review for a more complete description of this method, and page 339 for a contradiction to this result). The methodol-

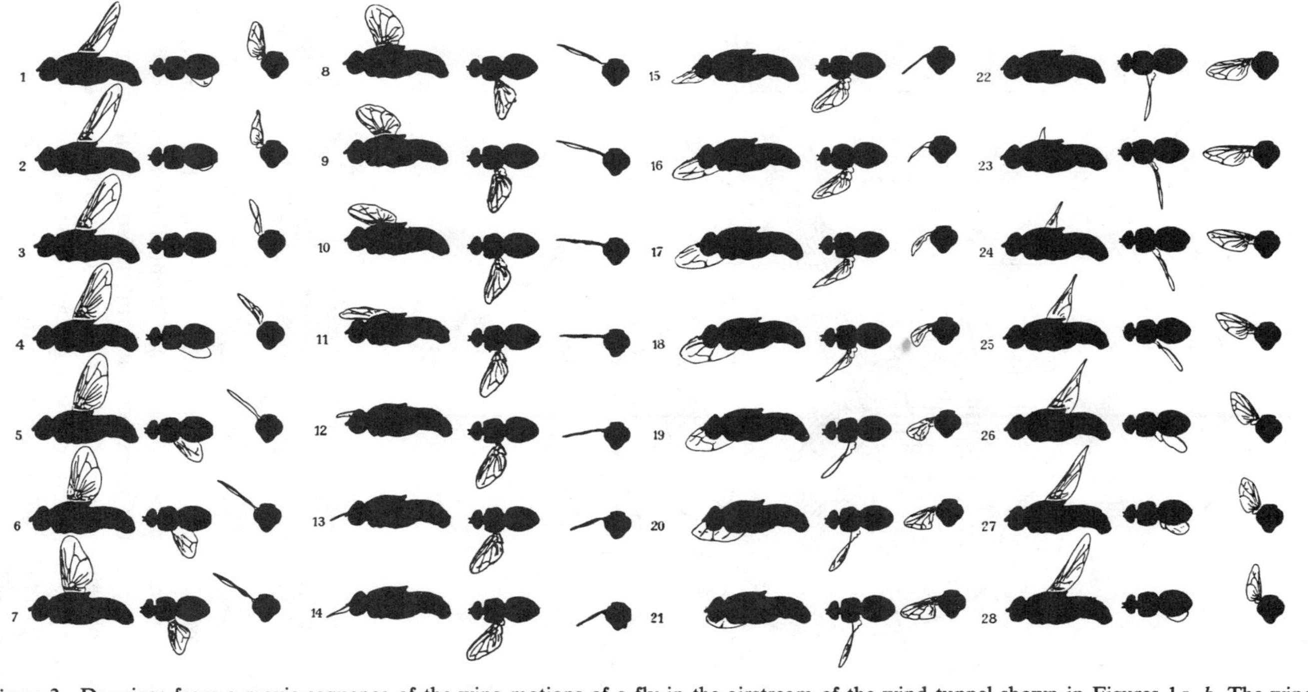


*Figure 2* Drawings from a movie sequence of the wing motions of a fly in the airstream of the wind tunnel shown in Figures 1*a*, *b*. The wing downstroke starts in picture 3 and continues through 16. At this point, the wing begins to rotate so that, on the upstroke, a favorable lift and thrust can still be generated. Rotation back to the best angle for the downstroke begins at 27 or 28 and continues through 1 and 2 at which time the sequence repeats. The time between each picture is 1/3200 s. Reproduced with permission from Nachtigall (1966).

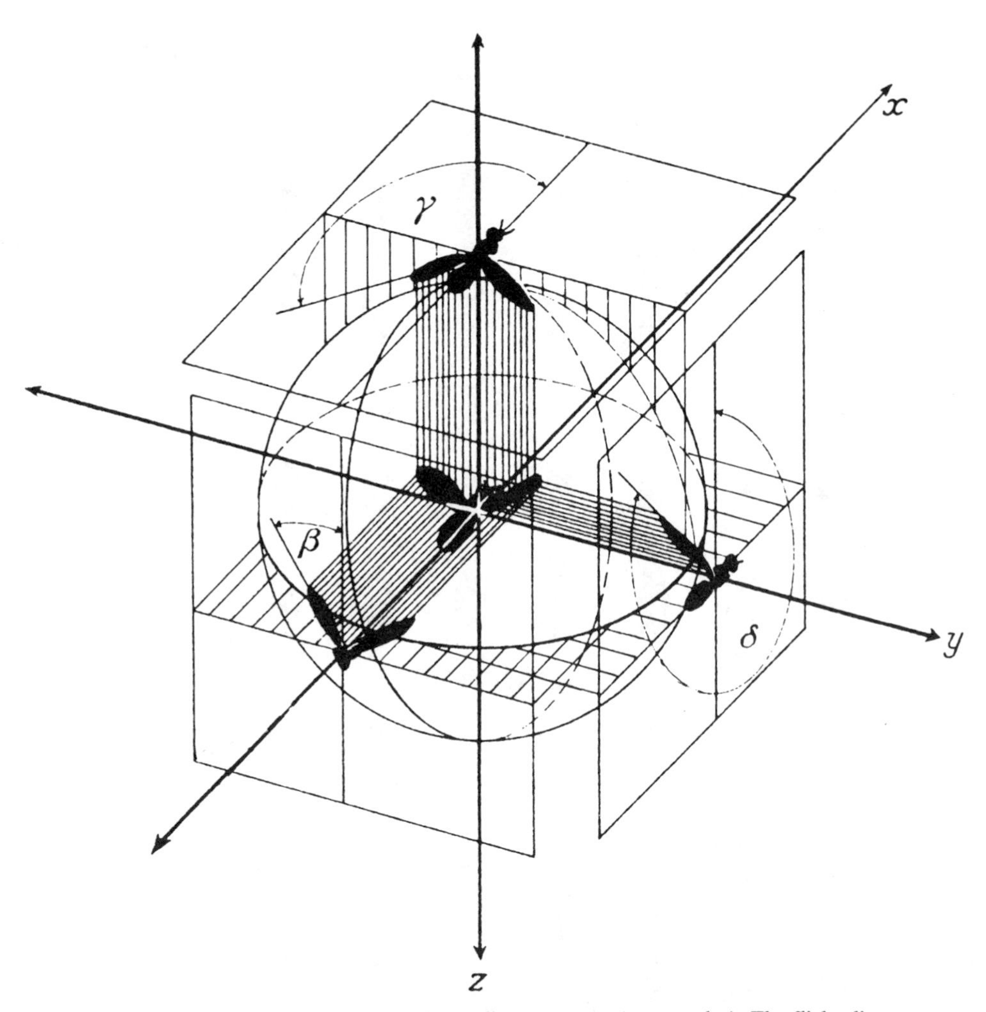


*Figure 3* Insect-centered orthogonal coordinate systems ($x$, $y$, and $z$). The flight direction is $x$. The three angles $\beta$, $\gamma$, and $\delta$ are those between a line drawn from wing root to wing tip and the $yz$, $xy$, and $xz$ planes, respectively. Reproduced with permission from Nachtigall (1966, 1974).

ogy of other investigators using tethered insects has been similar in spirit, if not in minute detail, to Nachtigall's and the reader is referred to the original papers for more complete information, although some small amount of detail is sprinkled throughout the following sections.

The photography of insects in free flight and the resultant analysis of wing motion is hampered by an inability to view the motion from several different angles and, hence, determine angles of attack and wing velocities with precision. The method is also useful only for hovering or

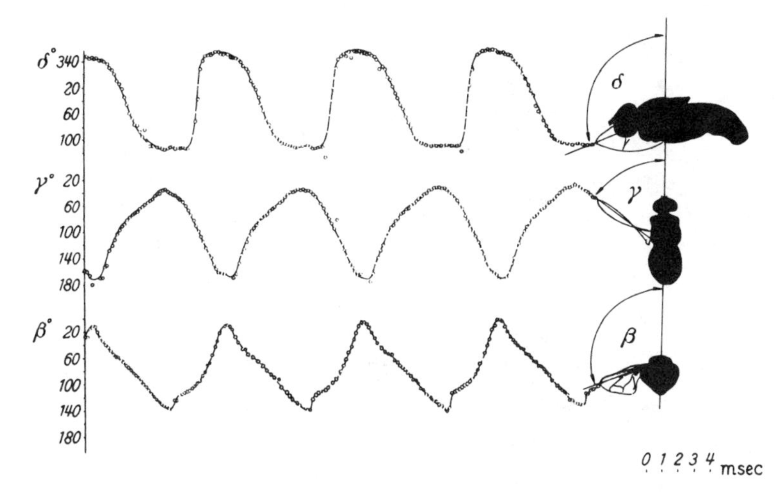


*Figure 4* The variation of $\beta$, $\gamma$, and $\delta$ during flight, in particular showing the fact that the downstroke takes place much more slowly than the wing upstroke. Time between points is 1/6400 s. The fly *Phormia* was flying steadily in an airstream of velocity 2.75 m/s. Reproduced with permission from Nachtigall (1966, 1974).

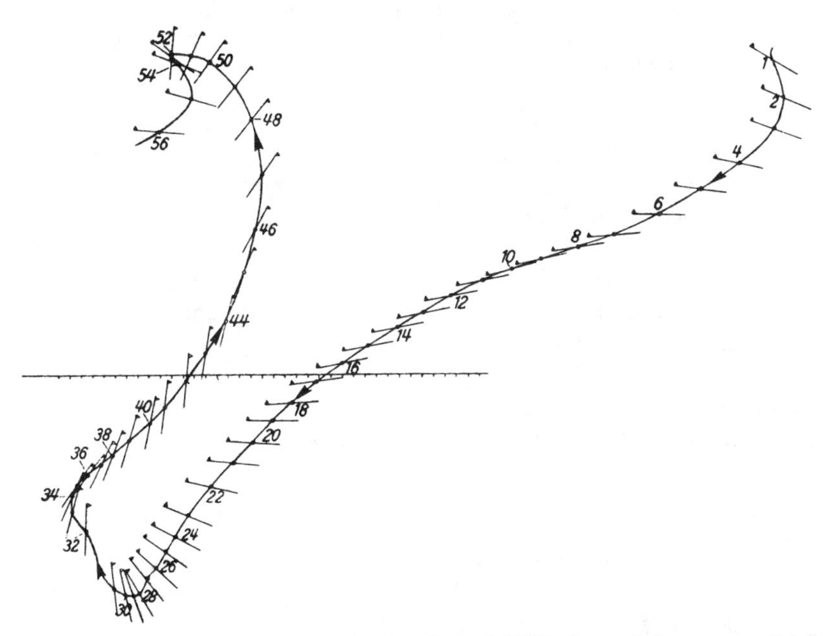

*Figure 5* The motion of the wing tip of a fly at 1/6400-s intervals in a space-related frame of reference with the insect moving from right to left. The plane at the wing tip is shown as a series of short lines and the angle they make with the wing path is the local angle of attack ($\alpha$) of the wing. Wing rotation prior to and during the upstroke is clear (points *28* through *51*) while the rotation back to the angle for the downstroke takes place very rapidly (points *51–54*). The small triangle is placed on the upper side of the leading edge and is filled in during the downstroke. The distance between points on the horizontal axis is the distance moved by the insect during 1/6400 s. Reproduced with permission from Nachtigall (1966, 1974).

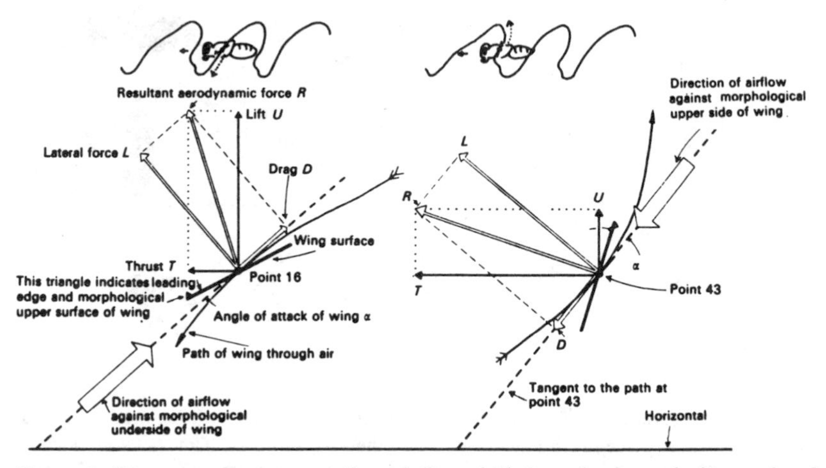


*Figure 6* Diagrammatic representation of the calculation of wing velocity, angle of attack, and forces at two representative points of Figure 5, point *16* of the downstroke and point *43* of the upstroke. In particular, during the upstroke the wing path and angle of attack are such that both a favorable thrust and lift are still generated on a quasi-steady basis. Reproduced with permission from Nachtigall (1974).

near-hovering insects, since the small fields of view and high framing rates, used to gain resolution, make it a matter of pure chance to capture a suitable image of a rapidly moving object. Excellent hovering sequences have been reported by Weis-Fogh (1973), Ellington & Runnells (1976), and Norberg (1975), among others. The best example of a published sequence of wing motions, by Weis-Fogh (1973), has already been reproduced in many places and need not be repeated. However, the excellent moving picture film, "High Speed Research Films of Free Flying Insects" by C. P. Ellington and G. G. Runnells of the Zoology Department at Cambridge University, is highly recommended to anyone interested in the subtleties of wing motion in a variety of insects.

## *Steady Aerodynamic Analysis of Unsteady Forward Flight*

Although in this section we emphasize the analysis of flapping flight, it is probably best to begin with a short discussion of the one case where the steady analysis is unquestionably correct. This occurs when the insect is gliding or soaring at constant velocity, i.e. when $n=0$, and hence $N=0$ and $k=0$. A straightforward analysis has been presented by Lighthill (1977) for birds, but the ideas are unchanged when applied to insects. During gliding flight, one balances lift and drag forces to the components of the insect's weight perpendicular and along a flight path oriented at an angle $\Lambda$ to the horizontal. This angle is then the inverse tangent of the ratio of drag to lift. Also, using standard relationships for lift and frictional, form, and induced drag, it is possible to determine, approximately, a speed at which the insect's drag is a minimum and speeds at which insects can soar at fixed angle of attack, etc. Jensen (1956), in his monumental analysis of the flight characteristics of the desert locust (*Schistocerca Gregaria*), devotes a short section to its gliding and soaring ability, presenting a measured lift-drag polar of the whole (dead) insect in gliding trim and concludes that a value of $\Lambda = \tan^{-1} 1/6.5 = 8.7°$ at 3.5 m/s is "best" for the desert locust.

It is also Jensen's (1956) analysis of flapping, forward flight that we follow here in order to discuss the typical assumptions made in the steady-state approach to the problem. The experiments with a tethered desert locust performed by Weis-Fogh (1956) and Jensen (1956) are similar to those of Nachtigall (1966) outlined in the previous section. In fact, the motion of both wings, relative to still air, is virtually identical to that shown in Figure 5, except that Jensen also observed a distortion of the fore-wing surface that is especially profound at the end of the downstroke and during the backward-moving upstroke.

In the second set of experiments, a single wing was mounted directly on a force-measuring balance and placed in the boundary layer of a wind tunnel (whether or not the boundary layer was laminar is not discussed) at various values of wing angle of attack $\alpha$ and sweep-back angle $\gamma$ (called $\beta$ in the original paper) in such a way that the variations in dynamic pressure along the span of the wing simulated those experienced by a wing rotating about a fixed point in a uniform flow (also see Newman et al 1977). The lift-coefficient versus drag-coefficient polar with $\alpha$ as parameter was produced for both a straight, uncambered wing and for the types of wing camber associated with the end of the downstroke and the upstroke (Figures 7 and 8), as well as for values of sweep-back angle appropriate to the range of angles found in flight. A similar curve was produced for the hind wing. From the curves of wing motion, similar to Figure 5, the velocity and angle of attack of several sections along the wing could be calculated at each of the various phases of the flapping cycle (Figure 6). It was then possible, using the lift-drag polars (e.g. Figures 7 and 8) to calculate the instantaneous forces experienced by the wings (as in Figure 6). No attempt was made to account for interaction effects between the independently moving pairs of wings. Jensen found that the averaged lift force calculated in this way balanced the weight of the insect and the thrust balanced the drag to within experimental accuracy. Subsequently, this result has been used to justify the use of steady aerodynamics in other problems (e.g. Pringle 1957) for which the value of $k$ is about the same as it is for the

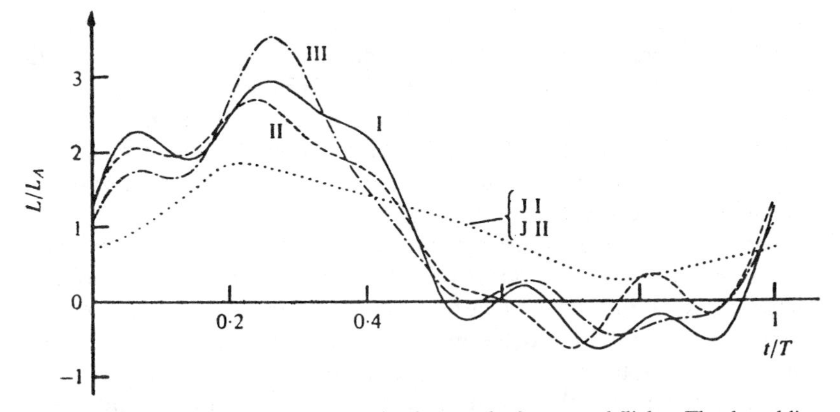


*Figure 9* Variations in lift during the wing beat cycle, for normal flights. The dotted line represents Jensen's (1956) results from flights JI and JII. The other curves are from flights I, II and III of Cloupeau et al (1979). For the abscissa, $t$ is time and $T$ the period of wing motion. For the ordinate, the lift ($L$) is normalized by the average lift over one cycle ($L_A$). On the average, the magnitude of lift variations observed in flights I–III is about twice that obtained from detached wings in steady flow conditions (JI and II). Reproduced with permission from Cloupeau et al (1979).

desert locust, i.e. $k=0.8$ at 2/3 of the wing semi-span.[1] The calculated instantaneous lift as a function of time is shown in Figure 9, together with measurement of this quantity by Cloupeau et al (1979), which we discuss further in the following section.

## *Measurements of Unsteady Lift*

The remarkable results of Jensen (1956) had not been subjected to a rigorous, independent check until experiments were performed recently by Cloupeau, Devillers & Devezeaux (1979) and Buckholtz (1980), although the importance of unsteadiness had been anticipated by the measurements of Bennett (1966, 1970) for both actual insect flight and a mechanical simulation.

Cloupeau et al (1979) used the same species of insect as Weis-Fogh & Jensen (1956) and a similar experimental arrangement except that, by designing and building a force balance with a high natural frequency, they were able to measure total instantaneous forces directly. From measured wing motions, they also calculated the instantaneous inertial

[1]One famous counterexample concerns calculations made by the aerodynamics staff of a well-known Southern California airplane company which revealed that the common bumblebee could not possibly fly. We presume that they not only ignored unsteady contributions to the lift but, more critically, were unaware of the more subtle aspects of wing motions that have been discovered since that time. See also Nachtigall (1974, p. 137).

forces required to accelerate the wings through their observed motion, which were then subtracted from the measured total force to give the aerodynamic contribution. The values of this latter quantity (Figure 9) are the most revealing. They show that, typically, the maximum, instantaneous lift is twice that found by Jensen and that it even becomes slightly negative during the upstroke. In this case also, the average lift equalled the insect weight.

Buckholtz (1980) also measured instantaneous forces using an ingenious force balance but, unfortunately, neither separated the relative magnitudes of inertial and aerodynamic forces nor made an explicit attempt to be certain that the mean horizontal and vertical forces were zero. He did note, however, that the wave forms of the instantaneous forces were remarkably free from harmonics of the fundamental wing frequency despite previous observations of higher harmonics in the wing motions themselves as photographed and measured by Nachtigall (1966). He also confirmed that the vertical forces in the liftward direction were a maximum during the downstroke and that thrust was maximum during the upstroke.

From these results, especially those of Cloupeau et al (1979), it appears that unsteady aerodynamic effects are important when $k$, for example, is of order one, as it is for a large number of insects. At which value of this, or any other appropriate parameter, such effects will become unimportant is unknown at the moment. A number of possible unsteady-flow phenomena are discussed qualitatively by Nachtigall (1979), including vortex production during wing rotation and that due to the motion of flexible wings, and the possibility of the generation of large lift during rapid angle-of-attack changes, among others. Which, if any, of these effects are important in actual flight has not been determined and a study of these questions is a very fruitful avenue for future research.

## THE FLUID DYNAMICS OF HOVERING FLIGHT

One limiting case that has excited a great deal of activity occurs when $V \to 0$ and $k \to \infty$ and the insect hovers at a fixed point in space. The important differences between the various classes of wing motion exhibited in photographic studies have been well summarized by Weis-Fogh (1973, 1975), Lighthill (1975), and Ellington (1978, 1980), among others, and will be mentioned only briefly here.

### *Normal Hovering*

In the most commonly observed pattern, the wing stroke is approximately horizontal. Such cases have been studied extensively by Weis-

Fogh (1973, 1975, 1977), who called the pattern normal hovering. Hummingbirds (Greenwalt 1960) and the majority of insects (Weis-Fogh 1973) are included in this group. The wing sections move in an almost horizontal figure-of-eight motion and rotate, or flip over, through large angles at the ends of the stroke so that the wings operate at a positive, or favorable, angle of attack over most of their travel (shown diagrammatically in Figure 10). These motions induce a large vertical velocity or downwash (Lighthill 1973, Figure 9) which Ellington (1980) has shown is effectively cancelled by the vertical wing velocities of the figure-of-eight pattern and results in a velocity that is essentially horizontal relative to the moving wing.

Weis-Fogh (1973), in an extensive study of this mode of hovering in which he used an extension of the steady-aerodynamics principles introduced by Walker (1925), Holst & Kuchemann (1941, 1942), and Osborne (1951), among others, showed that the average lift coefficients required to balance the insect's weight are moderate for most of the insects within this grouping, and within the limits to be expected, e.g. 0.5

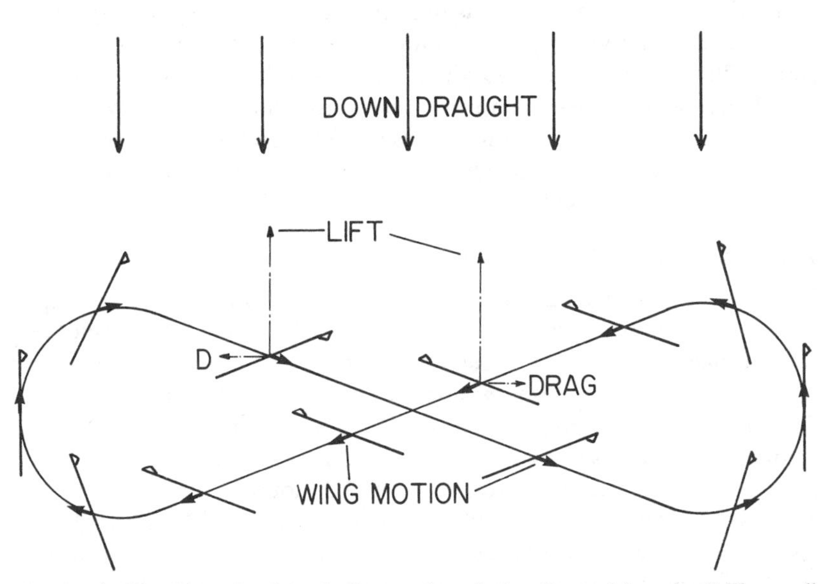


*Figure 10* Diagrammatic view of wing motions during "normal hovering." The small triangles are placed on the upper leading edge of the wing and one can see that the strong rotation before the return stroke (left-to-right) places the wing in a position to provide positive lift at all times on a quasi-steady basis. The effectiveness of this motion is easily demonstrated by the reader while floating in the water and moving his hands in the indicated manner. However, this method is rarely used in such cases since it is not the most efficient technique for the human hand, which is so stiff that either edge can be used as a leading edge. Insect and bird wings do not have this characteristic, of course, and the observed flip appears to be the best way to overcome this deficiency.

to 1.6, although this latter value is somewhat high. As Ellington (1980) and R. A. Norberg (1975) point out, such a result, while consistent with steady principles, does not exclude the possibility that unsteady effects create much larger and smaller instantaneous forces than those found by this analysis and that they ultimately cancel out over the whole wing cycle, so that in this case consistency with the results of steady theory is fortuitous. That unsteady effects must be important is shown in a preliminary study of the flip process in Maxworthy (1979, Figure 13) in which the vorticity that must be shed as the wing slows down and flips over interacts with the wing as it moves in the opposite direction to produce a positive lift. [Also see Ellington (1980, Figure 3) and Nachtigall (1979, Figure 2).] This latter study also suggests that the Wagner effect (Wagner 1925), by which the asymptotic lift of such an impulsively started wing or aerofoil is reached only slowly, may be greatly modified by a leading-edge separation bubble that is formed during the time it takes for the wing to move about one chord length. This bubble is not shed during the subsequent motion due to the combined effects of unsteadiness and a three-dimensional flow that removes vorticity from the upperside of the wing as it swings in an arc about its attachment point. This separation bubble gives the wing a large effective camber and thickness and greatly improves its performance, especially at low Reynolds numbers. We return to these points in more detail when we discuss the clap-and-fling mechanism of lift generation for which case the results are quite clear.

## *The Clap-and-Fling Mechanism*

This pattern of wing motion was discovered and named by Weis-Fogh (1973) while studying the chalcid wasp, *Encarsia Formosa*. The motion is, superficially, a small modification of the normal hovering mode, in that at one extreme of the motion, the wings clap together and then fling open, rather than flipping over as shown in Figure 10. The modified sequence is shown diagrammatically in Figure 11 viewed from behind the insect. Drawings from high-speed motion pictures are shown in Weis-Fogh (1973) and reproduced in Weis-Fogh (1975) and Lighthill (1975, 1977). Weiss-Fogh (1973) calculated that the minimum mean lift coefficient for this insect was far too high to be explained by conventional steady-state aerodynamic principles. Even after correction of his value by Ellington (1975), who took into account the extra wing area represented by fringe hairs or setae that increase this area by about 40% at the moderately low Reynolds numbers ($\dot{\omega}c^2/\nu$; $\dot{\omega}$ is a measure of the wing angular-rotation rate and its exact numerical value depends on the particular part of the wing motion being considered) of order twenty at

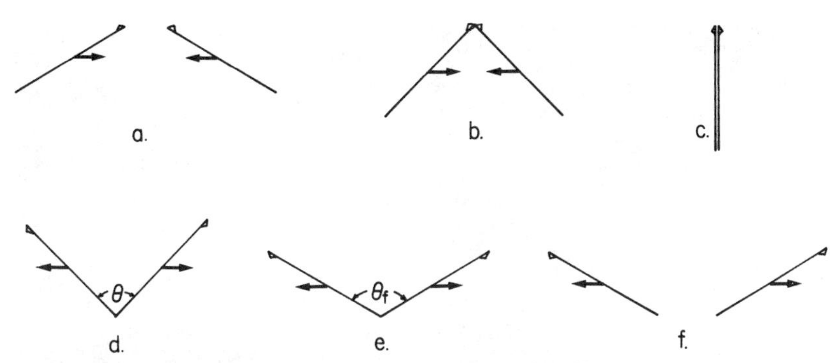


*Figure 11* The clap-and-fling mechanism. In *a–c* the leading edges come together and the gap between the wings closes to form an outflow and a downward-moving vortex. In *d–f* the wings rotate around the trailing edge, parting along the leading edge initially. As discussed in the text, the vortices produced by these unsteady motions are the result of large forces applied to the fluid by the wing.

which *Encarsis Formosa* hovers, the lift coefficient of 1.6 is at the upper limit of acceptability. Consequently, Weis-Fogh (1973) and Lighthill (1973) devised an ingenious explanation for the large forces that appear to exist during the fling, based on an inviscid source flow into the opening **V** of the two wings (Figures 11*c–e* and 12*a–c*). By the time the state of Figure 11*e* is reached and the wings began to part (Figure 11*f*) the circulation around each one of them is equal to $\Gamma$, and it was assumed that this would result immediately in a large lift, corresponding to $\rho \Gamma U_0$ per unit length ($U_0$ is the velocity at which the wings part relative to still air), without the need to continuously shed trailing vorticity and for the lift to build slowly to its asymptotic value as in the classical Wagner (1925) problem.

This model largely ignores the critical role played by viscous effects although Lighthill (1973) anticipated the possibility of leading-edge separation modifying the values of $\Gamma$ calculated from the inviscid, unseparated case. Maxworthy (1979), using two mechanical models, clarified the role of this separation and how it can account for the large lift force or, more critically, large lift coefficient that is needed to balance the insect weight. Flow visualization of small particles in a two-dimensional flow which models this process is shown in Figure 12. Maxworthy (1979) showed that the circulation contained in this separation bubble, also shown in the numerical computation of Haussling (1977), is approximately three times that calculated from the inviscid theory at the instant when the wings begin to part, Figure 12*c*. The force calculated from consideration of the time taken to develop the impulse contained in these vortices was actually far larger than required

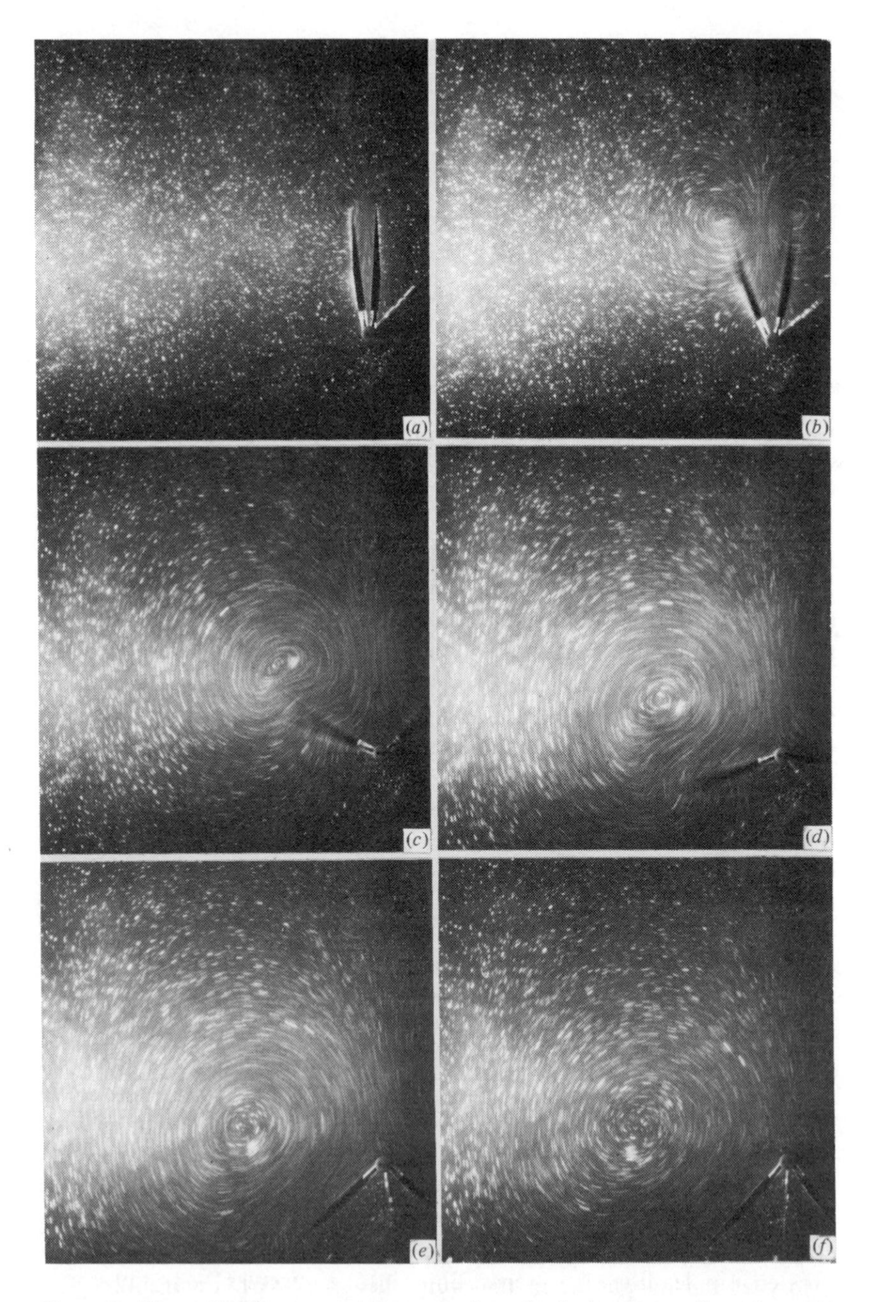


*Figure 12* The vortical flow created by an opening pair of wings in two dimensions. The flow is made visible by photographing small particles suspended in the fluid. The Reynolds number based on wing chord and opening velocity is about 30 in this sequence. Reproduced from Maxworthy (1979).

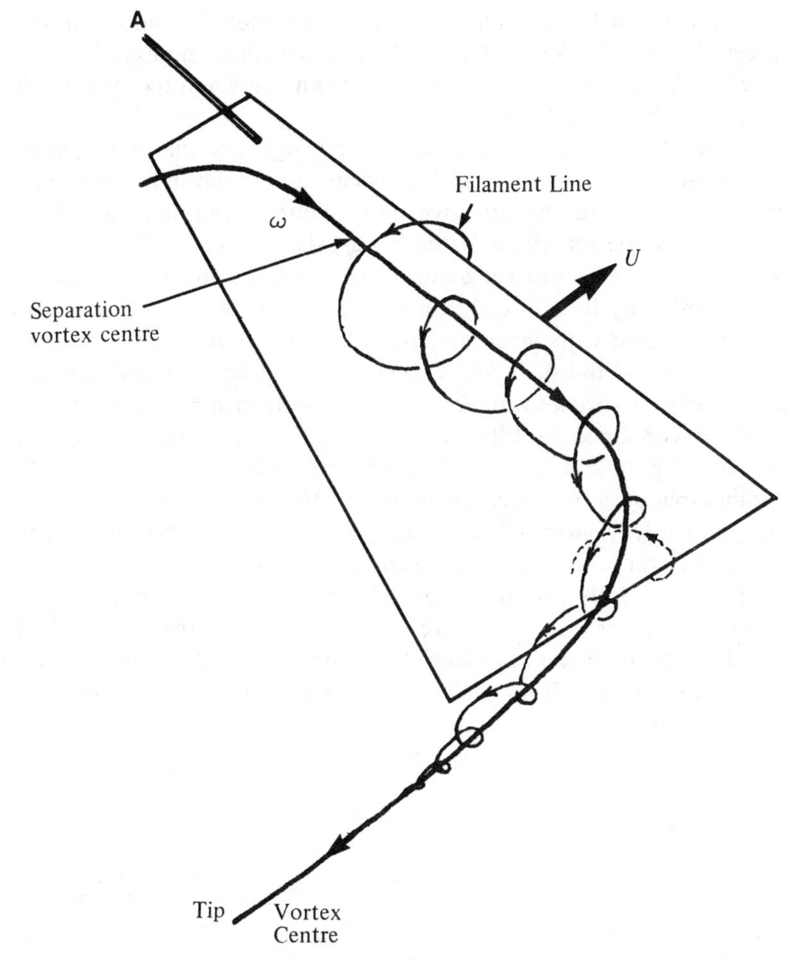


*Figure 13* The opening wing rotates about the wing root at *A*. The resulting pressure gradient along the length of the wing creates a flow that removes the vorticity formed by leading-edge separation and deposits it into a tip vortex. Thus, a quasi-steady vortex appears over the wing and favorably changes its effective camber and thickness. Reproduced from Maxworthy (1979).

and is presently being checked experimentally. Using a three-dimensional model he also showed that as the wings part (Figure 11*f*) a flow is quickly set up along the span of the wing by pressure gradients required to balance centrifugal forces, and this removes the vorticity[2] created by leading-edge separation, into a circular wing-tip vortex

[2] This three-dimensional flow is critical in preventing a buildup of vorticity over the wing, which under two-dimensional-flow conditions would be shed dramatically, as in a Kármán vortex street, with an associated loss of lift.

(Figure 13), which eventually evolves into a vortex ring surrounding the insect (Figure 14). Again, the creation of the fluid impulse of this ring over the time scale of wing opening gives an average force large enough to explain the natural observations.

As an alternative approach, Maxworthy suggested that the separation bubble shown in Figure 13 quickly reaches an equilibrium configuration and by increasing the effective wing camber and thickness greatly improves the performance of the wing. Also, since the flow is already separated, this effective wing can operate at large angles of attack with concommitantly large lift coefficients. Such effects are not uncommon in aerodynamic experience and include those found in the flow over delta wings and the so-called Kasper wing (e.g. Kruppa 1977). Some of these ideas on vortex formation in the two-dimensional case are similar to those contained in Ellington (1980), who also discussed the clap process (Figure 11*b*, *c*). In Figure 15, we show an unpublished flow photograph of a low-Reynolds-number two-dimensional clap which, independently, confirms Ellington's picture of a vortex-formation process that again results in a large estimated lift force.

Ellington (1980) also noted several insects that perform a near clap-and-fling, that is, the wings are separated by less than about half a chord length at closest approach. He speculated on the flow likely to be found in this case, but no studies have yet confirmed the details of these streamline patterns.

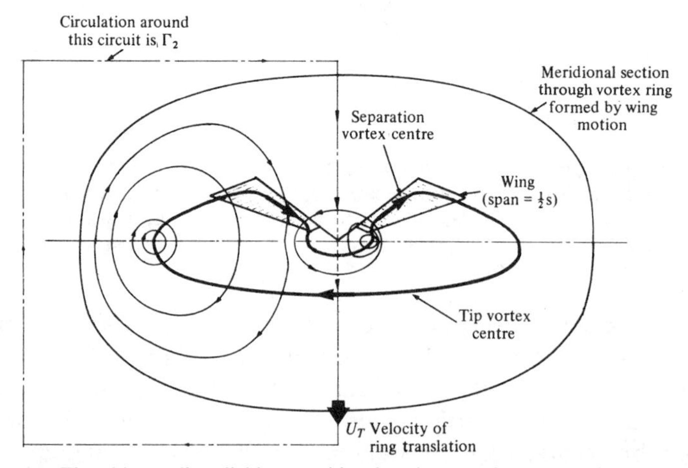


*Figure 14* The ultimate flow field created by the wing-opening process is a vortex ring of large impulse that moves downwards as the last member of a chain of rings produced by prior wing motion. Reproduced from Maxworthy (1979).

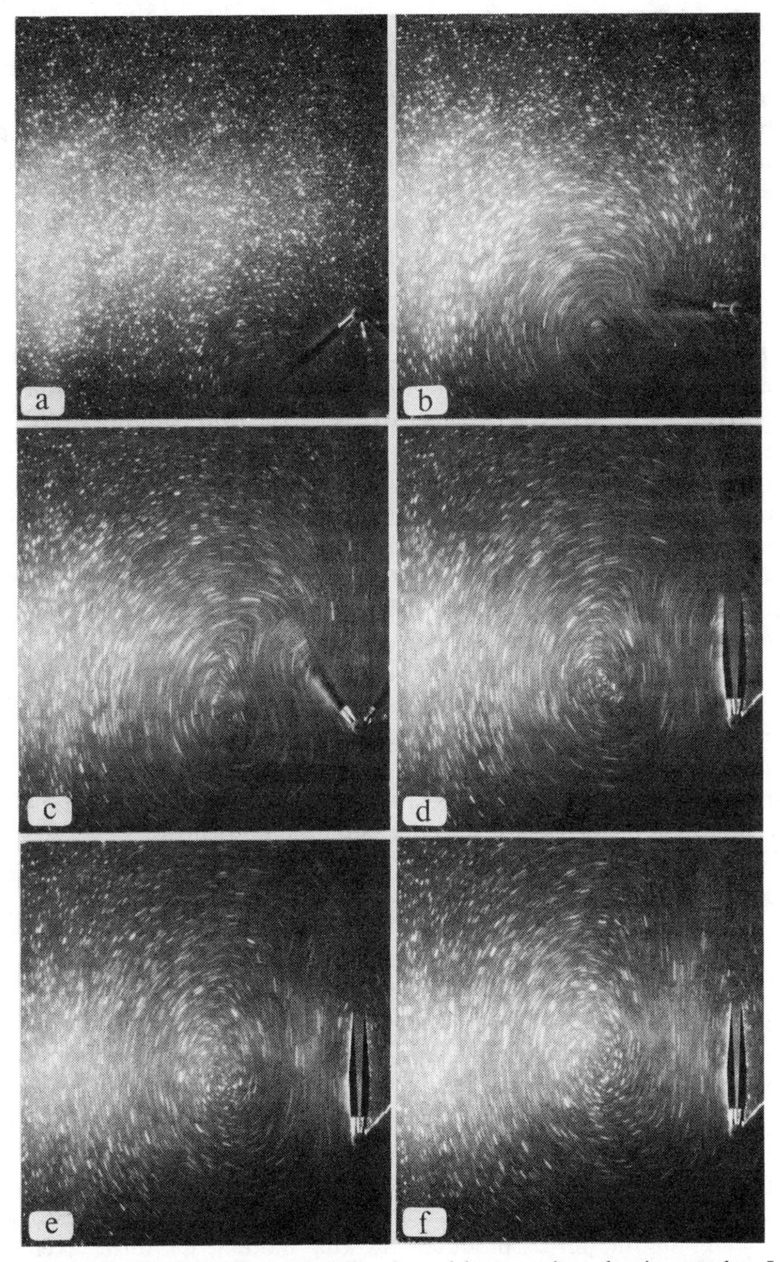


*Figure 15* The low-Reynolds-number flow formed by two wings clapping together. In this case, for experimental convenience, the vortex pair shown moves upwards, representing a downward force on the wings. To reproduce the motion created by an actual insect in flight, this set of photographs should be viewed with the page turned upside down. Unpublished photograph of T. Maxworthy.

### *Inclined Wing-Stroke Plane*

The true hover-flies (Weis-Fogh 1973) and small passerine birds and bats hover with the wing-stroke plane inclined at an angle of between 30° and 40° to the horizontal. Dragonflies hover with this angle at about 60° (Norberg 1975). It appears that, in general, the upstroke generates a very small force because either the wings rotate strongly to create a near-zero angle of attack or, in birds and bats especially, the wings flex to present a reduced surface area to the flow.

### *Vertical-Stroke Plane*

This type of motion is typical of the *Papilionoidea* or butterflies, and has been discussed in some detail by Ellington (1980) and Maxworthy (1979) and also observed by Chance (1975). The flow pattern formed by the opening wings is identical to that of the initial phase of the fling process (Figures 11*d* and *e*) discussed earlier, except that the final opening angle $\theta_f$ is typically 270°, although, at the later stages of their motion, the wings do *not* part (as they do in Figure 11*f*). We can again use the flow visualization of Figure 12 to indicate the type of flow field that is formed, while Ellington (1980) has successfully visualized the vortices produced by a tethered moth (see Chance 1975, also).

The vortices formed by such flow separation can be modeled as vortex rings or pairs and used to successfully calculate the unsteady lift forces required for takeoff. A satisfactory model to explain the ability of these insects to hover has not been forthcoming, since it is not yet clear how the wing upstroke is modified so as to produce a negative lift that is smaller than the positive lift produced on the downstroke.

## CONCLUDING REMARKS

By concentrating in some detail on the importance of unsteady flow in insect flight, we have ignored many other important topics that have been studied by many outstanding scientists. We apologize to them and to our readers for these omissions and hope that others will review these features in the near future. They might include questions of flight at very low Reynolds number, as in the *Thysanoptera*, or fringe-winged insects, the wings of which consist of a central spar supporting thin hairs, or *setae*, and no membraneous surface at all; interaction effects between pairs of wings beating out of phase; the role of wing flexibility in modifying the forces produced by wing motion; the form taken by the wake behind flapping insects (or birds); questions of the energy and power required for flight, etc., etc.

This work was supported internally at the University of Southern California.

*Literature Cited*

Bennett, L. 1966. Insect aerodynamics: Vertical sustaining force in near-hovering flight. *Science* 152:1263–66

Bennett, L. 1970. Insect flight: Lift and rate of change of incidence. *Science* 167:177–79

Buckholtz, R. H. 1980. Measurements of unsteady periodic forces generated by the Blowfly flying in a wind tunnel. *J. Exp. Biol.* In press

Chance, M. A. C. 1975. Air flow and the flight of the noctuid moth. In *Swimming and Flying in Nature. Vol. 2*, ed. Wu, Y.-T., Brokaw, C. J., Brennen, C., pp. 829–43. New York/London: Plenum

Cloupeau, M., Devillers, J. F., Devezeaux, D. 1979. Direct measurements of instantaneous lift in desert locust; comparison with Jensen's experiments on detached wings. *J. Exp. Biol.* 80:1–15

Ellington, C. P. 1975. Non-steady-state aerodynamics of the flight of *Encarsia Formosa*. In *Swimming and Flying in Nature. Vol. 2*, ed. Wu, Y.-T., Brokaw, C. J., Brennen, C., pp. 783–96. New York/London: Plenum

Ellington, C. P. 1978. The aerodynamics of normal hovering flight: Three approaches. In *Comparative Physiology—Water, Ions and Fluid Mechanics*. Cambridge Univ. Press

Ellington, C. P. 1980. Vortices and hovering flight. In *Proc. Conf. on Unsteady Effects of Oscillating Animal Wings*, Saarbrucken, West Germany, November 1977, ed. W. Nachtigall

Ellington, C. P., Runnells, G. G. 1976. High speed research films of free flying insects. Dept. Zoology, Univ. Cambridge.

Greenwalt, C. H. 1960. *Hummingbirds*. New York: Doubleday

Haussling, H. J. 1977. Boundary-fitted coordinates for accurate numerical solution of multi-body flow problems. Rep. of David W. Taylor Naval Ship Res. and Dev. Cent., Bethesda, Maryland 20084

Holst, E. v., Küchemann, D. 1941. Biologische und aerodynamische Probleme des Tierfluges. *Naturwissenschaften* 29:348–62

Holst, E. v., Küchemann, D. 1942. Biological and aerodynamical problems in animal flight. *J. R. Aero. Soc.* 46:39–56

Jensen, M. 1956. Biology and physics of locust flight III. The aerodynamics of locust flight. *Philos. Trans. R. Soc. London, Ser. B* 239:511–52

Kruppa, E. W. 1977. A wind tunnel investigation of the Kasper vortex concept. *Paper* 77-310. AIAA 13th Ann. Meet., Washington, DC, Jan. 1977

Lighthill, M. J. 1973. On the Weis-Fogh mechanism of lift generation. *J. Fluid Mech.* 60:1–17

Lighthill, M. J. 1975. Aerodynamic aspects of animal flight. In *Swimming and Flying in Nature. Vol. 2*, ed. Y-T Wu, C. J. Brokaw, C. Brennen, pp. 423–91. New York/London: Plenum

Lighthill, M. J. 1977. Introduction to the scaling of aerial locomotion. In *Scale Effects in Animal Locomotion*, ed. T. J. Pedley, pp. 365–404. London/New York: Academic

Maxworthy, T. 1979. Experiments on the Weis-Fogh mechanism of lift generation by insects in hovering flight. Part I. Dynamics of the fling. *J. Fluid Mech.* 93: 47–63

Nachtigall, W. 1966. Die Kinematik der Schlagflugelbewegungen von Dipteren. Methodische und analytische Grundlagen zur Biophysik des Insektenflugs. *Z. Vgl. Physiol.* 52:155–211

Nachtigall, W. 1974. *Insects in Flight*. New York: McGraw-Hill

Nachtigall, W. 1979. Rasche Richtungsanderung und Torsionen schwingender Fliegenflugel und Hypothesen über zugeordnete instationare Stromungseffekte. *J. Comp. Physiol.* 133:351–55

Newman, B. G., Savage, S. B., Schouella, D. 1977. Model tests on a wing section of an *Aeschna* dragonfly. In *Scale Effects in Animal Locomotion*, ed. T. J. Pedley, pp. 445–77. London: Academic

Norberg, R. A. 1975. Hovering flight of the dragonfly *Aeschna Juncea L.*, Kinematics and aerodynamics. In *Swimming and Flying in Nature, Vol. 2*, ed. Y-T Wu, C. J. Brokaw, C. Brennen pp. 763–81 New York/London: Plenum.

Pringle, J. W. S. 1957. *Insect Flight*. Cambridge Univ. Press

Osborne, M. F. M. 1951. Aerodynamics of flapping flight with applications to in-

sects. *J. Exp. Biol.* 28:221–45
Vogel, S. 1966. Flight in *Drosophila*. I. Flight performance of tethered flies. *J. Exp. Biol.* 44:567–78
Vogel, S. 1967a. Flight in *Drosophila*. II. Variations in stroke parameters and wing contour. *J. Exp. Biol.* 46:383–92
Vogel, S. 1967b. Flight in *Drosophila*. III. Aerodynamic characteristics of fly wings and wing models. *J. Exp. Biol.* 46:431–43
Wagner, H. 1925. Uber die Entstehung des dynamischen Auftriebes von Tragflügeln. *Z. Angew. Math. Mech.* 5:17–35
Walker, G. T. 1925. The flapping flight of birds. *J. R. Aero. Soc.* 29:590–94
Walker, G. T. 1927. The flapping flight of birds II. *J. R. Aero. Soc.* 31:337–42
Wood, T. 1970. A study of the instantaneous air velocities in a plane behind the wings of certain *Diptera* flying in a wind tunnel. *J. Exp. Biol.* 52:17–25
Weis-Fogh, T. 1956. Biology and physics of locust flight. II. Flight performance of the desert locust (*Schistocerca Gregaria*). *Philos. Trans. R. Soc. London Ser. B* 239:459–510
Weis-Fogh, T. 1973. Quick estimates of flight fitness in hovering animals including novel mechanisms for lift production. *J. Exp. Biol.* 59:169–230
Weis-Fogh, T. 1975. Flapping flight and power in birds and insects, conventional and novel mechanisms. In *Swimming and Flying in Nature, Vol. 2*, ed. Y-T. Wu, C. J. Brokaw, C. Brennen, pp. 729–62. New York/London: Plenum
Weis-Fogh, T. 1977. Dimensional analysis of hovering flight. In *Scale Effects in Animal Locomotion*, ed. T. J. Pedley, pp. 405–20. New York/London/San Francisco: Academic
Weis-Fogh, T., Jensen, M. 1956. Biology and physics of locust flight. I. Basic principles in insect flight. A critical review. *Philos. Trans. R. Soc. London Ser. B* 239:415–58

*Ann. Rev. Fluid Mech. 1981. 13:351–78*


# FIELD-FLOW FRACTIONATION (POLARIZATION CHROMATOGRAPHY)

*8180

*E. N. Lightfoot, A. S. Chiang, and P. T. Noble*
Department of Chemical Engineering, University of Wisconsin, Madison, Wisconsin 53706

## 1 INTRODUCTION

*Field-flow fractionation* is a term that has come to denote a narrow class of separations processes dependent to an unusual degree on nonuniform fluid motion. We shall be primarily concerned in this review with describing the nature of field-flow fractionation, or *polarization chromatography* as we prefer to call it, but we begin on a more general note. We wish here to orient the reader by introducing a morphological framework for the description of separations processes, by showing the somewhat specialized position of polarization chromatography within this framework, and, finally, by defining some concepts essential to understanding the discussions of later sections.

For our purposes it is most convenient to base discussion on the evolution of a differential feed pulse in a space-time continuum; such a process is in turn most easily visualized in two spatial dimensions as shown in Figure 1.

This figure is drawn for two solutes, species $i$ and $j$, introduced at zero time to the point $(x_0, y_0)$, and the heavy black lines show evolutions of the *centers of mass* of these two species. Closed surfaces, or *f-envelopes*, indicated by the lighter lines, are built from closed curves drawn at various times as shown. Each such curve is a line of constant solute concentration enclosing the same fixed fraction of solute. For example, the f-envelopes shown may be considered to enclose 95% of each species at all times. Then as the f-envelopes cease to overlap, here at about $t_3$,



0066-4189/81/0115-0351$01.00

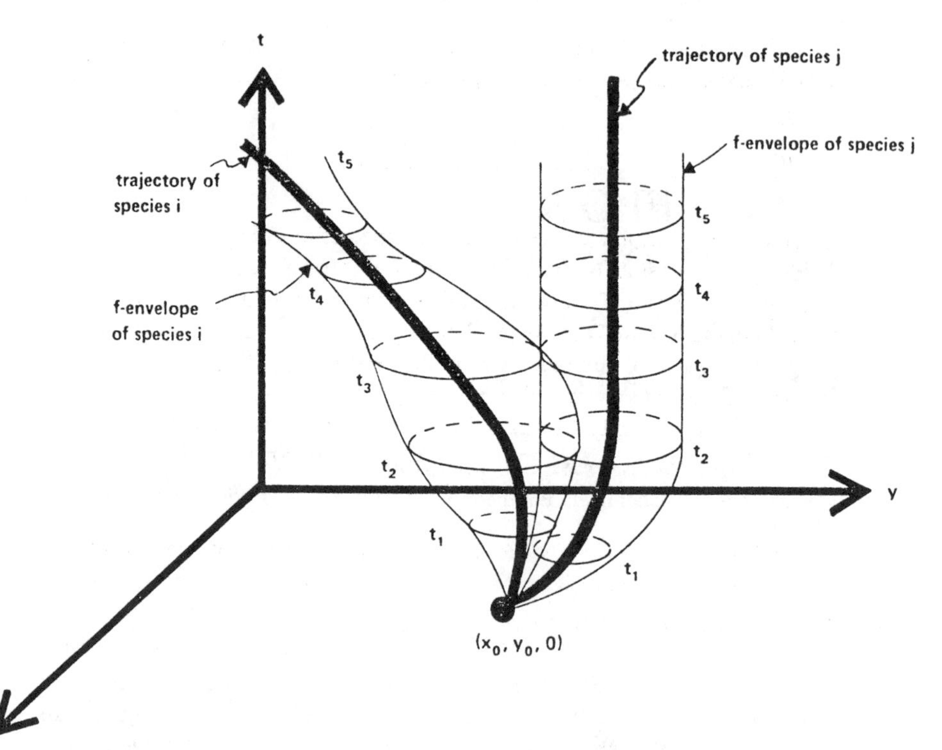


*Figure 1* Trajectories and f-envelopes for separating species *i* and *j* from a common differential feed pulse (Lee et al 1977).

95% of each solute can be obtained, with no more than 5% contamination by the other, simply by collecting each from the appropriate spatial region.

In practice it is not possible to concentrate feed solutes at a point, but if the "lateral" dimensions of the f-envelope at the time of solute withdrawal are large compared to those of the feed, the latter may often be considered as vanishingly small. Such processes are known as "chromatographic" separations, and Figure 1 may be considered to represent a rather general form of *two-dimensional chromatography*[1].

[1]This term reflects early use of this technique to fractionate natural pigments, for example from leaf extracts. These separations were made in glass columns with packings of white granular materials upon which the separation developed with the formation of brightly colored bands.

If separation takes place in a true continuum it is usually possible to describe the transport of each species by an equation of the form

$$\underbrace{\frac{\partial c_i}{\partial t} + (\bar{\mathbf{v}}_i \cdot \nabla c_i)}_{(1)} - \underbrace{\mathfrak{D}_{im}\nabla^2 c_i}_{(2)} + \underbrace{c_i(\nabla \cdot \bar{\mathbf{v}}_i)}_{(3)} = 0. \qquad (1)$$

Here $c_i$ is local molar concentration of species $i$, and $\bar{\mathbf{v}}_i$ is the observable velocity of $i$ resulting from directed fields or forces[2] and bulk fluid motion; $\mathfrak{D}_{im}$ is the effective mass diffusivity of $i$ through the mixture of species present, and $\nabla$ is the gradient operator; $t$ is time.

The trajectory of the center of mass depends only upon *term 1* in the above equation, as averaged over the system volume: the center of mass moves at unit velocity through time and at *the mass average velocity*

$$\langle \bar{\mathbf{v}}_i \rangle = \int_\tau c_i \bar{\mathbf{v}}_i \, d\tau \Big/ \int_\tau c_i \, d\tau \qquad (3)$$

through space. Here $\tau$ is system volume. Calculation of the trajectories of the centers of mass is the simplest and also most important task in describing a separation. We are normally only interested in processes in which the centers of mass separate: so-called first-moment separations. Polarization chromatography is of this type.

The extent of separation depends strongly on the shapes of the f-envelopes, and these are determined by *terms 2 and 3*. These two are, however, quite different in character.

*Term 2*, which describes such dispersive effects as Brownian motion, always leads to expansion of the envelopes. In the absence of confining walls and changes in $\langle \bar{\mathbf{v}}_i \rangle$ both dispersion and *resolution* (see Lee et al 1977) increase with the square root of time or distance traveled by the center of mass; this is characteristic of simple chromatographic separation, including field-flow fractionation.

*Term 3* describes what is known as *focusing* or *anti-focusing*: a contraction or expansion of the f-envelopes resulting from *spatial gradients* in $\bar{\mathbf{v}}_i$. Thus if $(\nabla \cdot \bar{\mathbf{v}}_i) < 0$ the f-envelopes for species $i$ contract, and the scale of

[2] One such force of particular interest to us is that producing *electrophoresis*, the migration of charged solute species—ions–in an electric field:

$$\bar{\mathbf{v}}_i^{(\mathrm{el})} = m_i \mathbf{E}, \qquad (2)$$

$$m_i \doteq \mathfrak{D}_{im} \nu_p (\mathcal{F}/\mathcal{R}T).$$

Here $\bar{\mathbf{v}}_i^{(\mathrm{el})}$ is the electrophoretic contribution to species velocity, $m_i$ is the *electrophoretic mobility*, and $\mathbf{E}$ is the electric field. Other symbols are defined immediately after Equation (5) of the next section.

spatial distribution of $i$ is decreased; species $i$ is said to be focused, and in general separation is enhanced.

Such focusing can be produced by *programming* in space: changing the bulk flow velocity, or strengths of force fields acting on solutes, with position. For axial flow in a cylindrical system such as that shown in Figure 2 focusing can be produced by *ultra-filtration*: movement of solvent outward across the confining walls without corresponding loss of solute. Success here requires that the wall be selectively soluble to the various species present, e.g. *permselective*. Temporal programming or changing $\bar{\mathbf{v}}_i$ with time is also useful and has an effect in the time domain very similar to focusing.

Most chromatographic operations, and all that we shall consider here, take place via such predominantly axial motion in a cylindrical system and for most purposes can be considered *one-dimensional*. For these, Equation (1) may be contracted by averaging all terms over the interior cross section or *lumen* of the flow system, to obtain

$$\frac{\partial \bar{c}_i}{\partial t} + \langle \bar{v}_{iz} \rangle \frac{\partial \bar{c}_i}{\partial z} - \varepsilon_{im} \frac{\partial^2 \bar{c}_i}{\partial z^2} + \bar{c}_i \frac{\partial \langle \bar{v}_{iz} \rangle}{\partial z} = 0 \tag{4}$$

where $\bar{c}_i$ and $\langle \bar{v}_{iz} \rangle$ are now averaged quantities and $\varepsilon_{im}$ is a *dispersion coefficient* which is process as well as state dependent (see Lee et al 1977); $z$ is the direction of the primary flow.

Equation (4) is the description normally used by chromatographers, and the primary concern is to produce as large differences in $\langle \bar{v}_{iz} \rangle$ as possible between the various species present. This is normally accomplished by partially filling the lumen of the column with a stationary granular *packing* exhibiting selective affinity for the solutes to be separated. The solutes tend to distribute rapidly between the fixed and moving phases, and those most strongly attracted by the solid move most slowly through the column. Polarization chromatography is unique in not using a packing.

A particularly exciting recent development, called *affinity chromatography*, uses highly species-specific sorbent species, or *ligands*, for retarding individual solutes. This technique is not directly available to polarization chromatography as no packing is used in this process. However, selected solute species can be reversibly bound, or *tagged*, with ligands to change their properties in the moving phase, and to produce a result much like that of affinity chromatography.

We now turn our attention to polarization chromatography itself. The reader desiring a more extensive background is referred to Giddings (1966), Lee & Lightfoot (1976), and Lee et al (1977).

# 2 POLARIZATION CHROMATOGRAPHY

## *Nature*

Field-flow fractionation (FFF), or polarization chromatography[3] (PC), is a single-phase analogue of conventional chromatography in which retardation is produced primarily by concentration polarization within a nonuniform flow field. Its basic nature can perhaps best be grasped by considering a specific example, and we choose here retardation of a charged protein by *electrical field-flow fractionation* or *electropolarization chromatography* (EPC).

In this process, described schematically in Figure 2, a protein is concentrated into a lens-shaped region adjacent to the inner wall of a permselective hollow fiber by a balance between electrophoretic migration and concentration diffusion. This is indicated by the shading in the top view of Figure 2*a*. The protein is also moved axially by convection, in the suspending buffer solution which is pumped down the fiber lumen, in the $z$-direction as shown. The axial-velocity profile is nonuniform, and is low in the region near the wall where the protein is concentrated; hence, mean axial velocity of the protein is less than that of the suspending buffer.

Chromatographic separation is obtained by introducing short feed pulses, containing mixtures of proteins that are polarized, hence retarded, to differing degrees.

The apparatus consists essentially of a permselective hollow fiber filled and surrounded by a suitable buffer and positioned in a transverse electric field. Some means of pumping buffer through the fiber lumen, of introducing feed pulses at the inlet, and of monitoring or separating solutes passing through the outlet, are also necessary. The fiber wall must be almost completely impermeable to protein, but at the same time it must offer only a modest resistance to small ions whose passage is essential to maintaining the intra-luminal field. As a practical matter, it should also have only a small hydraulic permeability to water, as ultra-filtration across the fiber wall can have a major disturbing effect on separation.

This simple example illustrates most of the essential features of field-flow fractionation, detailed immediately below:

1. Polarization within a single phase by a balance between some form of migration and either Brownian motion or concentration diffusion. The dynamic polarizations characteristic of FFF or PC are

[3] The reasons for using two terms for polarization processes are historical, and are discussed in Section 3, Part 2.

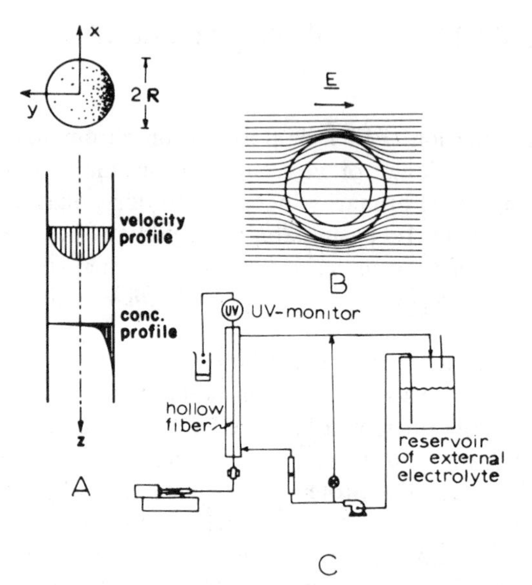


*Figure 2* Hollow-fiber EPC: (*A*) coordinate system, convection-velocity profile, and equilibrium-concentration profile in the hollow fiber, with the dotted region representing the lens-shaped concentrated protein region; (*B*) distortion of the electrical force lines due to the presence of an isotropic cylindrical duct; (*C*) sketch of the experimental setup.

contrasted in Figure 3 with the equilibrium-dominated polarizations of more conventional chromatography.

2. Nonuniform flow perpendicular to the direction of polarization. Three means of producing such nonuniform flows are shown in Figure 3.
3. A polarizing barrier impermeable to the solutes being separated but not to the polarizing force.

   In the case of EPC, mass must be transferred across this barrier to permit transmission of the polarizing force. This is undesirable and can be avoided in some other types of FFF.
4. Rigid boundaries accurately defining the polarization and flow fields.

   The hollow fiber of circular cross section shown in Figure 2 offers the advantage of automatically maintaining its desired shape provided only that luminal pressures exceed those outside the fiber.

   Close-spaced parallel membranes with inert side walls confining the flow are widely used, and annular systems consisting of coaxial cylindrical surfaces are also promising. Both are potentially more efficient than the hollow fiber in permitting use of more solute per unit volume at equal retardation. However, both present more

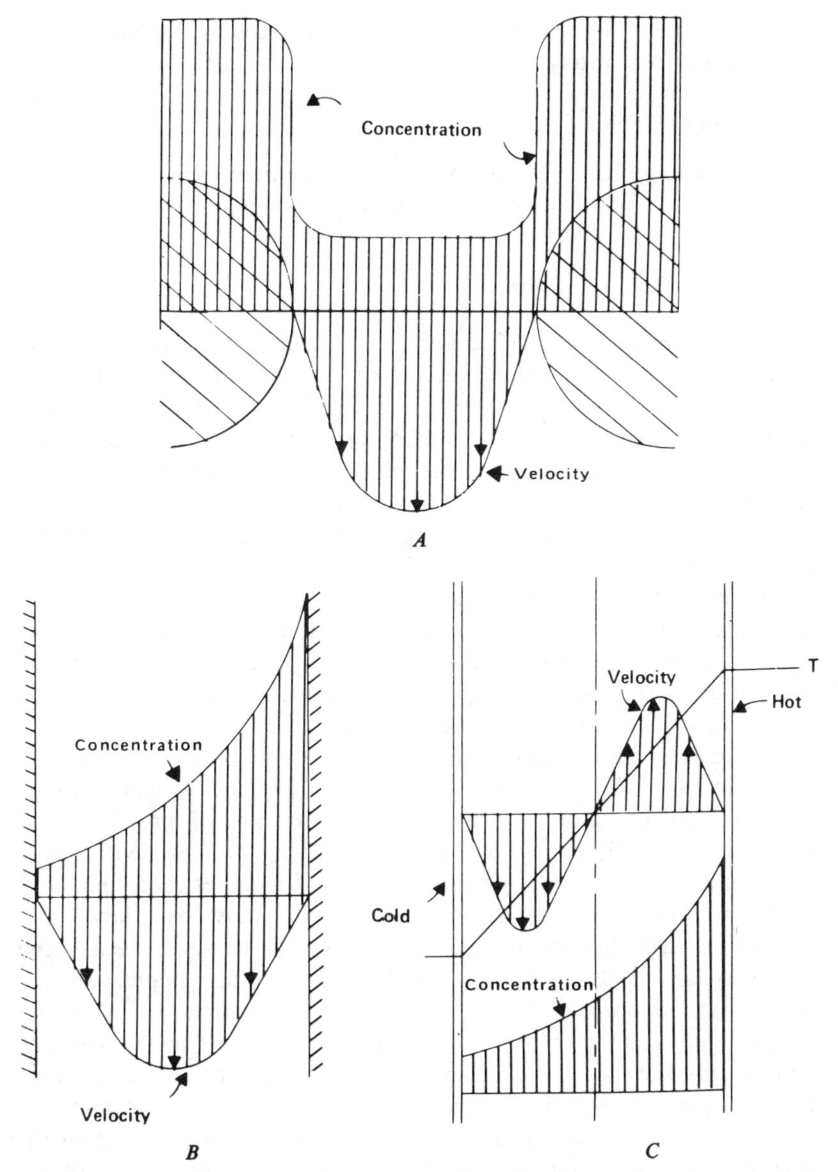


*Figure 3* Convection-augmented separations: (*A*) forced-convection thermodynamic polarization separation; (*B*) forced-convection dynamic polarization separation; (*C*) free-convection dynamic polarization separation (Lee et al 1977).

severe manifolding problems, and slit-flow geometries are hard to maintain accurately in practice: any nonuniformities in spacing can be shown to cause marked axial dispersion of solute via a generalization of Taylor diffusion.

The above four characteristics provide a convenient basis of classifying FFF processes, which we shall follow in this review.

## *Historical Development*

OLD RELATED PROCESSES The concept of combining intraphase concentration polarization with nonuniform convection as a separations technique is an old one, and it does not seem feasible to seek its true origin. There are, however, a number of well-known processes worth citing to put the purely chromatographic applications, of primary interest here, in proper perspective.

Among the oldest and most important are settling operations for removing particulates from flowing fluids and to classify particulates on the basis of size. In the simplest situations, solids are deposited by gravitational settling from fluids flowing in horizontal channels. Many variants exist, using both gravitational and centrifugal fields, in the mineral-dressing industry (Taggert 1953), and these are related to some of the particulate separations discussed below.

The thermo-gravitational Clausius-Dickel columns (see Grodzka & Facemire 1977), the diffusional analogue of Kirkwood & Brown (1952), and electro-decantation systems for concentration of proteins (Bier 1967) are all closely related to FFF. These, along with sweep diffusion, once seriously considered for fractionation of isotopes (Shachter et al 1965), are used on occasion but are for the most part laboratory curiosities. Centrifugally induced polarization is, however, being used on a very large scale for the fractionation of uranium isotopes, as is the related nozzle process. These differ from FFF primarily in *process morphology*, i.e. in the pattern of solute concentrations in a space-time continuum. They may also differ in the *magnitudes* of key parameters, for example solute and system size. All could logically be included under the general heading of FFF, however, in that they are fractionations produced by interaction of a polarizing field with a nonuniform flow.

PROCESSES OF DIRECT INTEREST In practice the terms FFF and PC have become associated with the *chromatographic mode* of operation: transient fractionations of a spatially concentrated feed mixture in which the centers of mass of the various species move in the same direction but at different speeds. We prefer the term polarization chromatography because it includes just those chromatographic operations in which retardation results from concentration polarization.

J. C. Giddings was the first to suggest such polarization-induced chromatographic separations, in 1966, and similar suggestions have since been made independently by others (Berg & Purcell 1967a, Lee et al 1974). The first implementation appears to have been the retardation of small polystyrene beads via thermal diffusion in both coiled tubes and rectangular channels by Thompson et al (1967). Berg & Purcell (1967a) developed a model for FFF starting from equilibrium sedimentation theory in 1967 and quickly confirmed their predictions experimentally (Berg, Stewart & Purcell 1967, Berg & Purcell 1967b). An approximate nonequilibrium theory was presented by Giddings in 1968, which enabled prediction of retention and, to a lesser extent, dispersion. Since that time, Giddings and his associates have produced a large number of useful papers introducing new polarizing forces, techniques, and mathematical analyses. These are listed in Table 1, along with the contributions of others. A review of the early FFF work can be found in Grushka et al (1973) and, in particular for Giddings' work, in Giddings et al (1976e).

Kesner et al (1976) were the first to demonstrate the use of electrophoretic forces, which show particular promise for the fractionation of proteins.

Lee et al (1974) first showed that the concentration polarization accompanying ultra-filtration of proteins could be utilized for chromatographic separations. Shortly thereafter, the use of hollow fibers for electropolarization chromatography was introduced by Reis & Lightfoot (1976).

In 1976 Lee & Lightfoot presented the first comprehensive quantitative description of FFF in slit flow. This work provides formally complete descriptions of retention and dispersion under both transient and asymptotic conditions, and it presents both the most complete analytic results available and the simplest to use. It does, however, assume constant transport properties and neglects secondary flows. Other mathematical descriptions for these idealized conditions, by Krishnamurthy & Subramanian (1977) and by Jayaraj & Subramanian (1978), followed shortly. In 1976 and 1978 Reis et al developed a corresponding description for dispersion in the much more formidable system of hollow fibers with planar fields.

Takahashi & Gill (1980), in a three-dimensional dispersion analysis, found the inert side walls of rectangular channels to increase dispersion to a remarkable extent, and Shah et al (1979) showed the concentration dependence of transport parameters also to have a large effect.

**Table 1.** Field-flow fractionation processes organized in terms of polarization forces

| Polarizing force | Researchers | Date | Techniques |
|---|---|---|---|
| Electrical | Caldwell et al | 1972 | Flexible membrane planar channel of proteins 1972 |
| | Reis & Lightfoot | 1976 | Hollow fiber EPC of proteins, programmed field |
| | Giddings et al | 1976f | Rigid membrane, planar channel of proteins |
| Flow-field | Lee et al | 1974 | Hollow fibers of dextran, proteins |
| | Lee & Lightfoot | 1976 | Mathematical model gives complete solution, planar channel of proteins |
| | Giddings et al | 1976c | Planar channel of proteins & polystyrene latex beads |
| | Giddings et al | 1977b | Viruses |
| | Giddings et al | 1977d | Use as a dialyzer |
| | Giddings et al | 1978b | Separation of water soluble polymers |
| | Giddings et al | 1978c | Separation of colloidal silica |
| | Giddings et al | 1979b | Programmed flow field |
| | Doshi & Gill | 1979 | Channel with one permeable wall |
| | Liauh | 1979 | Porous glass capillary |
| Centrifugal | Berg & Purcell | 1967b | Experiments with viruses 12000g |
| | Giddings et al | 1974 | Latex particles—1400g programmed rotor & solvent density |
| | Giddings et al | 1975d | Viruses at 400 g |
| Gravitational | Berg et al | 1967 | With polystyrene latex beads and 1 or 2 wall channels (with or without free surfaces) |
| | Giddings | 1978 | Glass beads—steric FFF |
| | Myers & Giddings | 1979 | Glass beads—continuous steric FFF |
| Thermal | Thompson et al | 1967 | Coiled capillaries & planar channels, polystyrene beads |
| | Thompson et al | 1969 | Programmed field |
| | Giddings et al | 1978d | Surface barriers for retentions enhancement |
| | Giddings et al | 1979a | Thermo-gravitational FFF, superposition of axial flow fields |
| Activity Gradient | Giddings et al | 1977a | Concept & proposed design for proteins |

[a] For more information about each process, please refer to Caldwell et al (1979), Gajdos & Brenner (1978), Giddings (1973a, b), Giddings et al (1975b), Gill & Sankarasubramanian (1970), Martin & Giddings (1980), Subramanian et al (1978), Yang et al (1977a, b).

At present, much attention is being given to discrepancies between the predictions of available quantitative models and actual observation. Giddings and his associates have generally found observed retardations to be in reasonably good agreement with expectation for a variety of physical situations (Giddings et al 1976d, f, Myers et al 1974) but dispersion has generally been larger than predicted. Lightfoot and his group, working primarily with electropolarization of proteins, have had a much more complicated experience. Retardation varies markedly with both composition and ionic strength of the buffers used and with the nature of the confining membrane. Reasonable agreement with expectation is typically observed at low field strengths, but at high fields retardation may be either much less than expected (Chiang et al 1979) or much greater (*electro-retention*; Reis & Lightfoot 1976). Dispersion decreases with increase in polarization, as required by available theory, but more slowly than expected.

FFF lends itself readily to modifications such as field programming (Giddings et al 1976b, 1979b, Yang et al 1974, Reis & Lightfoot 1976) and to superposition of fields. It is, for example, reported by Giddings that the free convection in vertically oriented thermal-diffusion polarized columns tends to enhance resolution (Giddings et al 1979a).

Introduction of grooves into the channel wall increases retention, sample size, and selectivity, but at the expense of much increased dispersion (Giddings et al 1978d). Use of large, settling, solute particles, especially in the presence of grooves, allows the wall effects of hydrodynamic exclusion to dominate as in hydrodynamic chromatography (Silebi & McHugh 1979). Giddings & Myers (1978) have labeled this limiting case, steric FFF.

## 3 POLARIZATION MECHANISMS AND THE DIFFUSIONAL BASES FOR FIELD-FLOW FRACTIONATION

The requirements for a quantitative description of FFF are essentially the same as for any other separation process. Thus for a system of $N$ identifiable species the following independent relations must be satisfied simultaneously:

$N$ species-continuity equations or mass balances,

$(N-1)$ translational-diffusion equations or particulate equations of motion,

Necessary initial conditions, boundary conditions, and equations of state.

For orientable molecules or particles one must also satisfy rotational-diffusion equations as discussed by Brenner & Cundiff (1972). These are, however, of limited present interest and will not be further discussed. The use of electromagnetic forces also requires solution of Maxwell's equations, at least in principle, but normally one can get by quite well with simple approximations. In electrical-polarization processes, for example, it is normally sufficient to assume a uniform unidirectional current density throughout the flow field of interest.

The behavior of small suspended particles differs surprisingly little from that of true molecular solutes under conditions of interest to us, and we shall base our discussion primarily on the latter. The motion of particulates will, however, be discussed briefly at the end of Section 3, at the end of the next section.

## *Selective Motion and the Generalized Stefan-Maxwell Equations*

Any mechanism producing net motion of solutes relative to the solution as a whole can be used to produce polarization, and a convenient reliable basis for describing such motion is provided by the generalized Stefan-Maxwell equations (Lightfoot 1974). We shall be primarily interested here in the motion of polarizing solutes, $p$, relative to a major component, the solvent, which we shall designate as $s$. For this situation it is convenient to write

$$cx_s x_p \mathbf{v}_p = cx_p \Bigg\{ \underset{1.}{x_s \mathbf{v}_s} - \mathcal{D}_{ps} \Bigg[ \underset{2.}{\nabla \ln \gamma_p} + \underset{3.}{\left( c\bar{V}_p - M_p \frac{c}{\rho} \right) \frac{\nabla P}{c\mathcal{R}T}} + \underset{4.}{\nu_p \frac{\mathcal{F}}{\mathcal{R}T} \mathbf{E}} + \underset{5.}{k_T \nabla \ln T} - \underset{6.}{(\tilde{\mathbf{g}}_p)_{\text{net}}} \Bigg] - \underset{7.}{\sum_{\substack{j=1 \\ \neq p,s}}^{N} \frac{\mathcal{D}_{ps}}{\mathcal{D}_{pj}} x_j (\mathbf{v}_j - \mathbf{v}_p)} \Bigg\} - \underset{8.}{c\mathcal{D}_{ps} \nabla x_p}. \tag{5}$$

Here

$\mathbf{v}_p, \mathbf{v}_s, \mathbf{v}_i$ = (macroscopically observable) velocities of solute of particular interest, solvent, and other solutes, respectively,

$x_i$ = mole fraction of species,

$\mathcal{D}_{ij}$ = the diffusion coefficient describing interaction of species $i$ and $j$,
$\gamma_i$ = mean thermodynamic activity coefficient of species $i$, calculated for existing concentration and any arbitrary standard temperature and pressure,
$\bar{V}_i$ = partial molal volumes of species $i$,
$M_i$ = molecular weight of species $i$,
$P$ = hydrostatic pressure,
$c$ = total molar concentration or molar density (approximately 55 g-moles per liter for aqueous solutions),
$\rho$ = solution mass density,
$\mathcal{R}$ = international gas constant,
$T$ = absolute temperature,
$\nu_i$ = electrophoretically[4] effective molar charge of species $i$ (effective number of unbalanced protons),
$\mathcal{F}$ = Faraday's constant ($9.652 \times 10^4$ abs-coulombs/g-equivalent),
$\mathbf{E}$ = electrostatic field (e.g. volts/cm),
$k_T$ = thermal diffusion ratio,
$\tilde{\mathbf{g}}_i$ = net body force per mole acting on species $i$, other than electrophoretic forces already considered in term 4,
$\nabla$ = gradient operator denoting magnitude and direction of rates of change of indicated parameters with position.

The mole fraction of solvent $x_s$ will often be very close to unity and under these circumstances need not be written explicitly.

Equation (5) is complex in detail, but it tells a basically simple story: the velocity of polarizing solute $p$ is just the sum of effects numbered 1 through 8, which are described in detail below. We note here that Equation (5) may be condensed to the simpler approximate form

$$\mathbf{N}_p \equiv c x_p \mathbf{v}_p \doteq \underset{\text{A}}{c x_p (\mathbf{v}_s + \mathbf{v}_F)} - \underset{\text{B}}{c \mathcal{D}_{ps} \nabla x_p} \tag{6}$$

which is useful for describing polarization-based chromatographic separations. It is, in fact, the justification for grouping them as a coherent class of separations. In writing Equation (6) we have

1. omitted explicit inclusion of solvent mole fraction $x_s$, which is normally close to unity,
2. noted that terms 2 through 6 are all multiplied by $x_p$ and are themselves not strongly dependent upon it.

[4] For a macromolecule such as a protein, $p$, the electrophoretic mobility $m_p = \nu_p \mathcal{D}_{ps} / \mathcal{R}T$.

We therefore choose to define

$$\mathbf{v}_F \equiv -\mathcal{D}_{ps}\left[\nabla \ln \gamma_p + \left(\frac{\bar{V}_p}{\bar{V}_m} - \frac{M_p}{M_m}\right)\frac{\nabla P}{c\mathcal{R}T} + \left(\frac{\nu_p \mathcal{F}}{\mathcal{R}T}\right)\mathbf{E} + k_T \nabla \ln T - (\tilde{\mathbf{g}}_i)_{\text{net}}\right] - \sum_{\substack{j=1 \\ \neq p,s}}^{N} \frac{\mathcal{D}_{ps}}{\mathcal{D}_{pj}} x_j(\mathbf{v}_j - \mathbf{v}_p)$$

$$\equiv \text{field-induced migration velocity of species } p. \tag{7}$$

We shall see below that all polarization-induced chromatographic separations can be explained to a usually satisfactory degree using Equation (6).

1. Polarization proceeds perpendicular to the main flow, i.e. in the $x$-$y$ plane of Figure 2, very nearly to a balance between terms A and B. Term A is the sum of convection and migrational motion, and term B represents Brownian dispersion.
2. Motion of solute $p$ down the column is primarily by convection at very nearly the solvent velocity $v_s$.

Classification of these separations then is according to the nature of contributions to $\mathbf{v}_F$ and the importance of that component of $\mathbf{v}_s$ in the direction of polarization.

We now proceed to identify these contributions:

1. Ultra-filtration or flow of solvent across the confining wall. Polarization is induced by horizontal components of $\mathbf{v}_s$, i.e. by *convection*.
2. Chemo-diffusion produced by spatial gradients in the activity coefficient of the polarizing species. This effect is normally small and is seldom used in single-phase systems. It may, however, be viewed as the primary driving force for *interphase* segregation and thus provides a link between polarization and more conventional chromatography.
3. Pressure diffusion produced by gradients in hydrostatic pressure $P$.
4. Electro-diffusion produced by gradients in electrostatic potential.
5. Thermal diffusion produced by temperature gradients.
6. Forced diffusion produced by a variety of electromagnetic forces. Magnetophoresis and dielectrophoresis produced by gradients in magnetic and electric fields are the simplest, but others also occur. A variety of short-range forces is of interest here in producing *adsorption* phenomena, which may be viewed as a limiting case of polarization separations.

7. Sweep diffusion produced by selective diffusional interaction between the polarizing solutes and a sweeping agent. Similar diffusional interactions between pairs of polarizing solutes are negligible for reasons discussed further below.
8. Brownian dispersion, which always opposes polarizing forces and provides the balance to them needed for effective separation. Where the chemo-diffusion effect is small it may be combined with Brownian dispersion. If the solution may be considered effectively binary,

$$D_{ps}(\nabla x_p + x_p \nabla \ln \gamma_p) \doteq \mathcal{D}_{ps} \nabla x_p \tag{8}$$

where

$$\mathcal{D}_{ps} = (1 + \partial \ln \gamma_p / \partial \ln x_p) D_{ps} \tag{9}$$

is the familiar binary diffusion coefficient normally used by chemists and engineers.[5]

It may now be seen that Equation (5) forms one basis for the classification scheme of Section 2. We shall use it for quantitative description of these processes in Section 3, Part 4.

Particulate suspensions can frequently be treated as true solutions by merely substituting the *number density* of particles $n_p$ for molar concentration and estimating diffusivity of particles relative to the fluid by

$$\mathcal{D}_{pF} = \kappa T / 6\pi\mu R_p f$$

where $\kappa$ is the Boltzmann constant, $\mu$ the solvent viscosity, and $R_p$ is particle radius. The quantity $f$ is a shape factor, needed for nonspherical particles and available in such standard sources as Tanford (1961). Additional corrections are needed for aerosols if their diameter approaches the mean free path of suspending gas (Perry & Chilton 1973). The formal description of thermal-diffusion coefficients and multicomponent interactions with surrounding fluids are available in standard references on kinetic theory, e.g. Hirschfelder et al (1954). Particle-gas interactions are best understood, and this literature is accessible, via Evans et al (1962). Recent experimental results are accessible via TFFF references cited in this review.

[5] $\mathcal{D}_{ps}$ is much more composition-dependent than $D_{ps}$ and, in multicomponent mixtures, can even be negative. Such occurrences form the basis for chemo-diffusion–based polarization processes and, for these, use of Equations (8) and (9) is clearly not advisable.

For small spherical particles initially at rest in a quiescent fluid, motion is described by

$$\sum_j \mathbf{F}_{pj} = \left(\frac{\kappa T}{\mathcal{D}_{pF}}\right)(\mathbf{v}_p - \mathbf{v}_F) + \kappa T \nabla \ln n_p$$

1. 2.

$$+\left(\frac{\rho_p^\circ - \rho_F^\circ}{\rho_F^\circ}\right) V_p \nabla P_\infty + \left(\rho_p^\circ + \frac{1}{2}\rho_F^\circ\right) V_p \frac{d\mathbf{v}_p}{dt}$$

3. 4.

$$+6\sqrt{\pi\mu\rho_F^\circ}\, R_p^2 \int_0^t \frac{\mathbf{v}_p'(\tau)}{\sqrt{t-\tau}}\, d\tau. \tag{10}$$

5.

Here

$\sum_j \mathbf{F}_{pj}$ = the sum of electro-magnetic forces acting on the particles,

$\rho_i^\circ$ = density of species $i$,

$V_p$ = volume of particle,

$P_\infty$ = pressure far enough from any particle to be uninfluenced by its hydrodynamic boundary layer.

Thermal diffusion and selective interaction with individual constituents of the fluid have been ignored in this highly simplified equation. The individual terms represent

1. viscous interaction (drag) between particle and fluid (Stokes' law),
2. Brownian motion (concentration diffusion),
3. pressure diffusion,
4. acceleration,
5. Basset forces (effect of past motion on viscous drag).

Terms 4 and 5 are relaxation phenomena and have no counterpart in our continuum description.

Other potentially useful polarizing forces also exist for particles and appear not to have been used for polarization chromatography to date. These include migration of both rigid and deformable particles in nonuniform shear fields (Bird et al 1977) and the Uhlenhopp effect (Bird 1979, Dill & Zimm 1979).

This last is a most unusual effect in that it depends on simultaneous extension and curvature of long flexible molecules in a curvilinear flow field and resultant migration toward the center of rotation. It is a very powerful polarizing force for nucleic acids and could be applied to an FFF morphology with annular geometry.

## *Conservation Relations, Boundary Conditions, and Equations of State*

In principle, one must solve the $N$ nonreactive continuity equations,

$$\partial c_i/\partial t + (\nabla \cdot c x_i \mathbf{v}_i) = 0, \tag{11}$$

for an $N$-species mixture simultaneously along with necessary boundary conditions, constraints, and equations of state. This has been done for electropolarization chromatography by Shah et al, but even these authors introduced a large number of simplifications, some of doubtful validity.

It is our belief that careful multicomponent analyses are required for those processes in which mass is transferred across the polarization barriers, e.g. ultra-filtration, electrophoresis, and sweep-diffusion–induced separations. Such analyses should, in fact, be given a high priority.

In practice, however, only the polarizing solutes are normally considered explicitly, and their motion is typically described by pseudo-binary expressions in which interactions between the polarizing species are neglected. Relative velocities of polarizing species are normally small enough that neglect of such interactions appears reasonable. Furthermore, the polarizing field is effectively perpendicular to the principal flow direction ($z$-direction in Figure 2) because of the large length-to-width aspect ratios required for successful operation. Finally, in all quantitative analyses except those of Shah et al (1979) for EFFF and Giddings et al (1976a) for TFFF, the polarizing fields of Equation (5), terms 1 through 7, have been assumed position-independent. This is not necessary and is probably not advisable in view of the large changes of physical properties characteristic of the polarization zone. In electrophoretic polarization of proteins, for example, both the effective protein charge $\nu_p$ and the solution conductivity, which determines $\mathbf{E}$ to a large extent, are quite composition dependent. All of these assumptions are, however, necessary in order to obtain the closed-form descriptions favored by most investigators to date.

Under these simplified conditions all forms of FFF are described by the same partial-differential equation,

$$\begin{aligned}\frac{\partial c_p}{\partial t} \doteq & -v_m \frac{\partial c_p}{\partial z} + \mathcal{D}_{pm} \frac{\partial^2 c_p}{\partial z^2} \\ & - \frac{\partial}{\partial x} c_p v_x + \frac{\partial}{\partial x} \mathcal{D}_{pm} \frac{\partial c_p}{\partial x} \\ & - \frac{\partial}{\partial y} c_p v_y + \frac{\partial}{\partial y} \mathcal{D}_{pm} \frac{\partial c_p}{\partial y}\end{aligned} \tag{12}$$

where

$c_p$ = molar concentration of polarizing solute,

$v_m = v_m(x, y)$ = axially directed velocity of the fluid mixture as a whole,

$\mathbf{v} = \delta_x v_x + \delta_y v_y$ = the $x$ and $y$ components of $\mathbf{v}_s + \mathbf{v}_F$,

$\mathcal{D}_{pm}$ = effective binary diffusivity of polarizing solute through the fluid mixture $\doteq D_{ps}$.

The underlined term describing diffusion in the primary flow direction is seldom important. The bulk of axial dispersion takes place through interaction of the nonuniform axial velocity $v_m$ and lateral diffusion, a type of convective dispersion quite similar to Taylor dispersion.

In most modeling efforts the variation of the polarization velocity $\mathbf{v}$ and of $\mathcal{D}_{pm}$ in the $x$ and $y$ directions is also neglected, but this degree of simplification can lead to serious errors.

Both initial and boundary conditions are of critical importance, from a physical as well as mathematical point of view.

It is normally assumed that solute enters the polarization zone as a sharp pulse uniformly distributed over the cross section. In this case, at $t=0$,

$$c = (c_0 L)\,\delta(z) \tag{13}$$

where $c_0 L$ is a weighting function providing a mass balance, and $\delta$ is the Dirac impulse function. In practice, the pulse is neither so concentrated nor so uniformly distributed. Retardation of solutes is not always large, and column volumes can be small. Hence, the feed may take up an appreciable fraction of column length.

From a physical standpoint, lateral transients, resulting from the lateral migration of solute from its initially wide distribution, can be important: migration time can be an appreciable fraction of solute residence time. The most complete numerical study of this effect, by Jayaraj & Subramanian (1978) for a planar channel, predicted a skewed axial solute distribution.

End effects are normally of little importance, however, and one can write without serious error

$$c_p,\ \partial c_p/\partial z \to 0 \qquad \text{for} \quad z \to \pm\infty. \tag{14}$$

These boundary conditions considerably simplify the determination of concentration profiles.

Finally, one normally assumes the lateral boundaries to be impenetrable to the polarizing solute:

$$(\mathbf{n} \cdot [\mathbf{v} c_p - \mathcal{D}_{pm} \nabla c_p])|_s = 0. \tag{15}$$

Here **n** is a local unit normal to the lateral bounding surface $s$, and $\nabla$ is the gradient operator.

Equations of state differ for each application, and we can only say here that variations in carrier-solvent viscosity (Hovingh et al 1970), as well as all transport properties, can be important.

## *Macroscopic Descriptions and System Behavior*

The most reliable and convenient quantitative descriptions of FFF are those of Lee & Lightfoot (1976) for slit flow, of Takahashi & Gill (1980) for rectangular channels, and of Reis & Lightfoot (1976) for Poiseuille flow, all obtained from generalizations of the Taylor-Gill-Subramanian dispersion model (Gill & Sankarasubramanian 1970) through using the generalized Sturm-Liouville theory of Ramkrishna & Amundson (1974). Their results can be most conveniently summarized in terms of area-mean compositions and the one-dimensional heat conduction equation,

$$\partial\langle c_p\rangle/\partial t+(\langle v_m\rangle/r_p)\partial c_p/\partial z=\varepsilon_p\partial^2\langle c_p\rangle/\partial z^2 \tag{16}$$

where

$$\langle Q\rangle=\frac{1}{S}\int_s Q\,ds=\text{area-mean value},$$

$$r_p=r_p(t)=\text{retardation coefficient for species } p,$$

$$\varepsilon_p=\varepsilon_p(t)=\text{effective axial diffusivity for species } p.$$

Complete description of transient behavior is provided in the above references.

Both $r_p$ and $\varepsilon_p$ approach asymptotes for strong polarization, which are useful for estimating the potential of polarization-based chromatographic separations. Asymptotic retardations are given by

$$(r_p)_\infty=\text{Pé}/3 \qquad \text{(slits)} \tag{17}$$

$$=\text{Pé}/4 \qquad \text{(tubes)} \tag{18}$$

where

$$\text{Pé}=\frac{v_0B}{\mathscr{D}}=\text{polarization Péclet number} \tag{19}$$

and $B$ is the radius of lumen or the half-width of a planar channel; $v_0$ is the maximum axial velocity. (The ratio $v_0$ to $\langle v\rangle$ equals 2 for the circular lumen or 3/2 for the planar channel.) The solute retardation is expressed by Giddings in terms of retention ratio, $R$:

$$R=\frac{1}{r_p}. \tag{20}$$

In the chromatography literature, the height of an equivalent theoretical plate, $H$, is commonly used to express dispersion. In terms of Giddings' notation, $H$ is calculated by the following equation:

$$H = 4B^2\lambda^2\psi\langle v\rangle R/\mathcal{D} + \frac{2\mathcal{D}}{\langle v\rangle} \tag{21}$$

where

$$\lambda = \frac{1}{2\text{Pé}}, \tag{22}$$

$$\psi \doteq 4(1-6\lambda) \text{ for small } \lambda. \tag{23}$$

The plate height can be related to $\varepsilon_p$ by

$$\frac{HR}{B} = \frac{2\varepsilon_p}{\langle v\rangle B} \tag{24}$$

where the right-hand side is equivalent to a reciprocal, effective, axial Péclet number. For large Pé the dispersion coefficient for both geometries is approximately

$$\varepsilon_p = \frac{8(v_0 B)^2}{\mathcal{D}\text{Pé}^4}\left(1 - \frac{5}{\text{Pé}} + \frac{7}{\text{Pé}^2}\right). \tag{25}$$

These results must be applied cautiously in slit flow, because numerical calculations by Takahashi & Gill (1980) indicate surprisingly large side effects for finite horizontal-aspect ratios.

# 4 EXPERIMENTAL OBSERVATIONS

Because of space limitations, we must content ourselves here with summary comments relating to retention and dispersion for selected polarization processes.

## *Thermal Diffusion*

Only polymeric solutes have been tested to date. These separate well, but as yet only in organic solvents (Myers et al 1974). Cross-channel temperature differences ranged from 50 to 158°C (Giddings et al 1975a). Improvements in technique have reduced separation time from about 50 hours (Thompson et al 1969) to less than one hour (Giddings et al 1978a). Retention and dispersion are predictable when corrections are made for temperature dependence of transport properties (see Figure 4). Reliability of this process permits useful measurements of thermal diffusivities, $\mathcal{D}_{ps}k_T$ in our nomenclature (Giddings et al 1970).

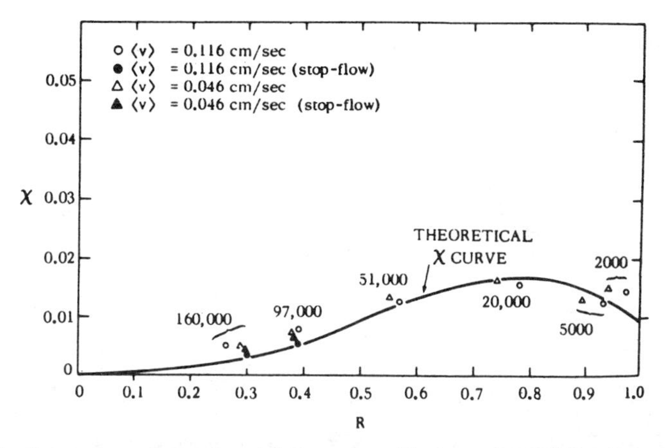


*Figure 4* Comparison of experimental dispersion with theory for TFFF, where $\chi=\lambda^2\psi R$ and is proportional to the first term in the plate height (Equation 21). The temperature difference across the channel was 50°C. The molecular weight of the polystyrene samples is given in the figure. Stop-flow experiments involved stopping the axial flow after sample injection in order to reach a steady-state profile at the entrance (Smith et al 1977)

## *Centrifugal and Gravitational Polarization*

Fractionation of viruses and polymer beads has been achieved with centrifugal fields of 50–500 g (Giddings et al 1975d); larger particles have been separated by gravitational fields in a manner similar to conventional settling operations (Berg, Purcell, & Stewart 1967). As shown in Figure 5, the retardation coefficient $r_p$ is linearly dependent upon field strength. Dispersion is larger than expected despite allowances for end effects, transients, and polydispersity of the feed. Care with channel design is necessary to minimize secondary flows (Giddings et al 1974). This process has been used to estimate the molecular weights of viruses (Giddings et al 1975d).

## *Ultra-Filtration–Induced Polarization*

A wide variety of macromolecules and particles, including proteins, synthetic polymers, and viruses, has been examined using this process. Fractionating power is demonstrated in Figure 6. Retardation is linearly dependent on polarizing fluid velocity, and observed dispersion is greater than predicted. Cross-flow velocities from $10^{-5}$ to $10^{-2}$ cm/sec have been used. Excess dispersion results at least in part from secondary flows. In addition to large-scale circulations, believed to be present, are short-wavelength disturbances (length scales of the order of millimeters)

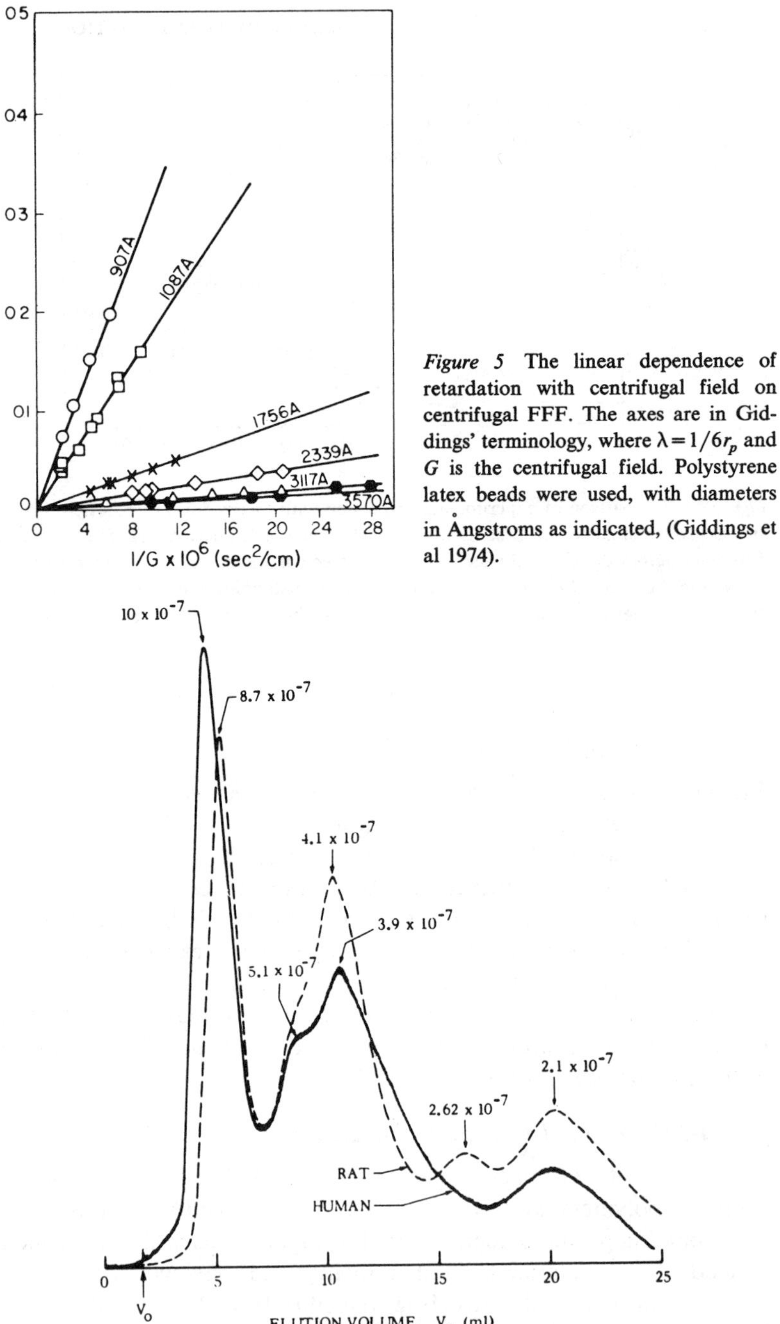


*Figure 5* The linear dependence of retardation with centrifugal field on centrifugal FFF. The axes are in Giddings' terminology, where $\lambda = 1/6r_p$ and $G$ is the centrifugal field. Polystyrene latex beads were used, with diameters in Ångstroms as indicated, (Giddings et al 1974).

*Figure 6* Fractionation of 15 to 20 μl samples human and rat plasmas by a flow FFF column with 1.68 ml void volume. Cross-flow velocities are $2.3 \times 10^{-2}$ cm/min, and calculated diffusion coefficients (cm$^2$/sec) are indicated for the various elution volumes (Giddings et al 1977c).

visible as parallel ridges on the polarizing membrane for colored proteins such as hemoglobin (H. L. Lee, unpublished, Madsen 1977). This technique has been used to determine diffusivities for solutes and suspensions ranging in size from protein molecules to micro-spheres.

## *Electrophoretic Polarization*

This process has been used primarily for retardation of proteins in both tubular and slit geometries using (estimated internal) electric fields up to approximately 20 v/cm. As shown in Figure 7, an approximately linear dependence of retardation on calculated field strength has generally been observed for modest fields (Giddings et al 1976f, Lightfoot et al 1980). This process is, however, inherently much more complex than those cited above, involving relatively long-range interactions of charged hydrophilic colloids with each other, a surrounding electrolyte, and a probably charged polarizing membrane. As yet, neither observed retention nor dispersion is in good agreement with available theory, especially at high fields.

Reliable prediction of macroscopic behavior requires knowing the concentration dependence of effective solute charge and solution conductivity as well as the viscosity and solute diffusivities needed for

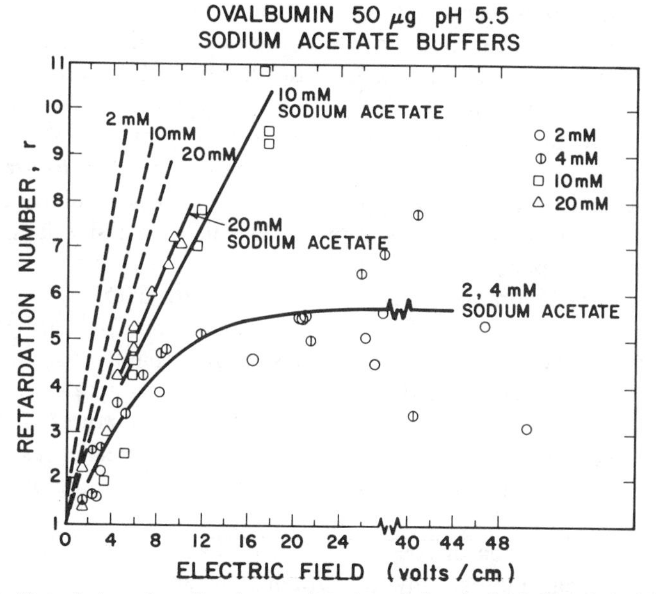


*Figure 7* Retardation of ovalbumin as a function of electric field. The dashed lines are theoretical predictions using Equation (17), and the mobility and diffusivity of the protein are for the limiting case of infinite dilution. The three dashed lines correspond to three values of the ionic strength used in correcting mobility to the proper buffer strength, using Henry's equation (Lightfoot et al 1980).

description of all polarization processes. In addition, one must understand the complex diffusional interactions taking place, both in solution and the "pores" of the membrane; these may introduce electrokinetic effects not specifically considered (Kesner et al 1976).

These effects should be more pronounced at lower ionic strengths and higher fields. Reis & Lightfoot (1976) first reported anomalous but beneficial enhanced retardation under these conditions, which they termed *electro-retention*. This finding has since been confirmed (Chiang et al 1979), and a variety of anomalous behavior patterns has been observed by the Wisconsin group: retentions may be either much higher or much lower than predicted. Dispersion is always larger than expected.

This process is most promising for analytical and, perhaps, commercial (Lightfoot et al 1980) separations. Much work remains, however, even to elucidate the salient characteristics of this flexible but complex process.

## 5 SUMMARY: THE NATURE AND UTILITY OF FFF

The term field-flow fractionation is used to denote polarization-induced chromatographic separations in which the fractionation process is confined to a single phase. Slow solid and interphase diffusional processes are avoided, along with the irreversible adsorption common in more conventional chromatography. In practice, FFF proves most useful for polymers and small particles, and it is closely related to fractional sedimentation of macroscopic particles. It appears particularly promising for such sensitive materials as proteins, which are easily damaged by interphase transport.

It compares favorably with gel-permeation chromatography in terms of resolution, capacity, and molecular-weight range (Giddings et al 1975c, Giddings 1976, 1979), and it has similar characteristics. This is primarily a high capacity, flexible, separation of moderate resolution, well suited to preliminary fractionation of complex mixtures. The bases of separation are variations in such transport properties as electrophoretic mobility and diffusivity, which tend not to be highly species-specific.

The greatest degree of specificity will probably be found in electropolarization chromatography of proteins, where pH, ionic strength, dielectric constant, and other less easily characterized properties affect each solute differently. The use of affinity tagging with highly charged ligands offers the possibility of greatly enhanced specificity and should be explored. EPC is, however, highly complex and not yet well understood.

The polarization process itself tends to reduce convective dispersion, and resolution should be superior to that of conventional chromatography with comparable retardation and with particle diameters equal to polarization channel width. The small dimensions, and hence high resolution, of high-pressure liquid chromatography have, however, yet to be matched. It has also proven difficult to achieve high retardations, but substantial progress is being made with thermal FFF (Giddings et al 1980c, d).

Promising applications in determination of transport properties and for both analytical and preparative scale separations have been reviewed on at least two occasions (Giddings et al 1978e, 1980b; see also Myers et al 1980, Giddings et al 1980a, e). Preliminary experimental results and economic analysis (Lightfoot et al 1980) confirm earlier suggestions (Lightfoot et al 1977) that electropolarization is promising for commercial-scale fractionation of blood plasma and other biologically active proteins.

*Literature Cited*


Berg, H. C., Purcell, E. M. 1967a. A method for separating according to mass a mixture of macromolecules or small particles suspended in a fluid. *Proc. Nat. Acad. Sci.* 58:862–69

Berg, H. C., Purcell, E. M. 1967b. A method for separating according to mass a mixture of macromolecules or small particles suspended in a fluid. III. Experiments in a centrifugal field. *Proc. Nat. Acad. Sci.* 58:1821–28

Berg, H. C. Purcell, E. M., Stewart, W. W. 1967. A method for separating according to mass a mixture of macromolecules or small particles suspended in a fluid, II. experiments in a gravitational field. *Proc. Nat. Acad. Sci.* 58:1286–91

Bier, M., ed. 1967. *Electrophoresis: Theory, Methods, and Applications*, Vol. II. New York: Academic

Bird, R. B. 1979. Use of simple molecular models in the study of the mechanical behavior of solutions of flexible macromolecules. *J. Non-Newtonian Fluid Mech.* 5:1–12

Bird, R. B., Armstrong, R. C., Hassager, O. 1977. *Dynamics of Polymeric Liquids*, Vol. 1. New York: Wiley

Brenner, H., Cundiff, D. W. 1972. Transport mechanics in systems of orientable particles. *J. Colloid Interface Sci.* 41(2):228

Caldwell, K. D., Kesner, L. F., Myers, M. N., Giddings, J. C. 1972. Electrical field-flow fractionation of proteins. *Science* 176:296–98

Caldwell, K. D., Nguyen, T. T., Myers, M. N., Giddings, J. C. 1979. Observations on anomalous retention in steric field-flow fractionation. *Sep. Sci. & Tech.* 14:935–45

Chiang, A. S., Kmiotek, E. H., Langan, S. M., Noble, P. T., Reis, J. F. G., Lightfoot, E. N. 1979. Preliminary experimental survey of hollow-fiber electropolarization chromatography (electrical field-flow fractionation) for protein fractionation. *Sep. Sci. & Tech.* 14(6):453–74

Dill, K. A., Zimm, B. H. 1979. A rheological separator for very large DNA molecules. *Nucleic Acids Res.* 7(3):735–49

Doshi, M. R., Gill, W. N. 1979. Pressure field-flow fractionation or polarization chromatography. *Chem. Engrg. Sci.* 34:725–31

Evans, R. B. III, Watson, G. M., Mason, E. A. 1962. Gaseous diffusion in porous media II. Effect of pressure gradients. *J. Chem. Phys.* 36(7):1894–1902

Gajdos, L. J., Brenner, H. 1978. Field-flow fractionation: extensions to nonspherical particles and wall effects. *Sep. Sci. & Tech.* 13(3):215–40

Giddings, J. C. 1966. A new separation

concept based on a coupling of concentration and flow nonuniformities. *Sep. Sci.* 1(1):123–25

Giddings, J. C. 1968. Nonequilibrium theory of field-flow fractionation. *J. Chem. Phys.* 49(1):81–85

Giddings, J. C. 1973a. Parameters for optimum separations in field-flow fractionation. *Sep. Sci.* 8(5):567–75

Giddings, J. C. 1973b. The conceptual basis of field-flow fractionation. *J. Chem. Ed.* 50(10):667–69

Giddings, J. C. 1976. Field-flow fractionation extending the molecular weightrange of liquid chromatography to one trillion ($10^{12}$). *J. Chrom.* 125:3–16

Giddings, J. C. 1978. Displacement and dispersion of particles of finite size in flow channels with lateral forces. Field-flow fractionation and hydrodynamic chromatography. *Sep. Sci. & Tech.* 13(3):241–54

Giddings, J. C. 1979. Field-flow fractionation of polymers: one-phase chromatography. *Pure & Appl. Chem.* 51:1459–71

Giddings, J. C., Myers, M. N. 1978. Steric field-flow fractionation: a new method for separating 1 to 100 μm particles. *Sep. Sci. & Tech.* 13(8):637–45

Giddings, J. C., Hovingh, M. E., Thompson G. H. 1970. Measurement of thermal diffusion factors by thermal field-flow fractionation. *J. Phys. Chem.* 74:4291–93

Giddings, J. C., Yang, F. J., Myers, M. N. 1974. Sedimentation field-flow fractionation. *Anal. Chem.* 46:1917–24

Giddings, J. C., Smith, L. K., Myers, M. N. 1975a. Thermal field-flow fractionation: Extension to lower molecular weight separations by increasing the liquid temperature range using a pressurized system. *Anal. Chem.* 47:2389–94

Giddings, J. C., Yoon, Y. H., Caldwell, K. D., Myers, M. N., Hovingh, M. E. 1975b. Nonequilibrium plate height for field-flow fractionation in ideal parallel plate columns. *Sep. Sci.* 10(4):447–60

Giddings, J. C., Yoon, Y. H., Myers, M. N. 1975c. Evaluation and comparison of gel permeation chromatography and thermal field-flow fractionation for polymer separations. *Anal. Chem.* 47:126–31

Giddings, J. C., Yang, F. J., Myers, M. N. 1975d. Application of sedimentation field-flow fractionation to biological particles: molecular weights and separation. *Sep. Sci.* 10:133–49

Giddings, J. C., Caldwell, K. D., Myers, M. N. 1976a. Thermal diffusion of polystyrene in eight solvents by an improved thermal field-flow fractionation methodology. *Macromolecules* 9:106–112

Giddings, J. C., Smith, L. K., Myers, M. N. 1976b. Programmed thermal field-flow fractionation. *Anal. Chem.* 48:1587–92

Giddings, J. C., Yang, F. J., Myers, M. N. 1976c. Flow field-flow fractionation: A versatile new separation method. *Science* 193:1244–45

Giddings, J. C., Yang, F. J., Myers, M. N. 1976d. Theoretical and experimental characterization of flow field-flow fractionation. *Anal. Chem.* 48:1126–32

Giddings, J. C., Myers, M. N., Yang, F. J., Smith, L. K. 1976e. Mass analysis of particles and macromolecules by field-flow fractionation. *Colloid and Interface Science* 4:381–98. New York: Academic

Giddings, J. C., Lin, G. C., Myers, M. N. 1976f. Electrical field-flow fractionation in a rigid membrane channel. *Sep. Sci.* 11:553–68

Giddings, J. C., Yang, F. J., Myers, M. N., 1977a. Criteria for concentration field-flow fractionation. *Sep. Sci.* 12(4):381–93

Giddings, J. C., Yang, F. J., Myers, M. N. 1977b. Flow field-flow fractionation: new method for separating, purifying and characterizing the diffusivity of viruses. *J. Virology* 21:131–38

Giddings, J. C., Yang, F. J., Myers, M. N. 1977c. Flow field-flow fractionation as a methodology for protein separation and characterization. *Anal. Biochem.* 81:395–407

Giddings, J. C., Myers, M. N., Yang, F. J. 1977d. The flow field-flow fractionation channel as a versatile pressure dialysis and ultrafiltration cell. *Sep. Sci.* 12:499–510

Giddings, J. C., Myers, M. N., Lin, G. C., Martin, M. 1977e. Polymer analysis and characterization by field-flow fractionation (one-phase chromatography). *J. Chromatogr.* 142:23–38

Giddings, J. C., Martin, M., Myers, M. N. 1978a. High speed polymer separations by thermal field-flow fractionation. *J. Chromatogr.* 158:419–35

Giddings, J. C., Lin, G. C., Myers, M. N. 1978b. Fractionation and size distribution of water soluble polymers by flow field-flow fractionation. *J. Liq. Chromatogr.* 1:1–20

Giddings, J. C., Lin, G. C., Myers, M. N. 1978c. Fractionation and size analysis of colloidal silica by flow field-flow fractionation. *J. Colloid Interface Sci.* 65:67–78

Giddings, J. C., Smith, L. K., Myers, M. N. 1978d. Surface barriers for retention enhancement in field-flow fractionation. *Sep. Sci. & Tech.* 13(4):367–85

Giddings, J. C., Myers, M. N., Moellmer, J. F. 1978e. Fine-particle separation and characterization by field-flow fractionation. *J. Chromatogr.* 149:501–17

Giddings, J. C., Martin, M., Myers, M. N. 1979a. Thermogravitational field-flow fractionation: an elution thermogravitational column. *Sep. Sci. & Tech.* 14:611–43

Giddings, J. C., Caldwell, K. D., Moellmer, J. F., Dickinson, T. H., Myers, M. N., Martin, M. 1979b. Flow programmed field-flow fractionation. *Anal. Chem.* 51:30–33

Giddings, J. C., Graff, K. A., Myers, M. N., Caldwell, K. D. 1980a. Field-flow fractionation: potential role in the analysis of energy related materials. *Sep. Sci. & Tech.* 15

Giddings, J. C., Myers, M. N., Caldwell, K. D., Fisher, S. R. 1980b. Analysis of biological macromolecules and particles by field-flow fractionation. *Methods Biochem. Anal.* 26:79–136

Giddings, J. C., Martin, M., Myers, M. N. 1980c. High resolution polymer separations in a 4-pass hairpin thermal field-flow fractionation column. *J. Polymer Sci.* Submitted

Giddings, J. C., Myers, M. N., Janca, J. 1980d. Retention characteristics of various polymers in thermal field-flow fractionation. *J. Chromatogr.* 186:37

Giddings, J. C., Myers, M. N., Caldwell, K. D., Pav, J. W. 1980e. Steric FFF as a tool for the size characterization of chromatographic supports. *J. Chromatogr.* 185:261

Gill, W. N., Sankarasubramanian, R. 1970. Exact analysis of unsteady convective diffusion. *Proc. R. Soc. London Ser. A* 316:341–50

Grodzka, P. E., Facemire, B. 1977. Clausius-Dickel separation: a new look at an old technique. *Sep. Sci.* 12(2):103–69

Grushka, E., Caldwell, K. D., Myers, M. N. Giddings, J. C. 1973. Field-flow fractionation. *Sep. Purif. Methods* 2:127–51

Hirschfelder, J. O., Curtiss, C. F., Bird, R. B. 1954. *Molecular Theory of Gases and Liquids*. New York: Wiley

Hovingh, M. E., Thompson, G. H., Giddings, J. C. 1970. Column parameters in thermal field-flow fractionation. *Anal. Chem.* 42:195–203

Jayaraj, K., Subramanian, R. S. 1978. On relaxation phenomena in field-flow fractionation. *Sep. Sci. & Tech.* 13(9):791–817

Kesner, L. F., Caldwell, K. D., Myers, M. N., Giddings, J. C. 1976. Performance characteristics of electrical field-flow fractionation in a flexible membrane channel. *Anal. Chem.* 48:1834–39

Kirkwood, J. G., Brown, R. A. 1952. Diffusion-convection. A new method for the fractionation of macromolecules. *J. Am. Chem. Soc.* 74(4):1056–58

Krishnamurthy, S., Subramanian, S. R. 1977. Exact analysis of field-flow fractionation. *Sep. Sci.* 12:347–49

Lee, H. L., Lightfoot, E. N. 1976. Preliminary report on ultrafiltration-induced polarization chromatography—an analog of field-flow fractionation. *Sep. Sci.* 11:417–40

Lee, H. L., Reis, J. F. G., Dohner, J., Lightfoot, E. N. 1974. Single-phase chromatography: solute retardation by ultrafiltration and electrophoresis. *AIChE J.* 20:776–84

Lee, H. L., Lightfoot, E. N., Reis, J. F. G., Waissbluth, M. D. 1977. The systematic description and development of separations processes. In *Recent Developments in Separation Science*, ed. N. N. Li. 3:1–69. CRC Press

Liauh, W. C. 1979. *Mass transfer in laminar flow field with ultrafiltration and its application in polymer fractionation*. PhD thesis. Penn. State Univ., University Park, Pa.

Lightfoot, E. N. 1974. *Transport Phenomena and Living Systems*. New York: Wiley

Lightfoot, E. N., Reis, J. F. G., Bowers, W. F., Lustig, H. M. 1977. Polarization chromatography and allied separation processes. In *Proc. Int. Workshop Technology for Protein Separation and Improvement of Blood Plasma Fractionation*, ed. H. E. Sandberg, pp. 463–89. *NIH #78-1422*

Lightfoot, E. N., Noble, P. T., Chiang, A. S., Czernicki, A. B., Moschella, A., Bowers, W. F. 1980. The potential of electropolarization chromatography for large-scale fractionation of blood plasma. In symposium *Processing and Fractionation of Blood Plasma*, AIChE Society June, 1980

Madsen, R. F. 1977. *Hyperfiltration and Ultrafiltration in Plate-and-Frame Systems*. New York: Elsevier

Martin, M., Giddings, J. C. 1980. Retention and non-equilibrium band broadening for a generalized flow profile in field-flow fractionation. *J. Phys. Chem.* Submitted

Martin, M., Myers, M. N., Giddings, J. C. 1979. Non-equilibrium and polydispersity peak broadening in thermal field-flow fractionation. *J. Liq. Chromatogr.* 2:147–64

Myers, M. N., Giddings, J. C. 1979. A continuous steric FFF device for the size separation of particles. *Powder Tech.* 23:15–20

Myers, M. N., Caldwell, K. D., Giddings, J. C. 1974. A study of retention in thermal field-flow fractionation. *Sep. Sci.* 9:47–70

Myers, M. N., Graff, K. A., Giddings, J. C. 1980. Application of field-flow fractionation to radioactive waste disposal. *Nucl. Chem.* In press

Perry, R. H., Chilton, C. H., eds. 1973. *Chemical Engineer's Handbook.* New York: McGraw-Hill 5th ed.

Ramkrishna, D., Amundson, N. R. 1974. Stirred pots, tubular reactors, and self-adjoint operators. *Chem. Eng. Sci.* 29:1353–61

Reis, J. F. G., Lightfoot, E. N. 1976. Electropolarization chromatography. *AIChE J.* 22(4):779–85

Reis, J. F. G., Ramkrishna, D., Lightfoot, E. N. 1978. Convective mass transfer in the presence of polarizing fields: dispersion in hollow fiber electropolarization chromatography. *AIChE J.* 24:679–86

Shachter, J., Von Halle, E., Hoglund, R. L. 1965. Diffusion separation methods. In *Encyclopedia of Chemical Technology*, ed. E. A. Parolla, 7:91–175. New York: Wiley

Shah, A. B., Reis, J. F. G., Lightfoot, E. N. 1979. Modeling electroretention of proteins during electropolarization chromatography. *Sep. Sci. & Tech.* 14:475–97

Silebi, C. A., McHugh, A. J. 1979. The determination of particle size distributions by hydrodynamic chromatography: an analysis of dispersion and methods for improved signal resolution. *J. Appl. Polym. Sci.* 23(6):1699–1721

Smith, L. K., Myers, M. N., Giddings, J. C. 1977. Peak broadening factors in thermal field-flow fractionation. *Anal. Chem.* 49:1750–56

Subramanian, R. S., Jayaraj, K., Krishnamurthy, S. 1978. Note on the interpretation of some field-flow fractionation experiments. *Sep. Sci. & Tech.* 13(3): 273–76

Taggart, A. F. 1953. *Handbook of Mineral Dressing.* New York: Wiley

Takahashi, T., Gill, W. N. 1980. Hydrodynamic chromatography: three dimensional laminar dispersion in rectangular conduits with transverse flow. *Chem. Eng. Commun.* In press

Tanford, C. 1961. *Physical Chemistry of Macromolecules.* New York: Wiley

Thompson, G. H., Myers, M. N., Giddings, J. C. 1967. An observation of a field-flow fractionation effect with polystyrene samples. *Sep. Sci.* 2(6):797–800

Thompson, G. H., Myers, M. N., Giddings, J. C. 1969. Thermal field-flow fractionation of polystyrene samples. *Anal. Chem.* 41:1219–22

Yang, F. J., Myers, M. N., Giddings, J. C. 1974. Programmed sedimentation field-flow fractionation. *Anal. Chem.* 46: 1924–30

Yang, F. J., Myers, M. N., Giddings, J. C., 1977a. High-resolution particle separations by sedimentation field-flow fractionation. *J. Colloid Interface Sci.* 60: 574–77

Yang, F. J., Myers, M. N., Giddings, J. C. 1977b. Peak shifts and distortion due to solute relaxation in flow field-flow fractionation. *Anal. Chem.* 49:659–62

*Ann. Rev. Fluid Mech. 1981. 13:379–97*


# FRAZIL ICE IN RIVERS AND OCEANS

*8181

*Seelye Martin*
Department of Oceanography, University of Washington, Seattle, Washington 98195

## INTRODUCTION

Frazil ice is the terminology for small discs of ice measuring 1–4 mm in diameter and 1–100 $\mu$m in thickness that form in turbulent, slightly supercooled water (Kivisild 1970). Because of the surface properties of these discs, once they form, they rapidly aggregate together and adhere to foreign material in the water. In rivers these crystals, which form at rapids and other areas of open water where their production rate can be as large as $10^6$ $m^3$ $day^{-1}$, cause serious problems with hydroelectric facilities. These problems include the reduction of available head by 25%, the blocking of turbine intakes, the blockage of hydroelectric reservoirs, and the freezing open of gates.

Because of the economic importance of these problems, river frazil ice is the subject of many papers. Recent reviews of river frazil ice include Michel's (1971) comprehensive survey, Osterkamp (1978), and Ashton (1978). Also, the *Proceedings on Ice Problems of the International Association for Hydraulic Research* (Int. Assoc. Hydraul. Res. 1970, 1975, 1978) contain many excellent papers on frazil ice. Finally, Hobbs (1974) gives a comprehensive review of the literature on the physical properties of ice, much of which is relevant to the frazil crystal.

In the ocean, frazil ice forms during winter both at the ice edge and in the large regions of open water within the pack ice called polynyas, in particular, in near-shore regions where the predominant winds blow the ice away from shore so that a large expanse of seawater at its freezing point is exposed to the cold air. Ice production proceeds rapidly in these polynyas and is accompanied by an outflow of saline water into the ocean. Because ocean frazil ice has become a source of economic concern only in the last decade with the onset of oil and gas development in the Arctic, the literature on ocean frazil ice is sparse and the



0066-4189/81/0115-0379$01.00

work on frazil ice has been almost entirely observational. Previous to the past decade, Zubov (1943) provides the only historical review.

The present review article primarily concentrates on the work of the last fifteen years. Specifically, the next or second section discusses the nucleation and multiplication of the basic frazil disc, and the surface properties of the discs that cause them to sinter together into clusters. The third section then discusses the generation of frazil ice in rivers, the flow of a suspension of the discs, the increase in size of the suspended material through sintering, and the subsequent deposition of the frazil ice under a solid cover. The fourth section is a parallel discussion for ice in the ocean.

## THE CRYSTAL PROPERTIES

### *Supercooling and the Crystal Shape*

In rivers frazil-ice production is a transient phenomenon that occurs in turbulent supercooled water. According to Carstens (1966) frazil ice grows in open water in the following way: first, because of wind and

*Figure 1* Photograph through crossed polaroids of a suspended solution of frazil-ice crystals; the photograph covers 25 mm in the vertical (from Martin & Kauffman 1981).

radiative cooling, the water temperature drops about 0.01–0.1 degrees below the freezing point; then the water abruptly fills with numerous frazil discs; finally, the water temperature following the mass of crystals increases on the order of minutes toward the freezing point.

Numerous other field observations of the formation of river frazil ice such as those of Michel (1971), Arden & Wigle (1973), and Osterkamp (1978) all report that frazil formation begins for supercoolings on the order of 0.1 degrees with the preferred crystal shape being a 1–5 mm diameter circular disc. As an example, Figure 1 shows a photograph taken by Martin & Kauffman (1981) of frazil discs of various orientations suspended in a 10 mm thick water layer, where the basic disc has a 1 mm diameter and a 1–10 $\mu$m thickness.

In his review of the preferred shapes of the ice crystals that form in supercooled water, Hobbs (1974, pp. 572–77) confirms the above observations. He shows that in fresh water that is supercooled by 0.1 to 0.3 degrees, then is nucleated by the introduction of either a very cold metal rod or snow crystals, the preferred growth habit of the resultant crystals is the circular disc. Lesser amounts of needles and semicircular discs also form. Arakawa (1955, also shown in Hobbs 1974, Figure 9.1) shows that the disc develops into a six-pointed stellar dendrite as it grows in radius. This dendritic form also affects the sintering process discussed below.

For a specific example of frazil ice formation, Wigle (1970) and Arden & Wigle (1973) describe a study in the Niagara River, which is 29 km long, has a mean depth of 4 m, and lacks a seasonal ice cover. Here frazil production on cold, clear nights causes an abrupt reduction in the river flow that can be as large as 25 percent. Through use of instrumentation such as underwater lights, collection trays, and thermistor arrays, the authors observed on cold, clear nights that frazil formation began at supercoolings of 0.05 degrees. Underwater observations showed that the frazil ice resembled "a driving snowstorm, as seen through the headlight beams of a moving automobile at night" (Arden & Wigle 1973, p. 1299). As time progressed, the entire river depth became supercooled and the frazil discs began to adhere to the rocks, gravel, and sediment of the river bottom, as well as to each other. When the crystals adhered to sands and sediments, they picked up the adhered material and moved with the river. When the crystals adhered to rocks and gravel, the crystals remained on the bottom then grew in size both from the adherence of additional crystals and from heat transfer to the supercooled water to form what is called "anchor ice." If the anchor ice grew large enough, the rocks and gravel also rose off the bottom and moved with the flow. In the morning following a period of frazil growth, the authors found that ice floes, discolored by entrained material, covered the river surface. The authors feel that the combination of the anchor ice growth, the suspended frazil ice, and ice floating on the surface caused the abrupt reduction in the river flow rate.

## *Nucleation and Multiplication of Frazil Ice*

The way in which frazil nucleates and multiplies in turbulent water has been a source of controversy; namely, do the ice crystals form spontaneously in the water column, or can new crystals only grow from ice already present in the water? Hobbs (1974) shows that for liquid water to freeze or to overcome the free-energy barrier associated with a change of phase to ice, the water, even if contaminated with clays or biological material, requires a supercooling on the order of 10 degrees. Since the observed supercooling in rivers is 0.01 to 0.1 degrees, homogeneous or spontaneous nucleation probably rarely occurs in nature. Therefore, the formation of frazil crystals must proceed from heterogeneous nucleation, or from the introduction of an ice crystal from outside the water.

Osterkamp (1977) reviews various natural seeding mechanisms for the initiation of frazil ice. These include crystals introduced into rivers as snow or frost crystals, water droplets on a $\mu$m scale splashing into the air then freezing and re-entering the water, and growth from natural ice at the riverbanks. As evidence of natural seeding, Osterkamp reports finding numerous ice crystals in air samples taken at night above arctic streams.

Once an ice crystal is introduced to the supercooled water, two experiments on supercooling and turbulence show that there is a rapid multiplication of ice nucleii through what is called "collision breeding." Garabedian & Strickland-Constable (1974), working with 0.5 liters of distilled and de-ionized water, which was supercooled in a cooling bath and stirred at 800 rpm, find that for supercooling of 0.1 degrees 44 crystals formed from the original unbroken seed in five seconds. Müller & Calkins (1978) find similar results for an oscillating grid in a one-liter tank, and use their data to estimate heat transfer to the nucleii. They also show that for experiments beginning with the same initial supercoolings the rate of frazil-ice production increases with the turbulence levels. In both cases, because the multiplication process is microscopic, the actual mechanism is unclear, but apparently the collision of the crystal with both the walls of the tank and other nucleii leads to the rapid crystal multiplication.

Garabedian & Strickland-Constable (1974) also found that for a supercooling of 0.5 degrees, a stirring rate of 900 rpm, and the addition of two 3-mm glass beads to the tank, no nucleation occurred even after stirring for one hour. Müller & Calkins (1978) also observed in their experiment that even for supercoolings of one degree, spontaneous nucleation never occurred; crystal growth took place only when an ice crystal was added to the agitated solution. The conclusion of these authors and Hanley (1978) is that the supercooled water must be seeded for frazil production to begin. In summary, the laboratory studies suggest that for both rivers and oceans, frazil-ice production begins with the addition to turbulent supercooled water of outside ice crystals which rapidly multiply into frazil ice through collision breeding.

### *Sintering*

Unfortunately, once the disc-shaped crystals appear in the supercooled water, they do not remain as separate crystals. Rather they rapidly sinter together into larger groups on the scale of 3–10 mm across called "flocs." Osterkamp (1978, Figure 1*b*) shows from field observations a photograph of these "flocs," and the crystal clusters in Figure 1 show a laboratory example. Once the flocs form, they too sinter together into larger chunks so that the size of the suspended material in the water increases with time. Thus, sintering causes both a rapid increase in the size of the suspended material and, as Martin & Kauffman (1981) show for the ocean case at least, a viscosity that increases as the rate of shear decreases for water-frazil suspensions.

To understand the physics of sintering, Hobbs (1974, Chapter 6) both reviews earlier work and summarizes current work on the surface properties of ice. This work shows that ice crystals, both alone and in contact with each other, adjust their shapes such that their surface free energy tends toward a minimum. For example, experiments show that the point of contact between two ice spheres is thermodynamically unstable. Because a chemical-potential gradient exists between the point of contact and the unstressed ice surface, there is a transfer of material to the contact point so that a neck forms between the two spheres. Extrapolation of the data in Hobbs' Figure 6*b* shows, for two spheres in air with radii of curvature of 10 $\mu$m, the time for the neck to grow to 1/4 of the sphere diameter is 6 s; for 1 $\mu$m the time is $10^{-2}$ s, even though Hobbs states that plastic flow of the ice will affect the results for the 1 $\mu$m spheres. This work implies that for discs with radii of order 1 mm and thicknesses of 1–10 $\mu$m, in contact in the manner shown in Figure 2, bonds between the discs form quickly. Further, because of the

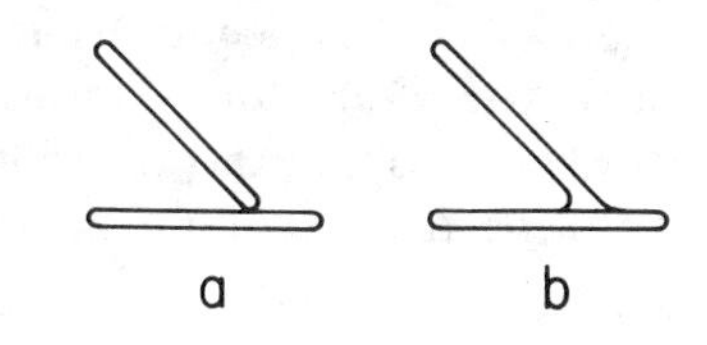


*Figure 2* The sintering together of two frazil discs: (*a*) at the moment of contact; (*b*) a short time later. See text for additional description.

radius-of-curvature decrease at the edge, these sintering times may be even shorter for discs that have evolved into stellar dendrites.

On a longer time scale, the frazil disc is by itself unstable. To minimize the surface free energy, the crystal shape over a period of days tends toward a sphere. Hobbs (1974, Chapter 6.10.1) discusses the analogous evolution of snow crystals, and Michel (1966) shows photographs of evolved frazil ice. This process, which tends to strengthen snow, may also strengthen deposits of frazil ice.

## RIVER FRAZIL ICE

River frazil ice occurs in two kinds of rivers. First, it occurs in those that remain open throughout the year such as the Niagara (Arden & Wigle 1973) and the rivers of southern Sweden (Larsen & Billfalk 1978). Second, in very cold climates frazil ice also forms in rivers where only sections with high velocities and strong rapids remain open, such as the Thjorsa river system in Iceland which feeds the Burfell power project (discussed in Rist 1970, Carstens 1968, and Carstens 1970b) and the La Grande River in Canada (reviewed in Michel 1978). For hydroelectric plants on the Swedish rivers Larsen & Billfalk (1978, page 235) observe that "the most common cause of problems was the occurrence of frazil," which clogged intake trash racks with ice and ice-carried rocks and boulders, one of which weighed 30 kg, and caused in one instance a gate to freeze open, resulting in a runaway turbine.

For both kinds of rivers, hydrologists concerned about power production use a simple classification of frazil ice to predict when it will be hazardous, and an empirical condition to predict as a function of river velocity when frazil ice will form. For the first, Devik (cited in Carstens 1966) divides frazil ice into "active" and "passive" ice. Active frazil, which occurs in supercooled water, adheres to almost all foreign material and causes the serious problems in hydroelectric plants. Passive frazil occurs when the surrounding water warms up to the freezing point; passive frazil is not sticky and, as Carstens (1968) shows, is handled hydraulically as an inverse sediment problem by the proper design of spillways and settling basins.

For the second, Carstens (1970a) gives the empirical condition used to predict frazil formation in Norwegian rivers. For cold air temperatures and a river surface velocity $V$, the condition is as follows: For $1.2 > V > 0.6$ m s$^{-1}$, frazil ice forms and accumulates on the surface; for $V > 1.2$ m s$^{-1}$, the frazil forms and is suspended in the water column. Carstens further adds that this criterion applies over depth variations of a factor of 5, so that it cannot be scaled as a Froude-number criterion. This criterion is in rough agreement with the laboratory experiments of Hanley & Michel (1977), who set 0.24 m $s^{-1}$ as the threshold velocity for frazil formation.

## *Evolution of Frazil Ice*

For both active and passive conditions, sintering causes the frazil discs to stick to each other. This leads to a rapid growth in the size of the suspended ice. Michel (1966, 1971) and Osterkamp (1978) use the following classification with photographic illustrations for the evolution of river frazil ice: 1. the previously described discs; 2. "flocs," which are collections of discs with scales of 5–100 mm; 3. "pans," which accumulate on the surface with diameters of order 1 m and thicknesses of 0.1–0.5 m; and 4. floes, with diameters of 1–30 m and thicknesses of 0.5–5 m. As Carstens (1968) observes, this rapid growth in the size of the suspended particles greatly complicates any modeling of frazil ice.

In an open river the ice is swept downriver where it accumulates against a boom or solid ice cover. As Michel (1966) states, this accumulation can lead to a 10–40 km-per-day growth of the solid ice sheet upstream. This growth continues until the ice sheet reaches the foot of a high-velocity rapid. As reviewed in Michel (1971), rapids serve as "ice factories." For the Russian Angara River described in Michel (1971, p. 66), ice-production rates during winter in regions of open water are of order $1-100\times10^6$ $m^3$ $day^{-1}$. The currents at the foot of the rapids carry this enormous mass of ice under the solid ice cover, where it accumulates in large, irregularly shaped billows. Dean (1977) found from observations in the St. Lawrence River that these billows are a mixture of frazil ice and larger pieces of solid ice. Figures 3 and 4 show examples of these accumulations; Figure 3 shows the frazil distribution in the intake pond of the Gamlebrofoss power plant in Kongsberg, Norway (Tesaker 1975), and Figure 4 is from the La Grande River in northern Quebec (Michel 1978).

The Gamlebrofoss observations in Figure 3 show the frazil-ice deposition pattern in both plan and cross-sectional views for 13 January 1972, where the frazil ice forms in the rapids upstream of the intake pond. Tesaker estimates that the volume of the under-ice deposition is $4\times10^4$ $m^3$, or about 20% of the volume of the pond, and, from the two cross-sectional profiles, that the frazil ice blocked about 60% of the normally ice-free area. Tesaker also observed for both 1971 and 1972 that the frazil ice appeared to be deposited in an inverted sediment fan. Second, the La Grande River observations from the winter of 1972–1973 by Michel & Drouin (cited in Michel 1978) show a 16-km-long under-ice

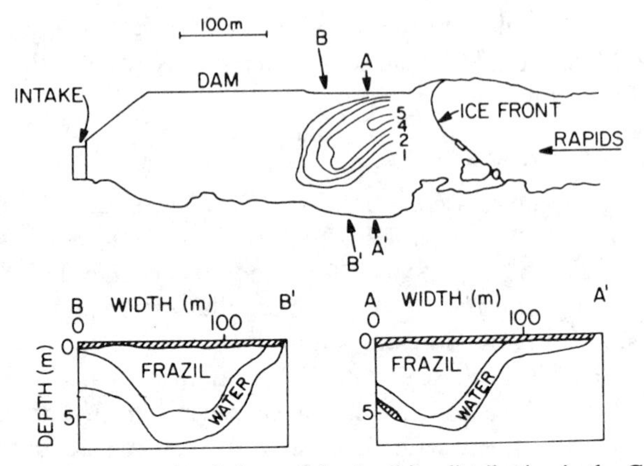


*Figure 3* Plan and cross-sectional views of the frazil-ice distribution in the Gamlebrofoss intake pond on 4 February 1971. Contours on plan view show frazil depth in m; the lines *AA′* and *BB′* indicated on the plan view show location of cross sections (redrawn from Tesaker 1975).

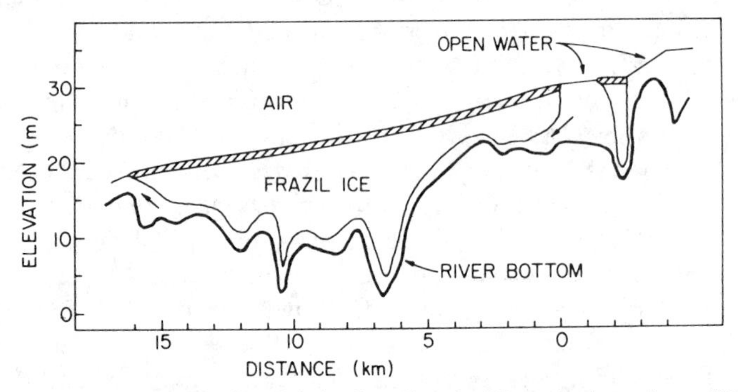


*Figure 4* Frazil-ice distribution in the La Grande River during winter 1972–1973 (redrawn from Michel 1978).

frazil deposition with an approximate volume of $5 \times 10^7$ m$^3$. This deposition began with the formation of a solid ice cover away from the rapids, followed by the accumulation of the frazil ice beneath the solid cover, beginning downstream and progressing upstream. As Figure 3 shows, the frazil ice filled most of the river depth. Michel (1978) further states, although not from the La Grande River observations, that frazil accumulations in low-velocity zones gradually increase the head loss so that in some cases the frazil-producing rapids are flooded out, a solid ice cover forms over the rapids, and the process advances upstream.

When the frazil ice generated in a rapid encounters a solid ice sheet, many investigators (as reviewed in Michel 1978 and Ashton 1978) use a Froude-number criterion to define the conditions under which this ice will flow beneath the solid cover. This criterion depends on the ice particle size and porosity, as well as water depth and velocity, and thickness and porosity of the ice edge. Michel (1978) and Tesaker (1975) have also had some success with the use of a Froude-number condition to predict the maximum thickness of the ice deposition beneath the cover. Finally, Michel (1978) and Tesaker (1975) briefly describe how the combination of solar radiation and increased water flow in the spring leads to the erosion of the deposited billows; they also point out that further observational and theoretical work is needed on frazil-ice decay.

### *A Model Experiment*

Carstens (1968) carried out the only model experiment on frazil-ice flow that is available in the literature. In his study, he built a hydraulic model of the Burfell, Iceland, power project, in which he assumed frazil ice could be treated as a suspension of buoyant sediments. He modeled the frazil behavior in spillways and inverse settling ponds through use of polyethylene shavings with a specific gravity of 0.92, even though, as he states, these shavings do not sinter together. Later, Carstens (1970b) briefly compares the flow in his model studies with that in the completed Burfell project. His photographs of the frazil-ice flow in the actual project show the existence of abrupt transition lines from fluid to solid behavior which did not occur in his model. In their experiments on wave absorption by ocean frazil ice discussed in the next section, Martin & Kauffman (1981) observed similar transitions, which were caused by the sintering-induced nonlinear viscosity of frazil ice. This nonlinear viscosity may also be the cause of the transition lines in the river-ice case, which suggests that a viscous model of frazil-ice flow will be of value to hydrologists.

### *Heat-Transfer Models*

For the calculation of frazil-ice production in rivers, Freysteinsson (1970) used the Rymsha-Donchenko model as described and elaborated on by Dingman et al (1968). Dingman et al (1968) use this model for calculations of the length of ice-free water produced in a river by the discharge from a thermal power plant, and obtain a fair agreement between the model and observation. Freysteinsson (1970) finds that this model gives good agreement with heat-flux measurements on the Thjorsa River System. Larsen (1978), however, carries out a term-by-term critical review of the different models that exist for radiative, evaporative, and turbulent atmospheric heat transfer. From observations made in rapids, which suggest that the frazil production rate is several times greater than predicted, he concludes that the effect of turbulent water flow must also be included in these heat-transfer models. In summary, the field appears to lack a satisfactory theoretical model for the prediction of frazil ice production in a turbulent water flow.

## OCEAN FRAZIL ICE

### *Where It Occurs*

Historically, the terminology for ice-disc formation is slightly different for the ocean than for rivers. For example, Armstrong et al (1966) classify oceanic ice discs into "frazil ice," or a light suspension of the individual platelets, and "grease ice," which is the official World Meteorological Organization terminology for a dense slurry of ice platelets on the ocean surface. The term grease ice is old whaling terminology which refers to the greasy appearance that the slurry gives the surface by damping the capillary waves. In the polar ocean frazil- or grease-ice formation takes place in at least four different situations: in regions of open water called leads and polynyas; at the interface between two fluid layers, each at their freezing point and with different salinities; adjacent to ice shelves and icebergs; and from the drainage of cold dense brine from sea ice into the underlying water.

LEADS AND POLYNYAS Frazil ice forms when cold winds blow across regions of open water that is at its freezing point. Here the cold air and wind waves combine to cool and agitate the surface water such that frazil ice forms. Dunbar & Weeks (1975), Martin et al (1978), and Martin & Kauffman (1981) review frazil-ice formation in regions of open water, where regions with width scales less than 100 m are called "leads" and larger regions are called "polynyas." In leads, the above observations show that the frazil crystals form throughout the open water, then a combination of wind and waves herds the crystals downwind to pile them up to depths of 0.1–0.3 m at the end of the lead. As this process continues, the ice cover advances laterally in the upwind direction until the open-water area is reduced to the point that frazil formation ceases. In large polynyas, a Langmuir circulation herds the frazil ice into streaks parallel to the wind and piles it up downwind to depths of order 1 m. Figure 5 shows a photograph of these Langmuir streaks in a large polynya south of Nome, Alaska, on 5 March 1978. At this time, the air temperature was $-20°C$ and the wind velocity was 15 $m\ s^{-1}$, parallel to the streaks, and blowing toward the large floes. The predominant wavelength in the photograph is 6 m; the floes are about 100 m across. The photograph clearly shows the Langmuir streaks and the piling up of the grease ice to measured depths of order 1 m against the edges of the large floes. Again, the observations suggest that the

*Figure 5* Oblique aerial photograph from 150 m of grease-ice formation off Nome on 5 March 1978; see text for further description (from Martin & Kauffman 1981).

frazil-ice crystals form in the open water, then are herded both into the Langmuir streaks and downwind.

BETWEEN WATERS OF DIFFERENT SALINITIES A second well-documented frazil formation mechanism occurs in the polar ocean when a layer of fresh melt water at its freezing point lies over a layer of seawater at its freezing point. This occurs in the mouths of arctic rivers (McClimans et al 1978), and in the polar summer when fresh melt water accumulates under the pack ice to form inverted melt ponds (Martin & Kauffman 1974). In both cases the temperature difference between the 0°C fresh water and the −1.6°C seawater leads to the production of both supercooled water and frazil ice. For the second case, Martin & Kauffman (1974) show that a 0.2–0.3-m-thick fresh-water ice layer grows during the polar summer under the melting pack ice. Lyons et al (1971) show that this process also contributes to the ice growth *from the bottom* of about $10^2$ km of the Ward Hunt Ice Shelf, which serves as a partial dam across Disraeli Fjord. In this case, fresh glacial melt water accumulates behind the dam, then flows out beneath it over the colder arctic seawater. The heat transfer between the two layers leads to the production of fresh-water ice. As McClimans et al (1978) state, the process is understood for the case of no relative velocities between the two layers, while the nature of this process in a velocity shear remains to be investigated.

ICE SHELVES AND ICEBERGS Another source of oceanic frazil ice occurs in the seawater adjacent to ice shelves and icebergs. Here, frazil ice may form in two ways: first, by direct cooling of the seawater from the cold ice; second, as Foldvik & Kvinge (1974, 1977) show both theoretically and observationally, by the upward movement of a seawater parcel that is at its freezing point. Because of the freezing-point depression with depth, as the parcel rises it becomes supercooled; if nucleation occurs in the rising water then under certain conditions the ice-water buoyancy increases, so that the parcel continues rising. To estimate the amount of ice generated by this process, Robin (1979) shows from Foldvik & Kvinge's data that if a 10-m-thick water layer at its freezing point moves upward at 10 mm $s^{-1}$ over a 100 m elevation change, $10^3$ $m^3$ $yr^{-1}$ of ice is generated per unit-meter width of ice shelf. These calculations suggest that ice shelves contribute to the frazil ice observed around Antarctica.

BRINE DRAINAGE The possibility also exists that the frazil ice observed beneath a solid ice cover forms locally. Lewis & Lake (1971) and Lewis & Milne (1977) suggest that the slow drainage of cold dense brine from the sea ice, which occurs over its growing season, leads to growth of the frazil-ice platelet layers. Observations summarized in Dayton & Martin (1971) and Lewis & Milne (1977) show that this brine leads to the growth under a solid ice sheet of ice stalactites which form around the brine outflow; Martin's (1974) laboratory study of stalactites shows that as much as half of the ice represented by the cold brine flows into the underlying water as frazil crystals, instead of going into stalactite growth. The sparse number of stalactites observed in the field, however, suggests that brine drainage cannot account for the large mass of observed frazil ice.

### *Underwater Observations*

The underwater observations in the Arctic and Antarctic are very similar to the river observations. Dayton et al (1969) observed in McMurdo Sound large billows of frazil platelets measuring 1–4 m in thickness under the solid ice cover (Figure 6). They also observed that on some days in winter the water filled with small ice spicules, and that anchor ice formed down to depths of 33 m both on the sea bottom and on lines hanging in the water. Both the ice spicules and the anchor ice

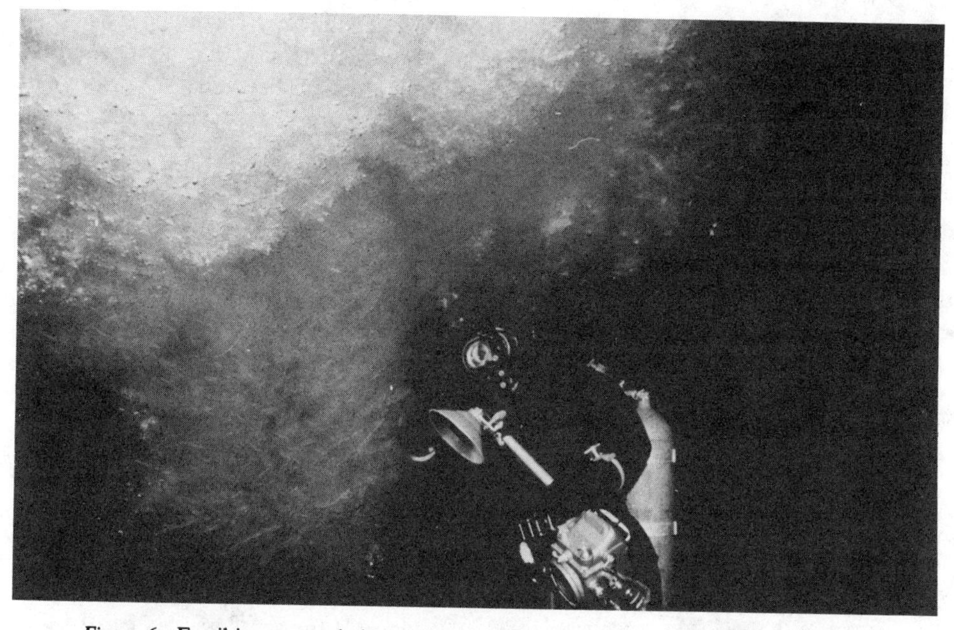

*Figure 6* Frazil-ice accumulation under the pack ice of McMurdo Sound (photograph courtesy Paul Dayton).

appeared suddenly, after the passage of days with no frazil formation. The formation of this frazil layer may be caused by the nearby McMurdo Ice Shelf.

For the sea ice adjacent to the coast at the Russian Antarctic station Mirny, there are also many reports of frazil-ice formation both on the surface in the large polynyas that form behind islands and grounded icebergs by the katabatic winds, and under the ice in large billows. Morecki (1965) describes observations made in 1958 of under-ice frazil billows measuring 1–5 m thick, and the growth of anchor ice down to depths of 5–50 m on lines hanging in the water column under a solid ice cover on different days throughout the winter. He also reports supercooling of the water column on the order of 0.1 degrees and attributes the frazil growth to supercooled water generated at the nearby ice shelves.

Baranov et al (1968) describe similar observations made in 1963. They attribute the frazil growth to the heat sinks provided by the nearby ice shelves and grounded icebergs, and by the large offshore polynyas which opened and closed throughout the winter in response to the katabatic winds. Finally, Cherepanov & Kozlovskii (1973) review previous observations and attribute the growth of the under-ice billows to ice formed in the persistent offshore polynyas which is redistributed by the ocean currents. They state that the under-ice dispersal of frazil depends on local currents, and in regions such as the Molodeshnaya Road where the current velocity is zero, frazil-ice billows are not observed.

In the Arctic, there have also been several recent diving surveys under the ice offshore of Prudhoe Bay, Alaska, in Stefansson Sound. From such surveys, K. Dunton and E. Reimnitz (unpublished, 1980) observed large frazil-ice billows under the solid ice cover at three weeks after freeze-up in both 1978 and 1979. They also observed that the individual frazil crystals in the billows were covered with a fine sediment layer. Because both 1978 and 1979 freeze-up at Prudhoe Bay took place during severe storms, they attribute the billows to frazil ice that formed during the storms, after which the billows froze into place.

All of the frazil ice observations described above can be explained by heat losses to either adjacent polynyas or ice shelves, or by the formation of the billows during freeze-up. Since the field observations (Dayton & Martin 1971, Lewis & Milne 1977) show that the ice volume in the observed stalactites is several orders of magnitude smaller than the ice in the frazil billows, it seems unlikely that the billows form locally from brine drainage. Therefore, the literature suggests that the winter formation of frazil ice in the ocean takes place in the same way as frazil formation in rivers. Namely, the ocean supercools on the order of 0.1 degree by cooling from either a polynya or possibly an ice shelf, then nucleation occurs from introduction of outside ice crystals. The frazil then forms and either is herded by the wind and currents into billows that freeze into place, or is carried under the ice cover by currents.

## *Viscous Properties*

To study the viscosity of frazil-ice slurry, Martin & Kauffman (1981) carried out in a cold room a wave-tank study of wave absorption by grease ice. Figure 7 shows a schematic diagram of the waves propagating into the ice; the surface waves herd the grease ice to the far end of the tank where the radiation stress creates the thickness increase with distance of the grease-ice layer.

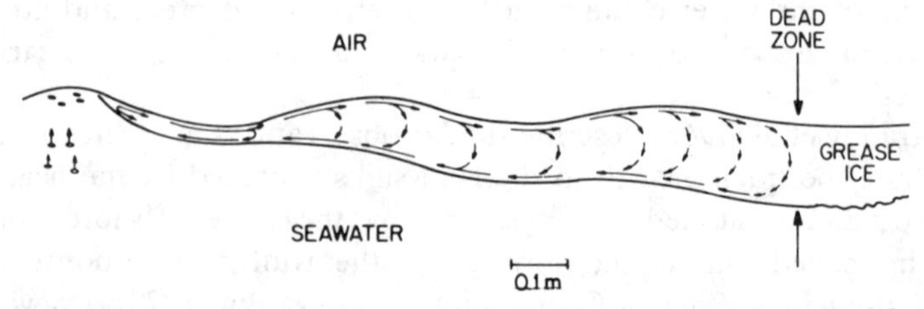


*Figure 7* Schematic drawing of waves propagating into grease ice from the laboratory experiments of Martin & Kauffman (1981); see text for additional discussion.

The figure also shows that the grease ice divides into regions of liquid and solid behavior separated by an abrupt transition zone measuring 5–10 mm wide, which Martin & Kauffman term the "dead zone." Ahead of the dead zone, the waves propagate as heavily damped water waves, where the wave attenuation generates the mean velocities shown by the arrows within the grease ice; behind the dead zone, the waves propagate as elastic waves. Also ahead of the dead zone, when the grease-ice depth is greater than $k^{-1}$ where $k$ is the wavenumber of the incident wave, the wave amplitude decays linearly such that the decay slope $\alpha$ obeys

$$\alpha \cong \tfrac{1}{4}(a_0 k)^2, \tag{1}$$

where $a_0$ is the wave amplitude ahead of the grease ice. Related observations showed that the relative volume of the ice crystals varied from 18–22% at the leading edge to 32–44% at the dead zone, where the largest concentrations in the experiment occur at the dead zone for large values of $a_0$.

Martin & Kauffman explain the wave-decay observations in terms of a yield-stress-viscosity model,

$$\mu = \frac{b^2}{\gamma}, \tag{2}$$

where $\mu$ is the effective dynamic viscosity, $\gamma$ is the scalar shear rate, and $b^2$ is the yield-stress coefficient. From their data,

$$b^2 = \tfrac{1}{4}kS, \tag{3}$$

where $S$ is the radiation stress of the incident wave.

Physically, the cause of this shear-dependent viscosity, which increases as the shear rate decreases, is the combination of sintering and concentration changes. The sintering creates bonds between the crystals and the increase in concentration with radiation stress creates more bonds per unit volume. These bonds, which present resistance to the shear, create the nonlinear viscosity. A similar study by Uzuner & Kennedy (1974) on the response to stress of a field of ice chunks shows that bonds form between the chunks by both sintering and regelation. Their results give a qualitatively similar strength increase with increases in concentration and decreases in the shear rate.

From qualitative experiments with fresh-water frazil slurries, Martin & Kauffman also found that fresh-water crystals tended to form stronger bonds than crystals in salt water. The difference between the two cases is probably due to the absence of a salt-water coating on the fresh-water crystals. One effect of this difference is that the large ice flocs that form under water in fresh water do not appear in the salt-water experiments.

## *Thermal Properties*

Because frazil ice forms in open water and is herded either downwind or into Langmuir streaks, the frazil growth rate is much greater than that of an undisturbed solid ice sheet. Although there are no systematic quantitative measurements of the growth rate, Martin & Kauffman (1981) found in the laboratory that for a large-amplitude wave field and an air temperature of −20°C the equivalent thickness of 0.1 m of solid ice grew in one hour, while for an undisturbed ice cover where the thickness increased by conduction, a 0.1 m ice growth required 24 hours. In the wave field, ice growth took place both in the open water and in the grease ice ahead of the dead zone. In this part of the grease ice, the stirring induced by the wave oscillations and the mean circulation kept the temperature of the ice surface warm. For air temperatures of −20 to −30°C with air blowing over the surface, the ice surface temperature ahead of the dead zone was only $10^{-1}$ to $10^{-2}$ degrees colder than the water temperature below the grease ice. Behind the dead zone, Martin et al (1978) show that the surface temperature decreased slowly with time, and grease ice solidified into the pancake ice. These observations ahead of the dead zone imply that the heat transfer in this region is of the same order as the open-water heat transfer.

## *Oceanic Importance*

The above work implies as Martin & Kauffman (1981) show in the Bering Sea that polynyas where wind or currents periodically sweep away or break open the ice cover and replace it with frazil ice are important regions of ice production. In these polynyas there is the possibility of the growth of the equivalent thickness of 2 m of solid ice in 20 hours; whereas in the high Arctic, 2 m of undisturbed ice growth requires an entire season. Further, K. Aagaard (in preparation) shows that the rapid ice growth in the polynya that forms in the Chukchi Sea off Cape Lisburne and Point Hope, Alaska, is accompanied by an oceanic salt flux that generates a dense outflow along the sea bottom.

Frazil ice also plays a geological and biological role in the ocean. First, K. Dunton and E. Reimnitz (unpublished, 1980) observed in Stefansson Sound, Alaska, that a fine sediment layer coated each deposited frazil crystal. They speculate that frazil ice may serve as both a scouring and sediment-transfer agent. Second, Dayton et al (1969) describe the biological role of frazil ice in McMurdo Sound. They found that the frazil-ice billows serve to shelter biological colonies and that anchor ice both transports species from the sea bottom to the pack-ice bottom, and renders certain shallow-water regions uninhabitable. For example, they found that the sponge colonies do not exist above the maximum depth of anchor-ice formation. These diverse examples suggest that frazil ice is important in determination of the geology, biology, and physical oceanography of the shallow-water regions of the polar oceans.

## SUMMARY

The present review discusses the problems of frazil-ice formation, flow, and deposition in rivers and oceans. In rivers, four research areas require attention. First, a model of frazil-ice production as a function of atmospheric and river-flow parameters needs to be constructed and confirmed in the laboratory and field. Second, the increase in the size and density of the suspended ice aggregates with time and the related viscosity of the suspension should be determined as a function of local shear and concentration. In particular, we need models of ice sintering and deposition under active and passive conditions, and related models for anchor-ice growth from both sintering of additional crystals to the deposited ice and from heat transfer to the surrounding supercooled water. Third, on a longer time scale we need models of the metamorphosis of single ice crystals and the effect of this shape change on the strength of the deposited frazil ice. Finally, we need additional studies on frazil-ice decay from flowing warm water and solar radiation.

Our needs in the ocean are similar. First, we need to clarify the mechanism of frazil-ice production; frazil ice is obviously produced in open water, while the literature suggests that frazil production also occurs at ice shelves and by brine drainage. The latter two mechanisms require field studies to determine their importance; the polynyas require a heat-transfer model (for calculation of the ice-production rate) as well as a related salt-flux model. Finally, the oceanic case also requires studies similar to those described for river ice on frazil-ice viscosity, deposition, and decay.

### ACKNOWLEDGMENTS

The author thanks Professors J. F. Kennedy, B. Michel, and T. E. Osterkamp for their help and encouragement in assembling the references for this paper, Ms. M. Peacock for her great help with the editing and bibliography, and Professor L. Larsen both for reading this work in draft and for organizing a timely conference on frazil-ice observations in Prudhoe Bay, Alaska. The author gratefully acknowledges the support of the Office of Naval Research under Task No. NR307-252,

Contract No. N00014-76-C-0234. Contribution number 1161, Department of Oceanography, University of Washington, and Contribution number 541, Department of Atmospheric Sciences, University of Washington.

*Literature Cited*


Arakawa, K. 1955. The growth of ice crystals in water. *J. Glaciol.* 2:463–68

Arden, R. S., Wigle, T. E. 1973. Dynamics of ice formation in the upper Niagara River. *Int. Symp. Role of Snow and Ice in Hydrol., Banff, Alberta 1972*, 2:1296–1313. Geneva: UNESCO-WMO-IAHS. 1483 pp.

Armstrong, T., Roberts, B., Swithinbank, C. 1966. *Illustrated Glossary of Snow and Ice*. Cambridge: Scott Polar Res. Inst. 60 pp.

Ashton, G. D. 1978. River ice. *Ann. Rev. Fluid Mech.* 10:369–92

Baranov, G. I., Nazintsev, Yu. L., Cherepanov, N. V. 1968. *Tr. Sov. Antarkt. Eksp.* 38. English transl. Formation conditions and certain properties of Antarctic sea ice (according to observations of 1963). *Collected Papers Sov. Antarct. Exp.* 38:74–86

Carstens, T. 1966. Experiments with supercooling and ice formation in flowing water. *Geofys. Publ.* 26, No. 9:1–18

Carstens, T. 1968. Hydraulics of river ice. *Houille Blanche* 23(N.S.):271–84

Carstens, T. 1970a. Heat exchanges and frazil formation. See Int. Assoc. Hydraul. Res. 1970, paper 2.11. 17 pp.

Carstens, T. 1970b. Modelling of ice transport. See Int. Assoc. Hydraul. Res. 1970, paper 4.15. 9 pp.

Cherepanov, N. V., Kozlovskii, A. M. 1973. *Probl. Arktiki Antarkt.* 42:49–58. English transl. Classification of Antarctic sea ice by the conditions of its formation. *Problems of the Arctic and Antarctic* 42:61–73

Dayton, P. K., Martin, S. 1971. Observations of ice stalactities in McMurdo Sound, Antarctica. *J. Geophys. Res.* 76:1595–99

Dayton, P. K., Robillard, G. A., DeVries, A. L. 1969. Anchor ice formation in McMurdo Sound, Antarctica, and its biological effects. *Science* 163:273–74

Dean, A. M. 1977. Remote sensing of accumulated frazil and brash ice in the St. Lawrence River. *Rep. 77-8, Cold Reg. Res. Eng. Lab.*, Hanover, N.H. 19 pp.

Dingman, S. L., Weeks, W. F., Yen, Y. C. 1968. The effects of thermal pollution on river ice conditions. *Water Resource Res.* 4:349–62

Dunbar, M., Weeks, W. F. 1975. Interpretation of young ice forms in the Gulf of St. Lawrence using side-looking airborne radar and infrared imagery. *Rep. No. 337, Cold Reg. Res. Eng. Lab.* Hanover, N.H. 41 pp.

Foldvik, A., Kvinge, T. 1974. Conditional instability of seawater at the freezing point. *Deep-Sea Res.* 21:169–74

Foldvik, A., Kvinge, T. 1977. Thermohaline convection in the vicinity of an ice shelf. In *Polar Oceans*, ed. M. Dunbar, pp. 247–55. Calgary: Arctic Institute of North America. 681 pp.

Freysteinsson, S. 1970. Calculation of frazil ice production. See Int. Assoc. Hydraul. Res. 1970, paper 2.1. 12 pp.

Garabedian, H., Strickland-Constable, R. F. 1974. Collision breeding of ice crystals. *J. Crystal Growth* 22:188–92

Hanley, T. O'D. 1978. Frazil nucleation mechanisms. *J. Glaciol.* 21:581–87

Hanley, T. O'D., Michel, B. 1977. Laboratory formation of border ice and frazil slush. *Can. J. Civ. Eng.* 4:153–60

Hobbs, P. V. 1974. *Ice Physics*. London: Oxford Univ. Press. 837 pp.

Int. Assoc. Hydraul. Res. 1970. *Symp. on Ice and Its Action on Hydraulic Structures, Reykjavik, Iceland*. 64 papers.

Int. Assoc. Hydraul. Res 1975. *Proc. Third Int. Symp. on Ice Problems, Hanover, N.H.* 627 pp.

Int. Assoc. Hydraul. Res. 1978. *Proc. Fifth Int. Symp. on Ice Problems, Lulea, Sweden*. Part 2, 475 pp. Part 3, 284 pp.

Kivisild, H. R. 1970. River and lake ice terminology. See Int. Assoc. Hydraul. Res. 1970, paper 1.0. 14 pp.

Larsen, P. 1978. Thermal regime of ice covered waters. See Int. Assoc. Hydraul. Res. 1978, Part 3:95–117

Larsen, P., Billfalk, L. 1978. Ice problems in Swedish hydro power operation. See Int. Assoc. Hydraul. Res. 1978, Part 2:235–43

Lewis, E. L., Lake, R. A. 1971. Sea ice and supercooled water. *J. Geophys. Res.* 76:5836–41

Lewis, E. L., Milne, A. R. 1977. Un-

derwater sea ice formations. See Foldvik & Kvinge 1977, pp. 239–45
Lyons, J. B., Savin, S. M., Tamburi, A. J. 1971. Basement ice, Ward Hunt Ice Shelf, Ellesmere Island, Canada. *J. Glaciol.* 58:93–100
Martin, S. 1974. Ice stalactites: comparison of a laminar flow theory with experiment. *J. Fluid Mech.* 63:51–79
Martin, S., Kauffman, P. 1974. The evolution of under-ice melt ponds, or double diffusion at the freezing point. *J. Fluid Mech.* 64:507–27
Martin, S., Kauffman, P. 1981. A field and laboratory study of wave damping by grease ice. *J. Glaciol.* (In press)
Martin, S., Kauffman, P., Welander, P. E. 1978. A laboratory study of the dispersion of crude oil within sea ice growth in a wave field. *Proc. 27th Alaska Sci. Conf. Fairbanks, August 4-7, 1976*, 2:261–87.
McClimans, T. A., Steen, J. E., Kjeldgaard, J. H. 1978. Ice formation in fresh water cooled by a more saline underflow. See Int. Assoc. Hydraul. Res. 1978, Part 2: 20–29
Michel, B. 1966. Morphology of frazil ice. In *Physics of Snow and Ice: Proc. Int. Conf. Low Temp. Sci.*, ed. H. Ōura, 1: 119–128. Sapporo: Inst. Low Temp. Sci., Hokkaido Univ. 711 pp.
Michel, B. 1971. Winter regime of rivers and lakes. *Monogr. III-Bla, Cold Reg. Res. Eng. Lab.* Hanover, N.H. 131 pp.
Michel, B. 1978. Ice accumulations at freeze-up or break-up. See Int. Assoc. Hydraul. Res. 1978, Part 2:301–17
Morecki, V. N. 1965. Underwater sea ice. *Probl. Arktiki Antarkt.* 19:32–38. English transl. *Def. Res. Board Can., Transl. T 497 R*, 1968
Müller, A., Calkins, D. J. 1978. Frazil ice formation in turbulent flow. See Int. Assoc. Hydraul. Res. 1978, Part 2:219–34
Osterkamp, T. E. 1977. Frazil-ice nucleation by mass-exchange processes at the air-water interface. *J. Glaciol.* 19:619–25
Osterkamp, T. E. 1978. Frazil ice formation: a review. *J. Hydraul. Div. ASCE* 104:1239–55
Rist, S. 1970. Ice conditions in the Thjorsa river system. See Int. Assoc. Hydraul. Res. 1970, Paper 3.11. 6 pp.
Robin, G. de Q., 1979. Formation, flow and disintegration of ice shelves. *J. Glaciol.* 24:259–72
Tesaker, E. 1975. Accumulation of frazil ice in an intake reservoir. See Int. Assoc. Hydraul. Res. 1975, pp. 25–38
Uzuner, M. S., Kennedy, J. F. 1974. Hydraulics and mechanics of river ice jams. *Iowa Inst. Hydraul. Res. Rep. 161.* Iowa Inst. Hydraul. Res., Univ. Iowa. 158 pp.
Wigle, T. E. 1970. Investigations into frazil, bottom ice and surface ice formation in the Niagara river. See Int. Assoc. Hydraul. Res. 1970, Paper 2.8. 15 pp.
Zubov, N. N. 1943. *Arctic Ice*. English transl. 1963. U.S. Naval Oceanogr. Off. and Am. Meteorol. Soc. 491 pp.

*Ann. Rev. Fluid Mech. 1981. 13:399-423*


# CUP, PROPELLER, VANE, AND SONIC ANEMOMETERS IN TURBULENCE RESEARCH

*8182

*J. C. Wyngaard*
National Center for Atmospheric Research, Boulder, Colorado 80307

## INTRODUCTION

Micrometeorologists have traditionally used cup, propeller, and vane anemometers to measure wind speed and direction; sonic anemometers are newer, dating from the 1950s. Today all four types are widely used in research, with cup, propeller, and vane devices also being standard in operational applications.

One early research problem was the measurement of mean wind profiles in the "constant flux" layer, the first few tens of meters above the earth's surface. There the shearing stress $\tau$ varies little with height, so its value several meters above the surface can be taken as the surface stress $\tau_0$, which is a scaling parameter for surface-layer structure but also represents the drag on the atmosphere and hence has a strong influence on global circulation. Direct measurements of $\tau$ were not possible until fairly recently, but at that time it was known that near the surface the vertical gradient of the mean wind speed $U$ varies as

$$\frac{\partial U}{\partial z} = \frac{1}{kz}\sqrt{\frac{\tau_0}{\rho}} = \frac{u_*}{kz}, \tag{1}$$

where $\rho$ is the density, $u_*$ is the friction velocity, and $k \sim 0.4$ is the von Kármán constant. Thus (1) was used to deduce $\tau_0$ from mean wind-profile measurements.

Surface-layer researchers subsequently attempted to measure $\tau$ directly. Above the surface, $\tau$ is represented by the turbulent momentum flux $\rho\langle u_1 u_3\rangle$, where $u_1$ and $u_3$ are wind fluctuations in the streamwise and vertical directions and the angle brackets represent an average.



0066-4189/81/0115-0399$01.00

According to Kaimal (1979) the first direct measurements of $\rho\langle u_1 u_3 \rangle$ were made in 1926 by F. J. Scrase; soon after, the direct measurement of fluxes of momentum, heat, and moisture in the atmospheric boundary layer (ABL) became a central research goal in micrometeorology. Today, researchers routinely measure these fluxes with various combinations of cup, propeller, vane, and sonic anemometry together with turbulent temperature and humidity sensors.

The scope of micrometeorological research has broadened considerably in the past two decades, and now the ABL is a recognized focus of turbulence research. In a normal day over land the ABL ranges through shear-driven, convectively driven, and stably stratified states, each having different structure and dynamics; this provides turbulence research opportunities not readily available in laboratory flows.

With the discovery of the ABL by the turbulence community have come renewed attempts to use the hot-wire anemometer. However, the excellent spatial and temporal resolution of the hot wire is not generally needed except for dissipative range measurements, and its inherent sensitivity to temperature can be a severe drawback in the relatively low-speed ABL flows. Thus cup, propeller, vane, and sonic anemometers remain the principal types used in ABL research applications.

While the static, or steady-flow, performance of rotating anemometers is now well understood, their dynamic response, more important in turbulence research, is not. Space does not permit an exhaustive review such as that done by Corrsin (1963) on laboratory anemometry, and so I will concentrate on the dynamical aspects here, emphasizing the dynamical nonlinearities, which, while not yet well explored, are particularly important in turbulence research.

## CUP ANEMOMETERS

### *Some Response Considerations*

Early studies of cup anemometers dealt with their behavior in steady, nonturbulent flow. Patterson (1926) gave an interesting account of early attempts to calculate the ratio of wind speed to cup speed, called the "factor"; it was taken as three by Robinson, the designer in 1846 of the unit in most common use 80 years later. Patterson attempted to calculate the factor from his extensive wind-tunnel measurements of the torque on stationary and moving cups. In spite of his prodigious effort, however, the agreement between his calculations and the observed factors was only fair.

Researchers turned their attention to cup-anemometer dynamics when it was discovered that they "overspeed," or indicate erroneously high

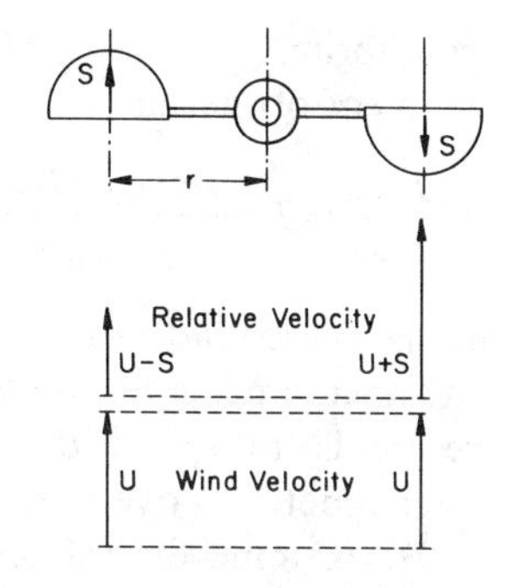


*Figure 1* Schematic plan view of a segment of an opposed-cup rotor.

mean values in gusty winds. We will illustrate this and other important aspects of cup-anemometer dynamics through a crude model.

Figure 1 shows a segment of an opposed-cup rotor at an instant when the cup arms are perpendicular to the wind direction. Let $c_d$ and $C_d$ be the drag coefficients of the front and rear cup faces, $A$ the cup area normal to the wind, and let the air and cup speeds be $\mathfrak{U}=U+u$ and $\mathfrak{S}=S+s$, the sum of mean and fluctuating parts. Then the torque $T$ is

$$T=\frac{\rho Ar}{2}\left[C_d(U+u-S-s)^2-c_d(U+u+S+s)^2\right]. \tag{2}$$

Patterson (1926) and later researchers have shown that $T$ actually varies strongly with angular position; however, for our purely illustrative purposes in this section we will neglect this angular dependence and assume that (2) represents the torque averaged over the basic period of cup rotation.

In a steady wind of speed $U$ with $u=0$, the torque vanishes (for a frictionless rotor) and the anemometer rotates steadily at peripheral speed $S=KU$ with $s=0$; $K$, the factor, is typically 0.3. Equation (2) gives for this steady case

$$C_d(U^2-2US+S^2)=c_d(U^2+2US+S^2). \tag{3}$$

Equations (2) and (3) can be combined to give

$$T=\frac{2\rho rC_d AKU^2}{(1+K)^2}\times\left[(1-K^2)\frac{u}{U}-(1-K^2)\frac{s}{S}+\left(\frac{u}{U}\right)^2+K^2\left(\frac{s}{S}\right)^2-(1+K^2)\frac{s}{S}\frac{u}{U}\right]. \tag{4}$$

Some of the nonlinear features of cup-anemometer dynamics are apparent from (4). Consider, for example, the initial response to a sudden change $u$ in wind speed, with the anemometer initially in equilibrium so that $s=0$. Then from (4) we have (with $I$ the polar moment of inertia)

$$\frac{I}{r}\frac{ds}{dt}=T=\frac{2\rho rC_{d}AKU^{2}}{(1+K)^{2}}\left[(1-K^{2})\frac{u}{U}+\left(\frac{u}{U}\right)^{2}\right]. \tag{5}$$

Equation (5) indicates that for finite-amplitude wind-speed fluctuations the anemometer responds to wind speed increases ($u>0$) faster than it responds to speed decreases. This is observed, and causes a cup anemometer to overspeed.

To examine the influence of anemometer inertia, consider the equation of motion with the crude expression (4) for torque in the limiting case of small fluctuations $u$ and $s$:

$$\frac{I}{r}\frac{ds}{dt}=\frac{2\rho rC_{d}AU(1-K^{2})}{(1+K)^{2}}[Ku-s]. \tag{6}$$

We rewrite this in the form

$$\frac{L}{U}\frac{ds}{dt}+s=Ku, \tag{7}$$

where $L=I/\rho r^{2}CA$ and $C$ is an effective drag coefficient, which also includes the numerical factors. $L$ has units of length and is called the "distance constant," since for a given anemometer it is fixed unless the air density changes or the speed range is so large that the drag coefficients change. Thus for small fluctuation levels (7) indicates a cup anemometer is a linear first-order system with a time constant $\tau$ inversely proportional to mean wind speed:

$$\tau\frac{ds}{dt}+s=Ku. \tag{8}$$

Cup anemometers in the linear limit are observed to follow (8), with the time constant proportional to $I/\rho r^{2}AU$ as predicted above.

While they respond primarily to the wind component in the cup plane, most cup anemometers also have a sensitivity to the velocity component perpendicular to the cup plane; we will refer to this as normal velocity sensitivity. This is usually measured by placing the anemometer in a low-turbulence-level wind tunnel operated at a speed $U$ and tilting the anemometer axis away from vertical by an angle $\theta$. Thus the wind speed in the cup plane is $U\cos\theta$, which would be the indicated speed if the anemometer had no normal velocity sensitivity. Figure 2*a* shows test results reported by MacCready (1966) for two cup anemometers; note that both have significant normal velocity sensitivity. The response asymmetry in $\theta$ is due to the effects of the anemometer housing. Kondo et al (1971) show similar results.

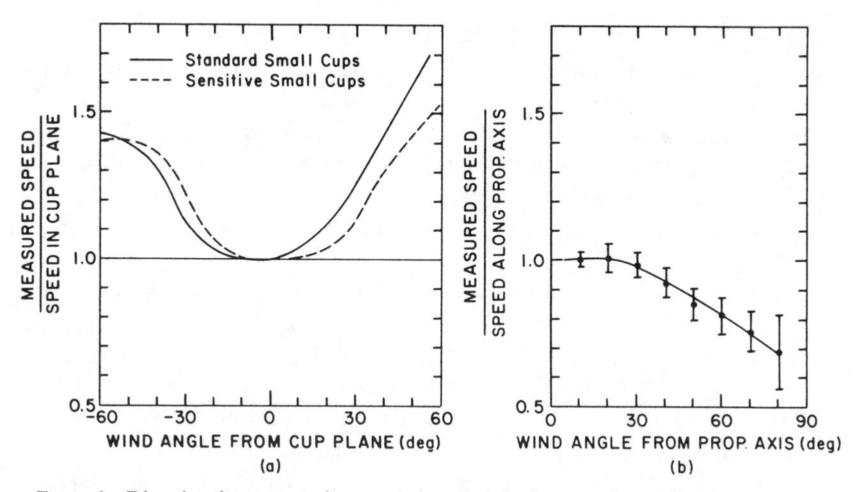


*Figure 2* Directional-response characteristics of typical cup and propeller anemometers. (*a*) cup response, from MacCready (1966). (*b*) Gill propeller response, from Pond et al (1979).

## *The Dynamic Equation*

The first published, systematic attempt to develop a finite-amplitude dynamic equation for cup anemometers seems to be that of Schrenk (1929). He noticed that the torque expression for the cup pair in the orientation shown in Figure 1 [our Equation (2)] can be written as

$$\frac{2T}{\rho \mathcal{S}^2 Ar} = (C_\mathrm{d} - c_\mathrm{d}) - 2(C_\mathrm{d} + c_\mathrm{d})\frac{\mathcal{U}}{\mathcal{S}} + (C_\mathrm{d} - c_\mathrm{d})\left(\frac{\mathcal{U}}{\mathcal{S}}\right)^2. \tag{9}$$

He then assumed that the torque averaged over one cycle also has the form of (9),

$$\frac{2T}{\rho \mathcal{S}^2 Ar} = \nu + \mu\left(\frac{\mathcal{U}}{\mathcal{S}}\right) + \lambda\left(\frac{\mathcal{U}}{\mathcal{S}}\right)^2, \tag{10}$$

where $\nu$, $\mu$, and $\lambda$ are undetermined coefficients. His next and most important step was to measure torque on a rotating cup wheel mounted in a wind-tunnel flow. He then fit the expression (10) to the results in order to evaluate $\nu$, $\mu$, and $\lambda$. His results are mainly of historical interest now, since $\nu$, $\mu$, and $\lambda$ depend on cup-wheel geometry and modern designs are different. However, his approach of directly measuring torque is an excellent technique for developing a dynamic equation for a cup anemometer.

Most other workers have attempted to derive dynamic equations by considering the forces on the individual cups. Some, for example, have started from the Brevoort-Joyner (1934, 1935) data on forces on stationary cups at various angles of incidence. This turns out to be an extraordinarily difficult path to take, in view of the myriad of subtle influences such as cup-wake effects. Ramachandran (1969) and Kondo et al (1971) have proceeded in this way. Acheson (1970) has instead followed Schrenk (1929) and correlated torque measurements with an expression of the form of (10), and Hyson (1972) has used a similar approach.

None of these efforts to develop a dynamic equation has dealt with normal velocity sensitivity, which judging from the data in Figure 2*a* should be important in cup-anemometer dynamics. In fact, Hyson (1972) concludes that the horizontal wind contribution to overspeeding in typical applications is only about 1%, so that if larger values of overspeeding occur they probably are due to w-effects.

Wyngaard, Bauman, & Lynch (1974) (hereafter called WBL) derived a general, nonlinear dynamic equation, which includes normal velocity sensitivity. Their approach is similar to Schrenk's in that it involves torque measurements on a rotating wheel, but their derivation of the dynamic equation from these measurements proceeds somewhat differently.

WBL used the quasi-steady assumption that the smoothed torque $\tilde{T}$ (that is, torque averaged over the basic period of cup rotation) depends only on $\mathfrak{U}$, the vertical velocity $\mathfrak{W}$, and $\mathbb{S}$. This implies that $\tilde{T}$ is zero under the equilibrium conditions of a steady horizontal wind:

$$\tilde{T}(U_0, 0, S_0) = 0, \tag{11}$$

where the subscript zero denotes the zero-torque condition. Thus WBL expressed $\tilde{T}$ as a three-variable Taylor series expansion about equilibrium:

$$\tilde{T} = \left.\frac{\partial \tilde{T}}{\partial S}\right|_0 s + \left.\frac{\partial \tilde{T}}{\partial U}\right|_0 u + \left.\frac{\partial \tilde{T}}{\partial W}\right|_0 w + \left.\frac{\partial^2 \tilde{T}}{\partial S^2}\right|_0 \frac{s^2}{2} + \left.\frac{\partial^2 \tilde{T}}{\partial U^2}\right|_0 \frac{u^2}{2} + \left.\frac{\partial^2 \tilde{T}}{\partial W^2}\right|_0 \frac{w^2}{2}$$

$$+ \left.\frac{\partial^2 \tilde{T}}{\partial U \partial S}\right|_0 us + \left.\frac{\partial^2 \tilde{T}}{\partial U \partial W}\right|_0 uw + \left.\frac{\partial^2 \tilde{T}}{\partial S \partial W}\right|_0 sw, \tag{12}$$

where $w$ is the vertical velocity fluctuation. Defining $u' = u/U_0$, $s' = s/S_0$, and $w' = w/U_0$ then leads to the dynamic equation

$$s' + \tau \frac{ds'}{dt} = a_1 u' + a_2 w' + a_3 s'^2 + a_4 u'^2 + a_5 w'^2 + a_6 u's' + a_7 u'w' + a_8 s'w', \tag{13}$$

where the time constant $\tau = -I(r\partial\tilde{T}/\partial S|_0)^{-1}$ can be shown to be the usual one having the form $L/U_0$. The definitions of the dimensionless coefficients $a_i$ follow from (12) and (13) and involve the torque derivatives at equilibrium; they are explicitly given by WBL.

WBL showed that (13) includes, as a special case where $w=0$, the form that follows from the torque expression (10) used by Schrenk. They found that the $a_i$ are subject to the constraints

$$a_1=1; a_3+a_4+a_6=0. \tag{14}$$

They also found that for symmetrical w-response (that is, a response in Figure 2*a* that is even in $\theta$) $a_2=a_7=a_8=0$, and that $a_5$ can be evaluated from the tilt test used to generate Figure 2*a*.

WBL presented measurements of the $a_i$ based on wind-tunnel measurements of the torque on a rotating cup anemometer. For a standard three-cup anemometer, they found

$$\begin{aligned} a_1&=1.03,\\ a_3&=-0.23,\\ a_4&=0.96,\\ a_5&=0.67,\\ a_6&=-0.73, \end{aligned} \tag{15}$$

with the other coefficients zero, within the experimental error. These results (15) are in good agreement with the constraints (14).

The WBL expression (12) for $\tilde{T}$ has all the terms that appeared in the illustrative model (4), but the $a_i$ are different from the coefficients in (4); for example, WBL found $a_3$ to be negative while (4) indicates it is positive. This suggests the hazards of attempting to derive a response equation through detailed consideration of the complex balance of forces on the rotating, interacting cups rather than measuring torque directly.

The WBL equation for their standard three-cup unit is

$$s' + \frac{\tau ds}{dt} = u' - 0.23s'^2 + 0.96u'^2 + 0.67w'^2 - 0.73u's'. \tag{16}$$

The presence of the $w'^2$ term, which represents normal velocity sensitivity, prevents $s'=u'$ from being a solution, even in the limit of fluctuations so slow that the time-lag term is negligible. In the general case with time lag and finite-amplitude $u'$, $s'$, and $w'$, the form of (16) indicates that nonlinear distortion of $s'$ can occur. We discuss some solutions of (16) in the next section.

## *Applications to Research*

Today's cup anemometers are quite different from those of Schrenk's time. They are also fairly easy to design and build, and many micrometerorologists do so. Examples are described by Jones (1965), Frenzen (1967), and Bradley (1969). Research models tend to have very-low-friction bearings, specially shaped, lightweight cups, and relatively small distance constants [Frenzen (1967) reports values of 0.3–0.6 m, but values for commercial units are typically 2–4 times larger] but can have widely varying normal velocity sensitivity. Frenzen (1967) minimizes this sensitivity with a "staggered six" unit, which has two three-cup wheels separated vertically and offset by 60°; Jones (1965) uses twelve slightly inclined cups. Figure 3 shows two current models of the Argonne

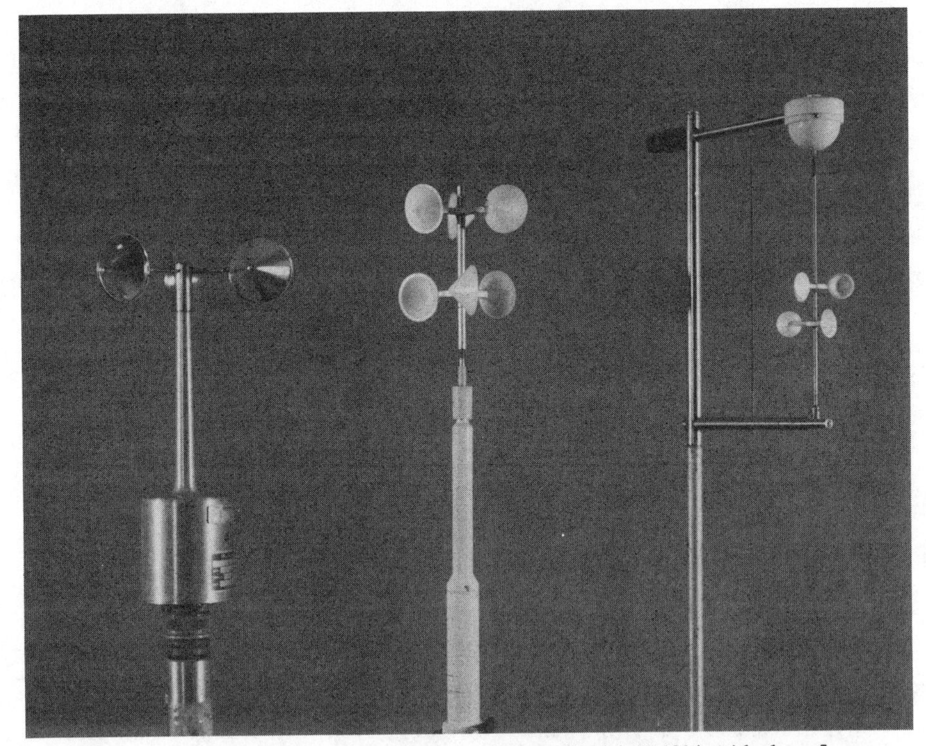

*Figure 3* Cup anemometers. *Left*: Weather Measure Model W-1034, with three 5-cm-diameter cups, distance constant 1.5 m. *Center*: Argonne National Laboratory (ANL) low-inertia (LI) model used (in conjunction with w-sensor) for momentum-flux measurements. Six 3.8-cm cups in two-deck array for cosine response; distance constant 0.35 m. *Right*: ANL/LI model for inertial-range measurements; six 2-cm cups, jewel bearings, distance constant 0.25 m. Paul Frenzen photo.

National Laboratory low-inertia cup anemometer and a standard instrument.

The two main research uses for cup anemometers are in mean wind-profile measurements and in turbulence measurements. The principal problem in the first application is overspeeding. If (as is usual) identical cup anemometers are used at different heights, the turbulence characteristics and hence the overspeeding errors will vary with height; thus the measured wind profile can be seriously distorted. To date, overspeeding has been documented primarily through wind-tunnel tests with strongly fluctuating sinusoidal or square-wave winds (Scrase & Sheppard 1944, Deacon 1951), and there is little published, direct evidence of its magnitude in ABL applications, partly because of the lack of another suitable anemometer for comparison. Izumi & Barad (1970) and Högström (1974) each report 10% cup-anemometer overspeeding in different surface-layer experiments, however.

Perhaps the strongest efforts to date to calculate overspeeding in ABL applications are those of Busch & Kristensen (1976) and Kaganov & Yaglom (1976). Each used the WBL response equation (16) for a typical three-cup unit. Busch & Kristensen time-averaged (16) to give an expression for the overspeeding $\langle s' \rangle$:

$$\langle s' \rangle = -0.23\langle s'^2 \rangle + 0.96\langle u'^2 \rangle + 0.67\langle w'^2 \rangle - 0.73\langle u's' \rangle \tag{17}$$

since $\langle u' \rangle = 0$ by definition. They used an approximate analytical technique to evaluate $\langle u's' \rangle$ and $\langle s'^2 \rangle$ and introduced measured wind statistics which enabled them to calculate the overspeeding for various surface-layer conditions. They found that under extreme conditions (a unit with a 20-m distance constant mounted at a height of 2 m over a relatively rough surface in convective conditions) the overspeeding was as large as 30% of the true mean. For a faster (smaller distance constant) unit the overspeeding could be much less. It would be desirable to confirm these results experimentally.

Kaganov & Yaglom (1976) give an exhaustive review of dynamic models, then adopt the WBL form (16) and a perturbation expansion to produce probably the most complete set of response calculations to date. They find overspeeding to be dominated by the normal velocity sensitivity and that it can reach the 10% levels reported by Izumi & Barad (1970) and Högström (1974) in field measurements. They also evaluate approximately the spectral distortion induced by the nonlinearity of the cup-anemometer response.

Like a vertically oriented hot-wire sensor, the cup anemometer responds to the horizontal speed; thus its fluctuating output is only approximately the streamwise wind fluctuation $u_1$. To illustrate, take an instantaneous anemometer signal that is indicating the instantaneous horizontal speed,

$$\mathcal{U} = U + u = \left[ (U_1 + u_1)^2 + u_2^2 \right]^{1/2}, \tag{18}$$

where $(U_1, 0)$ and $(u_1, u_2)$ are the mean and fluctuating horizontal wind vectors. Expanding the right side through second order and averaging gives

$$U = U_1 \left( 1 + \frac{\langle u_2^2 \rangle}{2U_1^2} \right), \quad u = u_1 + \frac{U_1}{2} \left[ \left( \frac{u_2}{U_1} \right)^2 - \left\langle \left( \frac{u_2}{U_1} \right)^2 \right\rangle \right]. \tag{19}$$

The first equation of (19) indicates the mean error, and the second the contamination of the fluctuating anemometer signal by the fluctuating lateral component. Both effects are also familiar in hot-wire anemometry.

In the linear first-order system approximation, a cup anemometer has a half-power frequency $\omega = U/L$. The distance constant $L$ of research units can be as small as 0.3–0.6 m, giving by Taylor's hypothesis a half-power streamwise wavenumber of $\omega/U = L^{-1} \sim 1.7\text{–}3.3$ rad m$^{-1}$, which is typically well into the inertial subrange at heights more than a few meters above the surface. Frenzen (1977) uses the Argonne low-inertia anemometers of Figure 3 for momentum-flux and inertial-subrange measurements.

## PROPELLER ANEMOMETERS

Most of today's propeller anemometers have helicoid rotors, and thus derive from the helicoid anemometer first described by W. H. Dines in 1887 (Gill 1973). According to Gill, Dines' anemometer was too complicated for general use and the helicoid was apparently not used again in anemometry until the Bendix Friez Aerovane was developed in the 1940s. In the 1960s Gill and MacCready independently perfected propeller anemometer–bidirectional wind vane units which have been fairly widely adopted.

Today the Gill unit (Holmes et al 1964) is perhaps the most common propeller anemometer used in research. Gill (1975) and Pond et al (1979) discuss its use as a turbulence sensor. Figure 4 shows some models currently manufactured by the R. M. Young Co., of Traverse City, Michigan.

Propeller anemometers, in first approximation, respond linearly to the velocity component normal to the plane of the propeller; we will call this the axial component. The propeller can be flat or helicoid, and most often has two or four blades. The entire unit can be mounted on a vane

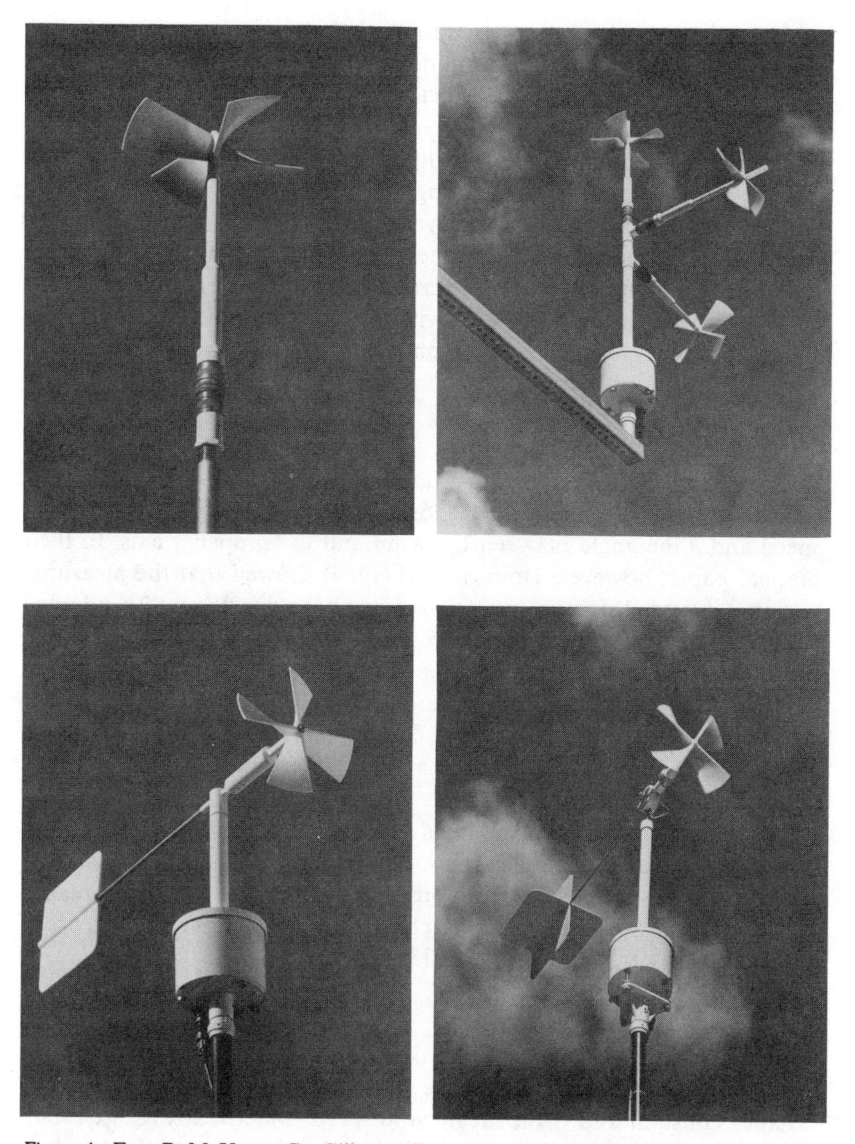

*Figure 4* Four R. M. Young Co. Gill propeller anemometers. *Upper left*: vertical-axis unit for measuring vertical velocity. *Upper right*: three-component anemometer. *Lower left*: propeller vane. *Lower right*: Propeller bivane. R. M. Young Co. photos.

to keep the propeller oriented into the wind; this is often preferred, since typical units have significant deviations from axial response. Other workers use fixed units and correct for the actual directional response characteristics.

The propeller anemometer has little competition as an inexpensive, rugged, simple sensor of fluctuating vertical velocity. When combined with sensors for fluctuating streamwise velocity, temperature, or humidity it has made possible the routine measurement of vertical turbulent fluxes. It is not without its idiosyncrasies, but fortunately many of these have been well documented; we will briefly survey the current understanding here. As with cup anemometers, the least-understood area is that of dynamic response.

### *Directional Sensitivity*

The propeller anemometer is, to a fair and widely used approximation, a "cosine-law" device; that is, it responds to $\mathcal{U}\cos\theta$, where $\mathcal{U}$ is the wind speed and $\theta$ the angle between the wind and the propeller axis. In their original paper, however, Holmes et al (1964) showed that the measured response departed significantly from cosine behavior. Figure 2*b*, adapted from Pond et al (1979), summarizes some measurements of the directional response of the Gill propeller anemometer. Note that it indicates a speed slower than the axial component, while typical cup anemometers (Figure 2*a*) indicate a speed larger than that in the cup plane. Monna & Driedonks (1979) present data for some other commercial units.

The departures from cosine behavior indicated by Figure 2*b* require correcting in research applications. The R. M. Young Co. boosts the output of its w-sensors by 25% in order to compensate for the decreased sensitivity near 90°, Figure 2*b*. More precise corrections are discussed by Drinkrow (1972) and Horst (1973a).

### *Frequency Response*

A propeller anemometer in an axial wind is usually treated as a first-order linear system with time constant $\tau = L/U$, where $L$ is the distance constant and $U$ the mean wind speed (MacCready & Jex 1964). To first approximation $L$ in axial winds is independent of $U$, which is why it rather than the time constant is usually quoted. $L$ is of the order of 1 m for four-blade, 23-cm-diameter Gill propeller.

This simple situation changes with off-axis winds, where the time constant is observed to vary with the wind angle (Camp & Turner 1970, Clink 1971, McBean 1972, Horst 1973a, Garratt 1975, Gill 1975, Pond et al 1979). However, in converting measured time constants to distance constants by multiplying by wind speed, some authors use the axial wind-velocity component and others the total wind speed; thus there is some confusion in the literature regarding the value of distance constants for off-axis winds. However, it is clear that for fixed wind speed the propeller time constant increases as $\theta$ increases. Garratt (1975), using the data of Hicks (1972), reports that for a Gill propeller the distance constant $U\tau$ (where $U$ is total wind speed) increases from 1.0 m at $\theta=0°$ to about 1.4 m at $\theta=60°$ and then sharply to about 3 m at $\theta=85°$, in fair agreement with the data of Gill (1975). Thus propeller response is much slower when used as a w-sensor (i.e. when the axis is vertical) than when used as a horizontal wind sensor.

## *Dynamical Response*

The propeller anemometer, like the cup anemometer, has inherently nonlinear dynamics. We illustrated the source of cup-anemometer nonlinearity with simple torque arguments based on Figure 1, finding that the responses to finite-amplitude wind-speed increases and decreases were different. A similar analysis of the propeller leads to the same result.

In principle, the propeller anemometer should overspeed, yet there is little reference to this in the literature; however, Brock (1973) has demonstrated propeller-anemometer overspeeding through wind-tunnel tests. Perhaps it is typically less serious than cup-anemometer overspeeding. Wieringa (1972) does mention a cup-anemometer overspeeding correction of 7% applied in a surface-layer experiment and indicates this was determined by comparison with propeller-speed data.

The dependence of the time constant on wind direction is evidence of dynamic nonlinearity of propellers. Further evidence for vertically mounted units is reported by Fichtl & Kumar (1974), who found that the time constant depends on the vertical velocity variance $\langle w^2 \rangle$, and by Francey & Sahashi (1979), who found that the measured power-spectral density of $w$ showed an $f^{-2.5}$ high-frequency falloff rather than the $f^{-2}$ of a first-order linear system. Another nonlinearity is the "dead zone" when used as a w-sensor. An axial wind of the order of 0.2 m s$^{-1}$ is required to overcome starting friction and turn the propeller, and this can cause a loss of small-amplitude w-fluctuations (Wieringa 1972, Horst 1973a).

Acheson (1970) has made one of the few attempts to formulate a nonlinear response equation for propeller anemometers. Adopting the experimental approach used by Schrenk (1929), he measured the torque on units rotating in axial flow in a wind tunnel and found basically the same response equation for cup and propeller anemometers. His solutions for idealized wind forcing and axial flow showed propeller-anemometer overspeeding. The WBL approach, which gives a general, nonlinear response equation for any flow direction, seems yet to be applied to propeller anemometers.

### *Applications to Research*

Propeller anemometers are perhaps the most common sensor for vertical velocity fluctuations and hence for turbulent-flux measurements. Because of their relatively slow w-response, however, they can fail to measure a large fraction of the flux if incorrectly used (McBean 1972, Hicks 1972, Garratt 1975, Kaimal 1975, Francey & Sahashi 1979, Pond et al 1979). Like other sensors, they are prone to vertical misalignment or "tilt," which can introduce spurious horizontal velocity contributions into the vertical velocity signal. Stringent leveling standards are required to minimize this (Pond 1968, Kaimal & Haugen 1969, Wieringa 1972).

In order to eliminate some of the nonlinearities associated with a w-sensing propeller, such as the dead zone and the strongly direction-dependent response time, some researchers prefer to use combinations of inclined units rather than vertical-horizontal arrays. Drinkrow (1972), Horst (1973a), and Pond et al (1979) discuss such applications. It has also been found that a vertical shaft extension on a w-propeller can increase the symmetry of the response (Hicks et al 1977). The same authors (see also Hicks 1972) have discovered that w-propellers are very sensitive to bearing friction, causing the performance as a w-sensor to deteriorate markedly with time. They have also experimented with faster propellers, but find this leads to increased sensitivity to bearing-friction effects.

## VANES

Vanes are often used to orient cup, propeller, or hot-wire anemometers into the wind, or used alone to indicate wind direction (see Figure 4). Figure 5 shows miniature vanes used on the NCAR research aircraft. They sense the direction components of the velocity relative to the nose boom (the attack and sideslip angles) to frequencies of 30 Hz at speeds of 70 m s$^{-1}$; hence, they resolve streamwise wavelengths as small as 2m.

In the schematic of the simple vane of Figure 6 (adapted from Wieringa 1967) the wind vector $\mathcal{U}$ makes an instantaneous angle $\beta$ to the vane. The vane force acts at the aerodynamic center, a distance $r_v$ from the axis of rotation. The vane velocity at the aerodynamic center is

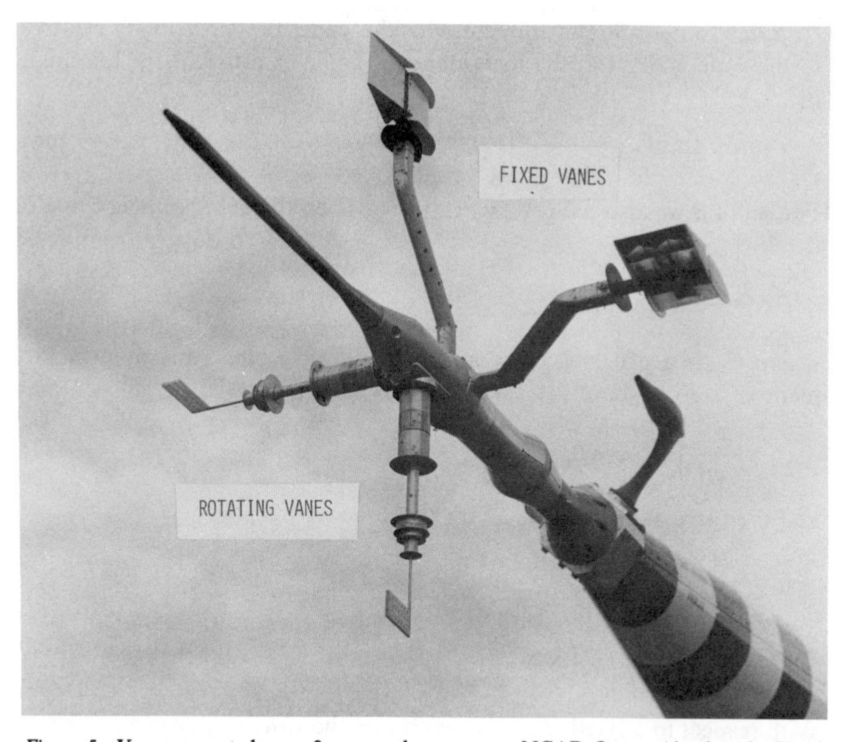


*Figure 5* Vanes mounted on a 2-m nose boom on an NCAR Queen Air aircraft. Fixed vanes have 7-cm leading edges, 300-Hz natural frequency; deflection sensed by strain gauges. Rotating vanes are 6.2 × 2.5 cm, 30-Hz natural frequency; position sensed by rotary differential transformer.

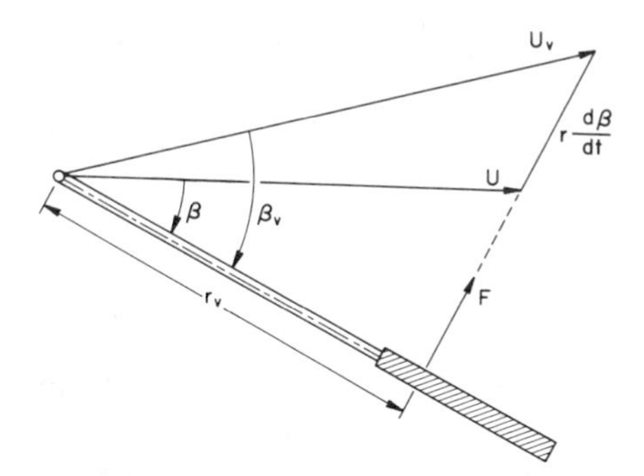


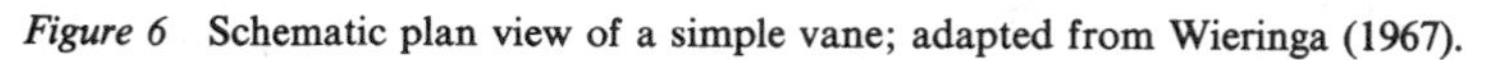

*Figure 6* Schematic plan view of a simple vane; adapted from Wieringa (1967).

$r_v d\beta/dt$. In coordinates moving with the vane, the air velocity relative to the vane is $\mathcal{U}_v$ and the instantaneous angle of attack $\beta_v$ is, for small $\beta$,

$$\beta_v \simeq \beta + \frac{r_v}{\mathcal{U}} \frac{d\beta}{dt}. \tag{20}$$

For small $\beta$ we also have $\mathcal{U}_v = |\mathcal{U}_v| \simeq \mathcal{U}$. Then the aerodynamic force $F$ is

$$F = \frac{\rho \mathcal{U}^2}{2} A C_L \cos \beta_v, \tag{21}$$

where $C_L$ is a lift coefficient and $A$ is the projected vane area in the plane of zero angle of attack. For small $\beta_v$,

$$C_L \simeq \left.\frac{\partial C_L}{\partial \beta}\right|_0 \beta_v, \ \cos \beta_v \simeq 1 \tag{22}$$

and the force $F$ and torque $T$ are

$$F \simeq \frac{\rho \mathcal{U}^2}{2} A \left.\frac{\partial C_L}{\partial \beta}\right|_0 \beta_v,$$

$$T \simeq \frac{\rho \mathcal{U}^2}{2} A \left.\frac{\partial C_L}{\partial \beta}\right|_0 r_v \left(\beta + \frac{r_v}{\mathcal{U}} \frac{d\beta}{dt}\right). \tag{23}$$

With respect to a fixed frame of reference, the vane-response equation is then

$$I \frac{d^2\beta}{dt^2} + \frac{\rho}{2} \mathcal{U} A \left.\frac{\partial C_L}{\partial \beta}\right|_0 r_v^2 \frac{d\beta}{dt} + \frac{\rho \mathcal{U}^2}{2} A \left.\frac{\partial C_L}{\partial \beta}\right|_0 r_v (\beta - \phi) = 0, \tag{24}$$

where $\phi$ is the wind direction. If a parameter $N$ is defined by

$$N = \frac{T}{\beta_v} = \frac{\rho \mathcal{U}^2 A}{2} \left.\frac{\partial C_L}{\partial \beta}\right|_0 r_v \tag{25}$$

then the response equation is simply

$$I \frac{d^2\beta}{dt^2} + \frac{r_v N}{\mathcal{U}} \frac{d\beta}{dt} + N(\beta - \phi) = 0. \tag{26}$$

The vane-response equation (26) is analogous to that for a damped spring-mass system undergoing linear motion. The parameter $N$ is analogous to the spring constant, while $r_v N/\mathcal{U}$ plays the role of the linear damping and $I$ replaces the mass. From (23) and (25) $N$ is the torque, per unit angle of attack, due to static deflection. The remaining torque is due to the vane motion, which causes the relative angle of attack seen by the moving vane to be $\beta_v$ instead of $\beta$; this induces what is called the damping torque. An "aerodynamic damping" $D$ is sometimes defined in the following way:

$$D=\frac{r_v N}{\mathcal{U}}. \tag{27}$$

Thus the equation of vane motion can be written

$$\frac{d^2\beta}{dt^2}+\frac{D}{I}\frac{d\beta}{dt}+\frac{N}{I}(\beta-\phi)=0, \tag{28}$$

which is also expressed as (MacCready & Jex 1964, Acheson 1970)

$$\frac{d^2\beta}{dt^2}+2\zeta\omega_n\frac{d\beta}{dt}+\omega_n^2(\beta-\phi)=0. \tag{29}$$

The natural frequency $\omega_n$ and damping $\zeta$ are, from (25) and (27)–(29),

$$\omega_n=\left(\frac{N}{I}\right)^{1/2}=\left[\frac{\rho\mathcal{U}^2 A\left.\dfrac{\partial C_L}{\partial\beta}\right|_0 r_v}{2I}\right]^{1/2}, \tag{30}$$

$$\zeta=\frac{D}{2(IN)^{1/2}}=\left[\frac{\rho r_v^3 A\left.\dfrac{\partial C_L}{\partial\beta}\right|_0}{8I}\right]^{1/2},$$

from which we see that $\omega_n$ varies linearly with wind speed $\mathcal{U}$ while $\zeta$ is a property of the vane alone, providing that the speed range is small enough that the lift characteristics do not change. Thus we can write $\omega_n$ as

$$\omega_n=\frac{\mathcal{U}}{\lambda};\quad \lambda=\left[\frac{2I}{\rho A\left.\dfrac{\partial C_L}{\partial\beta}\right|_0 r_v}\right]^{1/2}, \tag{31}$$

where $\lambda$ is a natural response length of the vane, independent of the wind speed. The parameter $\lambda_n=2\pi\lambda$ is often called the "natural wavelength."

The vane-response equation (29) is linear for fixed wind speed $\mathcal{U}$, making its solution straightfoward. MacCready & Jex (1964) discuss linear vane response in great detail. Here we simply note that if the wind-angle forcing is a stationary random variable we can write

(Batchelor 1960)

$$\phi(t)=\int e^{i\omega t}\,dE_{\text{in}}(\omega), \tag{32}$$

$$\beta(t)=\int e^{i\omega t}\,dE_{\text{out}}(\omega),$$

and from (29) we have

$$(-\omega^2+2i\zeta\omega_{\text{n}}\omega+\omega_{\text{n}}^2)\,dE_{\text{out}}=\omega_{\text{n}}^2\,dE_{\text{in}}. \tag{33}$$

The ratio of input and output power-spectral densities becomes

$$\frac{\Phi_{\text{out}}}{\Phi_{\text{in}}}=\frac{\langle dE_{\text{out}}\,dE_{\text{out}}{}^*\rangle}{\langle dE_{\text{in}}\,dE_{\text{in}}{}^*\rangle}=\frac{1}{\left[1-\left(\frac{\omega}{\omega_{\text{n}}}\right)^2\right]^2+4\zeta^2\left(\frac{\omega}{\omega_{\text{n}}}\right)^2}. \tag{34}$$

If the vane damping is small ($\zeta \ll 1$), (34) indicates the spectral response is sharply peaked for $\omega \sim \omega_{\text{n}}$. MacCready & Jex (1964) show that $\zeta \sim$ 0.5–0.7 is optimal, giving a good compromise between fast transient response and smooth spectral response.

In the general case the vane-response equation (29) can be written, by introducing (31),

$$\frac{d^2\beta}{dt^2}+\frac{2\zeta\mathcal{U}}{\lambda}\frac{d\beta}{dt}+\frac{\mathcal{U}^2}{\lambda^2}(\beta-\phi)=0. \tag{35}$$

If $\mathcal{U}$ fluctuates this equation is no longer simple to solve. Such stochastically forced equations are of current interest in their own right, but the only attempts to solve (35) in the context of anemometry seem to be those of Acheson (1970). He used a perturbation expansion to calculate the effects of speed fluctuations on the transfer function, which is (34) in the absence of fluctuations. He found that the fluctuations increased the magnitude of the transfer function at frequencies of the order of $\omega_{\text{n}}$. In view of the assumptions made in its derivation, however, it is not clear that (35) is appropriate for studies of vane response to finite-amplitude, three-dimensional turbulence.

The combination of a propeller or cup anemometer and a wind vane is a rather complicated dynamical system, although in the linear limit its response can be analyzed analytically. MacCready & Jex (1964) discuss this in some detail.

Chimonas (1980) describes an interesting application of a simple bivane, which records the instantaneous elevation angle $\phi=\tan^{-1}[(W+w)/(U+u)]$. He shows that when the mean vertical velocity $W$ is negligible the mean elevation angle $\langle\phi\rangle$ is, to second order, simply $\langle uw\rangle/U^2$; hence the bivane could be used to measure Reynolds stress.

Vanes that are constrained from rotating are also used. Lenschow (1971) describes one developed for aircraft mounting; it has faster response than a rotating vane and no bearing friction. Two of these fixed vanes are shown on their aircraft boom in Figure 5. Högström (1967) describes a fixed vane for tower applications.

# SONIC ANEMOMETERS

## *Introduction and History*

Sonic (also called acoustic) anemometers, in conjunction with modern developments in data processing and recording, have revolutionized surface-layer turbulence research. To a very accurate approximation a well-designed sonic anemometer measures the projection of the wind-velocity vector on the acoustic path, and is free of the nonlinearities, time lag, and most other deficiencies of cup, propeller, and vane anemometers. Its principal disadvantage is its high cost; a three-component unit is at least an order of magnitude more expensive than a three-component propeller anemometer, for example.

In this section we will briefly review the history and theory of operation of the sonic anemometer. A more complete survey was recently done by Kaimal (1979); much of the material here is taken from his paper.

Suomi (1957) was a pioneer in sonic anemometry. His device had a 1-m path; the transit-time difference of acoustic pulses traveling in opposite directions along this path gave the average velocity along the path direction. Later workers used continuous acoustic signals (Gurvich 1959, Bovsheverov & Voronov 1960, Kaimal & Businger 1963). The 1968 Kansas experiments (Haugen et al 1971, Haugen 1973) saw the first use of sonic anemometry in a large-scale field program. Three-component, pulse-type sonic anemometers with 20-cm paths were used at heights of 5.7, 11.3, and 22.6 m in a study of the structure and dynamics of the surface layer. Kaimal et al (1974) describe an improved device, an updated version of which will soon be commercially available from Applied Technology Corporation of Boulder, Colorado.

## *Theory of Operation*

Figure 7, from Kaimal (1979), shows the basic principle of the sonic anemometer. The wind vector **V** has components $V_d$ and $V_n$ along and normal to the acoustic path, and we assume that all velocities are averages along the paths. The transit times for two simultaneous pulses traveling from $T_1$ to $R_1$ and from $T_2$ to $R_2$ are, from the vector diagram

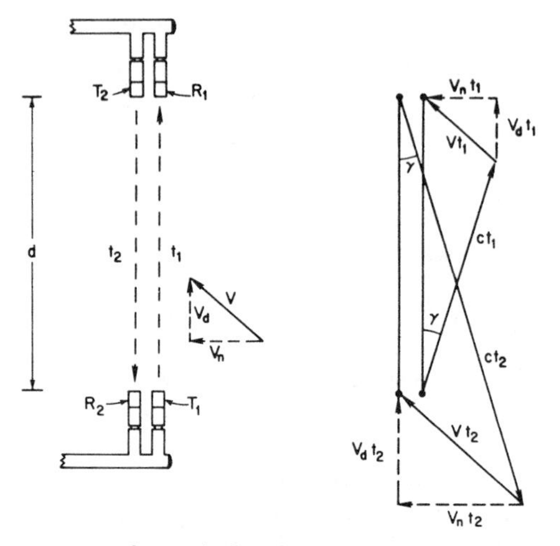


*Figure 7* Sound-ray vectors for a single-axis sonic anemometer, showing principle of operation; $T$ and $R$ represent transmitter and receiver, respectively. Adapted from Kaimal (1979).

in Figure 7,

$$t_1 = \frac{d}{c\cos\gamma + V_d}, \tag{36}$$

$$t_2 = \frac{d}{c\cos\gamma - V_d}.$$

Here $t_1$ and $t_2$ are the pulse transit times, $d$ is the path length, $c$ is the velocity of the sound in the air, and $\gamma = \sin^{-1}(V_n/c)$. Thus for $V^2 \ll c^2$ the transit-time difference is, from (36),

$$\Delta t = \frac{2d}{c^2} V_d. \tag{37}$$

Knowing the temperature $T$ gives $c^2$ from the relation $c^2 = kRT$, where $k$ is the ratio of specific heats and $R$ the gas constant. Thus, when $T$ is known, measuring $V_d$ becomes a matter of the measurement of the transit-time difference. For $d = 20$ cm this typically can be done with a resolution of about $10^{-7}$ s in commercial units, giving a velocity resolution of the order of 3 cm s$^{-1}$. Kaimal (1979) has discussed the influence of temperature and humidity fluctuations, which affect the sound speed $c$, on the velocity measurements; they are generally negligible.

## *Three-Dimensional Units*

According to Kaimal (1979), the problem of designing a three-dimensional sonic array with unobstructed exposure for all the axes is not a trivial one. Current models use a pair of horizontal axes separated by 120 degrees and one vertical axis; the three orthogonal velocity components are then retrieved from these outputs. However, the outputs must be monitored to ensure that the natural variations in wind direction do not exceed the allowable range of this array, which is of the order of 90 degrees in the horizontal. Figure 8 shows a three-axis sonic anemometer with 20-cm paths.

Current units with a pair of acoustic paths per axis typically have pulse rates on the order of 400 $s^{-1}$. Thus Kaimal (1979) points out that 10 or 20 successive measurements can be block-averaged to improve resolution and reduce spectral-aliasing effects, while leaving a signal of the order of 20 samples $s^{-1}$. Some units use only one transducer at each end of the path, acting alternately as a transmitter and a receiver; here

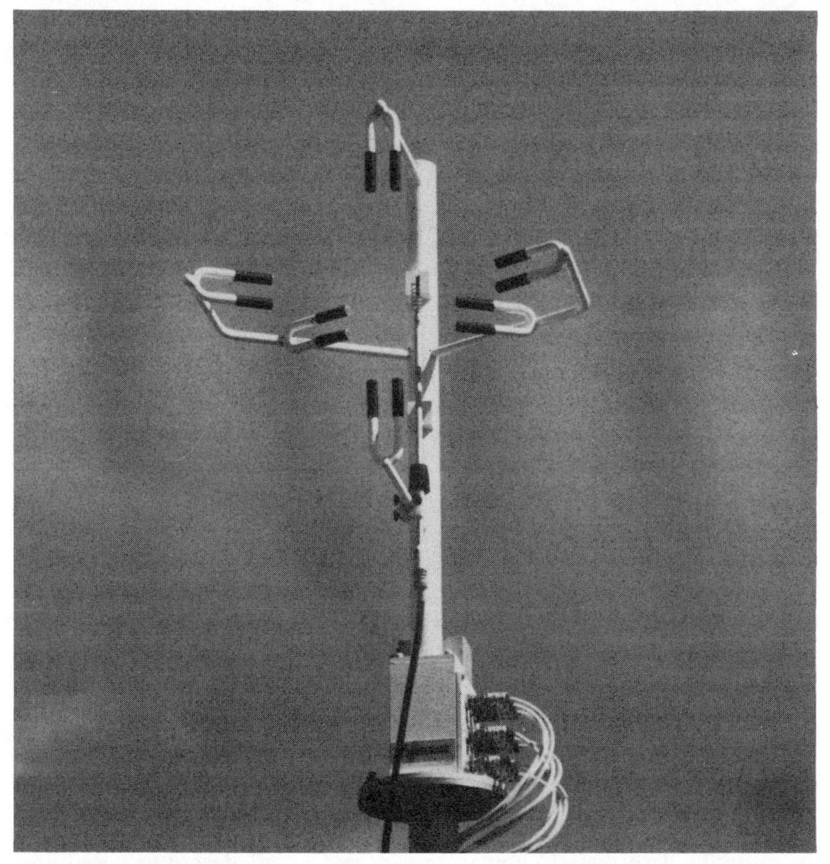

*Figure 8* A three-axis sonic anemometer (EG & G model 198-3) with 20-cm paths. J. C. Kaimal photo.

the sample rate must be less (no greater than 10 $s^{-1}$) to accommodate settling times in the transducers.

The spatial-resolution properties of three-component sonic arrays have been calculated in some detail by Kaimal et al (1968) and by Horst (1973b). They have calculated the effects of path length and path separation on measured horizontal and vertical velocity spectra, assuming the true spectra are inertial and isotropic. Their results indicate that the measured streamwise wavenumber spectra (inferred from frequency spectra through Taylor's hypothesis) of horizontal and vertical velocity components become distorted at streamwise wavenumbers $\kappa_1$ of the order of $d^{-1}$; thus the "distance constant" is effectively the path length, which at typically 20 cm is less than the distance constant of most cup or propeller anemometers. This gives sufficient spatial resolution to recover the vertical turbulent fluxes at heights of the order of 5 m and greater (Kaimal 1975).

Sonic-anemometer velocity measurements can be systematically underestimated because of the velocity defects in the transducer wakes along the acoustic paths. The severity of this effect depends on the array design. Kaimal (1979) cites some calibration data that suggest it could be important in very-short-path arrays, which tend to use transducers of the same diameter as longer-path units. However, there are as yet no published calculations or measurements of the influence of these errors on various turbulence statistics.

Kaimal's (1979) concluding remarks are appropriate here:

> The sonic anemometer with its rapid response, linear output and stable calibration has become a valuable tool for atmospheric research. It has provided high-quality turbulence data in a number of field experiments conducted during the last decade. A new three-axis configuration is currently being used for continuously monitoring the wind at eight levels on the Boulder Atmospheric Observatory's 300 m tower.
>
> Because of its complexity and high cost, the sonic anemometer is likely to remain a research instrument. Attempts are being made to reduce cost and to simplify its operation. Even if these attempts are successful, the sonic anemometer has an inherent limitation which precludes its use in adverse weather conditions. The instrument fails to respond in rain, wet snow, and heavy fog. Formation of water drops on the transducer temporarily affects its operation. Thus the sonic anemometer's main contribution will be in fair-weather observations of atmospheric turbulence, where it is still the instrument of choice among boundary-layer physicists.

## SUMMARY

The static response of cup and propeller anemometers to wind speed and direction is generally well understood. Their dynamic response can be broadly described in linear terms, but a full description requires nonlinear models. Much progress has been made in the last decade in formulating and studying nonlinear models of cup anemometers, but the extension to propellers has not been carried out although it is straightforward. There is little published work on the nonlinear dynamics of vanes.

Further studies of the nonlinear dynamics of cup, propeller, and vane anemometers would be most useful, particularly as researchers probe more deeply into ABL turbulence structure and require increasing fidelity from these commonly used sensors.

Sonic-anemometer response is inherently linear and hence better understood. The effect of temperature and water-vapor fluctuations on its measured velocities is well documented and generally negligible, and its spatial-resolution characteristics have been calculated in great detail. A remaining problem that deserves more attention is the influence of the transducer wakes on the measured velocities.

## ACKNOWLEDGMENTS

I am grateful to D. Acheson, F. Bradley, S. Corrsin, and J. C. Kaimal for a number of stimulating discussions on anemometry over the past decade. H. Baynton, F. Brock, P. Frenzen, J. C. Kaimal, D. Kipley, and D. Lenschow have been very helpful during the preparation of this review. Finally, I wish to thank P. Waukau for expertly typing several drafts of this manuscript. The National Center for Atmospheric Research is sponsored by the National Science Foundation.

*Literature Cited*

Acheson, D. T. 1970. Response of cup and propellor rotors and wind direction vanes to turbulent wind fields. *Meteorol. Monogr.* 11:252–61

Batchelor, G. K. 1960. *The Theory of Homogeneous Turbulence*. Cambridge Univ. Press. 197 pp.

Bovsheverov, V. M., Voronov, V. P. 1960. Acoustic anemometer. *Izv. Geophys. Series*. 6:882–85

Bradley, E. F. 1969. A small sensitive anemometer system for agricultural meteorology. *Agric. Meteorol.* 6:185–93

Brevoort, M. J., Joyner, U. T. 1934. Aerodynamic characteristics of anemometer cups. *NACA Tech. Note No. 489*

Brevoort, M. J., Joyner, U. T. 1935. Experimental investigation of the Robinson-type cup anemometer. *NACA Rep. No. 513*. 24 pp.

Brock, F. 1973. *Experimental analysis of propeller anemometer dynamic performance*. PhD thesis. Univ. Okla., Norman. 71 pp.

Busch, N. E., Kristensen, L. 1976. Cup anemometer overspeeding. *J. Appl. Meteorol.* 15:1328–32

Camp, D. W., Turner, R. E. 1970. Response tests of cup, vane, and propeller wind sensors. *J. Geophys. Res.* 75:5265–70

Chimonas, G. 1980. Reynolds stress deflections of the bivane anemometer. *J. Appl. Meteorol.* 19:329–33

Clink, W. L. 1971. Comment on "Response tests of cup, vane, and propeller wind sensors" by D. W. Camp, R. E. Turner, and L. P. Gilchrist. *J. Geophys. Res.* 76:2902

Corrsin, S. 1963. Turbulence: Experimental Methods. In *Handbuch der Physik*, ed. S. Flugge, C. Truesdell. 8(2):524–90. Berlin: Springer

Deacon, E. L. 1951. The over-estimation error of cup anemometers in fluctuating winds. *J. Sci. Instrum.* 28:231–34

Drinkrow, R. 1972. A solution to the paired Gill-anemometer response function. *J.*

*Appl. Meteorol.* 11:76–80

Fichtl, G. H., Kumar, P. 1974. The response of a propeller anemometer to turbulent flow with the mean wind vector perpendicular to the axis of rotation. *Boundary-Layer Meteorol.* 6:363–79

Francey, R. J., Sahashi, K. 1979. Gill propeller anemometer frequency response over the sea. *J. Appl. Meteorol.* 18:1083–86

Frenzen, P. 1967. Modifications of cup anemometer design to improve the measurement of mean horizontal wind speeds in turbulence. *Ann. Rep. Radiol. Phys. Div.*, Argonne Nat'l. Lab., *ANL-7360*, pp. 160–66

Frenzen, P. 1977. A generalization of the Kolmogorov–von Kármán relationship and some further implications on the values of the constants. *Boundary-Layer Meteorol.* 11:375–80

Garratt, J. R. 1975. Limitations of the eddy-correlation technique for the determination of turbulent fluxes near the surface. *Boundary-Layer Meteorol.* 8:255–59

Gill, G. C. 1973. The helicoid anemometer. *Atmosphere* 11:145–55

Gill, G. C. 1975. Development and use of the Gill UVW anemometer. *Boundary-Layer Meteorol.* 8:475–95

Gurvich, A. S. 1959. Acoustic microanemometer for investigation of the microstructure of turbulence. *Acoustic J.* (USSR) 5:368–69

Haugen, D. A., ed. 1973. *Workshop on Micrometeorology*, Am. Meteorol. Soc., Boston. 300 pp.

Haugen, D. A., Kaimal, J. C., Bradley, E. F. 1971. An experimental study of Reynolds stress and heat flux in the atmospheric surface layer. *Q. J. R. Meteorol. Soc.* 97:168–80

Hicks, B. B. 1972. Propeller anemometers as sensors of atmospheric turbulence. *Boundary-Layer Meteorol.* 3:214–28

Hicks, B. B., Frenzen, P., Wesely, M. L. 1977. Some preliminary results from the 1976 ITCE: I. Anemometry. *Ann. Rep., Radiol. & Environ. Res. Div.*, Argonne Nat'l Lab., *ANL-76-88*, pp. 127–32

Högström, U. 1967. A new sensitive eddy flux instrumentation. *Tellus* 19:230–39

Högström, U. 1974. A field study of the turbulent fluxes of heat, water vapour and momentum at a "typical" agricultural site. *Q. J. R. Meteorol. Soc.* 100: 624–39

Holmes, R. M., Gill, G. C., Carson, H. W. 1964. A propellor-type vertical anemometer. *J. Appl. Meteorol.* 3:802–4

Horst, T. W. 1973a. Corrections for response errors in a three-component propeller anemometer. *J. Appl. Meteorol.* 12:716–25

Horst, T. W. 1973b. Spectral transfer functions for a three-component sonic anemometer. *J. Appl. Meteorol.* 12:1072–75

Hyson, P. 1972. Cup anemometer response to fluctuating wind speeds. *J. Appl. Meteorol.* 11:843–48

Izumi, Y., Barad, M. L. 1970. Wind speeds as measured by cup and sonic anemometers and influenced by tower structure. *J. Appl. Meteorol.* 9:851–56

Jones, J. I. P. 1965. A portable sensitive anemometer with proportional d.c. output and a matching wind velocity-component resolver. *J. Sci. Instrum.* 42:414–17

Kaganov, E. I., Yaglom, A. M. 1976. Errors in wind-speed measurements by rotation anemometers. *Boundary-Layer Meteorol.* 10:15–34

Kaimal, J. C. 1975. Sensors and techniques for direct measurement of turbulent fluxes and profiles in the atmospheric surface layer. *Atmos. Tech.* 7:7–14. Available from NCAR, P. O. Box 3000, Boulder, Colorado 80307

Kaimal, J. C. 1979. Sonic anemometer measurement of atmospheric turbulence. *Proc. Dynamic Flow Conf., 1978*, pp. 551–65. P. O. Box 121, DK 2740 Skovlunde, Denmark

Kaimal, J. C., Businger, J. A. 1963. A continuous wave sonic anemometer thermometer. *J. Appl. Meteorol.* 2:156–64

Kaimal, J. C., Haugen, D. A. 1969. Some errors in the measurement of Reynolds stress. *J. Appl. Meteorol.* 8:460–62

Kaimal, J. C., Newman, J. T., Bisberg, A., Cole, K. 1974. An improved three-component sonic anemometer for investigation of atmospheric turbulence. In *Flow—Its Measurement and Control in Science and Industry* Vol. 1, pp. 349–59. Instrum. Soc. Amer.

Kaimal, J. C., Wyngaard, J. C., Haugen, D. A. 1968. Deriving power spectra from a three-component sonic anemometer. *J. Appl. Meteorol.* 7:827–37

Kondo, J., Naito, G., Fujinawa, Y. 1971. Response of cup anemometer in turbulence. *J. Meteorol. Soc. Japan.* 49:63–74

Lenschow, D. H. 1971. Vanes for sensing incidence angles of the air from an aircraft. *J. Appl. Meteorol.* 10:1339–43

MacCready, P. B. 1966. Mean wind speed measurements in turbulence *J. Appl.*

*Meteorol.* 5:219–25

MacCready, P. B., Jex, H. R. 1964. Response characteristics and meteorological utilization of propeller and vane wind sensors. *J. Appl. Meteorol.* 3:182–93

McBean, G. A. 1972. Instrument requirements for eddy correlation measurements. *J. Appl. Meteorol.* 11:1078–84

Monna, W. A. A., Driedonks, A. G. M. 1979. Experimental data on the dynamic properties of several propeller vanes. *J. Appl. Meteorol.* 18:699–702

Patterson, J. 1926. The cup anemometer. *Trans. R. Soc. Can.* Ser. III, 20:1–54

Pond, S. 1968. Some effects of buoy motion on measurements of wind speed and stress. *J. Geophys. Res.* 73:507–12

Pond, S., Large, W. G., Miyake, M., Burling, R. W. 1979. A Gill twin propeller-vane anemometer for flux measurements during moderate and strong winds. *Boundary-Layer Meteorol.* 16:351–64

Ramachandran, S. 1969. A theoretical study of cup and vane anemometers. *Q. J. R. Meteorol. Soc.* 95:163–80

Schrenk, O. 1929. Über die Trägheitsfehler des Schalenkreuz-Anemometers bei schwankender Windstärke. *Z. Tech. Phys.* 10:57–66

Scrase, F. J., Sheppard, P. A. 1944. The errors of cup anemometers in fluctuating winds. *J. Sci. Instrum.* 21:160–61

Suomi, V. E. 1957. Sonic anemometer. In *Exploring the Atmosphere's First Mile*, 1:356–66. New York: Pergamon

Wieringa, J. 1967. Evaluation and design of wind vanes. *J. Appl. Meteorol.* 6: 1114–22

Wieringa, J. 1972. Tilt errors and precipitation effects in trivane measurement of turbulence fluxes over open water. *Boundary-Layer Meteorol.* 2:406–26

Wyngaard, J. C., Bauman, J. T., Lynch, R. A. 1974. Cup anemometer dynamics. See Kaimal et al 1974, pp. 701–8

*Ann. Rev. Fluid Mech. 1981. 13:425–55*


# BROWNIAN MOTION OF SMALL PARTICLES SUSPENDED IN LIQUIDS



*W. B. Russel*
Department of Chemical Engineering, Princeton University, Princeton, New Jersey 08544

## INTRODUCTION

Intriguing random motions of small particles suspended in liquids were first reported by Robert Brown, a biologist, in 1828. Controversy concerning the origin of the motion persisted, however, for many decades stimulating a series of experiments by nineteenth-century scientists including Perrin (1910) and attracting notable theorists such as Einstein, Smoluchowski, Langevin, and Lorentz. This early work, reviewed by Nelson (1967), eventually confirmed the molecular nature of matter by relating the particle motion to the thermal fluctuations of molecules in the fluid.

This review will focus on more recent work concerning suspensions of rigid particles small enough to be affected by Brownian motion, but still large enough for the fluid to be treated as a continuum. These small dimensions, $\sim$1 nm–10 $\mu$m, render inertia negligible for steady motions, although acceleration must be retained in some cases because of the intrinsic transience of the movement. The dynamics of small molecules and polymers will not be discussed, except for the application of hydrodynamic theories to the former and the behavior of compact macromolecules such as globular proteins. Nor will long-range nonhydrodynamic interactions, such as electrostatic or dispersion forces, be treated explicitly since their effects are more quantitative than qualitative.

Even within these bounds there remain many interesting phenomena. Three general topics are examined in detail: Brownian motion of isolated particles, the effect of particle-particle and particle-wall interactions, and the role of Brownian motion in the rheology of suspensions.



0066-4189/81/0115-0425$01.00

Controversies surround several of the topics. Most arise from the molecular origin of Brownian motion which requires that the forces be appended to the usual continuum descriptions of fluid-particle systems. Of the several existing approaches we opt for the simplest while remaining consistent with more fundamental molecular theories. This review is intended to complement that of Batchelor (1976b).

## BROWNIAN MOTION OF ISOLATED PARTICLES

### *Langevin Descriptions of the Dynamics*

The Brownian motion of particles suspended in liquids can be described from an equation of motion balancing a random fluctuating force acting on the particle with its inertia and the fluid resistance. The approach of Langevin (1908), illustrated below for an isolated sphere, leads to the velocity-autocorrelation function characterizing the dynamics of a heavy sphere and the translational diffusion coefficient describing the net displacement. Recent work, noting the limitation of the conventional approach to heavy particles (Pomeau & Résibois 1975), has obtained the correct velocity autocorrelation for neutrally buoyant particles and verified that the diffusivity is unaffected. With this background the generalization to anisotropic particles and interacting spheres becomes straightforward.

The Langevin equation for a sphere of mass $m$ and radius $a$ with center at $\mathbf{x}$ and velocity $\mathbf{u}$,

$$m\frac{d\mathbf{u}}{dt} = -6\pi\mu a\mathbf{u} + \mathbf{F}(t), \tag{1}$$

includes the assumption that the forces that the fluid molecules exert on the particle can be separated into rapid fluctuations $\mathbf{F}(t)$, with time scales characteristic of molecular motion ($\sim 10^{-13}$ s for water), and a much slower viscous response characterized by the pseudosteady Stokes drag. The Brownian forces are random

$$\langle \mathbf{F}(t) \rangle = 0 \tag{2}$$

and uncorrelated on the time scales of particle motion, i.e.

$$\langle \mathbf{F}(t)\mathbf{F}(t+\tau) \rangle = \mathsf{F}\delta(\tau). \tag{3}$$

The assumption that at thermal equilibrium kinetic energy is partitioned equally among the three translational modes of the particle

$$\frac{1}{2}m\langle \mathbf{u}(t)\mathbf{u}(t) \rangle = \frac{1}{2}kT\mathbf{I} \tag{4}$$

serves to determine $\mathsf{F}$.

Successive integrations of (1) with $\mathbf{x}(0)=\mathbf{x}_0$ and $d\mathbf{x}/dt(-\infty)=0$ lead to $\mathbf{x}(t)$ from which the velocity autocorrelation function

$$\mathbf{R}(\tau)=\langle \mathbf{u}(t)\mathbf{u}(t+\tau)\rangle$$
$$=\frac{\mathsf{F}}{12\pi\mu am}\exp\left(-\frac{6\pi\mu a}{m}\tau\right) \tag{5}$$

follows. From (4)

$$\mathsf{F}=12\pi\mu akT\mathbf{I}, \tag{6}$$

comprising a fluctuation-dissipation theorem relating the strengths of the random Brownian fluctuations to the steady frictional forces and thereby reflecting their common origin in the interactions between the particles and the solvent molecules.

The autocorrelation function indicates that the energy imparted to a particle by each thermal impulse decays exponentially on the viscous time scale $m/6\pi\mu a$ ($\sim 10^{-9}$s for a neutrally buoyant 0.1 $\mu$m sphere in water). The subsequent random forcing has no coherent effect.

A parallel macroscopic analysis of the diffusion process in terms of Fick's law shows that

$$\lim_{t\to\infty}\frac{1}{2}\frac{d}{dt}\langle(\mathbf{x}-\mathbf{x}_0)(\mathbf{x}-\mathbf{x}_0)\rangle=\int_0^\infty \mathbf{R}(\tau)\,d\tau=D_0\mathbf{I}. \tag{7}$$

With (5) this provides the Stokes-Einstein relation $D_0=kT/6\pi\mu a$ for the diffusion coefficient of an isolated sphere.

The assumptions implicit in the Langevin equation have been established by Mazur & Oppenheim (1970) and Albers, Deutch & Oppenheim (1971). They began with the Liouville equation for both the particles and the fluid molecules and integrated over the positions and momenta of the latter to obtain the equation of motion for the particle. The conventional form (1) results when all relaxation times associated with the fluid are short compared with those of the particle. According to their papers this means $m/m_f \gg 1$ with $m_f$ the mass of a fluid molecule. In fact, the vorticity of the continuum fluid must diffuse faster than the particle loses inertia as well, i.e.

$$\frac{a^2}{\nu}\ll\frac{m}{6\pi\mu a},$$

indicating that the particle also must be much denser than the fluid. Thus, as noted by Hauge & Martin-Löf (1973), the appropriate limit is actually $m/m_f\to\infty$ with a fixed ratio of particle to molecular size. The normal Langevin formulation, therefore, does not apply to the neutrally buoyant particles of frequent interest.

This limitation was pointed out by Lorentz in 1911 (Hauge & Martin-Löf 1973) but attracted little attention until numerical simulations of molecular motions in liquids by Rahman (1964) and Alder & Wainwright (1967) produced velocity correlations with a long tail decaying as $t^{-3/2}$, rather than exponentially as predicted by (5). Alder & Wainwright (1970) recognized the hydrodynamic origin of the effect and solved the transient Navier-Stokes equations numerically to predict the asymptotic decay.

More recently three separate analyses have predicted correctly the full velocity correlation as well as the diffusion coefficient for comparable particle and fluid densities. Assumptions (2) and (3), concerning the Brownian forces, remain valid but the evaluation of $\mathbf{F}$ and the friction law requires some care. When $a^2/\nu \sim m/6\pi\mu a$, vorticity generated by the sudden acceleration, due to the Brownian impulse, diffuses away on the same time scale that the particle decelerates. As a result, the fluid inertia remains important, preventing the viscous force from reaching the pseudosteady limit before the motion dies away. Linearity is preserved, however, because the velocities remain small enough to render negligible the convective terms in the equations of motion. Hence classical solutions to the unsteady Stokes equations suffice for the time-dependent drag.

The approaches differ in the way they determine $\mathbf{F}$. Hauge & Martin-Löf (1973) substituted the unsteady drag for an incompressible fluid in equation (1) and calculated $\mathbf{F}$ from the theory of Brownian fluctuations in a fluid continuum. Hinch (1975) espoused a purely continuum approach but applied the Langevin equation to the fluid as well as the particles with the drag determined by the instantaneous stress field. This method accounts directly for thermal fluctuations in the fluid, thereby permitting $\mathbf{F}$ to be calculated from the equipartition of kinetic energy for both fluid and particles. The velocity autocorrelation for a neutrally buoyant particle resulting from both analyses

$$R(\tau)=\frac{2}{3}\frac{kT}{m}\operatorname{Im}\{\alpha e^{\alpha^2\tau}\operatorname{erfc}\alpha\sqrt{\tau}\} \tag{8}$$

with

$$\alpha=\sqrt{3}+i, \qquad \tau=\frac{3\nu t}{4a^2}$$

can be interpreted as the response of an initially stationary sphere to an impulse of magnitude $kT$ at $\tau=0$. The long-time limit

$$\lim_{\tau\to\infty} R(\tau)=\frac{1}{8\sqrt{3\pi}}\frac{kT}{m}\tau^{-3/2} \tag{9}$$

agrees with the molecular-dynamics simulations. The short-time limit

$$\lim_{\tau \to 0} R(\tau) = \frac{2}{3}\frac{kT}{m} \tag{10}$$

illustrates that the initially rapid acceleration causes the impulse to be distributed between the particle and the fluid added mass $\frac{1}{2}m$.

One mechanism by which the initial impulse, received solely by the particle, can be redistributed over the added mass was defined by Zwanzig & Bixon (1975) following earlier work by Chow & Hermans (1973). For a slightly compressible fluid, i.e. $a/c \ll a^2/\nu$, the velocity correlation falls from the true initial value $kT/m$ to (10) as

$$R(t) = \frac{2}{3}\frac{kT}{m}\left\{1 + \frac{1}{2}e^{-(3/2)(ct/a)}\left(\cos\frac{\sqrt{3}}{2}\frac{ct}{a} - \sqrt{3}\sin\frac{\sqrt{3}}{2}\frac{ct}{a}\right)\right\}. \tag{11}$$

This short initial transient, due to acoustic damping with no effect of viscosity, is followed by the slower viscous decay predicted by (8). The complete time dependence of the velocity correlation plotted in Figure 1 differs significantly from the exponential form predicted with the pseudosteady force law. Note also that only for $ac/\nu \gg 10^2$, e.g. spheres $>0.1\mu$ in water, do the acoustic and viscous time scales separate completely.

Substitution of the correct $R(t)$ for neutrally buoyant particles into (7) generates the same diffusivity as does the exponential form valid only for heavy particles. Indeed, as recognized by the above authors and

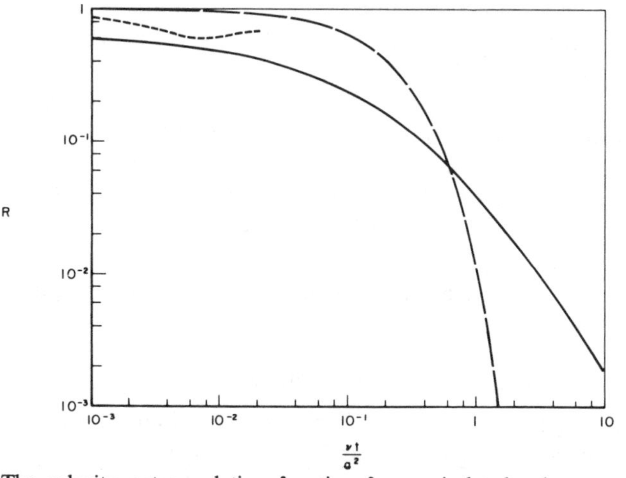


*Figure 1* The velocity autocorrelation function for an isolated sphere: —— with pseudosteady friction law, —— incompressible with time-dependent friction, ----- compressible but inviscid with $ac/v = 10^2$.

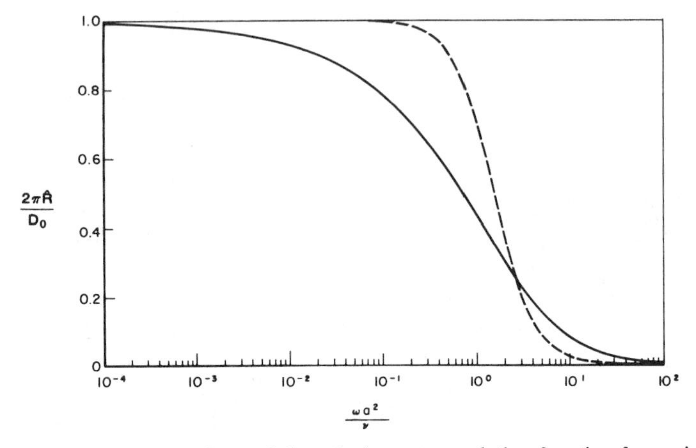


*Figure 2* The Fourier transform of the velocity autocorrelation function for an isolated sphere: _ _ _ _ _ with pseudosteady friction, —— compressible with time-dependent friction.

discussed by Batchelor (1976b), a different result would be difficult to reconcile. The physical explanation stems from the relation

$$\int_0^\infty R(t)\,dt = 2\pi \hat{R}(0) \tag{12}$$

where $\hat{R}(\omega)$ is the Fourier transform of the velocity correlation. Apparently the low-frequency components, representing almost steady motion, provide the largest net displacements and dominate the diffusion process. The two autocorrelation functions necessarily share a common zero-frequency limit (Figure 2) and therefore agree on the diffusion coefficient; the remainder of the spectra, while interesting, is superfluous.

These results indicate clearly that the diffusion coefficients for bodies of arbitrary shape, obtained directly from their pseudosteady hydrodynamic mobilities (Brenner 1974), predict accurately the increase in mean-square displacement and rotation for $t \gg l^2/\nu$ ($l$ = the characteristic dimension). Given the difficulty in measuring small spatial fluctuations, at least by light scattering, at time scales less than $10^{-10}$ s (Berne & Pecora 1976) this information fully characterizes the currently observable dynamics.

## *Diffusion in Biological Membranes*

Cell membranes are generally thought to be lipid bilayers of thickness ~2 nm which, though fluid, are $10^2$–$10^3$ times more viscous than the adjacent aqueous phases. Interest in their structure and function has stimulated measurements of rotational and translational diffusion for both the lipids themselves and large proteins intercalated in the bilayer (Edidin 1974). The data, when interpreted correctly, provide information on the membrane fluidity significant to the understanding of molecular clustering, rearrangement, and other aspects of biological organization. The intrinsic two-dimensionality of the process, however, modifies the hydrodynamic mobilities and complicates the interpretation in some subtle and interesting ways described below.

Saffman & Delbrück (1975) model the diffusant as a cylinder of radius $a$ spanning a planar membrane of thickness $h$. Since the membrane viscosity $\mu$ greatly exceeds that of the surrounding fluid, the molecule can protrude from the membrane without significant consequences. The problem presents some intriguing physics due to the two-dimensional nature of the motion within the membrane. The pseudosteady rotational mobility suffices to determine the rotational diffusion coefficient

$$D_r = \frac{kT}{4\pi\mu a^2 h} \tag{13}$$

but the translational mobility for Stokes flow in an infinite membrane is infinite. The authors propose several possible resolutions to the paradox based on additional constraints or physical processes: a finite membrane size, coupling with adjacent fluids of finite viscosity, or unsteady inertial effects within the membrane. They conclude that the correct result depends on the length scales characterizing the individual effects, with the second probably controlling for biological membranes.

A finite membrane forces the fluid velocity due to steady translation to be zero at the boundary rather than diverging logarithmically. For a characteristic membrane radius $R \gg a$ this leads to

$$D_0 = \frac{kT}{4\pi\mu h} \ln\frac{R}{a}, \tag{14}$$

introducing a weak but troublesome dependence on the membrane size plus a smaller geometrical effect.

The singular nature of the planar problem indicates that a small, but finite, viscosity in the adjacent fluids could generate a critical three-dimensionality. Saffman (1976) treated the surrounding phases as Newtonian with viscosity $\mu'$ and the membrane as having viscosity $\mu$ in the plane and an infinite viscosity in the normal direction because of the highly oriented molecular structure of thin membranes. When $\mu' a/\mu h \ll 1$ the adjacent fluids only influence the membrane flow in the outer region

$r \gtrsim h\mu/\mu'$. The resulting three-dimensional motion removes the logarithmic singularity, leaving a finite translational diffusivity

$$D_0 = \frac{kT}{4\pi\mu h}\left(\ln\frac{\mu h}{\mu' a} - 0.5771\right). \tag{15}$$

For large membranes bounded by truly inviscid fluids (perhaps gases) the mobility for steady translation, of course, becomes finite at finite Reynolds number, but the resulting velocity-dependent mobility

$$\frac{\ln\dfrac{4\mu}{\rho U a} - 0.0771}{4\pi\mu h}$$

invalidates the linear analysis leading to (7). Saffman (1976) avoided this difficulty by resorting to a Langevin analysis with unsteady, but linear, friction, much like the development of Zwanzig & Bixon (1975). The diffusivity, deduced from the indefinite form of (7), depends on time as

$$D_0 = \frac{kT}{4\pi\mu h}\left(\ln\frac{4\mu t}{\rho a^2} - 1.5771\right) \tag{16}$$

and diverges as $t \rightarrow \infty$ as expected from the steady solution.

For diffusion processes in biological membranes Saffman & Delbrück (1975) concluded that

$$\frac{R}{a} \sim 10^4, \qquad \frac{\mu t}{\rho a^2} \sim 10^{15}, \qquad \frac{\mu h}{\mu' a} \sim 10^2$$

so that the smallest diffusivity (15) controls. The theory has proven useful for interpreting experimental data (e.g. Wu et al 1979) but few sets are sufficiently complete to test it. As cited in the original paper, Cone (1972) and Poo & Cone (1974) measured both rotational ($\sim 5 \times 10^4$ $s^{-1}$) and translational ($\sim 4 \times 10^{-13}$ $m^2/s$) diffusivities of rhodopsin ($a \sim 2 \times 10^{-9}$ m) in disk membranes from the frog's retina ($\mu \sim 10^{-1}$ $Ns/m^2$, $\mu' \sim 10^{-3}$ $Ns/m^2$, $h \sim 2 \times 10^{-9}$ m). The observed ratio

$$\frac{D_0}{D_r} \sim 0.2 - 2.0 \times 10^{-17}\ m^2$$

brackets the value $1.6 \times 10^{-17}$ $m^2$ predicted by (13) and (15) although the individual values deviate significantly, perhaps because of the uncertainty in membrane viscosity. The corresponding ratio for the equivalent sphere in an infinite fluid,

$$\frac{4}{3}a^2 \sim 5 \times 10^{-17}\ m^2,$$

shows this theory, still used at times to interpret data, to be numerically inaccurate as well as physically inappropriate.

In addition to the effect on membrane-bound species, diffusion within biological membranes also may be significant within a larger context. Even with the lower diffusivity implied by $\mu/\mu' \sim 10^3$, the partitioning of a reactant into a membrane can greatly enhance the reaction rate due to the lower dimensionality of the diffusion process therein (Adam & Delbrück 1968) and the molecular orientation imposed by the planar geometry (Poo & Cone 1974).

## *Application of Hydrodynamic Theories to Molecular Motion*

The study of molecular motion in liquids is a venerable field amply reviewed in the chemical physics literature (e.g. Pomeau & Résibois 1975, Hynes 1977). The topic arises here because of the recurrent use, and frequent success, of hydrodynamic models in explaining Brownian motions of molecules ranging from argon in the condensed state to macromolecules in solution. Recently molecular dynamics simulations mentioned above and data obtained with new experimental techniques, such as depolarized light scattering (Bauer et al 1976), have demonstrated both the potential and some of the limitations of the approach, thereby generating further theoretical and experimental activity. This section will not delve into the detailed molecular theories now emerging but will review briefly the hydrodynamic theories and illustrate their comparison with experimental data.

Einstein's (1956) original predictions for the translational and rotational diffusion coefficients apply to spherical molecules with a no-slip boundary condition at the solvent-molecule interface. This assumption was immediately questioned with respect to events at the molecular scale (Sutherland 1905) and a slip condition proposed instead. The alteration increases the translational diffusivity only by a factor of 3/2 but reduces the rotational friction coefficient to zero allowing the sphere to reorient inertially with

$$D_r = \frac{3}{4\pi}\left(\frac{5kT}{2ma^2}\right)^{1/2}. \tag{17}$$

Indeed some spherical molecules approach this limit (Bauer et al 1974). Rotation of nonspherical molecules, however, drives fluid motion even with slip at the surface, but the friction coefficient remains considerably smaller than without slip, except for extreme geometries. The rotational diffusion coefficient thus provides a sensitive indication of the appropriate boundary condition once the molecular geometry and size are known.

The appropriate friction coefficients with slip were calculated by Hu & Zwanzig (1974) for spheroids and by Youngren & Acrivos (1975) for ellipsoids. For a general ellipsoid the coefficients characterizing rotation about each of the three principle axes are nonzero and unequal. With axial symmetry, however, two become equal and the third is zero. Only in the limits of a needle or a disk does the friction coefficient with slip equal that without slip.

Diffusion coefficients for the individual rotational modes can be extracted from reorientation times detected by depolarized light scattering and $^{13}C$ NMR according to

$$\tau = \frac{1}{6D_r} + \tau_0 \tag{18}$$

(Berne & Pecora 1976). Time scales measured for a variety of systems indeed vary linearly with the solvent viscosity as implied by (18) with an intercept $\tau_0$ which correlates with the inertial time scale (Bauer et al 1974). These results support the validity of the hydrodynamic model and have been compared with quantitative predictions based on known molecular volumes and shapes.

Data for small molecules, ranging in molecular volume from chlorine ($2.6 \times 10^{-3}$ nm$^3$) to valeric acid (0.209 nm$^3$), dissolved in low-viscosity organic liquids which do not exhibit strong solute-solvent interactions provides one limit. For these systems the no-slip theory predicts relaxation times two to ten times too long, but as illustrated in Figure 3 the theory for spheroidal particles with slip provides generally acceptable values. Two exceptions are the aromatics, benzene and nitrobenzene, which have geometries a bit too complex to be approximated by a spheroid. Youngren & Acrivos (1975) found that a more detailed representation of benzene as six hemispheres (hydrogen atoms) attached to an oblate spheroid (carbon ring) resulted in the much improved prediction denoted by the asterisk in Figure 3.

Solutions with strong solute-solvent interactions, such as hydrogen bonding, pose more difficult problems. The results depend strongly on the particular solvent with the no-slip boundary condition apparently appropriate in some cases (Millar et al 1979, von Jena & Lessing 1979), while others fall intermediate between the two limits (Bauer et al 1974).

Ultimately, a satisfactory theory for the motion of small molecules must blend a detailed statistical-mechanical description of the short-time collisional dynamics with a hydrodynamic formulation of the subsequent large-scale motions (Hynes 1977). Only then will the true nature of these apparent transitions from no-slip to slip behavior with decreasing molecular size and decreasing intermolecular attraction be fully understood.

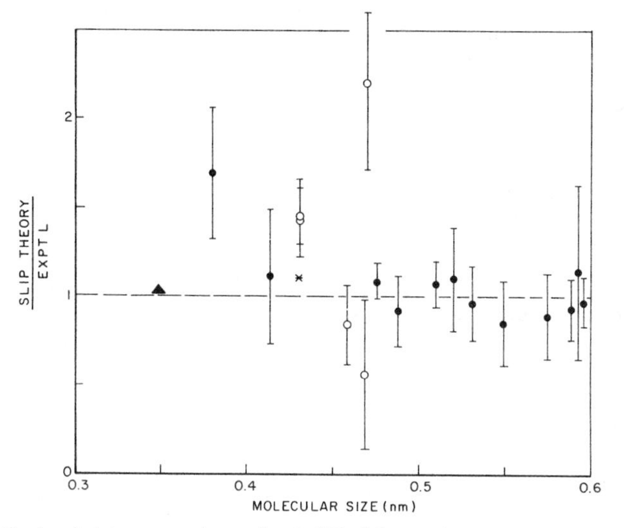


*Figure 3* Ratio of the measured rotational diffusivity to that predicted for an ellipsoid with perfect slip as a function of molecular size: ○ nitrobenzene and benzene (Bauer et al 1974); ● benzene (Youngren & Acrivos 1975); ▲ chlorine (Topalian et al 1979).

# EFFECTS OF INTERACTIONS

## *Generalized Description of the Dynamics*

The Brownian motion of particles suspended at concentrations for which hydrodynamic interactions become significant has attracted considerable attention in recent years (e.g. Ermak & McCammon 1978, Hess & Klein 1978). Two complications enter the Langevin formulation, even in the low-frequency limit with pseudosteady hydrodynamics. Clearly, the friction coefficients—now configuration dependent—couple the motions of the interacting particles. Less obvious is the coupling between the fluctuating Brownian forces at separations on the order of the particle size. The general analysis for $N$ interacting spheres in a volume $V$ sketched below represents a generalization of several existing treatments, illustrating the continuum-mechanics approach to the problem and the differences from the single particle limit.

The coupled Langevin equations for $N$ identical spheres without external couples, written in matrix form, become

$$m\frac{d\mathbf{u}}{dt} = -\mathbf{Z}\cdot\mathbf{u} + \mathbf{F} \tag{19}$$

with

$$\mathbf{u} = 0 \text{ at } t = -\infty$$

and $\mathbf{x}=\mathbf{x}_0$ at $t=0$. The matrices

$$\mathbf{u}=\{\mathbf{u}_i\}$$

$$\mathsf{F}=\{\mathbf{F}_i\}\ i=1,\ldots,N \tag{20}$$

account for the velocities and the Brownian forces on all $N$ particles while the elements of the generalized friction tensor

$$\mathbf{Z}=\{\zeta_{ij}\}\ i,j=1,\ldots,N \tag{21}$$

determine the force on the $i$th particle due to the velocity of the $j$th. Each component depends on all the positions $\mathbf{x}_k(k=1,\ldots,N)$, thereby coupling the set of equations. Only for well-separated spheres, i.e.

$$\lim_{\substack{|\mathbf{x}_k-\mathbf{x}_i|\to\infty \\ k\neq i}} \zeta_{ij}=\begin{cases} 0 & j\neq i \\ 6\pi\mu a\mathbf{I} & j=i \end{cases}, \tag{22}$$

do the particles move independently. The Brownian forces remain random

$$\langle\mathsf{F}\rangle=0$$

and uncorrelated in time

$$\langle\mathsf{F}(t)\mathsf{F}^T(t')\rangle=\mathsf{F}_0\delta(t-t'), \tag{23}$$

while the kinetic energy imparted to the particles is partitioned equally among the translational modes as

$$\frac{1}{2}m\langle\mathbf{u}\mathbf{u}^T\rangle=\frac{1}{2}kT\mathbf{I}. \tag{24}$$

This system can be solved exactly as was the equation for a single particle, provided the configuration $\{\mathbf{x}_i\}$ does not change significantly on the viscous time scale. As a result

$$\mathsf{F}_0=2kT\mathbf{Z}$$

and

$$\mathbf{R}(\tau)=\langle\mathbf{u}(t)\mathbf{u}^T(t+\tau)\rangle=\frac{kT}{m}\exp\left\{-\mathbf{Z}\frac{\tau}{m}\right\} \tag{25}$$

so that

$$\mathbf{D}=\int_0^\infty \mathbf{R}(\tau)d\tau=kT\mathbf{Z}^{-1} \tag{26}$$

where $\mathbf{Z}^{-1}=\{\omega_{ij}\}$ is the generalized mobility tensor. From (25) one can readily verify that the change in the relative position, $r_{ij}=|\mathbf{x}_i-\mathbf{x}_j|$, of any two particles during the viscous relaxation time

$$\frac{\langle(\Delta r_{ij})^2\rangle}{\langle r_{ij}^2\rangle}\sim\frac{2(mkT)^{1/2}}{3\pi\mu a}$$

remains small for all conditions of interest (Batchelor 1976a).

The generalized fluctuation-dissipation theorem in (25) reveals two interesting facets of the Brownian forces in a coupled system. Since

$$\langle \mathbf{F}_i(t)\mathbf{F}_j(t')\rangle = 2kT\zeta_{ij}(\mathbf{x}_k)\delta(t-t') \tag{27}$$

the forces acting on two interacting particles ($i \neq j$) are coupled and their magnitudes ($i=j$) depend on the configuration $\{\mathbf{x}_k\}$. Other derivations, perhaps more fundamental but also more involved, starting with the Liouville equations for the particles and the molecules arrive at the same conclusion (Lax 1966, Zwanzig 1969, Deutch & Oppenheim 1971).

The elements of **D** comprise multicomponent diffusion coefficients

$$\mathbf{D}_{ij} = kT\omega_{ij}(\mathbf{x}_k) \tag{28}$$

indicating that the flux of a particular particle may be hindered by hydrodynamic interactions, since $\mathbf{D}_{11} \leqslant kT/6\pi\mu a\mathbf{I}$, but also becomes coupled to the diffusion of its neighbors since $\mathbf{D}_{ij} \neq 0$ for $i \neq j$.

The tracer or self-diffusion coefficient, characterizing the wandering of a tagged particle in a uniform environment of untagged neighbors, reflects the former effect. For untagged spheres at volume fractions $\phi \ll 1$ without long-range interaction potentials, the configurational average of (28) yields

$$\langle \mathbf{D}_{11}\rangle = \frac{kT}{6\pi\mu a}(1-1.83\phi)\mathbf{I} \tag{29}$$

(Batchelor 1976a, Anderson & Reed 1976a). Figure 4 illustrates the uncertainties which plague comparisons of this result with measurements for small macromolecules such as bovine serum albumin. The

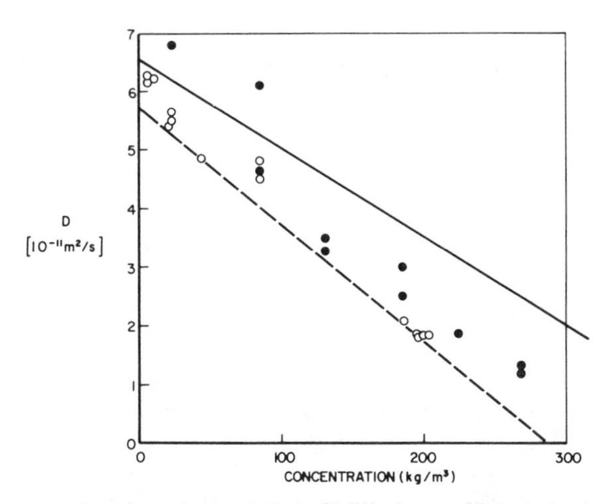


*Figure 4* Concentration dependence of the self-diffusion coefficients for bovine serum albumin in water: • pH = 4.9 (isoelectric point) (Keller et al 1971); ○ 1.5 M NaCl (Kitchen et al 1976); ----- Equation (29) with $a$ = 3.75 nm; —— Equation (29) with $a$ = 3.28 nm.

effective hydrodynamic radius must be extracted from the data, which introduces considerable uncertainty into the slope for the range of $a$ ~3.28–3.75 nm found in the literature. In addition, weak long-range forces, which may persist even at the isoelectric point or high salt concentrations, can alter the slope significantly (Anderson & Reed 1976a). Nonetheless, the results provide reasonable corroboration of the theory.

## *Equilibrium Analysis of Gradient Diffusion*

Diffusion coefficients, but not the detailed dynamics of colloidal particles, can be derived from the analysis of an equilibrium system in which an external potential acts on the particles to create a nonuniform concentration. In his classic paper Einstein obtained the diffusivity for an isolated sphere by balancing the flux due to the external potential against diffusion down the concentration gradient and recognizing the gradient in chemical potential as the appropriate driving force for the latter. Recently Batchelor (1976a) extended Einstein's thermodynamic argument to finite concentrations; his detailed calculation for pair interactions between hard spheres corrects the diffusion coefficient for hydrodynamic and potential interactions in the dilute limit. The alternative statistical-mechanical approach sketched below (Russel et al 1980) reaches the same conclusion without invoking thermodynamic arguments about the effective driving force for diffusion. The resulting generalized Stokes-Einstein relation, valid at arbitrary concentrations, directly relates the gradient diffusion coefficient to the sedimentation coefficient and the osmotic compressibility.

Consider a closed system of $N$ spheres suspended in a Newtonian liquid and subjected to the external potential $U(x)$. At equilibrium the sedimentation and diffusion fluxes must balance locally as

$$-D\nabla n(\mathbf{x})-\frac{K(\phi)}{6\pi\mu a}n\nabla U(\mathbf{x})=0. \tag{30}$$

In addition, the Boltzmann distribution

$$P_N=\frac{1}{Q}\exp-\frac{1}{kT}\left\{V(\mathbf{x}_1,\ldots,\mathbf{x}_N)+\sum_{i=1}^{N}U(\mathbf{x}_i)\right\} \tag{31}$$

determines the probability of finding the $N$ particles in configuration $\{\mathbf{x}_k\}$ in terms of the interparticle potential $V$ and the external potential. $Q$ normalizes the distribution so that

$$\int P_N d^3\mathbf{x}_1\ldots d^3\mathbf{x}_N=N!,$$

while

$$n(\mathbf{x}_1)=\frac{1}{(N-1)!}\int P_N d^3\mathbf{x}_2\ldots d^3\mathbf{x}_N \tag{32}$$

is the local concentration. The sedimentation coefficient $K(\phi)$ in principle can be calculated from $P_N$ and creeping flow solutions for $N$ interacting spheres; exact $O(\phi)$ corrections to Stokes' law are available for hard-sphere repulsions (Batchelor 1972) and longer-range potentials (Anderson & Reed 1976b).

The diffusivity $D$ follows from (30) once $\nabla n(\mathbf{x})$ has been related to $\nabla U$. For slowly varying potentials one can show that

$$\nabla n(\mathbf{x}_1)=-\frac{n(\mathbf{x}_1)}{kT}\nabla U(\mathbf{x}_1)\left\{1+4\pi n(\mathbf{x}_1)\int_0^{\infty} r^2[g(r)-1]\,dr\right\}. \tag{33}$$

where

$$g(r)=\frac{1}{n^2(\mathbf{x}_1)}\frac{1}{(N-2)!}\int \exp\left(-\frac{V}{kT}\right)d^3\mathbf{x}_3\ldots d^3\mathbf{x}_N \tag{34}$$

is the radial distribution function without the external potential. The bracketed term in (33) can be identified as the osmotic compressibility $kTdn/d\pi$ (Reed & Gubbins 1973) leaving

$$D=\frac{K(\phi)}{6\pi\mu a}\frac{d\pi}{dn}. \tag{35}$$

This result is rigorous and exact for a slowly varying external potential, i.e. $a\nabla U/U \ll 1$. The independent roles of hydrodynamic and thermodynamic forces in the diffusion process are best illustrated with the rigorous results available in the dilute limit. For hard-sphere repulsions, i.e.

$$V=\begin{cases} 0 & r_{12}>2a, \\ \infty & r_{12}<2a. \end{cases} \tag{36}$$

Batchelor (1976a) notes that

$$\frac{d\pi}{dn}=kT(1+8\phi)$$

and

$$K(\phi)=1-6.55\phi \tag{37}$$

so that $\dfrac{D}{D_0}=1.45\phi.$

The thermodynamic enhancement represents an osmotically driven expansion into a region of lower pressure and slightly overcompensates for the hydrodynamic retardation of the sedimentation process. Both effects depend strongly on the interparticle potential; long range repulsions increase the osmotic compressibility but decrease the sedimentation coefficient (Reed & Anderson 1980). The net effect can be dramatic increases in the diffusivity (Anderson et al 1978). The physical situation, therefore, differs markedly from self-diffusion in which neighboring spheres merely provide a passive resistance to motion.

The hard-sphere theory has proven controversial for two reasons. Part of the confusion arises from the published predictions for the $O(\phi)$ coefficient in $D/D_0$ ranging from $-2.6$ to $+8.0$ (Pyun & Fixman 1964, Altenberger & Deutch 1973, Phillies 1973, Anderson & Reed 1976a, Harris 1976). Recently Felderhof (1978) and Wills (1979) independently confirmed (37) and, more important, discussed at length the approximate hydrodynamics and, in some cases, incomplete physics responsible for the other values.

Attempts to verify (37) experimentally have been confounded by residual effects of long-range interparticle potentials with small macromolecules such as bovine serum albumin (Fair & Jamieson 1980) and the difficulty in interpreting dynamic light-scattering data for large particles.

The data of Newman et al (1974) appears to be free of both problems. Their aqueous solutions of fd bacteriophage DNA were quite monodisperse with a molecular weight of $1.86 \pm 0.06 \times 10^6$ and a hydrodynamic

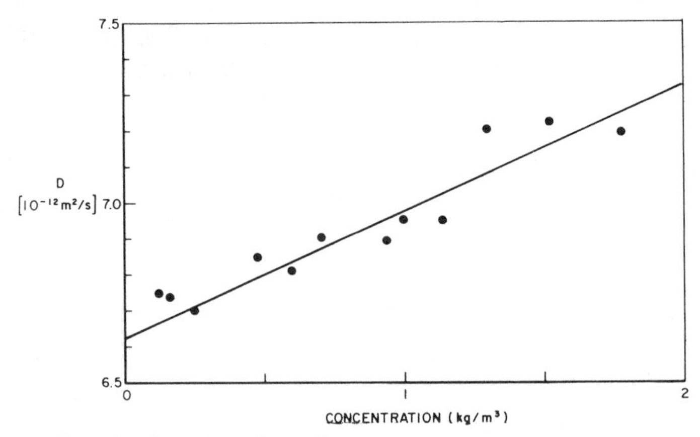


*Figure 5* Concentration dependence of the gradient-diffusion coefficient for fd bacteriophage DNA in water: ● light scattering data of Newman et al (1974); —— Equation (37).

**Table 1** $O(\phi)$ coefficients (from Newman et al 1974)

| | Hard sphere theory | Experiment |
|---|---|---|
| $K$ (sedimentation) | $-6.55$ | $-6.7 \pm 0.8$ |
| $\frac{1}{kT}\frac{d\pi}{dn}$ (osmotic compressibility) | 8.0 | $7.6 \pm 3.9$ |
| $\frac{D}{D_0}$ (diffusion) | 1.45 | $1.2 \pm 0.4$ |

radius of $31.6 \pm 0.6$ nm. They measured independently the sedimentation and diffusion coefficients and the osmotic pressure as functions of concentration. The dynamic light-scattering data for $D$ in Figure 5 was independent of scattering angle since the wavelength of light significantly exceeded the molecular size. As illustrated in Table 1 the observed concentration dependence of all three quantities fell within experimental error of the hard-sphere predictions. The osmotic pressure establishes that long-range interparticle forces are negligible and the sedimentation coefficient verifies the hydrodynamic theory, leaving the diffusion data as a clear confirmation of the generalized Stokes-Einstein equation, at least for the dilute limit.

## *Dynamic Light Scattering*

The development of dynamic light scattering as a rapid and accurate technique for measuring translational and rotational diffusion coefficients of colloidal particles and dissolved macromolecules has greatly advanced the study of Brownian motion. In addition to making basic studies of the phenomena easier, the technique has proven quite valuable for the characterization of biological and synthetic macromolecules with respect to molecular size and shape (e.g. Bloomfield 1977, McDonnell & Jamieson 1977).

Since suspended particles or macromolecules generally scatter far more light than does a low-molecular-weight fluid, the scattering intensity directly reflects their size, shape, and dynamical properties. For example, the spectrum of light scattered from a volume containing $N$ identical rigid spheres of mass $m$ (Berne & Pecora 1976)

$$I(\mathbf{k}, \omega) = Nm^2 P(k) \operatorname{Re} \int_0^\infty G(\mathbf{k}, t) e^{-i\omega t} \, dt \tag{38}$$

depends on the single-particle scattering function $P(k)$ and the coherent-structure factor

$$G(\mathbf{k}, t) = \frac{1}{N} \langle n^*(\mathbf{k}, t) n^*(\mathbf{k}, 0) \rangle . \tag{39}$$

Here

$$n^*(\mathbf{k},t)=\int n(\mathbf{x},t)e^{-i\mathbf{k}\cdot\mathbf{x}}\,d^3\mathbf{x} \tag{40}$$

is the spatial transform of the fluctuating number density and

$\mathbf{k}$ = the wave vector of the scattered light,

$|k|=\frac{4\pi}{\lambda}\sin\frac{\theta}{2}$,

$\lambda$ = wavelength of the light in the medium,

$\theta$ = scattering angle relative to the incident beam.

Conventional total-intensity light scattering measures

$$\int I(\mathbf{k},\omega)d\omega=m^2P(k)\langle n^{*2}(\mathbf{k},0)\rangle, \tag{41}$$

thereby detecting for a known concentration only the particle size. The full spectrum, on the other hand, also contains information on the dynamics of the concentration fluctuations. Several books (Cummins & Pike 1974, Chu 1974) describe the experimental techniques. This section will focus on the theoretical problem of relating $n^*(\mathbf{k},t)$ to the thermodynamical and mechanical properties of the suspension.

The Fourier transform of the usual conservation equation

$$\frac{\partial n^*(\mathbf{k},t)}{\partial t}=-i\mathbf{k}\cdot\mathbf{J}^*(\mathbf{k},t) \tag{42}$$

suffices to determine $n^*$ once the transform of the flux $\mathbf{J}^*$ is specified. For noninteracting spheres Fick's law yields

$$\mathbf{J}^*=-i\mathbf{k}D_0n^* \tag{43}$$

leading to

$$G(\mathbf{k},t)=e^{-k^2D_0t} \tag{44}$$

and

$$I(\mathbf{k},\omega)=Nm^2P(k)\frac{D_0k^2}{\omega^2+D_0^2k^4}. \tag{45}$$

The former indicates that concentration fluctuations with length scale $k^{-1}$ disappear on the diffusion time scale $(k^2D_0)^{-1}$. Optical mixing or beating techniques which detect decay times of $\sim 10^{-6}$ s and slower, therefore, can measure diffusion coefficients for particles larger than a few nanometers.

At finite concentrations, however, (44) fails because of interactions between particles. As noted in the previous section, the diffusion coefficient becomes concentration dependent, a moderate effect for hard-sphere interactions but a dramatic one for strong electrostatic repulsions. In addition, experiments with large particles (Colby et al 1975, Fijnaut et al 1978) or long-range electrostatic effects (Schaefer & Berne 1974, Brown et al 1975) reveal decay times for the correlation function that deviate from the expected $k^{-2}$ dependence. The effective diffusion coefficient depends, therefore, on the wavelength of the fluctuation as well as the mean concentration, considerably complicating the interpretation.

Such observations stimulated several attempts to generalize the theory to include interactions (Altenberger & Deutch 1973, Harris 1976, Ackerson 1976, 1978, Felderhof 1978, Wills 1979, Altenberger 1979). As with the theories mentioned in the previous section the results vary considerably, in part due to different levels of approximation for the hydrodynamics. As a result opinions differ about which coefficient (gradient diffusion or self-diffusion) is measured by dynamic light scattering. Altenberger (1979) approaches the correct answer via a complex route, apparently failing only in the detailed hydrodynamics, and also discusses the shortcomings of the previous efforts. Our much simpler approach outlined below (Russel & Glendinning 1980) follows the conventional treatment of the light scattering based on (38)–(40) and employs the exact two-sphere hydrodynamics to predict the concentration and wave-number dependence of the effective diffusion coefficient for hard spheres in the dilute limit.

The transformed flux $\mathbf{J}^*$ in (42) arises from diffusion down concentration gradients with spatial periodicity $\mathbf{k}$. As for the $ak \ll 1$ limit discussed in the previous section, the effective diffusivity can be derived for an equilibrium system, rather than the more complex non-equilibrium one implied by (42). This requires the construction of a spatially periodic potential $U(\mathbf{x}) = U^*(\mathbf{k})e^{i\mathbf{k}\cdot\mathbf{x}}$ to balance the diffusion flux and maintain steady concentration variations of the same wavelength. Then

$$\mathbf{J}^* = -i\mathbf{k}D(\mathbf{k},\phi)n^*(\mathbf{k}) = \frac{K(\mathbf{k},\phi)}{6\pi\mu a}n_0 i\mathbf{k}U^*(\mathbf{k}) \tag{46}$$

provided the deviations from the mean number density $n_0$ remain small. Now $K(\mathbf{k},\phi)$ must be calculated and $n^*(\mathbf{k})$ related to $U^*(\mathbf{k})$.

For long wavelengths the sedimentation coefficient must asymptote to (37). Then all particles have the same velocity, and hydrodynamic interactions produce three distinct $O(\phi)$ effects on the motion of an individual particle: a passive retardation by force-free neighbors ($-1.83\phi$), a backflow and local pressure gradient from the net motion of all the particles $(-5.0\phi)$, and enhancement due to nearby particles moving in the same direction $(+0.28\phi)$. As $ak$ becomes $O(1)$ the large-scale motion is suppressed by the periodicity, and the velocities of nearby particles shift out of phase. As a result, the last two contributions decay to zero in an oscillatory fashion leaving, for $ak \gg 1$, no effect of the motion of other particles but only the passive resistance of neighbors $(-1.83\phi)$ as in the self-diffusion process.

The gradient in the mean number density can be evaluated from a Boltzmann distribution reflecting the small, spatially varying potential. The resulting proportionality constant between $n^*(\mathbf{k})$ and $-U^*(\mathbf{k})/(n_0 kT)$ is the static-structure factor

$$S(k) = 1 + 4\pi n_0 \int_0^\infty (g-1)\frac{\sin kr}{kr} r^2\, dr \tag{47}$$

which for $k=0$ reduces to the osmotic compressibility. For hard spheres

$$S(k) = 1 + \frac{3\phi}{(ak)^2}\left(2\cos 2ak - \frac{\sin 2ak}{ak}\right); \tag{48}$$

as with the sedimentation coefficient the $O(\phi)$ term decays from $-8\phi$ for $ak \ll 1$ to zero for $ak \gg 1$ in an oscillatory fashion. The final result for the effective diffusion coefficient from (46) and (48) is

$$D(k,\phi) = \frac{kT}{6\pi\mu a}\{1 + K_D\phi\} + O(\phi^2); \tag{49}$$

where $K_D$ is the function of $ak$ plotted in Figure 6.

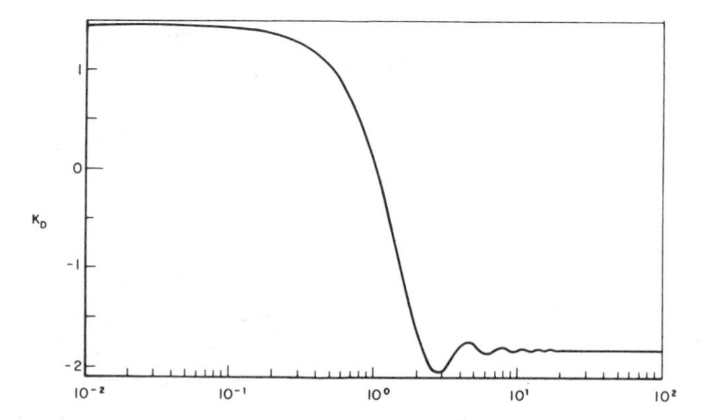


*Figure 6* The concentration dependence of the effective diffusion coefficient as a function of the wavenumber of the gradients.

Finally, the conservation equation (42) can be integrated to determine the correlation function

$$G(\mathbf{k},t)=S(k)e^{-k^2D(k,\phi)t} \tag{50}$$

and the spectrum

$$I(\mathbf{k},\omega)=Nm^2P(k)S(k)\frac{k^2D(k,\phi)}{\omega^2+k^4D^2(k,\phi)}. \tag{51}$$

Clearly, at finite concentrations dynamic light scattering only detects gradient diffusion when $ak \lesssim 0.2$ while for $ak \gtrsim 4$ the process closely resembles self-diffusion. For the visible spectrum and scattering angles between 45° and 135° these translate into $a \lesssim 20$ nm and $a \gtrsim 0.20$ $\mu$m, respectively.

While the theory seems to answer some troublesome questions, no definitive data are available for testing the predictions. Figure 7 shows data from Fijnaut et al (1978) for polymethylmethacrylate spheres with $a=0.12$ $\mu$m in benzene. The comparison is only qualitative because most of the volume fractions lie far beyond the range of validity of (49); in addition, the highly swollen latices may deviate significantly from hard-sphere behavior. Nonetheless, the trends are reasonably consistent with the predictions.

The hydrodynamic portion of the theory is complete but the light-scattering analysis remains less certain. Berne & Pecora (1976) carefully

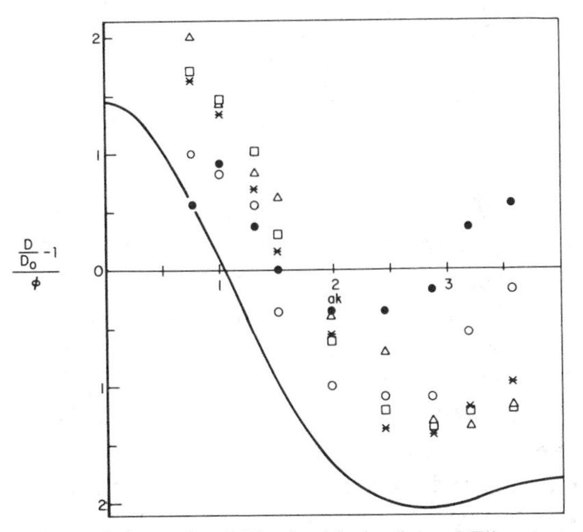


*Figure 7* Comparison of theory for $D(k,\phi)$ with the data of Fijnaut et al (1978): $\phi$, (●) 0.06, (○) 0.11, (∗) 0.25, (□) 0.33, (△) 0.50.

note that conventional treatments of dynamic scattering pertain only to particles small relative to the wavelength of light and neglect multiple scattering. For independent particles the latter is certainly negligible and the single-particle scattering function $P(k)$, available from Mie theory (Kerker 1969), accounts for the former effect. But when $\phi \sim O(1)$ and $ak \sim O(1)$, exactly the region of interest, multiple scattering may be significant and unaccounted for within the current theory.

### *Hindered Diffusion in Pores*

Several transport processes of biological or industrial interest involve diffusion and convection of small particles or macromolecules through packed beds, small capillary tubes, or membranes containing small pores. For steady-state transport through porous membranes, either physiological or artificial, the key questions center on the variation in the rejection characteristics and flux with the flow rate, particle size, and membrane structure (Bean 1972). For intrinsically transient separation techniques such as gel permeation and hydrodynamic chromatography (Silebi & McHugh 1978), the performance depends on the propagation time and dispersion of a pulse. One idealized model, applicable in some degree to both processes, considers the movement of spheres of radius $a$ through a cylindrical pore of radius $r_0$ (Anderson & Quinn 1974, Brenner & Gaydos 1977). In fact, thin sheets of track-etched mica (Bean 1972) with very uniform pores provide model membranes having almost exactly this geometry.

For dilute concentrations within the pores, i.e. $\phi < 2a/3r_0$ (Anderson & Quinn 1974), the conservation equation for the particles is (Zwanzig 1969, Brenner & Gaydos 1977, Felderhof 1978)

$$\frac{\partial P}{\partial t} + \nabla \cdot \mathbf{U}(\mathbf{x}) P = \nabla \cdot \mathbf{D}(\mathbf{x}) \cdot \left\{ \nabla P + \frac{P}{kT} \nabla V \right\}. \tag{52}$$

The equation holds both within and without the pore with $P$, the single-particle probability density or the local number density, becoming spatially uniform in the surrounding fluid. Within a pore, hydrodynamic interactions with the wall retard the motion, reducing the sphere velocity $\mathbf{U}(\mathbf{x})$ below the local fluid velocity and generating the anisotropic diffusivity tensor $\mathbf{D}$ with elements $D_{ij}(\mathbf{x}) \lesssim D_0$. Potential interactions with the pore wall produce the force $-\nabla V$ on the particle.

The role of entrance and exit effects can be illustrated by scaling the radial position on $r_0$ and the axial position on the membrane thickness $L$ within the pore so that

$$\left(\frac{L}{r_0}\right)^2 \frac{\partial P}{\partial \tau} + \mathrm{Pe}\, u \frac{\partial P}{\partial z} = \left(\frac{L}{r_0}\right)^2 \frac{1}{r} \frac{\partial}{\partial r}\left( r d_{\perp} e^{-V/kT} \frac{\partial}{\partial r}\left(P e^{V/kT}\right)\right) + d_{\parallel} \frac{\partial^2 P}{\partial z^2} \tag{53}$$

but on $r_0$ in the vicinity of the pore mouths so that

$$\frac{\partial P}{\partial \tau}+\mathrm{Pe}^*\nabla\cdot\mathbf{u}P=\nabla\cdot\mathbf{d}e^{-V/kT}\cdot\nabla Pe^{V/kT}. \tag{54}$$

The pore and entrance Peclet numbers are related by $\mathrm{Pe}=u_b L/D_0=(L/r_0)\mathrm{Pe}^*$ with $u_b$ the average fluid velocity in the pore. Within the pore the dimensionless particle velocity $u$ and diffusivities, $d_\perp$ and $d_\parallel$, and the interaction potential depend only on $r$, but outside $\mathbf{u}$, $\mathbf{d}$, and $V$ vary in two dimensions.

For long pores $P\exp(V/kT)=n(z)$ at steady state within the pore. The axial flux then consists of an $O(D_0\exp[-V_0/kT]\Delta n/L)$ contribution from diffusion and $O(u_b\exp[-V_0/kT]\Delta n)$ from convection. Here $V_0$ denotes the characteristic potential within the pore and $\Delta n$ the concentration difference through the pore.

The concentrations within the entrance and exit regions are invariably assumed to be in equilibrium with the bulk, i.e. $P\exp(V/kT)=n^*=$ constant (Anderson & Quinn 1974). According to (54) this requires that convection into the pore be weak, leaving diffusion to accommodate the axial flux through the pore. Equating the internal and external fluxes determines

$$\frac{\Delta n^*}{\Delta n}\sim\frac{r_0}{L}\frac{1+\mathrm{Pe}}{1+\mathrm{Pe}^*},$$

establishing that $n^*\sim$ constant and equilibrium is approached if $r_0/L\ll 1$ along with $\mathrm{Pe}^*\ll 1$.

With closely spaced pores, rather than isolated ones as considered above, the external transport process becomes one-dimensional and controlled by the convective mass-transfer coefficient $k$, leading to

$$\frac{\Delta n^*}{\Delta n}\sim\frac{fD_0}{kL}(1+\mathrm{Pe})$$

with $f$ the void fraction of the membrane. For thin highly porous membranes, entrance effects thus can be significant even with $r_0/L\ll 1$ (Malone & Anderson 1978).

Anderson & Quinn (1974) examined steady-state transport through a long pore with equilibrium at the ends, neglecting dispersive effects arising from coupling between radial diffusion and the parabolic velocity profile. Then $P\exp(V/kT)=n(z)$ for all Pe so that integration of (53) over the cross section gives

$$\mathrm{Pe}\langle u\rangle\frac{dn}{dz}=\langle d_\parallel\rangle\frac{d^2n}{dz^2} \tag{55}$$

with $n=n_0$ at $z=0$ and $n=n_1$ at $z=1$. The dimensionless particle velocity

$$\langle u\rangle = \frac{\int u e^{-V/kT}\,da}{\int e^{-V/kT}\,da} \tag{56}$$

and pore diffusivity

$$\langle d_{\|}\rangle = \frac{\int d_{\|} e^{-V/kT}\,da}{\int e^{-V/kT}\,da}$$

are normalized to characterize the motion of an average particle within the pore. The dimensionless axial flux, normalized on diffusion over the full cross section, is

$$\begin{aligned} J_z &= \Phi\left(-\langle d_{\|}\rangle \frac{dn}{dz} + \mathrm{Pe}\langle u\rangle n\right) \\ &= \Phi\langle \mathrm{Pe}\rangle\langle d_{\|}\rangle \frac{e^{\langle \mathrm{Pe}\rangle} n_0 - n_1}{e^{\langle \mathrm{Pe}\rangle} - 1} \end{aligned} \tag{57}$$

where $\langle \mathrm{Pe}\rangle = \mathrm{Pe}\langle u\rangle / \langle d_{\|}\rangle$; the steric factor

$$\Phi = \int e^{-V/kT}\,da \tag{58}$$

relates the average concentration in the pore mouths to the bulk fluid concentrations $n_0$ and $n_1$.

Brenner & Gaydos (1977) treated the unsteady situation including dispersive effects by extending the classical Taylor-Aris theory to allow for the variable mobilities and interaction potential crucial to pore transport. Their result for the dispersion coefficient

$$D^* = D_0\left(\langle d_{\|}\rangle + \frac{\mathrm{Pe}^{*2}}{48}\langle d_v\rangle\right) \tag{59}$$

indicates the effect to be significant for transient processes when $\mathrm{Pe}^* > 1$. At steady state, however, since $\mathrm{Pe} \gg \mathrm{Pe}^*$ the diffusion flux and also the effect of dispersion are then insignificant relative to convection (Anderson & Quinn 1974).

Brenner & Gaydos (1977) calculated all three hydrodynamic coefficients from the best available approximate solutions for a rigid sphere moving in a tube. Their asymptotic expansions for the mobilities, valid for $a/r_0 \ll 1$, matched exact solutions for motion near a plane wall in the inner region $(r_0 - r)/r_0 \ll 1$ with approximations from the method of reflections for $(r_0-r)/a \gg 1$. Without long range interaction potentials the sphere is merely excluded from $r_0-a<r<r_0$ and

$$\Phi\langle d_{\parallel}\rangle = 1+\frac{9}{8}\frac{a}{r_0}\ln\frac{a}{r_0}-1.539\frac{a}{r_0}+o\left(\frac{a}{r_0}\right)$$

$$\Phi\langle u\rangle = 1-7.90\left(\frac{a}{r_0}\right)^2+o\left(\frac{a}{r_0}\right)^2 \tag{60}$$

$$\Phi\langle d_v\rangle = 1-3.862\frac{a}{r_0}+14.40\left(\frac{a}{r_0}\right)^2+o\left(\frac{a}{r_0}\right)^2$$

with $\Phi=(1-a/r_0)^2$. The interactions with the wall reduce the diffusive flux substantially and the convective slightly, e.g. $\Phi\langle d_{\parallel}\rangle \sim 0.6$ and $\Phi\langle u\rangle \sim 0.9$ for $a/r_0 \sim 0.1$. The corresponding mean velocity $\langle u\rangle$ is increased, however, because of the exclusion of particles from the slowly moving fluid near the wall.

Several groups have attempted to test these predictions with the track-etched membranes and either bovine serum albumin or polystyrene latices. Significant hindrance has been observed but comparison with the theory remains inconclusive due to difficulties with adsorption onto the pore walls (Wong & Quinn 1976) and electrostatic interactions with the pore walls (Malone & Anderson 1978).

## THE EFFECT OF BROWNIAN MOTION ON THE RHEOLOGY OF SUSPENSIONS

Suspensions of particles in Newtonian liquids often exhibit markedly non-ideal rheological behavior. With larger particles for which hydrodynamic effects dominate, the stress may not be Newtonian and may vary with the history of the sample but must remain a linear function of the rate of strain, dependent only on the volume fraction and geometry of particles and their instantaneous configuration (Batchelor 1970). For smaller particles subject to Brownian motion, however, the constitutive relation becomes nonlinear in the rate of strain and shows a fading memory. The additional strain rate and time dependence both scale on the diffusion time $\mu a^3/kT$, with $a$ the characteristic dimension of the particle; the data of Krieger (1972) in Figure 8 for polystyrene spheres of several sizes in three different fluids illustrate the former scaling quite clearly. Other nonhydrodynamic forces, particularly long-range colloidal interactions, can affect the rheology significantly (Russel 1980) but will not be included here.

The physical mechanism responsible depends on the geometry of the particle. For example, a shear flow tends to orient a rigid rod in the

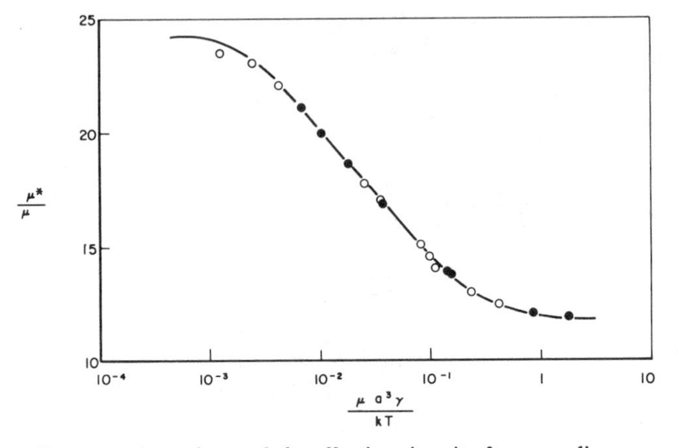


*Figure 8* Shear-rate dependence of the effective viscosity for monodisperse suspensions of polystyrene spheres at $\phi = 0.50$ in (—) water, (o) benzyl alcohol, (•): m-cresol (Krieger 1972).

direction of flow, causing gradients in the orientational distribution function and thereby generating an opposing rotational diffusion process. As a result Brownian motion is significant even at infinite dilution (Leal & Hinch 1971, Hinch & Leal 1972). With spheres, however, orientations are irrelevant and positions only become significant at finite concentrations. Then hydrodynamic and potential interactions can cause spatial nonuniformities in the probability density which drive translational diffusion. In both cases Brownian motion exerts two influences, a direct contribution to the bulk stress tensor from the Brownian torque or force acting on the particle and an indirect effect on the viscous stress through the orientational or spatial distribution function. Batchelor (1976b) recently has reviewed in detail the salient features of both problems. This section will concentrate on the nature of the direct Brownian stress and its role in the rheology of suspensions of spheres.

The existence of a direct thermodynamic contribution to the bulk stress in a suspension has been questioned at times (e.g. Leal & Hinch 1971). The conventional continuum formulation ignores fluctuating stresses due to thermal motion of the fluid and therefore misses the direct Brownian contribution. A thermodynamic stress, including the effect of Brownian motion as well as any interparticle forces or torques, must be appended to the bulk viscous stresses defined by Batchelor (1970). The derivation, generally attributed to Giesekus (1962) for rotational motion and outlined by Batchelor (1977) for translational diffusion of spheres, resembles the development of the stress tensor for the theory of rubber elasticity (Bird et al 1977).

For a volume $\mathcal{V}$ containing $N$ axisymmetric particles with positions $\mathbf{x}_i$ and orientations $\mathbf{e}_i$ the Helmholtz free energy $A$ depends on the partition function $P_N(\mathbf{x}_k, \mathbf{e}_k)$ and the total interaction potential $V(\mathbf{x}_k, \mathbf{e}_k)$ as

$$A = kT \ln P_N + V. \tag{61}$$

For an arbitrary homogeneous deformation $\mathbf{K}$ perturbing the system slightly from equilibrium

$$dA = \mathcal{V}\mathbf{K} : \langle \boldsymbol{\sigma} \rangle^{\text{thermo}} = kT d\ln P_N + dV. \tag{62}$$

With the chain rule for the differentiation and the relation between $\mathbf{K}$ and particle displacements, the latter becomes

$$\sum_{i=1}^{N} \left( \mathbf{x}_i \cdot \mathbf{K} \cdot \frac{\partial}{\partial \mathbf{x}_i} + (\mathbf{e}_i \mathbf{I} - \mathbf{e}_i \mathbf{e}_i \mathbf{e}_i) : \mathbf{K} \cdot \frac{\partial}{\partial \mathbf{e}_i} \right) (kT \ln P_N + V).$$

For arbitrary deformations $\mathbf{K}$ then

$$\begin{aligned} \langle \boldsymbol{\sigma} \rangle^{\text{thermo}} &= \frac{1}{\mathcal{V}} \sum_{i=1}^{N} \left( \mathbf{x}_i \frac{\partial}{\partial \mathbf{x}_i} + (\mathbf{e}_i \mathbf{I} - \mathbf{e}_i \mathbf{e}_i \mathbf{e}_i) \cdot \frac{\partial}{\partial \mathbf{e}_i} \right) (kT \ln P_N + V) \\ &= \frac{1}{\mathcal{V}} \sum_{i=1}^{N} (\mathbf{x}_i \mathbf{F}_i + \mathbf{L}_i \times \mathbf{e}_i \mathbf{e}_i) \end{aligned} \tag{63}$$

where $\mathbf{F}_i$ and $\mathbf{L}_i$ are the net nonhydrodynamic forces and torques, respectively, acting on the $i$th particle. Clearly, gradients in the partition function, due to either an applied flow or interparticle forces, generate bulk effects equivalent to those from steady forces,

$$-kT \frac{\partial \ln P_N}{\partial \mathbf{x}_i},$$

and torques,

$$-kT \mathbf{e}_i \times \frac{\partial \ln P_N}{\partial \mathbf{e}_i} \tag{64}$$

on the individual particles. The complete direct Brownian stress consists of this thermodynamic contribution plus the viscous stresses caused by these forces and torques.

Calculation of these stresses requires knowledge of $P_N$ in the presence of flow. For dilute suspensions of spheres without long range interaction potentials the conservation equation for the pair density $P_2$ is

$$\frac{\partial P_2}{\partial \tau} + \nabla \cdot \mathbf{u} P_2 = \nabla \cdot \mathbf{D} \cdot \nabla P_2 \tag{65}$$

with $\mathbf{D} = 2(\mathbf{D}_{11} - \mathbf{D}_{12})$ from (28) and $\mathbf{u}(\mathbf{r})$ the relative velocity of two spheres with separation $\mathbf{r}$ in a homogeneous shear flow $\mathbf{E}$. The most

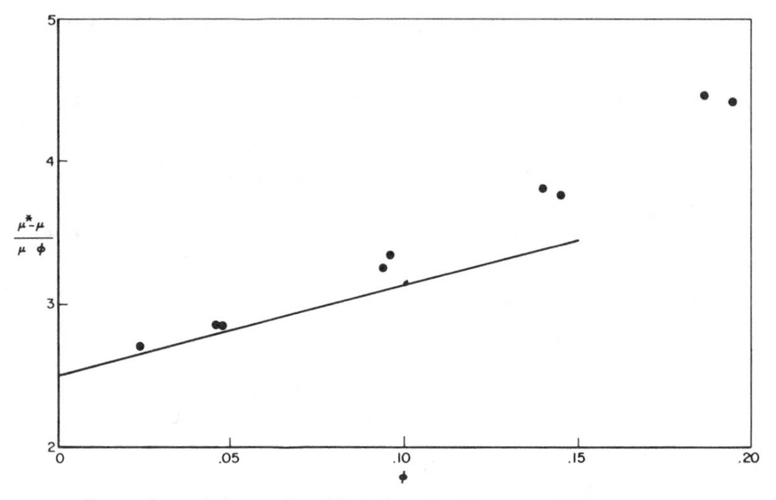


*Figure 9* Comparison of the predicted low-shear limiting viscosity for a dilute suspension of hard spheres (——) with data from Saunders (1961) (●) for polystyrene latices.

significant Brownian effects occur for $\mu a^3\gamma/kT \ll 1$ for which

$$P_2(\mathbf{r}) = n_0\left(1 - \frac{\mu a^3}{kT}\frac{\mathbf{r}\cdot\mathbf{E}\cdot\mathbf{r}}{r^2}f\left(\frac{r}{a}\right)\right). \tag{66}$$

The shear thus increases the pair density where $\mathbf{r}\cdot\mathbf{u}<0$ and reduces it where $\mathbf{r}\cdot\mathbf{u}>0$ causing Brownian forces which oppose the motion and increase the stress (Batchelor 1977).

Batchelor's result for the low-shear limiting viscosity

$$\frac{\mu^*}{\mu} = 1 + 2.5\phi + 6.2\phi^2 + O(\phi^3) \tag{67}$$

includes a direct Brownian contribution of $0.90\phi^2$. Unfortunately, the corresponding high-shear limit, where direct Brownian stresses become negligible, has not been determined for shear flow. An $O(\phi^2)$ coefficient less than 6.2 would clearly demonstrate that the decrease in viscosity with increasing shear rate illustrated in Figure 8 arises from the diminished role of Brownian motion. In Figure 9 (67) is plotted with the data of Saunders (1961) for monodisperse polystyrene latices in water. In view of the experimental difficulties in suppressing long-range repulsions without inducing aggregation and in measuring small viscosity increments accurately, the comparison is quite good.

## CONCLUDING REMARKS

In recent years several uncertainties in analyses of Brownian motion of small particles in liquids have been resolved. Existing hydrodynamic theories now appear to describe accurately the detailed motion and collective behavior of dilute suspensions of colloidal particles, compact macromolecules, and even small molecules, with a modified surface boundary condition. Long-range interaction potentials, which can play a dominant role, have yet to be included quantitatively in some cases. Likewise, the difficult problem of multiparticle interactions at higher concentrations remains. Nonetheless, the techniques for incorporating Brownian effects into continuum treatments of suspensions have been established.

## ACKNOWLEDGMENTS

I am indebted to E. J. Hinch for sharing his thoughts on Brownian motion with me and to my colleagues D. A. Saville and W. R. Schowalter for their support and comments.

*Literature Cited*

Ackerson, B. J. 1976. Correlations for interacting Brownian particles. *J. Chem. Phys.* 64:242–46

Ackerson, B. J. 1978. Correlations for interacting Brownian particles. II. *J. Chem. Phys.* 69:684–90

Adam, G., Delbrück, M. 1968. In *Structural Chemistry and Molecular Biology*, ed. A. Rich, N. Davidson, pp. 198–215. San Francisco: Freeman, 906 pp.

Albers, J., Deutch, J. M., Oppenheim, I. 1971. Generalized Langevin equations. *J. Chem. Phys.* 54:3541–46

Alder, B. J., Wainwright, T. E. 1967. Velocity autocorrelations for hard spheres. *Phys. Rev. Lett.* 18:988–90

Alder, B. J., Wainwright, T. E. 1970. Decay of the velocity autocorrelation function. *Phys. Rev. A* 1:18–21

Altenberger, A. R., Deutch, J. M. 1973. Light scattering from dilute macromolecular solutions. *J. Chem. Phys.* 59:894–98

Altenberger, A. R. 1979. On the wave vector dependent mutual diffusion of interacting Brownian particles. *J. Chem. Phys.* 70:1994–2002

Anderson, J. L., Quinn, J. A. 1974. Restricted transport in small pores. *Biophys. J.* 14:130–49

Anderson, J. L., Rauh, F., Morales, A. 1978. Particle diffusion as a function of concentration and ionic strength. *J. Phys. Chem.* 82:608–16

Anderson, J. L., Reed, C. C. 1976a. Diffusion of spherical macromolecules at finite concentration. *J. Chem. Phys.* 64:8240–50

Anderson, J. L., Reed, C. C. 1976b. In *Colloil Interface Sci.*, ed. M. Kerker. 4:502–12. New York: Academic. 587 pp.

Batchelor, G. K. 1970. The stress system in a suspension of force-free particles. *J. Fluid Mech.* 41:545–70

Batchelor, G. K. 1972. Sedimentation in a dilute dispersion of spheres. *J. Fluid Mech.* 52:245–68

Batchelor, G. K. 1976a. Brownian diffusion of particles with hydrodynamic interaction. *J. Fluid Mech.* 74:1–29

Batchelor, G. K. 1976b. Developments in microhydrodynamics. In *Theoretical and Applied Mechanics*, ed. W. Koiter, pp. 33–55. Amsterdam: North Holland. 260 pp.

Batchelor, G. K. 1977. The effect of Brownian motion on the bulk stress in a suspension of spherical particles. *J. Fluid Mech.* 83:97–117

Bauer, D. R., Brauman, J. I., Pecora, R. 1974. Molecular reorientation in liquids. Experimental test of hydrodynamic models. *J. Am. Chem. Soc.* 96:6840–43

Bauer, D. R., Brauman, J. I., Pecora, R. 1976. Depolarized light scattering from liquids. *Ann. Rev. Phys. Chem.* 27:443–64

Bean, C. P. 1972. In *Membranes—A Series of Advances*, ed. G. Eisenman. 1:1–54. New York: Dekker. 333 pp.

Berne, B. J., Pecora, R. 1976. *Dynamic Light Scattering*. New York: Wiley. 376 pp.

Bird, R. B., Hassager, O., Armstrong, R. C., Curtiss, C. F., 1977. *Dynamics of Polymeric Liquids. Vol 2 Kinetic Theory*. New York: Wiley. 250 pp.

Bloomfield, V. A. 1977. Hydrodynamics in biophysical chemistry. *Ann. Rev. Phys. Chem.* 28:233–59

Brenner, H. 1974. Rheology of a dilute suspension of axisymmetric Brownian particles. *Int. J. Multiphase Flow.* 1:195–341

Brenner, H., Gaydos, L. J. 1977. The constrained Brownian movement of spherical particles in cylindrical pores of comparable radius. *J. Colloid Interface Sci.* 58:312–56

Brown, J. C., Pusey, P. N., Goodwin, J. W., Ottewill, R. H. 1975. Light scattering study of dynamic and time averaged correlations in dispersions of charged particles. *J. Phys. A: Math. Gen.* 8:664–82

Chow, T. S., Hermans, J. J. 1973. Brownian motion of a spherical particle in a compressible fluid. *Physica* 65:156–62

Chu, B. 1974. *Laser Light Scattering.* New York: Academic. 317 pp.

Colby, P. C., Narducci, L. M., Bluemel, V., Baer, J. 1975. Light scattering measurements from dense optical systems. *Phys. Rev.* A12:1530–38

Cone, R. A. 1972. The rotational diffusion of rhodopsin in the visual receptor membrane. *Nature New Biol.* 236:39–43

Cummins. H. Z., Pike, E. R., eds. 1974. *Photon Correlation and Light-Beating Spectroscopy.* New York: Plenum. 584 pp.

Deutch, J. M., Oppenheim, I. 1971. Molecular theory of Brownian motion for several particles. *J. Chem. Phys.* 54:3547–55

Edidin, M. 1974. Rotational and translational diffusion in membranes. *Ann. Rev. Biophys. Bioeng.* 3:179–201

Einstein, A. 1956. *Investigations on the Theory of Brownian Movement.* New York: Dover. 122 pp.

Ermak, D. L., McCammon, J. A. 1978. Brownian dynamics with hydrodynamic interactions. *J. Chem. Phys.* 69:1352–60

Fair, B. D., Jamieson, A. M. 1980. Effect of electrodynamic interactions on the translational diffusion of bovine serum albumin at finite concentration. *J. Colloid Interface Sci.* 73:130–35

Felderhof, B. U. 1978. Diffusion of interacting Brownian particles. *J. Phys. A: Math. Gen.* 11:929–37

Fijnaut, H. M., Pathmamanoharan, C., Nieuwenhuis, E. A., Vrij, A. 1978. Dynamic light scattering from concentrated colloidal suspensions. *Chem. Phys. Lett.* 59:351–55

Giesekus, H. 1962. Elasto-viskose Flüssigkeiten, für die in stationären Schichtströmungen sämtliche Normalspannungskomponenten verschieden gross sind. *Rheol. Acta* 2:50–62

Harris, S. 1976. Diffusion effects in solutions of Brownian particles. *J. Phys. A: Math. Gen.* 9:1895–98

Hauge, E. H., Martin-Löf, A. 1973. Fluctuating hydrodynamics and Brownian motion. *J. Stat. Phys.* 7:259–81

Hess, W., Klein, R. 1978. Dynamical properties of colloidal systems. *Physica* 94A: 71–90

Hinch, E. J., Leal, L. G. 1972. The effect of Brownian motion on the rheological properties of a suspension of nonspherical particles. *J. Fluid Mech.* 52: 683–712

Hinch, E. J. 1975. Application of the Langevin equation to fluid suspensions. *J. Fluid Mech.* 72:499–511

Hu, C., Zwanzig, R. 1974. Rotational friction coefficients for spheroids with the slipping boundary condition. *J. Chem. Phys.* 60:4354–57

Hynes, J. T. 1977. Statistical mechanics of molecular motion in dense fluids. *Ann. Rev. Phys. Chem.* 28:301–21

Keller, K. H., Canales, E. R., Yum, S. I. 1971. Tracer and mutual diffusion of proteins. *J. Phys. Chem.* 75:379–87

Kerker, M. 1969. *The Scattering of Light and Other Electromagnetic Radiation.* New York: Academic, 414 pp.

Kitchen, R. G., Preston, B. N., Wells, J. D. 1976. Diffusion and sedimentation of serum albumin in concentrated solutions. *J. Polym. Sci. Symp.* 55:39–49

Krieger, I. M. 1972. Rheology of monodisperse latices. *Adv. Colloid Interface Sci.* 3:111–36

Langevin, P. 1908. Sur la théorie du mouvement brownien. *C. R. Acad. Sci.* 146:530–33

Lax, M. 1966. Classical noise IV: Langevin methods. *Rev. Mod. Phys.* 38:541–66

Leal, L. G., Hinch, E. J. 1971. The effect of weak Brownian rotations on particles in shear flow. *J. Fluid Mech.* 46:685–703

Malone, D. M., Anderson, J. L. 1978. Hindered diffusion of particles through small pores. *Chem. Eng. Sci.* 33:1429–40

Mazur, P., Oppenheim, I. 1970. Molecular theory of Brownian motion. *Physica* 50:241–58

McDonnell, M. E., Jamieson, A. M. 1977. Quasielastic light scattering measurements of diffusion coefficients in poly-

styrene solutions. *J. Macromol. Sci.—Phys.* B13:67–88
Millar, D. P., Shah, R., Zewail, A. H. 1979. Picosecond saturation spectroscopy of cresyl violet: rotational diffusion by a "sticking" boundary condition in the liquid phase. *Chem. Phys. Lett.* 66:435–40
Nelson, E. 1967. *Dynamical Theories of Brownian Motion*. Princeton Univ. Press. 142 pp.
Newman, J., Swinney, H. L., Berkowitz, S. A., Day, L. A. 1974. Hydrodynamic properties and molecular weight of fd bacteriophage DNA. *Biochemistry* 13: 4832–38
Perrin, M. J. 1910. *Brownian Motion and Molecular Reality*. London: Taylor & Francis. 93 pp.
Phillies, G. D. J. 1973. Effect of intermolecular interactions on diffusion. I. Two-component solutions. *J. Chem. Phys.* 60:976–82
Pomeau, Y., Résibois, P. 1975. Time dependent correlation functions and mode-mode coupling theories. *Phys. Rep.* 19c:64–139
Poo, M., Cone, R. A. 1974. Lateral diffusion of rhodopsin in the photoreceptor membrane. *Nature* 247:438–41
Pyun, C. W., Fixman, M. 1964. Frictional coefficient of polymer molecules in solution. *J. Chem. Phys.* 41:937–44
Rahman, A. 1964. Correlations in the motions of atoms in liquid argon. *Phys. Rev.* A136:405–11
Reed, C. C., Anderson, J. L. 1980. Hindered settling of a suspension at low Reynolds number. *AIChE J.* In press
Reed, T. M., Gubbins, K. E. 1973. *Applied Statistical Mechanics*. New York: McGraw-Hill. 506 pp.
Russel, W. B. 1980. A review of the role of colloidal forces in the rheology of suspensions. *J. Rheol.* 24:287–317
Russel, W. B., Glendinning A. B. 1980. The effective diffusion coefficient detected by dynamic light scattering. *J. Chem. Phys.* Submitted
Russel, W. B., Hinch, E. J., Rallison, J. 1980. On gradient diffusion of rigid spheres in concentrated suspensions. *J. Fluid Mech.* Submitted
Saffman, P. G. 1976. Brownian motion in thin sheets of fluid. *J. Fluid Mech.* 73:593–602
Saffman, P. G., Delbrück, M. 1975. Brownian motion in biological membranes. *Proc. Natl. Acad. Sci.* 72:3111–13
Saunders, F. L. 1961. Rheological properties of monodisperse latex systems I. Concentration dependence of relative viscosity. *J. Colloid Sci.* 16:13–22
Schaefer, D. W., Berne, B. J. 1974. Dynamics of charged macromolecules in solution. *Phys. Rev. Lett.* 32:1110–13
Silebi, C. A., McHugh, A. J. 1978. An analysis of flow separation in hydrodynamic chromatography of polymer latices. *AIChE J.* 24:204–11
Sutherland, G. B. B. M. 1905. A dynamical theory of diffusion for non-electrolytes and the molecular mass of albumin. *Philos. Mag.* 9:781–85
Topalian, J. H., Maguire, J. F., McTague, J. P. 1979. Liquid $Cl_2$ dynamics studied by depolarized Raman and Rayleigh scattering. *J. Chem. Phys.* 71:1884–88
von Jena, A., Lessing, H. E. 1979. Rotational diffusion anomalies in dye solutions from transient dichroism experiments. *Chem. Phys.* 40:245–56
Wills, P. R. 1979. Isothermal diffusion and quasi-elastic light-scattering of macromolecular solutes at finite concentrations. *J. Chem. Phys.* 70:5865–74
Wong, J. H., Quinn, J. A. 1976. In *Colloid and Interface Science*, ed. M. Kerker. V:169–80. New York: Academic. 507 pp.
Wu, E. S., Jacobson, K., Szoka, F., Portis, A. 1979. Lateral diffusion of a hydrophobic peptide in phospholipid multilayers. *Biochemistry* 17:5543–50
Youngren, G. K., Acrivos, A. 1975. Rotational friction coefficients for ellipsoids and chemical molecules with the slip boundary condition. *J. Chem. Phys.* 63:3846–48
Zwanzig, R. 1969. Langevin theory of polymer dynamics in dilute solution. *Adv. Chem. Phys.* 15:325–32
Zwanzig, R., Bixon, M. 1975. Compressibility effects in the hydrodynamic theory of Brownian motion. *J. Fluid Mech.* 69:21–25

*Ann. Rev. Fluid Mech. 1981. 13:457–515*


# ORGANIZED MOTION IN TURBULENT FLOW

*8184

*Brian J. Cantwell*
Department of Aeronautics and Astronautics, Stanford University, Stanford, California 94305

## I INTRODUCTION

In nearly every area of fluid mechanics, our understanding is limited by the onset or presence of turbulence. Although recent years have seen a great increase in our physical understanding, a predictive theory of turbulent flow has not yet been established. Aside from certain results that can be derived through dimensional reasoning, it is still not possible to solve from first principles the simplest turbulent flow with the simplest conceivable boundary conditions. Our continuing inability to make accurate, reliable predictions seriously limits the technological advancement of aircraft design, design of turbomachinery, combustors, mixers, and a wide variety of other devices that depend on fluid motion for their operation.

Anyone who is introduced to the subject of turbulence for the first time quickly encounters the decomposition of the unsteady flow first proposed by Osborne Reynolds in 1895. Various flow variables are divided into a mean and fluctuating part, and upon substitution into the Navier-Stokes equations the result is a system of equations identical in form to the original system except for convective stress terms, which arise from averaging products of velocity fluctuations. In order to close the system of equations, a second relation is needed between the convective stresses and the mean velocity field. Until recently, much theoretical and experimental effort was focused on finding relationships that could be applied to larger and larger classes of mean flows with the ultimate hope of finding a universal constitutive relation for "turbulent fluid." There was never any guarantee that such a relation actually exists and the goals of this effort remain largely unrealized. Hope for a universal turbulence model has been slowly replaced by the growing



0066-4189/81/0115-0457$01.00

realization that the formulation of an adequate theory will require a greatly improved understanding of the physics of turbulent motion.

Recently there has been an increased emphasis on direct analysis or measurement of complex unsteady flow fields under rather restricted conditions. Remarkably elaborate and detailed experimental and analytical studies have been made possible by advances in electronic and optical instrumentation and computational hardware and software. An extremely important element in current experimental research is a renewed emphasis on the use of flow visualization and a widespread awareness that flow visualization can play a very broad role in improving our physical understanding of complicated turbulent phenomena (Kline 1978). Today, flow visualization, as an experimental tool for identifying new flow structure and as a conceptual tool for reducing complex flow processes to simple pictures, lies at the heart of what may eventually become a genuinely new understanding of turbulent flow.

In order to understand current research trends, it is helpful to review the way in which our physical picture of turbulence has evolved over the past several decades. To an investigator of the 1920s or 1930s, turbulence was essentially a stochastic phenomenon having a well-defined and repeatable mean superimposed on a randomly fluctuating velocity field. The motion was characterized by a wide range of scales limited in size only by the overall dimensions of the flow. This picture of randomly interacting scales led to the semi-empirical theories of Prandtl (1925) and Taylor (1915, 1932) in which convective stresses were connected to the mean flow by an effective eddy viscosity[1] or mixing length. Although the concept of an eddy played a central role in constructing mathematical models of turbulence, it was essentially an abstraction and early work on turbulence is marked by a singular lack of any attempt to sketch or schematically visualize turbulent motion.

The complexity of inhomogeneous turbulence led Taylor (1935) to study a homogeneous and isotropic field of turbulence which is initiated by some uniformly distributed stirring mechanism and decays with time. Interest in the statistical theory of turbulence dominated the field throughout the 1940s, and the early development of this theory, though usually attributed primarily to Kolmogorov (1941), also includes important contributions from Onsager (1945), Von Weizsacker (1948), Kovasznay (1948), Heisenberg (1948a, b), and Chandrasekhar (1949). For a review and useful early perspective on the turbulence problem see Liepmann (1952).

Considerations of the spectral evolution of homogeneous and isotropic turbulence led to the important observation that the energy-containing structure of isotropic turbulence does not depend directly on

[1]A concept introduced by Boussinesq in 1877.

the value of the fluid viscosity. It was observed that if the Reynolds number is high enough to insure turbulence then the energy-containing flow structure is similar at all higher Reynolds numbers. In addition it was observed that if the Reynolds number is large enough, the zone of dissipation and the zone of production of turbulent energy will be widely separated in wave-number space. In this case the small-scale motion is in a state of local isotropic equilibrium. Implicit in the idea of local isotropy is the assumption that the direct coupling between large- and small-scale motions is weak and that the behavior of small eddies is universally the same from flow to flow.

A new element was added to the physical picture of turbulence by Corrsin (1943) and Townsend (1947) who showed that the outer edges of turbulent shear flows, specifically wakes and jets, are only intermittently turbulent. The intermittent nature of turbulence was confirmed by Corrsin & Kistler (1954) and Klebanoff (1954) and by the middle of the 1950s the physical picture of turbulence had pretty much evolved to the state illustrated by several examples in Figure 1. This picture, which is attributed primarily to Townsend (1956), combines the laminar superlayer of Corrsin & Kistler (1954) with a field of turbulent fluid of nearly uniform intensity. The dynamical characteristics of this field are similar to the characteristics of isotropic turbulence. The turbulent fluid is moved about by the slow convective motion of a system of large eddies whose dimensions are comparable to the width of the flow and much larger than the scale of the eddies containing most of the turbulent energy. Outside the turbulent interface, the motion is unsteady, irrotational potential flow induced by movement at the boundary.

Townsend viewed the large eddies and small-scale turbulence as the main features of what he called a double structure, and he emphasized the importance of the large eddies in controlling transport.[2] He also realized that the large eddies ought to have a quasi-deterministic form and attempted, really for the first time, to draw a picture of the large-eddy motion. Attempts to produce such a picture usually took the form of inferences based on the long-time-averaged spatial-correlation tensor

$$R_{ij}(\mathbf{x}, \xi) = \overline{u_i(\mathbf{x}, t)u_j(\mathbf{x}+\xi, t)} \tag{1}$$

measured in an Eulerian frame. Extensive correlation measurements by Grant (1958) in the far wake of a circular cylinder ($x/D=533$) were used by Payne & Lumley (1967) to produce the structure shown in

[2] Interestingly, the discussion of double structure does not appear in Townsend's second edition (1976). Apparently influenced by recent experimental evidence in boundary layers and channels, Townsend was moved to delete the discussion of double structure from the new text.

Figure 2*a*. More recently, Townsend (1970) inferred the double-roller large-eddy structure shown in Figure 2*b* for a general shear flow and the double-cone structure shown in Figure 2*c* for a wall-bounded shear flow (Townsend 1976).

This approach to the large-eddy structure through the Eulerian spatial-correlation tensor is rooted in the stochastic random-scales picture of "turbulent fluid" and suffers from a number of shortcomings. The first and most obvious is that there does not exist a unique relationship between the correlation tensor and the unsteady flow that produces it. A second shortcoming is that the correlated portion of the signal associated with the passage of an organized and repeatable motion will be degraded by other contributions to the total signal. This may lead to results that emphasize certain mechanisms while ignoring others, and that may be erroneous or incomplete even in a qualitative sense. Early correlation measurements taken in mixing layers failed spectacularly to reveal the large spanwise eddies now known to dominate this flow. A third shortcoming is that the averaging point does not propagate with the disturbances that are responsible for the correlated portion of the signal. A Lagrangian averaging process might reveal quite a different structure from the Eulerian average. A final shortcoming is that the correlation method leads to an incomplete picture of the flow with vortex lines left unclosed and schematic patterns of lines that often defy definition. In short, the method offers no information about how arrays of moving large eddies are laced together to complete the flow field.

Beginning in the early 1960s, experiments were performed that began to change the view of turbulence just summarized. The last twenty years of research on turbulence have seen a growing realization that the transport properties of most turbulent shear flows are dominated by large-scale vortex motions that are not random. The form, strength, and scale of these organized motions vary from flow to flow and methods used to identify them are as varied as the motions themselves.

## II ORGANIZED MOTION IN THE TURBULENT BOUNDARY LAYER

### *(a) Motivation*

It is remarkable that the earliest observations of organized motion were made in the turbulent boundary layer along a wall where the motion is most complex. At least part of the reason is simply that this is the flow that has historically received the greatest attention because of its technological importance and would therefore be the most likely one to reveal

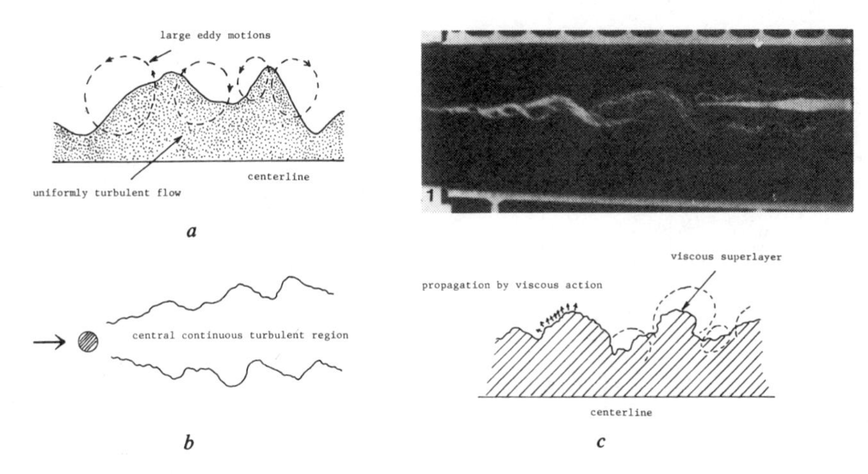


*Figure 1* Several conceptual views of turbulent flow. (*a*) Sketch of a jet flow from Townsend (1956); (*b*) and (*c*) Sketches of a wake flow from Hinze (1959).

its structure first. However, another reason has to be related to the fascinating and compelling mean-flow behavior that the boundary layer exhibits. To a good approximation, the mean velocity profile of a turbulent boundary layer may be divided into three parts (see Figure 3*b*).

*O–A*: Viscous sublayer $0 \leqslant y^+ \lesssim 7$

$$y^+ = u^+. \tag{2}$$

*A–B*: Buffer layer $7 \lesssim y^+ \lesssim 30$. Several relations are available for this region. An implicit formula due to Spalding (1961) which matches (2) and (4) is

$$y^+ = u^+ + e^{-Kc}\left\{e^{Ku^+} - 1 - Ku^+ - \frac{1}{2}(Ku^+)^2 - \frac{1}{6}(Ku^+)^3\right\} \tag{3}$$

*B–C–D*: Logarithmic and outer layer $30 \lesssim y^+ \leqslant \delta^+$. An empirical formula that works well for a variety of pressure gradients is (Coles 1956)

$$u^+ = \frac{1}{K}\ln y^+ + C + \frac{\Pi(x)}{K} 2\sin^2\left(\frac{\pi}{2}\frac{y^+}{\delta +}\right) \tag{4}$$

where

$$y^+ = \frac{yu^*}{\nu},\ \delta^+ = \frac{\delta u^*}{\nu}, \tag{5}$$

$$u^+ = u/u^* \tag{6}$$

and

$$u^* = \sqrt{\frac{\tau_\omega}{\rho}} \tag{7}$$

where

$$\tau_\omega = \mu \frac{\partial u}{\partial y}\bigg|_{y=0}. \tag{8}$$

The skin-friction coefficient is

$$C_f = \frac{\tau_\omega}{\frac{1}{2}\rho u_\infty^2} = 2\left(\frac{u^*}{u_\infty}\right)^2. \tag{9}$$

The profile (4) is determined by the two empirical constants $K$ and $C$ and the function $\Pi(x)$. Typical values are $K=0.4$ and $C=5.0$. The pressure gradient determines $\Pi(x)$. For $dP/dx=0$, $\Pi=0.62$. General features of the structure of the boundary layer are usually described in terms of wall variables $(u^*, \nu)$ or outer variables $(u_\infty, \delta)$, although it is well to keep in mind that they are not independent quantities. In particular, the boundary-layer thickness $\delta$ is determined from the other three. From (4)

$$\delta = \frac{\nu e^{\left(K\frac{u_\infty}{u^*} - KC - 2\Pi\right)}}{u^*}. \tag{10}$$

The first remarkable feature of the turbulent boundary layer is the universality of the near-wall ($y^+ \gtrsim 30$, $y^+/\delta^+ \ll 1$) behavior of Equation (4). Regardless of pressure gradient, wall roughness, or Reynolds number, the logarithmic dependence of $u$ on $y$ is observed.

The second remarkable feature is summarized by the results due to Klebanoff (1954) shown in Figure 3*a* which show that a sharp peak in the rate of production of turbulent energy (production $= -\langle u'v'\rangle \partial\bar{u}/\partial y$) occurs at the outer edge of the viscous sublayer. Measurements in pipe flow by Laufer (1954) show a similar effect. Integration over the thickness of the boundary layer leads to the result that the first 5% of the boundary layer contributes over half of the total production of turbulent energy. This important result was the primary motivation for the early work of Kline & Runstadler (1959) and later Kline et al (1967), and remains the primary motivation for much of the work on boundary-layer structure being carried on today.

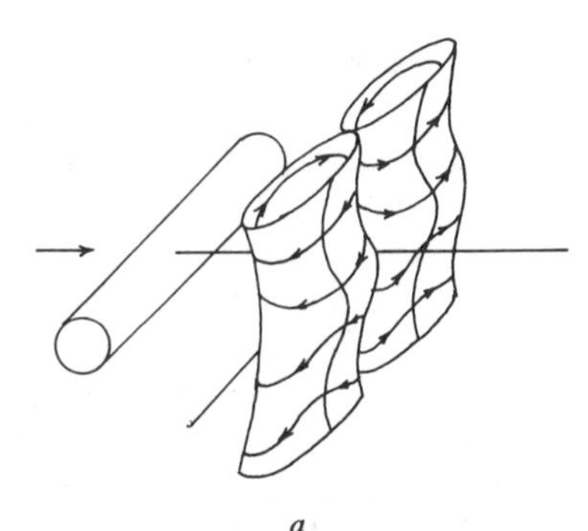

*a*

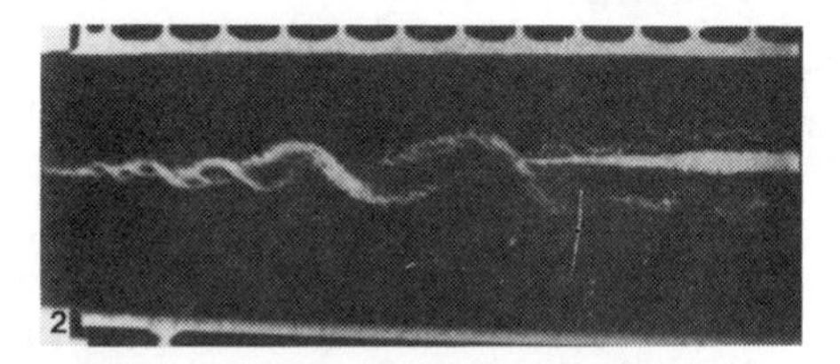


*Figure 2* Eddy structure inferred from correlation measurements for several flows. (*a*) Cylinder wake by Payne & Lumley (1967); (*b*) Inclined double-roller structure for general shear flow from Townsend (1970); (*c*) Double-cone structure for wall flow from Townsend (1976).

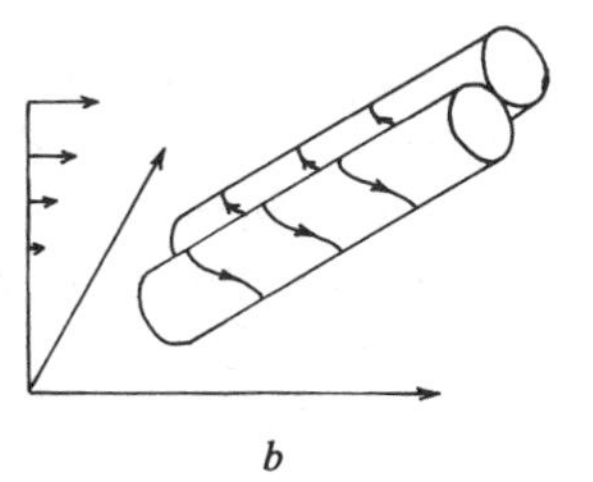

*b*

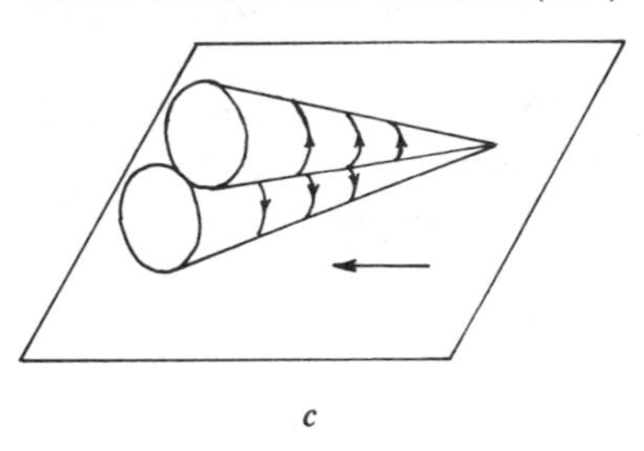

*c*

## (b) *Methods of Approach*

Efforts to isolate coherent structure in the boundary layer have followed two basic lines of approach. The first approach uses various modified forms of the correlation method used by Townsend. The main disadvantage of this method in most of its various forms is that the details of the organized flow pattern in physical coordinates are not determined. The main advantage is that the coherent structure is represented within a well-defined mathematical framework that allows quantitative statements to be made about its statistical properties.

The second line of approach makes use of various methods of flow visualization to make direct observations of complex unsteady turbulent motions. Here, flow visualization is used in the broadest sense to include conventional methods using hydrogen bubble, dye, shadowgraph, and Schlieren techniques as well as nonconventional methods based on conditionally averaged velocity measurements tied together to form a picture of the flow pattern.

An extension of the Eulerian spatial-correlation method to a time-space correlation

$$R_{ij}(\mathbf{x}, \boldsymbol{\xi}, \tau) = \overline{u_i(\mathbf{x}, t)u_j(\mathbf{x}+\boldsymbol{\xi}, t+\tau)} \tag{11}$$

was used by Favre et al (1957) to retain phase information about the flow structure in a turbulent boundary layer. The spatial separation of probes plus the time delay for maximum correlation leads to information about the propagation velocity of the structure. In general, they were able to show that the large-scale streamwise velocity fluctuations are convected with the local mean speed. By placing two probes at different distances from the wall they found that a positive time delay of the signal from the probe nearest the wall was required for a maximum of the correlation coefficient, with the time delay increasing with increasing distance between the probes. From this they concluded that the convected turbulent structure was inclined to the wall. In addition, they found that the scale of the correlation of streamwise velocity fluctuations was much larger in the streamwise than in the cross-stream direction.

There is a subtle difference between the correlation method used by Townsend and that of Favre. Townsend uses spatial correlations to infer a physical picture of the large-eddy flow pattern, whereas the space-time correlation map itself often becomes the "physical" picture of the flow.

### (c) *Near-Wall Studies*

Willmarth & Wooldridge (1962) used the space-time correlation to study pressure fluctuations at the wall under a thick turbulent boundary layer. Using two flush-mounted pressure sensors, they measured the correlation surface $R_{pp}(x, \tau)$ ($x$ is in the streamwise direction) shown in Figure 4*a*. This figure shows two propagating maxima; one of these propagates against the flow, which they attribute to acoustic disturbances propagating upstream from the wind-tunnel diffuser. The main result is the maximum in $R_{pp}$ leaning in the downstream direction which is associated with the propagation and decay of turbulent eddies. By filtering the measured $R_{pp}$ in several frequency bands, they deduced roughly $u_c/u_\infty = 0.83$ for large pressure-producing eddies and $u_c/u_\infty = 0.69$ for small pressure-producing eddies. In contrast to the results of Favre et al they found that the spatial extent of the pressure correlation is approximately the same in directions parallel and transverse to the stream.

In a series of following papers (Willmarth & Wooldridge 1963, Tu & Willmarth 1966, see also the review by Willmarth 1975), this picture was extended and refined. Willmarth & Wooldridge (1963) introduced a scheme called the vector field of correlations in which a field of vectors with components $R_{pu}$ and $R_{pv}$ ($p$ is wall pressure, $u$ and $v$ are velocity components at a fixed probe) is plotted in physical coordinates corresponding to the measuring points of the fixed velocity probe. The result

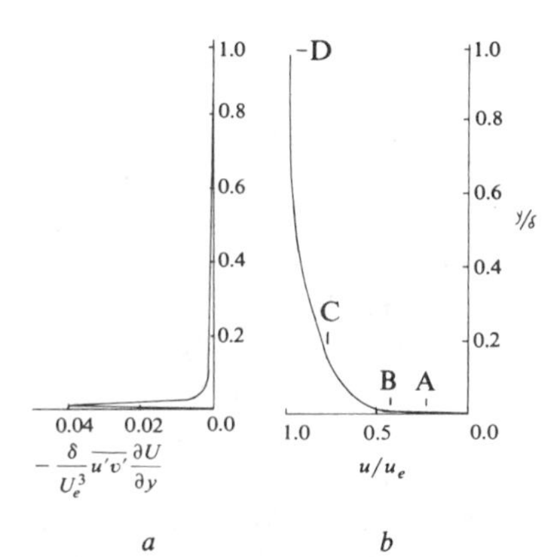


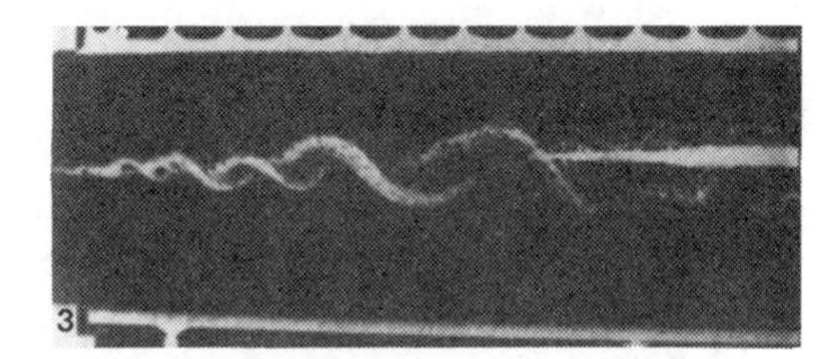


*Figure 3* Mean behavior of a turbulent boundary layer. (*a*) Normalized turbulence-energy-production rate per unit volume (Klebanoff 1954); (*b*) Typical mean-velocity profile; (*c*) Sublayer streaks at $R_\delta = 9000$; (*d*) Sublayer streaks at $R_\delta = 26000$.

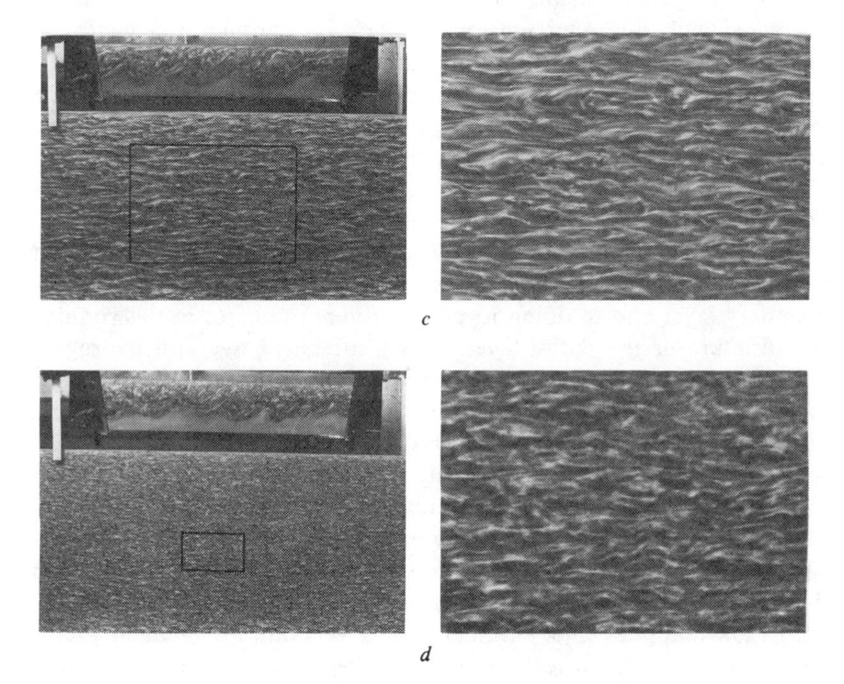


of such a correlation picture is shown in Figure 4*b*. Here the correlation map truly becomes a picture of the flow. In effect this picture represents a conditional average of the velocity field over an ensemble of flow disturbances and represents an early attempt to produce a picture of the velocity field associated with the organized structure in a turbulent boundary layer. Later work (Willmarth 1975) showed a strong correlation between streamwise velocity fluctuations and wall shear. In summary, three separate variables of the flow, the velocity field, wall pressure, and wall shear were all found to be highly correlated over a significant portion of the boundary layer.

Beginning in the late 1950s, a series of experiments was begun at Stanford using flow visualization to study the turbulent boundary layer. This effort culminated in the work by Kline et al (1967). Several new features of the flow in the near-wall region of the turbulent boundary layer were revealed. The flow in a low-Reynolds-number turbulent boundary layer was visualized by a hydrogen-bubble wire placed parallel to, and at various distances above, the wall. They found that even when the wire was deep in the viscous sublayer at $y^+ = 2.7$ the bubbles did not follow straight trajectories as they moved slowly along the plate, but rather they accumulated into an alternating array of high- and low-speed regions called "streaks." They observed that the streaks interacted with the outer portions of the flow through a sequence of four events: gradual outflow, liftup, sudden oscillation, and breakup. To the sequence of three events from liftup to breakup they applied the term "bursting." In addition, they found that a favorable pressure gradient ($dP/dx < 0$) tended to reduce the rate of bursting and an unfavorable pressure gradient ($d\rho/dx > 0$) tended to increase the rate and intensity of bursting. It was conjectured that the bursting phenomenon plays a dominant role in the production of turbulent energy, that it dominates the transfer process between inner and outer regions of the boundary layer and in doing so plays an important role in determining the structure of the entire layer. Using combined dye and hydrogen-bubble visualization plus hot-wire measurements, Kline et al were able to estimate various scales of motion associated with the streaks and bursts. They deduced from visual data that the average spanwise streak spacing (i.e. the distance for one full wavelength) for a smooth wall in all pressure gradients was approximately $\lambda_z^+ = \lambda_z u^* / \nu = 100$. The sequence of events associated with bursting was as follows: Initially the streak of hydrogen bubbles drifts slowly downstream and outward from the wall. When the streak reaches $y^+ = 8-12$ it begins to oscillate. This oscillation amplifies and terminates in a very abrupt breakup in the region $10 < y^+ < 30$. After the breakup the streak of bubbles is contorted, stretched, and ejected outward along an identifiable trajectory. They observed that beyond $y^+ = 40$ the ejected fluid moves at about 80% of the mean velocity in the outer part of the boundary layer. Putting all the visual and quantitative information together they constructed the schematic picture of the streak breakup process shown in Figure 5*a*. The various stages in the bursting process are summarized in Figure 5*b*.

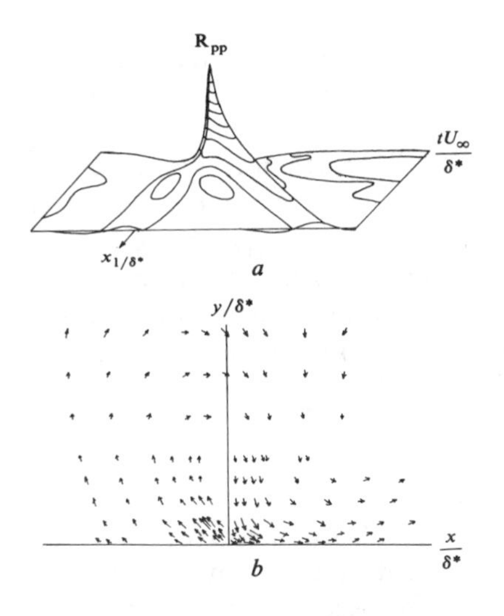


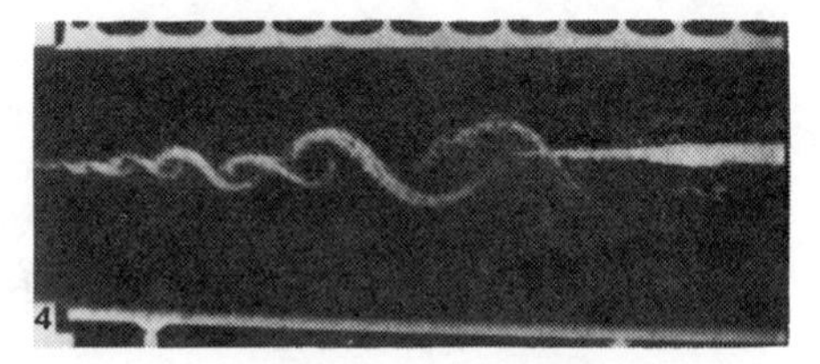

*Figure 4* Turbulent boundary-layer structure based on space-time correlations by Willmarth & Wooldridge (1963). (*a*) Correlation between two pressure transducers mounted flush with the wall; (*b*) Vector field of pressure-velocity correlations, magnitude of the vector at any point is $(R_{pu}^2 + R_{pv}^2)^{1/2}$, direction of the vector is $\tan^{-1}(R_{pv}/R_{pu})$.

Corino & Brodkey (1969) observed what they called ejections near the wall in fully developed high-Reynolds-number pipe flow. In contrast to Kline et al, they viewed the sublayer ($y^+ < 5$) as essentially passive with the ejections originating in a zone away from the wall between $y^+ = 5$ and $y^+ = 15$. It was noticed that the ejection always ended with an action called a sweep, which consisted of axial movement of upstream fluid sweeping out fluid from the previous ejection event. They examined the effect of Reynolds number on the frequency of the ejection process over a range from 2300 (laminar flow) to 50,000 and in general found that the number and intensity of events increased with increasing Reynolds number. Corino & Brodkey estimated that the ejections accounted for approximately 70% of the Reynolds stress measured by Laufer (1954).

These results were confirmed by Kim et al (1971) who showed that virtually all of the net production of turbulent energy in the range $0 < y^+ < 100$ occurs during bursts.

Willmarth & Lu (1972) studied the instantaneous $u'v'$ product near the wall and found very large values during bursting with rare events reaching $u'v' \sim 60\langle u'v' \rangle$ at $y^+ = 30.5$. They also found that large contributions to the Reynolds stress occurred during the sweep phase observed by Corino & Brodkey. A similar observation was also made by Grass (1971). Prior to this, most of the contributions to the Reynolds stress were assumed to occur during the outflow of low-speed fluid.

Taken together, these initial observations constitute a significant step, which has provided the inspiration for much of the work on turbulent boundary layers that has followed. They illuminate an important link between a quasi-deterministic, repeatable unsteady motion and the production and maintenance of mean turbulent transport.

In the initial studies of bursting the process was viewed as an essentially wall-bounded phenomenon with scales of motion determined from wall parameters $u^*$ and $\nu$. Thus it came as something of a shock when Rao et al (1971), after examining data over a fairly wide range of Reynolds number ($600 < R_\theta < 9000$),[3] showed that even in the wall layer the mean burst period scales with outer ($u_\infty$, $\delta$) rather than inner ($u^*$, $\nu$) variables with the mean dimensionless time between bursts given by

$$\frac{u_\infty T}{\delta} \approx 6. \tag{12}$$

Moreover, they found that the mean burst rate did not vary greatly with distance from the wall. Rao et al suggested that such bursts may be a general feature of *all* turbulent flows. They visualized large outer eddies scouring the slow-moving inner layer releasing bursts of turbulent energy by creating regions of intense shear in the inner layer by triggering local instabilities. The inner layer is seen as neither passive nor solely responsible for energy production, but as strongly interacting with the outer layer. They also suggest a mixed scaling with inner variables for the spanwise spatial scale and outer variables for the time between bursts which leads to $u_\infty u^*/F\delta^*\nu$ (where $\delta^*$ is displacement thickness and $F$ is the burst rate per unit span) as a quantity that is practically independent of Reynolds number. Aside from brief discussions in Kline et al and Rao et al, the remaining literature on this subject takes relatively little notice of the need for information on the scaling parameters for the spanwise spacing between bursts.

In an excellent and very extensive study, Grass (1971) used hydrogen-bubble data corrected for the lag effect due to the bubble-wire wake to measure structural features of turbulent flow over smooth, transitionally rough, and fully rough walls ($u^*k/\nu = 0.0$, 20.7, 84.7 where $k$ is the roughness height). He found that ejections and inrushes were present irrespective of the surface roughness. Grass suggests a universal ejection type of momentum-transport mechanism which extends across a major portion of the boundary-layer thickness. The mechanism is visualized as jets of low-momentum fluid ejected from the boundary region and randomly distributed with respect to position and time. He suggests further that the general ejection process is a common feature of the flow

[3] The range has since been extended by Narayanan & Marvin (1978) to $600 < R_\theta < 35{,}000$.

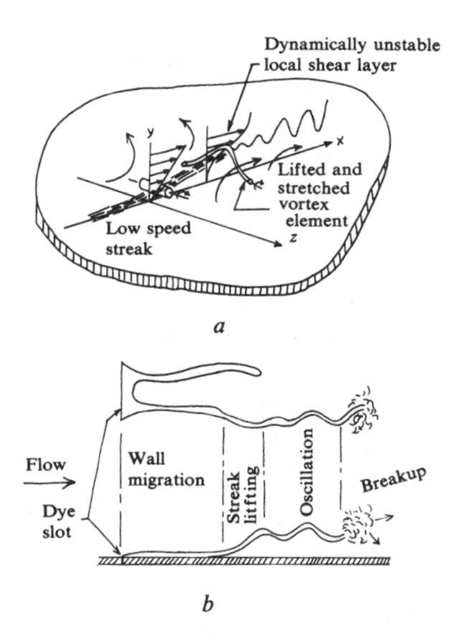


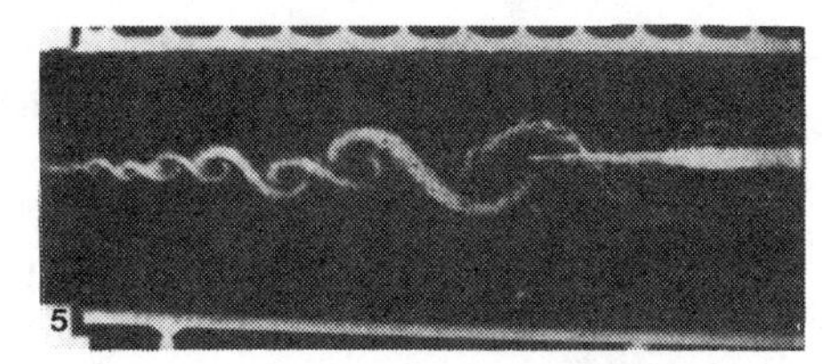


*Figure 5* Schematic views of near-wall turbulent boundary-layer structure based on direct observations. (*a*) Mechanics of streak breakup by Kline et al (1967), (*b*) Sequence of events in (*a*) by Kline (1978).

structure irrespective of boundary-roughness conditions. This is an important conjecture that raises some question regarding the precise role of sublayer streaks in the bursting process. The wall flow of a fully rough-wall boundary layer must be substantially different from that of a smooth-wall boundary layer. Yet the basic structure of the organized motion appears to be largely unchanged.

More recently, Blackwelder & Eckelmann (1979) have made a rather detailed study of the structure of wall streaks using heated wall elements to infer streamwise and spanwise vorticity at the wall (Figures 6*a* and *b*). They find the strength of streamwise vortices to be about one order of magnitude less than the mean spanwise vorticity. They identify the low-speed streak observed by Kline et al and others as the accumulation region between streamwise vortices where the vertical component of the secondary motion is directed away from the wall. They find the streamwise length of the vortices to be $\Delta x^+ \sim 1000$.

## (*d*) *Outer-Flow Studies*

As evidence for organized structure near the wall accumulated, attention began to focus on the flow in the outer part of the layer and the possible connection or interaction which may exist between the outer and wall layer.[4] Kovasznay et al used conditionally averaged space-time

[4]As a general rule, the term "wall layer" refers to $0<y^+<100$, which includes the viscous sublayer and a portion of the logarithmic region. The term "outer layer" refers to all the rest.

autocorrelations of several flow variables to draw a three-dimensional correlation map of the outer structure (Figure 7). They observed that the vorticity appeared to exhibit a discontinuity across the turbulent interface of the bulge whereas the velocity was continuous. In addition, they noticed that there was a considerable difference between the upstream-facing (back) and downstream-facing (front) portions of turbulent humps in the outer flow. They suggest, almost in passing, that if the flow were viewed in a coordinate system moving with the average convection velocity of the interface, fluid would appear to arrive at a stagnation point on the back of a bulge located at about $y/\delta=0.8$. This brief remark represents a most important observation which forms part of a common thread that runs through much of the recent literature on organized structure in turbulent flow; that it is the upstream-facing portion of the turbulent-nonturbulent interface that is most active, and that this activity is associated with a saddle-point flow in a convected frame of reference. A wide variety of flows, not just turbulent boundary layers, seem to exhibit this property. Observations of intense turbulent activity along upstream-facing interfaces may be found in Wygnanski & Champagne (1973; the turbulent slug in pipe flow), Falco (1977; turbulent boundary layer), Brown & Thomas (1977; turbulent boundary layer), Wygnanski et al (1976; turbulent spot), Cantwell et al (1978; turbulent spot), Gad-el-Hak & Blackwelder (1979; turbulent spot), and Cantwell & Coles (1980; cylinder near wake).

Kovasznay et al found that the individual bulges in the outer flow are correlated over $3\delta$ in the streamwise direction and $\delta$ in the spanwise direction. They conjecture that the bulges in the interface become passive and that only the birth of new ejected lumps (presumably from the wall) is the mechanism that maintains the Reynolds stress of the outer layer. They speculate that the bursts observed by Kline et al (1967) near the wall are responsible for the large-scale motion in the outer flow that they observe. They also conjecture that the large-scale wall-pressure fluctuation pattern may be caused by the same mechanism.

Blackwelder & Kovasznay (1972) extended the earlier work with measurements closer to the wall. They found that intense motions in the wall region remained strongly correlated out to $y/\delta=0.5$ confirming other observations that the disturbance associated with bursting extends across the entire layer.

In a pair of papers, Offen & Kline (1974, 1975) attempted to draw a kinematical description of the relationship between the inner and outer flow. They posed a causal relationship for the interaction between bursts and the flow in the logarithmic region that produces sweeps

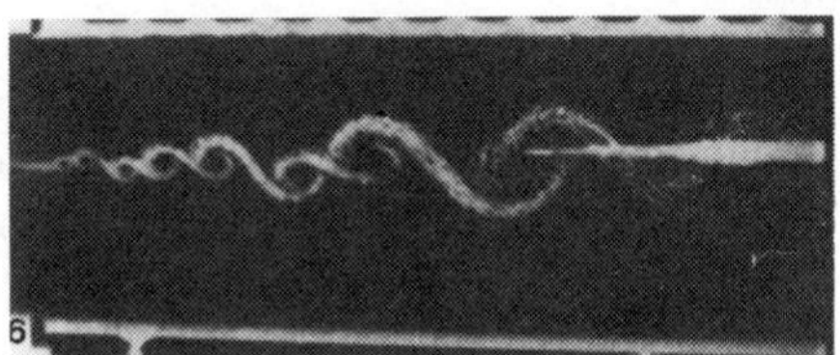

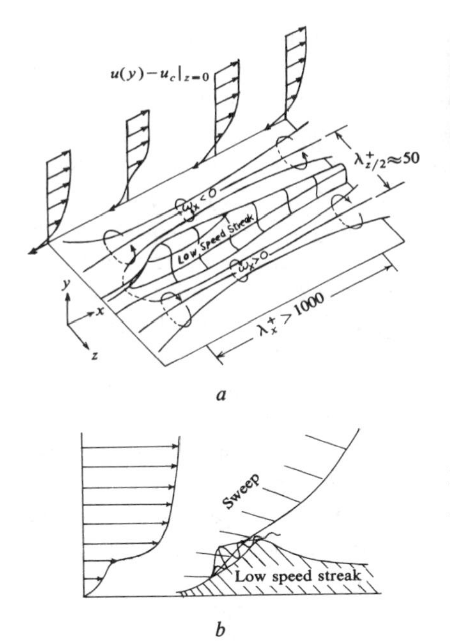


*Figure 6* Model of near-wall turbulent boundary-layer structure from Blackwelder (1978). (*a*) Counter-rotating streamwise vortices with the resulting low-speed streak; (*b*) Localized shear-layer instability between an incoming sweep and low-speed streak.

which, in turn, influence the generation of bursts farther downstream. Inspired partially by observations of vortex interactions in the plane mixing layer, they advanced the hypothesis that the bulges in the superlayer may be the consequence of vortex pairing between the vortices associated with two to four bursts.

Brown & Thomas (1977) correlated wall shear with velocity across the layer. They observed that the wall shear has a slowly varying part and a high-frequency part and that the two appear to be coupled. They established a line of maximum correlation which lay at an angle of 18° to the wall in the downstream direction and hypothesized that this was due to some organized structure at an oblique angle to the wall which produces a characteristic response in wall shear as it moves along the plate at about 0.8 $u_\infty$. They also observed a sharp step in the velocity which occurs at the trailing interface of the outer bulge. Falco (1977) combined visual and hot-wire observations in the outer layer. He also noticed considerable activity on the trailing interface of the outer bulge which he associated with Reynolds-stress-producing motions due to small-scale eddies in the outer layer with length scales on the order of 100 to 200 $\nu/u^*$. Brown & Thomas, Falco, and Blackwelder & Kovasznay draw very similar sketches of the organized structure (Figures 8*a*, *b*, and *c*).

In an extensive visual study, Smith (1978) used a moving frame of reference to study the interaction between inner and outer layers in a turbulent boundary layer at Reynolds numbers somewhat higher than the range studied by Falco. He observes the burst sequence to be related to the passage of a large-scale motion with a generally transverse rotation similar to observations by Nychas et al (1973). The large-scale motion is described as an agglomeration of smaller-scale vortical structures of varying sizes, strengths, orientations, and coherency with an overall spanwise rotation. Velocities of different structures within the large-scale motion varied from 0.7 $u_\infty$ to 0.95 $u_\infty$ with the center of rotation moving at about 0.8 $u_\infty$. In some of the observations of the interaction of wall-region fluid with the outer flow, free-shear-layer type vortical structures were observed to form in a region between $y^+ = 10$ and 40. He argues that the formation of these spanwise wavelike motions is a key mechanism in the entrainment process of the low-speed, wall-region fluid into the higher-speed sweep fluid and thus is a source of the high-Reynolds-stress production in the wall region during the bursting process.

Head & Bandyopadhyay (1978) used flow visualization and hot-wire measurements to draw quite a different picture of the turbulent boundary layer (Figure 9*a*). They make the point that Reynolds-number effects on the detailed boundary-layer structure are likely to be important and that experiments at $\mathrm{Re}_\theta < 1000$ (which covers about two-thirds of the literature on this subject) may give results that are quite unrepresentative of those at really high Reynolds numbers. They suggest that for values of $\mathrm{Re}_\theta$ in the range 1000–7000 the most characteristic feature of the boundary layer is not the existence of large-scale coherent motions, but of structures formed by the random amalgamation of features that are small in the streamwise direction but highly elongated along lines at about 40° to the surface. It is inferred that these represent hairpin vortices similar to the horseshoe vortices postulated by Theodorsen (1955). They suggest that the Reynolds-stress-producing outer eddies of Falco are in fact the tips of the hairpin vortices. This model has some features in common with some recent computations by Leonard (1979) of three-dimensional vortex motions in a developing turbulent spot (Figure 9*b*). Here, initially transverse vortex lines near a wall are perturbed and the perturbation is allowed to grow. Eventually a large-amplitude motion is observed in which the (now wavy) array of vortex lines assumes a shape similar to a family of Theodorsen vortices with their heads pointed downstream at an inclined angle to the plate. More recently, Perry et al (1980) have proposed a model of near-wall boundary-layer structure based on strings of Λ-shaped vortices similar

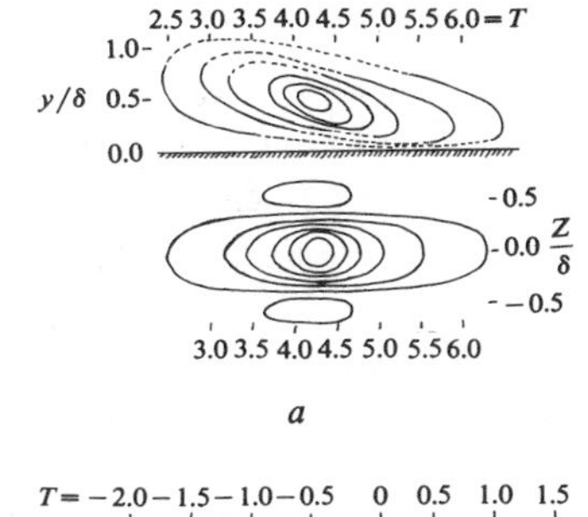


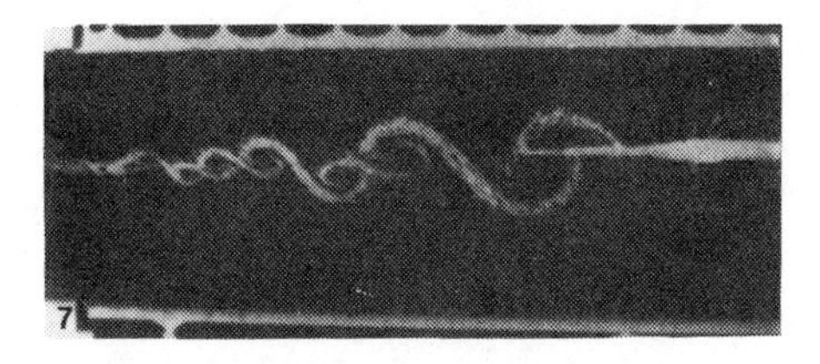

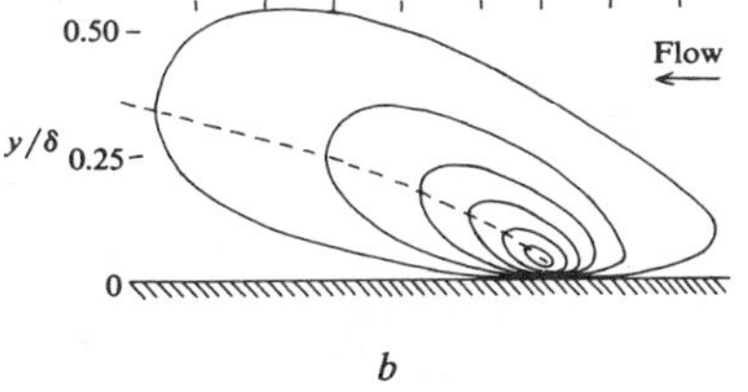


*Figure 7* Turbulent boundary-layer structure based on space-time correlations between the velocity at a fixed probe and the velocity at a probe that is positioned at various points in the flow. (*a*) Fixed probe at $y/\delta = 0.5$ from Kovasznay et al (1970); (*b*) Fixed probe at $y/\delta$ 0.03 from Blackwelder & Kovasznay (1972).

to those observed in a visual study of boundary-layer turbulence by Hama & Nutant (1963). Their model reproduces a number of the mean features of wall turbulence including the logarithmic velocity profile.

All of the investigations discussed thus far have involved observations in naturally occurring turbulent boundary layers. Several additional investigations should also be noted. Zilberman et al (1977) generated turbulent spots and allowed them to pass into a turbulent boundary layer that was tripped by a spanwise array of spheres. They were able to track the perturbation associated with the spot for approximately 70 turbulent-boundary-layer thicknesses in the streamwise direction. The convection speed of the disturbance was 0.9 $u_\infty$ and the main features of the spot structure imbedded in the boundary layer were in general agreement with other observations of the outer large structure.

Part of the difficulty in measuring the organized motion lies in the fact that it is extremely difficult to isolate and average. An interesting solution to this dilemma was proposed by Coles & Barker (1975). They created a synthetic boundary layer in water by using a series of controlled disturbances near the leading edge of a flat plate. These produced systematic moving patterns of turbulent spots in a laminar flow which, when averaged, gave a turbulent boundary layer with a standard mean-velocity profile. The idea here is to create a flow that simulates the naturally occurring flow but is much easier to study. This work has been continued in an air boundary layer by Savas (1979) who made extensive measurements of intermittency in the outer part of the layer. He found that the celerity of outer large eddies was 0.88. Savas observed an interaction phenomenon he called "eddy transposition." This involves the appearance and rapid growth of regions of new turbulence to the rear of the original spots and in the gaps between them. The original spots then decay and disappear. During the transposition process the number of large eddies is twice the normal value and these eddies form a honeycomb pattern of hexagons with empty centers. When transposition is complete, the original hexagonal pattern is restored with a substantial phase shift. Savas suggests that by fitting spots together in a close-packed hexagonal pattern spaced 2.5$\delta$ in the spanwise direction and about 8$\delta$ in the streamwise direction, one can create an optimal synthetic boundary layer.

### *(e) Other Wall-Bounded Flows*

A number of the works discussed in the last section have involved studies in fully developed pipe flow. However, no discussion of coherent motion in turbulent flow would be complete without some discussion of the remarkably complex transition in pipe flow. Motivated partially by the observations of Lindgren (1969), Wygnanski & Champagne (1973) undertook an extensive study of transition in a pipe over a Reynolds-number range from 2000 to 50,000. For smooth or only slightly disturbed inlets, transition occurs as a result of instabilities in the developing boundary layer long before the flow becomes fully developed in the pipe. This type of transition gives rise to turbulent slugs, which may occur at any Reynolds number greater than 3200. They occupy the entire cross-section of the pipe (Figure 10*a*) and grow in length as they proceed downstream. Although the lengths of individual slugs vary, an upper limit is of the order of the length of the pipe. The structure of the flow in the interior of a slug is identical to that in a fully developed turbulent pipe flow. Near the front and rear interfaces where the mean motion changes from a laminar to a turbulent state, the velocity profiles develop inflections. The total turbulence intensity near the interfaces is generally four to seven times higher than in the interior of the slug and may reach 15% of the velocity at the center of the pipe. The maxima in the Reynolds stresses occur at approximately the same location as the inflection points in the velocity profile. Slugs were never observed while the velocity profile was close to parabolic. The implication of this is that stability calculations in the developing region of the pipe are more relevant to the natural process of transition in a pipe than the analyses that are solely concerned with the stability of Poiseuille flow. Wygnanski & Champagne found a second type of intermittently turbulent flow which involved "puffs" of mixed laminar and turbulent flow in the fully developed portion of the pipe for $2000 < \mathrm{Re} < 2700$. Puffs are formed when a large disturbance is introduced at the inlet. While slugs are

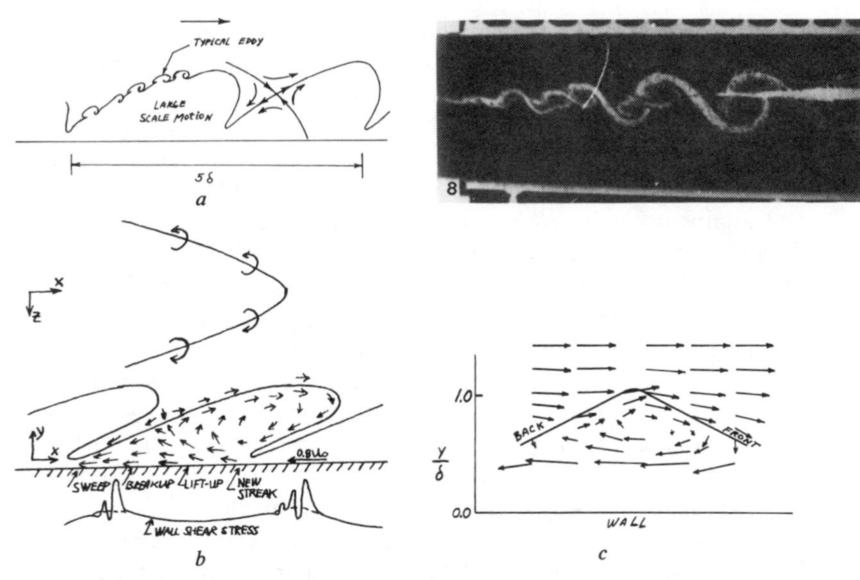


*Figure 8* Three conceptual models of the outer structure of a turbulent boundary layer. Each figure is referenced to an observer who moves at 0.8 $u_\infty$. (*a*) A composite of several figures from Falco (1977); (*b*) Outer flow plus wall shear from Brown & Thomas (1977); (*c*) From Blackwelder & Kovasznay (1972).

associated with transition from laminar to turbulent flow in the wall layer of the inlet flow, puffs apparently correspond to an incomplete relaminarization process. Occasionally puffs were observed to split and/or decay. Puffs can only be seen at $2000 \leqslant \mathrm{Re} \leqslant 2700$, whereas slugs occur at any $\mathrm{Re} > 3200$.

Wygnanski et al (1976) carried out a fairly extensive study of the puff. They were able to produce turbulent puffs that would maintain themselves indefinitely at Reynolds numbers around 2200. Nonequilibrium puffs at Reynolds numbers away from this value tend to grow and split or decay. The various regimes are as follows:

| | |
|---|---|
| $\mathrm{Re} \leqslant 2200$ | Puffs diminish with $x$. |
| $2200 \leqslant \mathrm{Re} \leqslant 2300$ | A single disturbance at the inlet gives rise to a single puff. No splitting is observed. The puff appears to be in equilibrium. |
| $2300 \leqslant \mathrm{Re} \leqslant 2600$ | Puffs grow with $x$. As Re increases, the number of puffs per disturbance increases reaching a maximum of four at $\mathrm{Re} = 2600$. |
| $2600 \leqslant \mathrm{Re} \leqslant 2800$ | A further increase of Re decreases the number of puffs until at $\mathrm{Re} \approx 2800$ a single slug of turbulence is observed. |
| $\mathrm{Re} \geqslant 3200$ | Only slugs are observed. |

Wygnanski et al (1976) made measurements of the equilibrium puff at Re=2220. A streamline picture of the puff in moving coordinates is shown in Figure 10*b*. In this frame of reference, the flow consists of two large flattened ring vortices which rotate in the same direction along with a small eddy in the vicinity of the rearward interface. The turbulent intensity in the puff increases gradually towards the rear of the puff and attains a maximum at the trailing edge. Near the leading edge of the puff there is no clear interface between laminar and turbulent flow, in contrast to the turbulent slug which is bounded by two well-defined interfaces spanning the entire cross section of the pipe.

Another curious, highly organized, wall-bounded flow is the spiral turbulence between two counter-rotating cylinders studied by Coles & Van Atta (1967). Under certain conditions the flow between two concentric, counter-rotating, cylinders consists of helical bands of alternating laminar and turbulent flow. The band of turbulence rotates at a speed roughly halfway between the speed of the inner and outer cylinders (Figure 10*c*). Fluid enters the turbulence at a severe angle, but leaves at a grazing angle so that the rate of detrainment of fluid passing from a turbulent to a laminar state is actually very low. In fact, the shape of the upstream-facing interface appears to be controlled primarily by the process of viscous decay. A similar observation was made by Wygnanski & Champagne in the turbulent slug in pipe flow. They found that a unique relationship existed between the velocity of the fluid and the velocity of the upstream-facing interface such that sudden relaminarization of turbulent fluid is prevented.

Another wall-bounded flow that has received considerable attention is the turbulent spot first studied by Emmons (1951) and Mitchner (1954). Shortly after these initial observations, Schubauer & Klebanoff (1955) defined celerities (0.55–0.9) for the upstream and downstream interfaces of the spot. Elder (1960) looked at the merging of spots and found a strongly nonlinear interaction in which one spot merges with another with very little loss of identity. Coles & Barker (1975) and Wygnanski et al (1976) made detailed measurements of the streamline pattern of an average spot. Both investigations concluded that the ensemble-averaged spot was essentially a single large horseshoe vortex superimposed on small-scale motions, which average out in the mean. The structure is not unlike a Theodorsen vortex, but larger. The extensive detail of the pictures drawn by Wygnanski et al of the puff and spot (Figures 10*b* and 11*a*) and Coles & Barker of the spot represent a significant advance over earlier attempts to visualize organized motion in turbulence. They also represent something of a retrenchment from the complexity of the turbulent boundary layer to flows where the

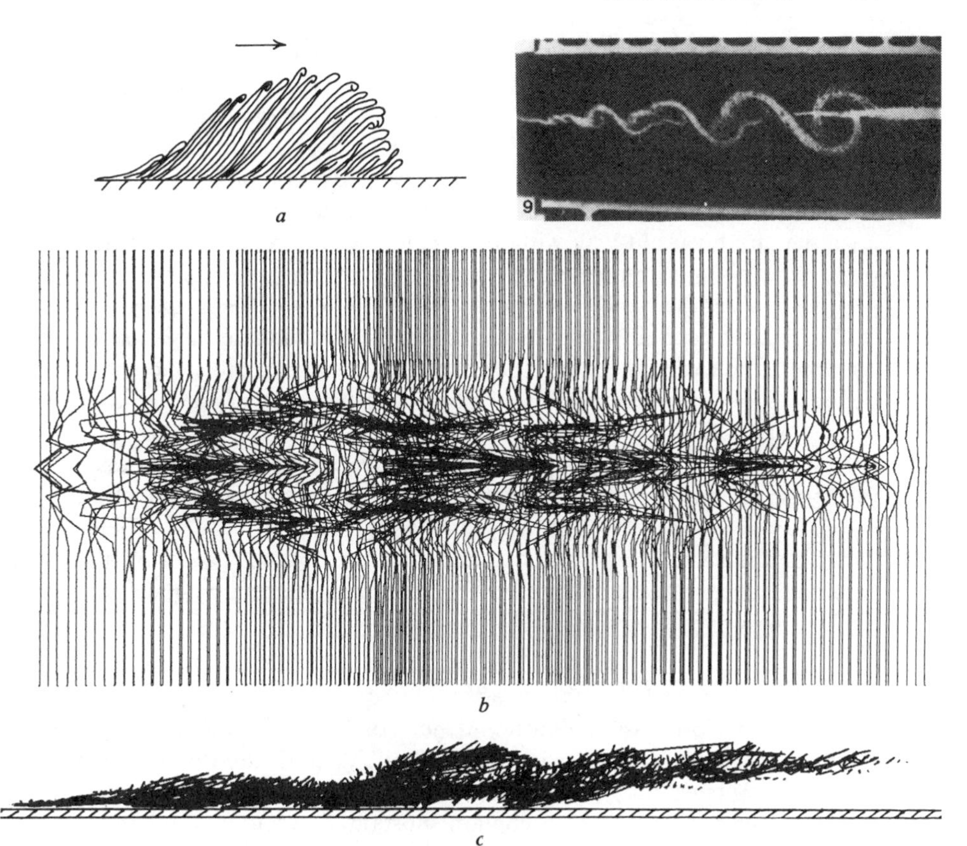


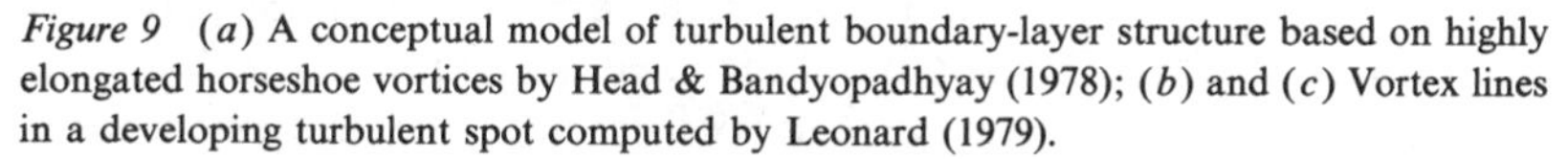


*Figure 9* (*a*) A conceptual model of turbulent boundary-layer structure based on highly elongated horseshoe vortices by Head & Bandyopadhyay (1978); (*b*) and (*c*) Vortex lines in a developing turbulent spot computed by Leonard (1979).

organized motion is much easier to isolate and identify. As for most of the major features of the spot, both investigations were in generally good agreement. There was, however, one major area of disagreement related to the celerity of the most energetic part of the motion. Space-time correlations near the wall led Wygnanski et al to a celerity of 0.65 $u_\infty$. Coles & Barker observed a conspicuous minimum in the streamwise velocity in the outer part of the spot. By timing the arrival of this minimum at several stations downstream of the spot origin they deduced a celerity of 0.83 $u_\infty$.

Cantwell et al (1978) carried out measurements on the plane of symmetry of the spot and made use of the approximately conical behavior of the spot to collapse their velocity data using similarity variables $\xi = x/u_\infty t$, $\eta = y/u_\infty t$ and $u/u_\infty = U(\xi, \eta)$, $v/u_\infty = V(\xi, \eta)$. In these coordinates, the equations for unsteady particle paths reduce to an autonomous system which can be integrated graphically by simply plotting isoclines. The result of this process is a diagram (Figure 11*c*) showing how the spot entrains free stream fluid. A useful property of the entrainment diagram is that it is invariant for all uniformly moving observers. As a result, structural features of the flow are brought out in a simple and invariant way without reference to the speed of a moving observer. The entrainment diagram for the spot includes four critical points, two saddles and two stable foci. The outer focus moves at $0.78\, u_{\infty}$ and is probably responsible for the velocity minimum observed by Coles & Barker. The inner focus moves at $0.65\, u_{\infty}$, the same speed deduced from the maximum space-time correlation measurements of Wygnanski et al.

More recently, Wygnanski et al (1979) have found that the laminar relaxation region following the spot is occupied by Tollmien-Schlichting instability waves induced by the fluctuating motions near the outer wings of the spot. They suggest that spot growth is controlled by the formation of new spots due to the breakdown of the trailing Tollmien-Schlichting waves.

### *(f) Summary of the Flow Structure*

We have seen a procession of different views of turbulent-boundary-layer structure deduced using a variety of methods. Each method has its own special advantages and disadvantages. The correlation approach leads to well-defined pictures which contain substantial amounts of quantitative statistical information. However, it tends to be unphysical and yields little information about the detailed structure of the flow. The direct approach using flow visualization is more physical but leads to conflicting results which are often difficult to interpret and organize. The problem here has often been compared to the fable about the five blind men and the elephant. Each man explores a small part of the animal and then makes conclusions about the nature of the whole beast. The fable points out the limits of human observation compared to the totality of facts required to make a correct conclusion.

Any attempt to summarize various data for organized structure in the turbulent boundary layer runs smack into a maze of ambiguous labels and conflicting definitions. Most observed quantities exhibit wide variation about a poorly defined mean. As a result common denominators are rare. Nevertheless, certain properties of the organized structure are beginning to be established.

There appear to be four main constituents of the organized structure. Nearest the wall is a fluctuating array of streamwise counter-rotating

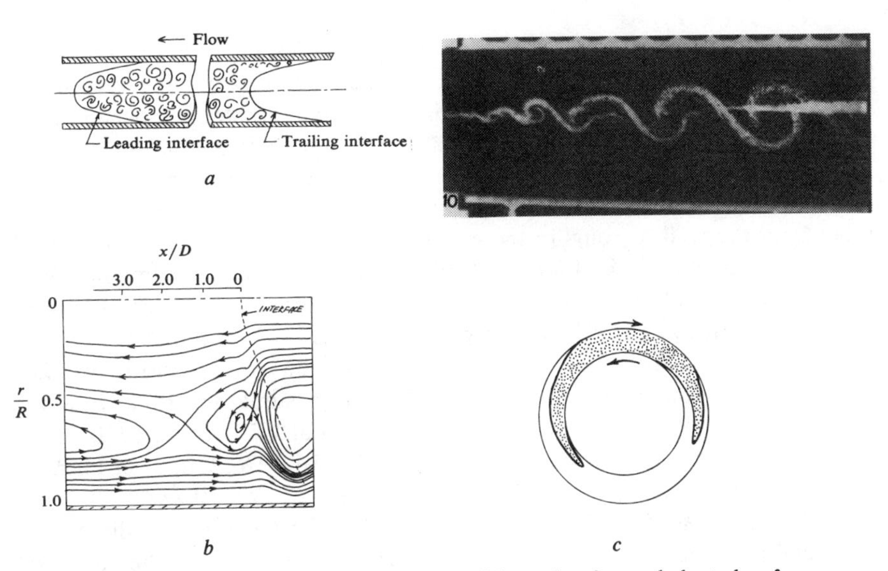


*Figure 10* Organized motion in pipe flow. (*a*) Schematic of a turbulent slug from Wygnanski & Champagne (1973); (*b*) Ensemble-averaged streamline patterns in a turbulent puff from Wygnanski et al (1976); spiral turbulence (*c*) from Coles & Van Atta (1967).

vortices. See Figures 3*c* and *d*. The vortices densely cover all parts of the smooth wall. Slightly above the streamwise vortices but still quite close to the wall is a layer that is regularly battered by bursts that involve very intense small-scale motions of energetic fluid. The outer layer is also occupied by intense small-scale motions. These are found primarily on the upstream-facing portions of the turbulent-nonturbulent interface; the backs of the bulges in the outer part of the layer. The outer small-scale motions are part of an overall transverse rotation with a scale comparable to the thickness of the layer. The various components, along with some notation, are summarized schematically in Figure 12. These components and fairly crude estimates of their scale, position, celerity, and lifetime are discussed below.

1. $\lambda_x$—Length of sublayer structure in the streamwise direction (Blackwelder 1978, Blackwelder & Eckelmann 1979, Praturi & Brodkey 1978). Observations vary from $\lambda_x = 100\ \nu/u^*$ to $\lambda_x = 2000\ \nu/u^*$ with $1000\ \nu/u^*$ as a best value. An issue here, which may account for some of the variation, is the distinction between sublayer streaks and sublayer longitudinal vortices. The consensus of data seems to be that streaks are the product of an accumulation process in regions that lie between streamwise vortices where there is an upwelling of fluid in the secondary motion (motion in a plane normal to the direction of flow). The

longitudinal scale of streaks might be quite different from the longitudinal scale of the secondary vortex motions that produce the accumulation. The exact form of the secondary streamline pattern associated with sublayer vortices, and how this pattern is matched to the more random secondary motions in the outer flow, is at present unclear.

2. $\lambda_y$—Vertical half scale of sublayer structure based on the distance from the wall to roughly the center of a streamwise vortex (Kline et al 1967, Bakewell & Lumley 1967, Blackwelder & Eckelmann 1979, Kovasznay et al 1970). Observations vary from $\lambda_y = 10\ \nu/u^*$ to $\lambda_y = 25\ \nu/u^*$ with 15 $\nu/u^*$ about average. Note that $y^+ = 30$ is about the vertical distance at which the law of the wall (Equation 4) begins to be valid.

3. $\lambda_z$—Spanwise (full wavelength) scale of sublayer structure (Kline et al 1967, Bakewell & Lumley 1967, Kim et al 1971, Grass 1971, Smith 1978, Praturi & Brodkey 1978, Cantwell et al 1978, Gupta et al 1971, Oldaker & Tiederman 1977, Hanratty et al 1977, Lee et al 1974, Willmarth & Yang 1970). This is probably the most universally agreed upon quantity in turbulent boundary-layer structure. The accepted mean value is $\lambda_z = 100\ \nu/u^*$. The statistics of this quantity are somewhat skewed, with the most probable value around 80 $\nu/u^*$. Some observations (Gupta et al 1971) indicate a possible dependence of $\lambda_z u^*/\nu$ on Reynolds number.

4. $b_x$, $b_y$, and $b_z$—Length scales of energetic near-wall eddies (Corino & Brodkey 1969, Dinkelacker et al 1977, Smith 1978, Sabot et al 1977). Here the correspondence between observations by different investigators becomes very difficult to pin down. However, it appears that in a region between $y^+ = 5$ and 40 very energetic parcels of fluid are observed to form through some mechanism of instability, which has yet to be fully identified. Most often the mechanism is described in terms of a fast inviscid oscillatory instability arising from a slowly varying instantaneous inflection point in the velocity profile (Figure 6*b*). Assigning scales to the parcel of fluid that participates in this process is very uncertain; however, typical estimates range from 20 $\nu/u^*$ to 40 $\nu/u^*$ for $b_x$ and from 15 $\nu/u^*$ to 20 $\nu/u^*$ for $b_y$. There is essentially no information on $b_z$ although the observations of Smith (1978) suggest that $b_z$ may be several times $b_x$ or $b_y$. Measurements based on wall pressure of the streamwise extent of energetic small-scale motions lead to somewhat larger estimates for $b_x$ on the order of 60 $\nu/u^*$ to 100 $\nu/u^*$.

5. $X_b$—The persistence distance of energetic near-wall eddies (Dinkelacker et al 1977, and Praturi & Brodkey 1978). Values of $X_b$ between $0.5\delta$ and $1.5\delta$ are suggested although the evidence is very sparse with wall-pressure measurements indicating the larger value.

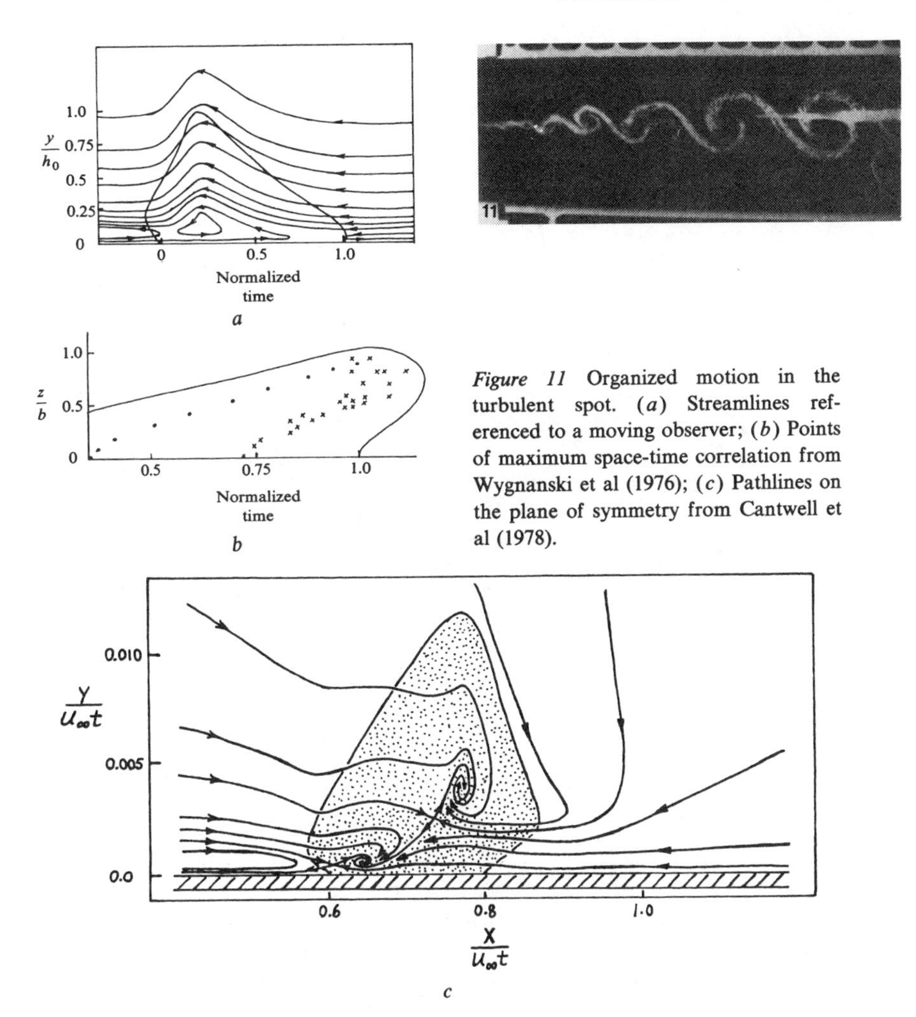


*Figure 11* Organized motion in the turbulent spot. (*a*) Streamlines referenced to a moving observer; (*b*) Points of maximum space-time correlation from Wygnanski et al (1976); (*c*) Pathlines on the plane of symmetry from Cantwell et al (1978).

6. $C_b$—The celerity of energetic near-wall eddies (Kline et al 1967, Dinkelacker et al 1977, Brown & Thomas 1977, Cantwell et al 1978, Willmarth & Wooldridge 1962, Wygnanski & Champagne 1973). Most data indicate a value of $C_b$ around $0.65\,u_\infty \pm 0.05\,u_\infty$. Wall-pressure data (Dinkelacker et al, 1977) indicate a much lower value around $0.2\,u_\infty$.

7. $l_x, l_y, l_z$—Length scales of energetic outer-flow eddies (Falco 1977, Zilberman et al 1977, Smith 1978). Called "typical eddies" by Falco, these outer-flow motions appear to be slightly flattened mushroom vortices with $l_x$ on the order of 200 $\nu/u^*$ and $l_y$ on the order of 100 $\nu/u^*$. There is no specific information on $l_z$; however, a value between

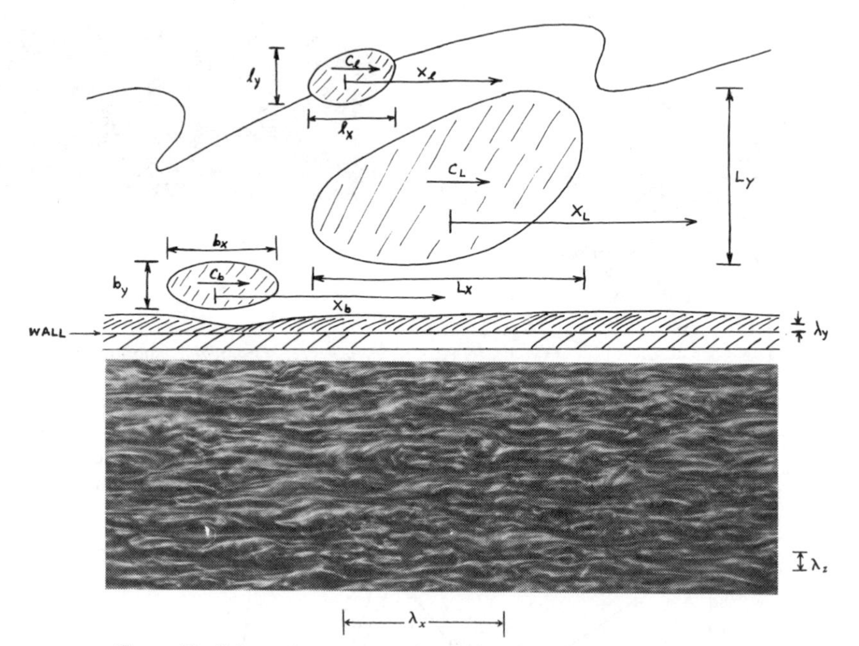


*Figure 12* Schematic summary of turbulent boundary-layer structure.

100 and 200 $\nu/u^*$ would be expected. Falco suggests that the typical eddies scale with the Taylor microscale and that they account for a significant fraction of the Reynolds stress in the outer flow. This is contrary to the commonly held belief that the Reynolds stresses are produced primarily by the largest eddies in the flow. The evidence thus far is relatively sparse and more measurements are needed over a wide Reynolds-number range to confirm these observations.

8. $X_l$—Persistence distance of energetic outer-flow eddies. These eddies appear to travel approximately five times their own streamwise length or about 1000 $\nu/u^*$ before losing their identity.

9. $C_l$—Celerity of energetic outer-flow eddies (Kovasznay et al 1970, Smith 1978, and Savas 1979). Typical values between 0.8 $u_\infty$ and 0.9 $u_\infty$ are observed.

10. $L_x$, $L_y$, $L_z$—Length scales of the large-scale motion in the outer flow (Kovasznay 1970, Dinkelacker et al 1977, Brown & Thomas 1977, Falco 1977, Zilberman et al 1977, Smith 1978, Head & Bandyopadhyay 1978, Praturi & Brodkey 1978, Willmarth & Wooldridge 1962). Most observations give a value for $L_x$ between $\delta$ and $2\delta$ at a height of about $0.8\delta$ above the surface. $L_x$ inferred from wall-pressure data is somewhat smaller with typical values between $0.5\delta$ and $\delta$. At $0.8\delta$ above the surface, the width $L_z$ of the outer large eddy appears to be between $0.5\delta$

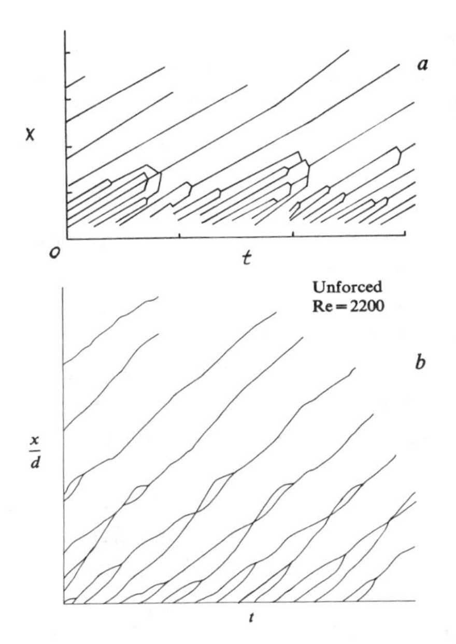


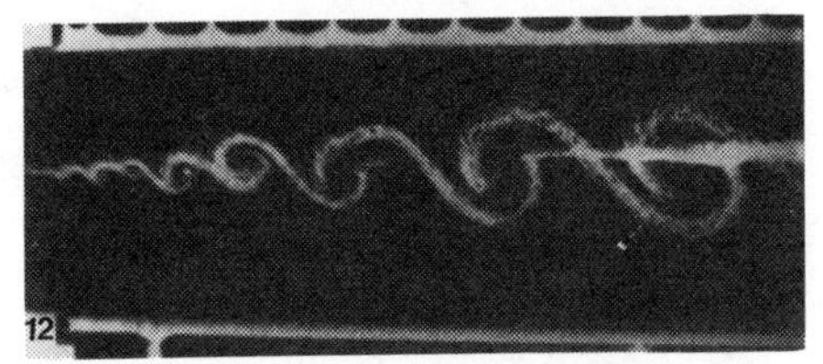


*Figure 13* Vortex trajectories. (*a*) In a plane mixing layer from Brown & Roshko (1974); (*b*) In the axysymmetric mixing layer in the initial region of a jet from Bouchard & Reynolds (1978).

and $\delta$ with the eddy centers spaced about 2.0 to 3.0$\delta$ apart in the spanwise direction.

11. $X_L$—Persistence distance of the large-scale motion in the outer flow (Kovasznay et al 1970, Dinkelacker et al 1977). Typical values are about 1.6$\delta$ and 2$\delta$ at a height of about 0.8$\delta$ above the plate.

12. $C_L$—Celerity of the large-scale motion in the outer flow (Kovasznay et al 1970, Dinkelacker et al 1977, Brown & Thomas 1977, Zilberman et al 1977, Smith 1978, Savas 1979, Cantwell et al 1978, Sabot et al 1977, Zakkay et al 1978, Willmarth & Wooldridge 1962, Coles & Barker 1975). A variety of measurements indicate a value between 0.8 and 0.9 $u_\infty$ at a height of about 0.8$\delta$. Some wall-pressure data indicate a somewhat lower value.

13. $T_B$—Period between bursts (Kline et al 1967, Kim et al 1971, Rao et al 1971, Kovasznay et al 1970, Brown & Thomas 1977, Falco 1977, Narayanan & Marvin 1978, Smith 1978, Savas 1979, Sabot et al 1977, Zakkay et al 1978). This is one of the more studied variables in turbulent boundary-layer structure. Early observations scaled $T_B$ with wall variables. Now it appears to be fairly well established that $T_B$ scales with outer variables and the generally accepted number is $T_B u_\infty/\delta \simeq 6$; however, there is a considerable amount of scatter about this value with a range from 2.5 to 10. It is also found that $T_B$ varies only slightly across the layer implying, in agreement with other observations, that the occurrence of a burst affects the entire layer.

14. $\beta$—Angle of maximum correlation (Corino & Brodkey 1969, Brown & Thomas 1977, Head & Bandyopadhyay 1978). Measurements correlating wall shear and streamwise velocity give $\beta = 18°$. Other observations of a characteristic lean angle for the large structure range from 8° to 40°.

15. $u'v'_{max}$—Maximum measured instantaneous $u'v'$ (Lu & Willmarth 1973, Falco 1977, Nychas et al 1973). In the outer portion of the flow $(y/\delta = 0.8)$, instantaneous values of $u'v'$ exceeding ten times the local mean have been observed. Near the wall $(y^+ \sim 30)$ values of $u'v'$ as large as 60 times the local mean have been observed.

16. $C_{pw_{max}}$—Maximum change in wall-pressure coefficient (based on free-stream velocity) during the passage of the organized motion (Dinkelacker et al 1977, Savas 1979, Cantwell et al 1978). Peak-to-peak variations on the order of 0.035 to 0.050 are observed.

17. $w'_x$—Root-mean-square streamwise vorticity fluctuations near the wall (Willmarth & Bogar 1977, Hanratty et al 1977). Typical values around one tenth the mean vorticity in the spanwise direction are observed.

18. $u^{+\prime}_{max}$—Maximum value of $\langle u'^2 \rangle^{1/2}/u^*$ (Coles 1978). This maximum occurs at about $y^+ = 15$. In an extensive survey of the literature, Coles collected a considerable body of data on fluctuations in the sublayer. The results indicate that over a wide range of Reynolds number $(100 < \delta^+ < 10^4)$, $u^{+\prime}_{max}$ has a nearly constant value of 2.75. At the same position $v^{+\prime}$ is about 0.6 and $w^{+\prime}$ is about 1.0. Using the "universal" numbers, $u^+/w^+ = 2.75$ as $y^+ \to 0$ and $\lambda^+ = 100$ plus Equation (3) matched to (4) and (2), Coles was able to produce a model of the secondary flow in the sublayer that gave excellent agreement with the collected measurements of $u^{+\prime}$, $v^{+\prime}$, $w^{+\prime}$, and $u'v'/\tau_w$ in the range $0 < y^+ < 15$.

## *(g) Discussion*

Very few issues regarding the organized structure in the turbulent boundary layer could be considered resolved.[5] It is clear from the work of Corino & Brodkey, Kim et al, Willmarth, Grass, and others, that most of the production of turbulent energy near the wall occurs during

[5] The generally disordered state of our understanding of turbulent boundary-layer structure is well recognized by workers in the field. A useful step toward improving this situation with many good remarks identifying resolved as well as unresolved issues may be found in the summary of a recent Michigan State workshop on structure by Kline & Falco (1980).

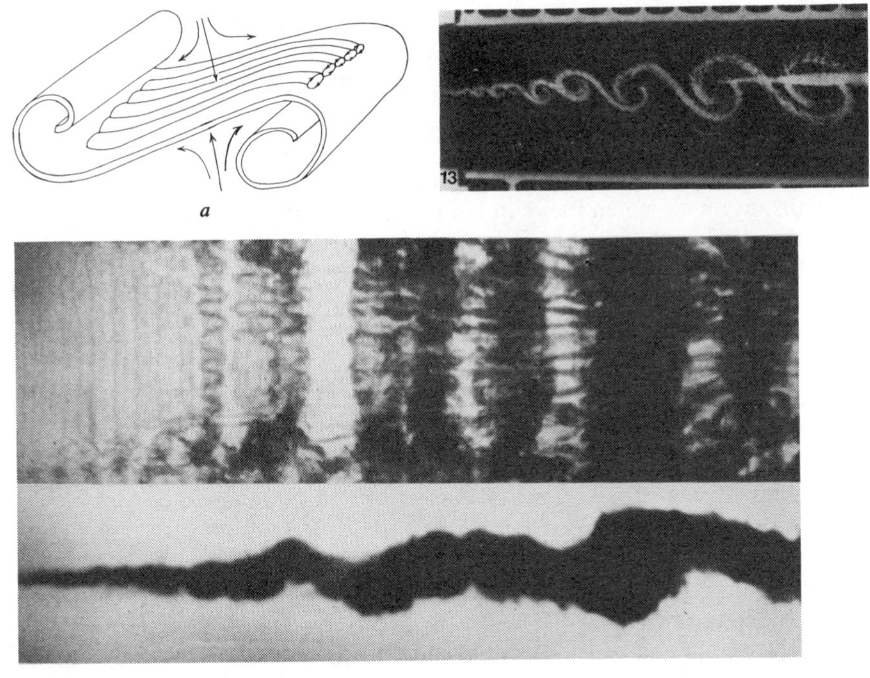


*Figure 14* Three-dimensional structure in a plane mixing layer. (*a*) A conceptual picture from Corcos (1979). (*b*) A chemically reacting layer from Briedenthal (1978).

bursts. However, the interaction process that creates conditions under which bursts occur is very far from understood. It appears fairly well established that the mean time between bursts scales with outer variables. This is essentially a statement about the spacing between bursts in the streamwise direction as well as the rate at which bursts are produced at the wall, propagate, and decay. However, very little is known about the scaling laws for the spacing of bursts in the spanwise direction.

There is a variety of views regarding the interaction between inner and outer layers. In each case a causal relationship is drawn between different aspects of the organized structure. However, the elliptic nature of incompressible fluid motion makes cause and effect exceedingly difficult to distinguish and the hope is that, once a true understanding is reached, the need for such a distinction will vanish. Central to most views of the interaction process is the idea that bursts are the result of an inviscid instability of the instantaneous streamwise velocity profile (see Figures 5*a* and 6*b*). Inflection points in the instantaneous profile at

$y^+$ between 30 and 50 have been observed by Kline et al 1967, Smith 1978, and Blackwelder & Kaplan 1971. One view of the interaction (Rao et al 1971, Laufer & Narayanan 1971, and Kovasznay et al 1970) sees bursting as an instability of the sublayer produced by the pressure field associated with the large-scale motion in the outer layer. In another view (Offen & Kline 1974, 1975), the emphasis is on flow disturbances which move in from the logarithmic region much closer to the wall. In this view, sweeping motions from the log region impress on the wall the temporary adverse pressure gradient required to bring about the streak lifting that precedes a burst.

Another aspect of the interaction problem regards the maintenance of the outer flow. The predominant view is that the outer flow is in some sense the wake formed by a composite of successive bursts near the wall. Offen & Kline view the bulges in the outer flow as the result of vortex pairing between eddies associated with two to four bursts. In contrast, the observations of Rao et al and Narayanan & Marvin that the mean time between bursts is nearly independent of distance above the wall, suggests a single structure that fills the layer.

Scaling parameters for the length scales of streamwise vortices near the wall are well established to be $\nu$ and $u^*$. This is essentially guaranteed by the universality of the logarithmic shape of the velocity profile near the wall [Equation (4) for $y^+/\delta^+ \ll 1$]. However, the scaling of mean times between bursts on outer variables, plus the fact that bursts account for most turbulence production near the wall, contradicts this universality. Somehow the effect of, say, pressure gradient or roughness on bursting is manifested only through its effect on $\tau_w$ without any residual effect on the details of the shape of the velocity profile.

Scaling parameters for structural features of the outer portion of the boundary layers are much more in doubt. This arises from the fact that the effect of the wall is felt throughout the layer [recall Equation (10)]. As a result, any eddy structure in the outer layer will exhibit a dependence on wall variables. To see this consider the scaling properties of the Taylor and Kolmogorov microscales in the outer portion of the boundary layer. If production and dissipation of turbulent energy scale together in this region, then dimensional analysis leads to

$$\frac{\lambda}{\delta} \sim \frac{1}{R_\delta^{1/2}} \text{ and } \frac{\eta}{\delta} \sim \frac{1}{R_\delta^{3/4}} \tag{19}$$

where $\lambda$ and $\eta$ are the Taylor and Kolmogorov microscales, respectively.[6]

[6]Unfortunately, $\lambda$ is also used to denote sublayer streak scales which do not necessarily correspond to a Taylor microscale. To maintain the distinction, we use subscripted $\lambda$'s ($\lambda_x, \lambda_y, \lambda_z$ to denote streak scales).

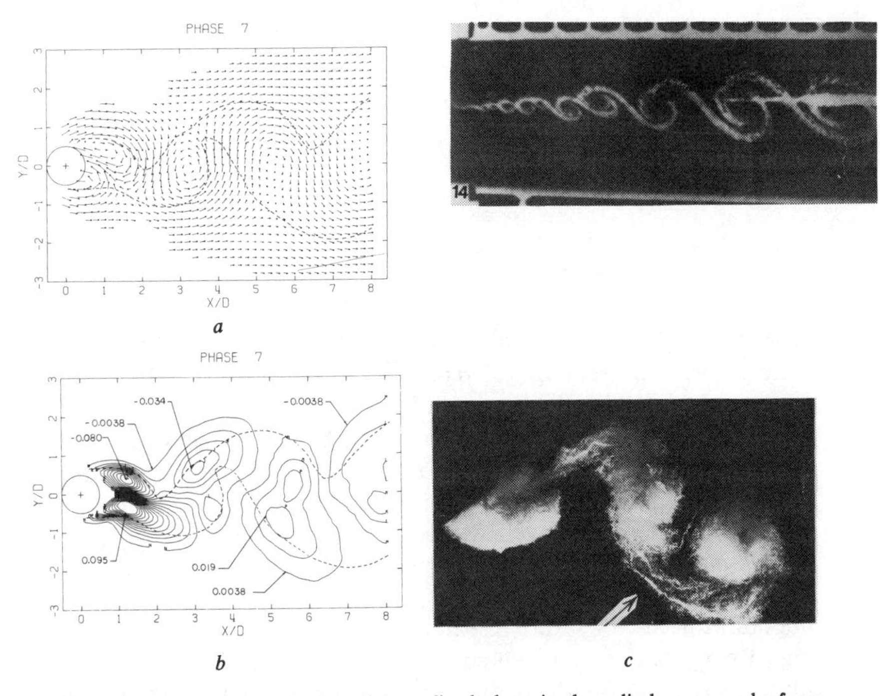


*Figure 15* Ensemble-averaged mean flow at fixed phase in the cylinder near wake from B. J. Cantwell and D. E. Coles (1980 in preparation). (*a*) Mean flow referenced to an observer who moves downstream at 0.75 $u_\infty$; (*b*) Shearing stress $\langle u'v' \rangle / u_\infty^2$ at constant phase; (*c*) Three-dimensional motions in the near wake of a flapped hydrofoil from Meijer (1965).

In a turbulent boundary layer we may write

$$\frac{\lambda u^*}{\nu} \sim \left(\frac{C_f}{2}\right)^{1/2} R_\delta^{1/2} \text{ and } \frac{\eta u^*}{\nu} \sim \left(\frac{C_f}{2}\right)^{1/2} R_\delta^{1/4}. \tag{20}$$

Making use of (10) we have

$$\frac{\delta u^*}{\nu} \sim e^{K\sqrt{\frac{2}{C_f}}}, \quad \frac{\lambda u^*}{\nu} \sim \left(\frac{C_f}{2}\right)^{1/4} e^{\frac{K}{2}\sqrt{\frac{2}{C_f}}},$$

$$\frac{\eta u^*}{\nu} \sim \left(\frac{C_f}{2}\right)^{3/8} e^{\frac{K}{4}\sqrt{\frac{2}{C_f}}}. \tag{21}$$

All three length scales depend exponentially on $1/C_f$, a quantity which increases very slowly with increasing $x$ along the layer. Over typical ranges of $C_f$ where observations are made both $\lambda u^*/\nu$ and $\eta u^*/\nu$

increase with decreasing $C_f$ although the dependence on a fractional power of $C_f$ mitigates the rate of increase somewhat. The dependence on $C_f$ is somewhat stronger when $\delta$, $\lambda$, and $\eta$ are normalized on $u_\infty$. Thus

$$\frac{\delta u_\infty}{\nu} \sim \left(\frac{2}{C_f}\right)^{1/2} e^{\kappa\sqrt{\frac{2}{C_f}}}, \quad \frac{\lambda u_\infty}{\nu} \sim \left(\frac{2}{C_f}\right)^{1/4} e^{\kappa\sqrt{\frac{2}{C_f}}},$$

$$\frac{\eta u_\infty}{\nu} \sim \left(\frac{2}{C_f}\right)^{1/8} e^{\kappa\sqrt{\frac{2}{C_f}}}. \tag{22}$$

Here the dependence on a power of $1/C_f$ leads to a faster rate of increase with decreasing $C_f$. Thus as far as eddy length scales are concerned, normalization with wall variables will invariably lead to a slower dimensionless rate of change with $C_f$ or, equivalently, with Reynolds number. No experiment to date covers a broad enough range of Reynolds number to properly resolve the scaling laws for the flow structure in the outer layer. Moreover, the mean time between bursts $u_\infty T/\delta$ may also exhibit a weak dependence on wall variables when the range of Reynolds numbers has been extended. One of the most important needs for future research is to extend current observations to high Reynolds numbers, not so much to identify new transport mechanisms as Head & Bandyopadhyay (1978) suggest, but rather to establish the scaling laws for the mechanisms that are observed to play an important role at low Reynolds numbers.

## III ORGANIZED MOTION IN FREE SHEAR FLOWS

In a study of transition in a laminar free shear layer, Freymuth (1966) noted the presence of highly regular vortex motions in the nonlinear stages of transition. He observed that the onset of subharmonic wavelengths was associated with an interaction he called "slip" in which two adjacent vortices rotate about a common axis and coalesce into a single structure. Downstream, the regular vortices appeared to give way to a chaotic motion. Traditionally, the breakdown to random, three-dimensional motion has been argued on the basis of vortex stretching along the principal axis of strain of the mean velocity field. Taking the curl of the momentum equation leads to the equation for vorticity

$$\frac{\partial \omega_i}{\partial t} + \frac{\partial}{\partial x_j}(u_j\omega_i) = \omega_i \frac{\partial u_i}{\partial x_j} + \nu \frac{\partial^2 \omega_i}{\partial x_j \partial x_j}. \tag{23}$$

The first term of the right-hand side of (23), the vortex stretching term,

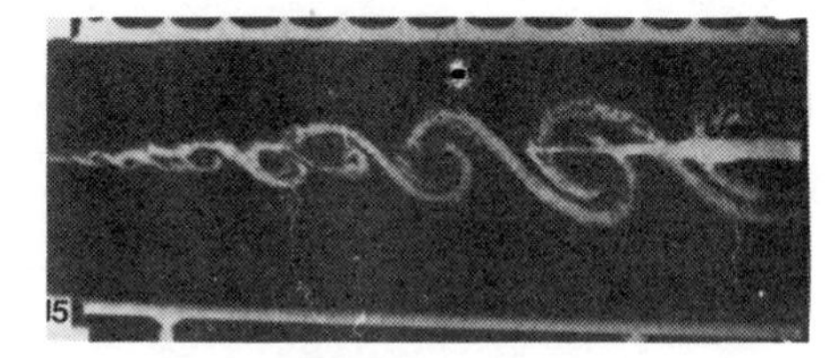


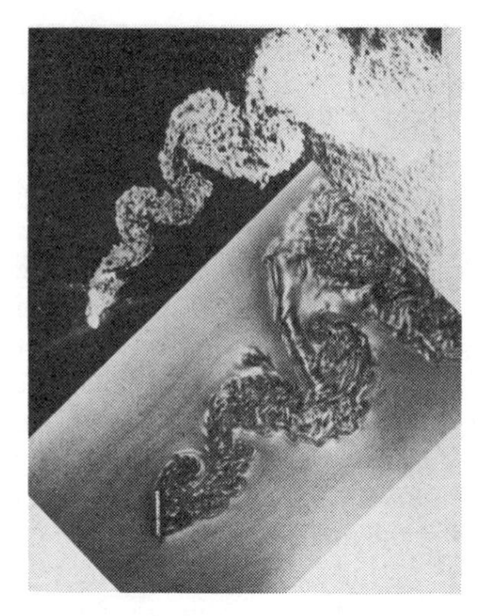

*Figure 16* Oil slick from the tanker, Argo Merchant, grounded off the Nantucket Shoals in 1976 (Re$\sim10^8$). Superimposed is the flow past an inclined flat plate at Re$=10^3$.

acts as a source term for vorticity. Any spanwise vorticity, it could be argued, would be quickly dominated by much more intense vorticity in the streamwise direction. Furthermore, the streamwise vortex motions would contain most of the turbulent energy and any transverse vorticity would contribute little to the dynamics of the flow.

In a study of the turbulent mixing layer, Brown & Roshko (1974) discovered that the layer was dominated by large-scale spanwise vortex motions. These motions originate in the transitional part of the layer; they do not vanish when smaller-scale turbulence sets in and they appear to remain as a permanent feature of the flow at all higher Reynolds numbers. Winant & Browand (1974) carried out a detailed study of the vortex pairing observed by Freymuth in a low-Reynolds-number shear flow. In pairing, adjacent vortices rotate about each other under their mutual induced velocity field. As the rotation progresses they amalgamate into a single vortex of larger scale. Winant & Browand suggest that pairing is the principal mechanism by which a shear layer grows. A similar pairing process was observed by Brown & Roshko at much higher Reynolds number. A sequence of pictures from a movie taken by the Caltech group showing a pairing event is shown in the upper right-hand corner of the pages of this review (see also Roshko 1976). An *x-t* diagram of eddy trajectories showing the amalgamation of two and sometimes three vortices is shown in Figure 13*a*. Figure 13*b* shows eddy trajectories in the initial shear layer of an axisymmetric jet measured by Bouchard & Reynolds (1978). Here, vortex centers can be observed to orbit about each other several times before coalescing.

The picture of the mixing layer which emerges from this is of a double structure composed of an array of moving and interacting large-scale spiral vortex motions superimposed on a background of finer-scale, presumably random, turbulence. Although the persistence of these eddies remains a subject of some controversy (Chandrasuda et al 1978), there is ample evidence that they are present and play an important role at Reynolds numbers far above those at which the mixing layer would be considered transitional (Dimotakis & Brown 1976). In some ways, recent observations confirm several of the ideas of Townsend who had proposed a similar double structure with fine-scale turbulence convected about by eddies whose size was comparable to the overall scale of the flow. However, the large eddies were only vaguely perceived as a field of randomly interacting, slow moving, three-dimensional motions. The observations in the mixing layer gave symmetry and form to the conceptual picture of the large-scale structure of turbulent flow. Suddenly it was feasible and reasonable to draw a picture of turbulence! The hand, the eye, and the mind were brought into a new relationship that had never quite existed before; cartooning became an integral part of the study of turbulence.

Observations of the plane mixing layer stimulated a renewed interest in modeling unsteady viscous flow using discrete vortex arrays. This method had been used by Abernathy & Kronauer (1962) to model the laminar vortex street in a two-dimensional wake. Now it was resurrected as a method for modeling the large-scale motion in turbulence.[7] The essential idea here is to lump the continuous field of vorticity into individual vortex elements. The time evolution of the flow is then simulated by solving a set of first-order ordinary differential equations

$$\frac{dx_i}{dt} = \sum_{\substack{j=1 \\ i \neq j}}^{N} u_j(x_i, t) \tag{24}$$

where $x_i$ is the coordinate of the $i$th vortex point and $u_j$ is the velocity induced at $x_i$ by the $j$th vortex point. Since the flow outside each vortex core satisfies Laplace's equation, the velocity induced at $x_i$ is found by superposition. The method has been used in both two and three dimensions to model a number of flows including vortex sheet roll-up (Chorin & Bernard 1973), mixing layers (Ashurst 1977), wakes (Clements 1973,

[7] In remarkable anticipation of later observations, Onsager (1945) used the Hamiltonian structure of the equations of motion to analyze probable states for an array of point vortices. When the "temperature" of the array is negative, point vortices of the same sign tend to organize into large compound vortices. For more discussion, see the recent review by Saffman & Baker (1979).

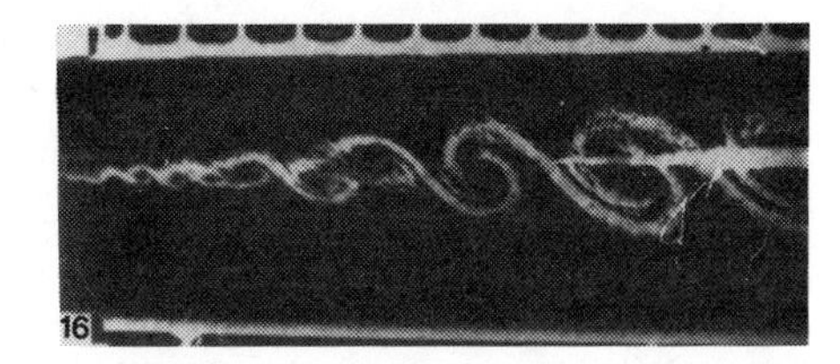

*Figure 17* Flow patterns in three-dimensional flows from Perry, Lim & Chong (1980). (*a*) Smoke geometry of a neutrally buoyant wake structure. (*b*) Schematic representation of streamlines in a three-dimensional wake. (*c*) Smoke rising from a cigarette. (*d*) Smoke rising from a chimney.

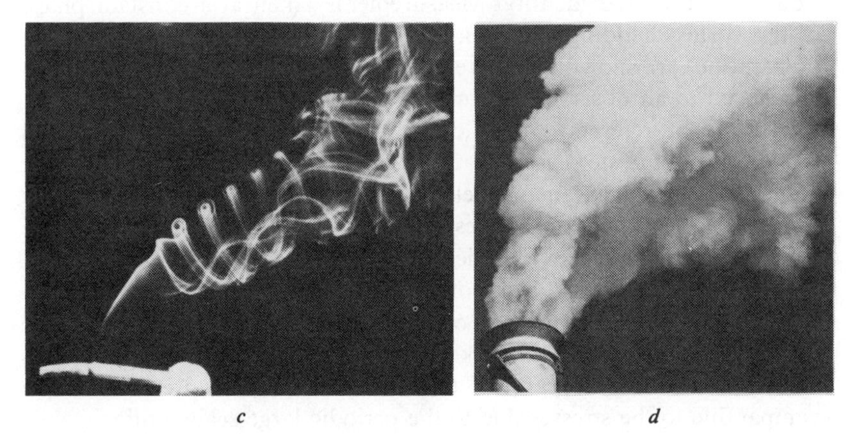


Sarpkaya 1975), and turbulent spots (Leonard 1979). Generally, the simulations reproduce the large-scale motions in these flows remarkably well. However, they tend to do less well at simulating the associated stresses. At least part of the reason for this appears to be due to the neglect of small-scale three-dimensional motions, which contribute significantly to the stress.

Observations of organized structure in free-shear flows are not limited to mixing layers. Crow & Champagne (1971) observed quasi-deterministic motions in the developed portion of an axisymmetric jet. Bevilaqua & Lykoudis (1971) observed dye entrained into a two-dimensional wake and concluded that the interface convolutions visible on photographs of wakes are the outer edges of large vortices and that it is vortex induction by these large eddies that is primarily responsible for the entrainment of laminar fluid by the wake.

All of the evidence discussed so far is for organized motion at the largest scale of the flow superimposed on random background turbulence. However, there is an increasing amount of evidence for a high degree of order at smaller scales. Observations of highly organized three-dimensional motions in the mixing layer have been reported by Briedenthal (1978) who examined a chemically reacting flow in water. Figure 14*b* shows simultaneous side and top views of the flow. The large transverse eddies are strung together by a spaghetti-like net of small-scale streamwise counter-rotating vortices not unlike the streamwise vortices found near the wall under a turbulent boundary layer.

A flow that has been known to be dominated by coherent vortex motions for a long while is the near wake of a bluff body. Here the eddies are produced in a constant, regular manner and, except for some dispersion, are not subject to pairing or any other strong interaction that would obscure their identity. Measurements taken at a constant phase of the vortex-shedding cycle (B. J. Cantwell and D. E. Coles 1980, in preparation) are shown in Figure 15. Figure 15*a* shows the velocity field referenced to an observer who moves downstream with the vortices. In this frame of reference, the flow is quasi-steady and the organized motion is manifested as a pattern of centers and saddles. Similar patterns of the instantaneous velocity field in the wake of a bluff body have been measured by Davies (1976; D-shaped cylinder), Owen & Johnson (1978; circular cylinder at $M=0.6$), and Perry & Watmuff (1979; three-dimensional wake of an ellipsoid).

Some indirect evidence for organized small-scale structure can be found in the cylinder near wake depicted in Figure 15*b*. The stresses associated with the background turbulence in this flow are found to be comparable to the stresses due to the periodic large-scale motion (taken as a fluctuation away from the globally averaged mean flow). The background normal stresses $\langle u'^2 \rangle$ and $\langle v'^2 \rangle$ show expected behavior with maxima near vortex centers and minima between vortices. However, the background shearing stress $\langle u'v' \rangle$ shown in Figure 15*b* achieves a maximum in the saddle region between the vortices. If one forms the

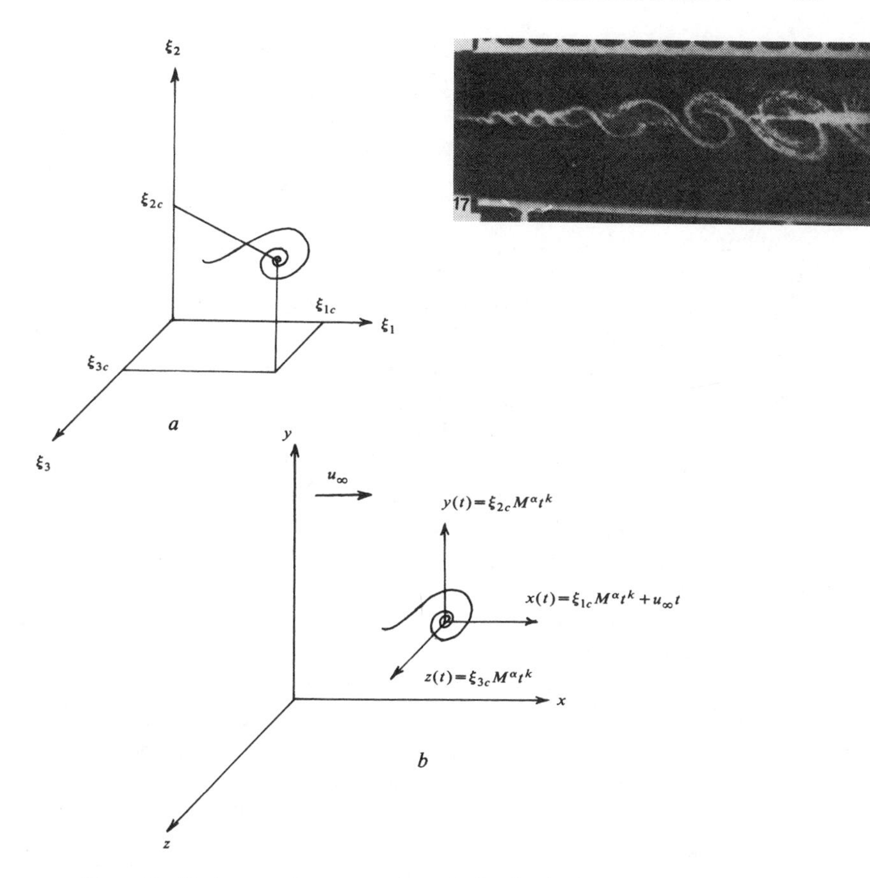


*Figure 18* Trajectory of a nonsteady critical point in physical coordinates.

correlation coefficient

$$R=\frac{\langle u'v'\rangle}{(\langle u'^2\rangle)^{1/2}(\langle v'^2\rangle)^{1/2}} \tag{25}$$

it is found that near a vortex center, where the vorticity and turbulent energy are at their maximum, $R$ is about 0.1, whereas, in the region of the saddle-point flow between vortices, where the background turbulent energy is at a minimum and where the transverse component of ensemble mean vorticity is nearly zero, $R$ is between 0.5 and 0.6. The background turbulence in this flow is neither small nor random. It has structure. Vortex stretching due to the straining motion at the saddle must lead to a substantial strengthening of the component of vorticity aligned with the diverging separatrix of the saddle. Some evidence for this effect is shown in the photograph of the wake of a supercavitating hydrofoil shown in Figure 15*c* where well-defined lines of cavitation bubbles connecting adjacent large eddies indicate a low-pressure zone associated with intense, highly stretched vorticity.

In a sense these observations also confirm some early views of turbulence in which the energy-containing eddies lie along the principal axis of strain of the mean velocity field. The main and very important difference is that the streamwise vortices convect with the flow responding, not to the stationary mean flow, but to the unsteady strain field imposed by the large eddies. Through the coupling between the large-scale motion and background turbulence the flow is differentiated into convecting regions of active and passive turbulence.

The question that immediately arises is, At what scale does organized motion in turbulent flow cease? Corcos (1979) has recently proposed a model for the plane mixing layer in which the flow is treated as essentially deterministic at all scales. The result is a cartoon of the mixing layer (Figure 14*a*), in which flattened streamwise vortices with thickness comparable to the Taylor microscale form in a perturbed saddle-point flow between adjacent large-scale vortices. The prospect raised here is that the physics of turbulent flow can be understood and modeled by considering an equivalent small number of complex laminar flows. If this is so, then many traditional ideas about turbulence, for example the concept of local isotropy and the cascade picture of turbulent energy exchange between large and small scales, need to be re-examined in light of the possibility that even very small-scale motions may be highly organized. Old ideas will not be discarded, but they will receive a new, more illuminating, physical interpretation.

An area of active current research, which has been stimulated by observations of organized motion in both wall-bounded flows and free-shear flows, is large-eddy simulation. The basic approach here (Reynolds 1976) is to compute large-scale unsteady motions explicitly while modeling only those motions that lie at scales below that which can be resolved by the computational grid. Several flows have been simulated thus far with encouraging results. Calculations of homogeneous and isotropic turbulence were used by Clark et al (1979) to generate "empirical" constants for Reynolds-stress models which previously could only have been determined from measurements of grid turbulence. Large-eddy simulation of turbulent channel flow by Moin et al (1978) reproduced mean velocity profiles (particularly the logarithmic portion near the wall) which were in good agreement with the measurements of Hussain & Reynolds (1975). In addition, the calculations

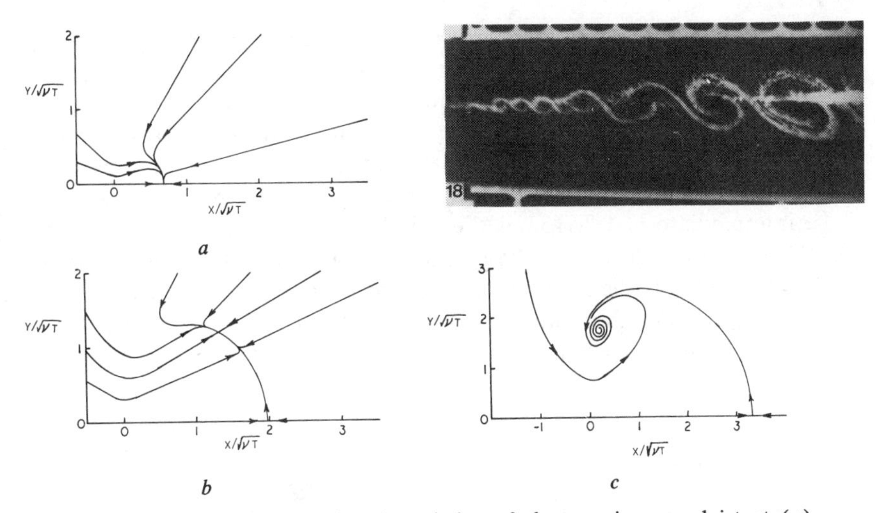


*Figure 19* Entrainment diagrams for the solution of the creeping round jet at (*a*) Re=2.0, (*b*) Re=8.0, (*c*) Re=20.0.

yielded data for quantities such as pressure and pressure-velocity correlations that cannot be measured experimentally.

At the heart of this approach is the assumption that unresolved sub-grid-scale fluctuations are only slightly nonisotropic and can be modeled in a universal way. The cutoff between large and small scales is essentially a function of the current state of computer technology. As machines become larger and faster (as they have and will become) the fraction of the spectrum of energy containing turbulent motions that can be computed explicitly will grow toward one. However, even the most optimistic forecaster of this evolution would concede that the technology-imposed cutoff between large and small scales will never approach the order of the Kolmogorov scale in a high-Reynolds-number, aerodynamically useful flow. A grid of this resolution in a given volume would require $N \cong \mathrm{Re}^{9/4}$ where $N$ is the number of grid points and Re is the Reynolds number of the flow in question based on global velocity and length scales. For a 4-cm chord compressor blade at 400 m/sec ($\mathrm{Re} \simeq 10^6$) $N$ equals $3 \times 10^{13}$. For a 200-cm airfoil at the same speed ($\mathrm{Re} \simeq 5 \times 10^7$) $N$ equals $2 \times 10^{17}$. However, estimates of the largest memory sizes available by the end of the century (Chapman 1979) do not exceed $N = 10^{11}$ (a four-order-of-magnitude increase in size over today's largest memories). The above estimate of $N$ is rather pessimistic. One may only need to resolve a Taylor microscale, in which case $N \sim \mathrm{Re}^{3/2}$. In addition, efficient grid-generating schemes can be used to reduce $N$ substantially.

In any case, for the foreseeable future, large-eddy simulation of high-Reynolds-number turbulent flows will have to stop at scales that will be several orders of magnitude greater than the smallest scales in the flow. In this respect the observations of small-scale organized motion raise a number of critical questions. Are these motions small in the sense of a Taylor or Kolmogorov microscale or are they really of intermediate scale? How is the coupling between large and small scales really accomplished? Is the motion at small scales the same from flow to flow? What is the maximum scale size at which local isotropy begins to be valid and will computers ever be able to see this scale for flows of engineering importance?

## IV TRANSITION AND THE CONTROL OF MIXING

More and more evidence has accumulated in the past few years that shows that virtually all turbulent shear flows are sensitive to transition, or more precisely, to perturbations applied during transition. Traditionally, it was believed that sufficiently far from their point of origin turbulent shear flows would reach an asymptotic state in which their rate of growth and decay would become independent of the manner in which the flow was started (see, for example, Liepmann 1962). The overwhelming body of data shows that once turbulence is established the overall properties of turbulent shear flows away from solid boundaries are virtually independent of viscosity. Although viscosity is essential to the creation of turbulence it seems to serve only to establish smaller and smaller scales of motion with more and more intense velocity gradients sufficient to dissipate the increased rate of energy input as the Reynolds number is increased. However, the Taylor microscale is usually associated, on dimensional grounds, with a velocity perturbation comparable to that associated with the largest eddies. It is not at all obvious that the overall behavior of the flow can depend so little on Reynolds number in the presence of such intense motions that depend strongly on Reynolds number.

The same data that demonstrates the Reynolds-number invariance of turbulence (illustrated by Figure 16) also shows very wide scatter (Roshko 1976). The mixing-layer spreading rates measured by Winant & Browand for $u_2/u_1=0.4$ at a streamwise Reynolds number of $10^4$ are in close agreement with the measurements of Spencer & Jones (1971) at a Reynolds number of $10^6$. In contrast, the spreading rate measured by Wygnanski & Fiedler (1970) at $\mathrm{Re}=5\times10^5$ differs by 30% from the value measured by Liepmann & Laufer (1947) for $u_2/u_1=0$ at a

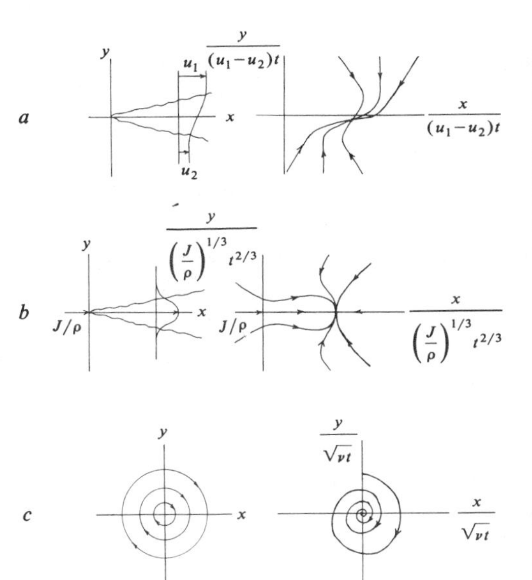


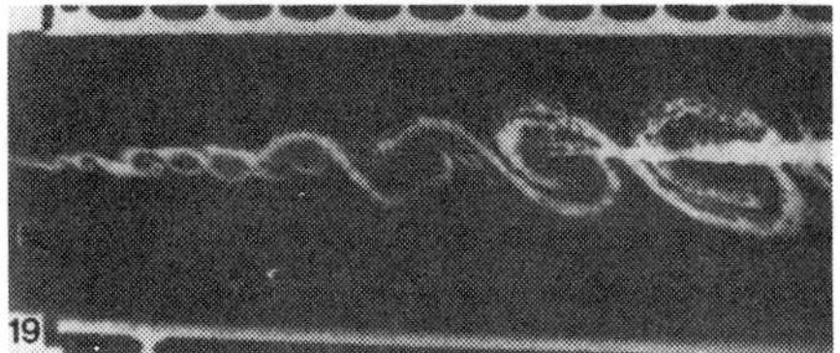


*Figure 20* Entrainment diagrams for several flows. (*a*) The stationary turbulent mixing layer; (*b*) The stationary plane turbulent jet; (*c*) The Oseen viscous vortex.

Reynolds number of $10^6$. Batt (1975) suggests that this is related to the presence or absence of a boundary-layer trip on the splitter plate ahead of the origin of the layer. As Roshko points out, much of the evidence suggests that any important effects of Reynolds number appear indirectly through conditions affecting transition rather than through the direct action of viscosity on the developing turbulent structure. In a recent evaluation of shear-layer data, Birch (1980) has suggested that much of the scatter in measurements of shear-layer spreading rate can be attributed to variations in the effective origin of the layer.

Effects of initial conditions have been observed by Leonard (1979) in numerical simulations of developing turbulent spots. In this study the spot is imbedded in an initially planar array of transverse vortex lines (Figure 9*b* and *c*). He finds that the downstream amplitude distribution of the distorted and stretched vortex lines appears to be directly related to the amplitude distribution of the initial disturbance.

Furuya & Osaka (1976) examined the effect of a distribution of roughness placed near the leading edge of a flat-plate zero-pressure-gradient turbulent boundary layer. Measurements of the spanwise distribution of momentum thickness showed a variation that was directly related to the spanwise variation of roughness. When the measurements were made several thousand boundary-layer thicknesses downstream, the same spanwise distribution of momentum thickness was found. Instead of decaying with distance, the amplitude of the variation had increased. In fact, the variation simply scaled with the boundary-layer thickness. Apparently the effect of roughness placed near the plate leading edge is to produce a permanent change in the large structure of the layer.

Other evidence that shear flows are sensitive to small perturbations may be found in the probe interference effects on shear layers observed by Hussain & Zaman (1978). They found that when a hot-wire probe was inserted into a laminar free-shear layer, it was found to induce stable edge-tone-like upstream oscillations. Their results suggest that a reference probe used near the origin of a free shear layer can alter the basic instability frequency of the layer, in turn influencing the downstream development of the coherent structure.

The sensitivity of shear flows to small perturbations applied during transition, together with the recognition that the developed flow is dominated by organized structures which appear to be remnants of transition, are the ingredients that suggest that some form of control over turbulent mixing may be possible. Oster et al (1977) were able to exert remarkable open-loop control on a plane mixing layer by flapping a small spanwise ribbon just downstream of the end of the splitter plate in the initial region of the layer. By adjusting the amplitude and frequency of the oscillation, they were able to halve or double the angle of spread of the layer. Under certain conditions the layer would spread for a while, stop spreading for a short distance, and then begin spreading again. The vibrating ribbon strongly influences the rate and phase at which vorticity is injected into the flow. The result is controlled vortex formation reminiscent of the "locking on" of vortex shedding observed to occur on vibrating bluff bodies (Griffin & Ramberg 1974). One can easily envision the use of effects of this kind in a closed-loop system where the flow is sensed and a perturbation is applied to produce some desired change in the flow. There is little doubt that the next few years will see a great deal of fruitful research in this area.

Are turbulent eddies like tennis balls or like medicine balls? It was once felt that an initially perturbed flow would rebound and eventually reach a unique asymptotic state. It now appears that an initially perturbed turbulent flow may remain permanently perturbed; that it may have an infinity of possible asymptotic states corresponding to an infinite variety of perturbed initial states. If this is so, then tremendous benefits may be realized through the control of mixing. However, it complicates our theoretical understanding. It is no longer enough to predict the angle of spread of a mixing layer; now we must be able to predict all possible angles of spread as a function of initial conditions.

This issue is still far from resolved and points up the need for standards for taking data. Present and future measurements need to be carefully and accurately scaled. In particular, it is necessary to distinguish between differences in flow conditions, methods of data presentation and reduction, and true differences in state.

## V ENTRAINMENT DIAGRAMS

A glossary of terms used to describe the coherent structure in turbulent flows would be long indeed. There are bursts, sweeps, streaks, typical eddies, streamwise vortices, transverse vortices, large eddies, etc. Every term is used to describe some structural feature of the motion. One of the central problems in current research is to relate these usually observed features to the usually not observed variables of fluid motion; streamlines, streaklines, pathlines, pressure, and their various derivatives.

The unsteady patterns of centers and saddles observed in conditionally averaged turbulent flow have much in common with the phase portraits of nonlinear dynamical systems. The connection is through the equations for particle paths,

$$\frac{dx_i(t)}{dt} = u_i(\mathbf{x}(t), t). \tag{26}$$

If a frame of reference can be found in which the flow is steady, then (26) reduces to an autonomous system with integral curves that coincide with the streamlines of the velocity field referred to the same frame. A number of authors have made use of this fact to explore the properties of solution trajectories in a variety of steady-flow situations. Oswatitsch (1958) and Lighthill (1963) classified certain critical points which can occur near a rigid boundary. Perry & Fairlie (1974) reviewed critical-point analysis in a general way and applied the technique to the problem of three-dimensional separation. They placed special emphasis on the fact that the method provides a wealth of topological language particularly well suited to the description of fluid-flow patterns. Recently Hunt et al (1978) applied critical-point theory to flow-visualization studies of bluff obstacles. More recently, Peake & Tobak (1980) reviewed the use of critical-point theory as it is applied to the study of three-dimensional vortex flows about various bodies in high-speed flow.

The success of critical-point theory for studying steady flow, coupled with the observations of organized spiral vortex motions in unsteady flows (critical-point-like motions) have led to a search for ways to extend the theory to unsteady flow. Perry & Lim (1978) used critical-point analysis to produce a qualitative description of the three-dimensional unsteady flow patterns in co-flowing jets and wakes (Figure 17*a* and *b*). Pullin (1978) studied the evolution of critical points in the unsteady streamline pattern associated with vortex-sheet rollup from the edge of an impulsively started plate.

The instantaneous streamline patterns used for these analyses provide a form of flow visualization combined with substantial amounts of quantitative information about the flow. Most importantly, they focus on the problem of connecting vortex structures together to complete the flow field. However, there are significant conceptual problems involved in interpreting unsteady streamline patterns as they relate to entrainment. In an unsteady flow, streamlines can move across fluid pathlines; thus the stream function provides little insight into the behavior of the fluid itself. Furthermore, structural features of the flow often remain hidden in a picture of instantaneous streamlines.

Particle trajectories plotted in physical coordinates also present similar conceptual difficulties. If the integration of the particle-path equations is carried out over a volume of particles, then each point in space will be traversed by an infinite set of trajectories, each with a different slope corresponding to the passage of particles through the point at successive instants of time. In addition, the pattern of particle paths, like the pattern of streamlines, depends on the frame of reference. For a recent appreciation of this problem see Lugt (1979).

Certain time-dependent flows can be reduced to a self-similar form. Such flows usually depend on one or at most two global parameters. In this case, some of the above objections can be removed by reducing the particle-path equations (26) to an autonomous system in similarity coordinates. Figure 11*c* is a diagram of particle trajectories in similarity coordinates used to analyze the flow structure on the plane of symmetry of a turbulent spot. Particle trajectories in similarity coordinates were used by Turner (1964) to analyze the flow pattern in a rising turbulent thermal which was modeled using an expanding Hill's spherical vortex. This method can be used to generate entrainment diagrams for a wide variety of shear flows, some steady, some unsteady, some laminar, and some turbulent. Some of the flows are governed by the Navier-Stokes equations, some by the boundary-layer equations, and some by the Oseen equation. In most cases the transformations involve simple stretchings, although more complex transformations involving uniform and logarithmic rotations and arbitrary nonuniform translations (Cantwell 1978) are allowed. In fact, it appears that virtually every incompressible viscous flow that we ordinarily think of as self-similar in

space (the Blasius boundary layer, the round jet, the plane turbulent mixing layer, etc) can be thought of as a special case of a more general unsteady flow that is self-similar in time with particle paths which reduce to an autonomous system. In the general case the flow is assumed to have been started at some initial time. Table 2 summarizes various similarity transformations and the reduced equations that may be derived from them. The basic similarity variable is

$$\xi_i = \frac{x_i - V_i t}{M^\alpha t^k} \tag{27}$$

where $M$ is an invariant of the motion with units $L^m T^{-n}$ and $\alpha$ and $k$ are chosen so that $M^\alpha t^k$ has the dimensions of a length. Thus

$$\alpha = 1/m, k = n/m. \tag{28}$$

Included are flows where there is a uniform velocity to the right in the $x$-direction so that $V = (u_\infty, 0, 0)$ and flows where there is no externally imposed velocity $V = (0, 0, 0)$. Consider the trajectory of a critical point shown schematically as a stable focus in Figure 18. If we take $\xi_1$ in the direction of the external flow, and $\xi_2$ and $\xi_3$ as cross-stream directions, then in physical coordinates, the trajectory of the critical point is given by

$$x_c = u_\infty t + \xi_{1c} M^\alpha t^k, y_c = \xi_{2c} M^\alpha t^k, z_c = \xi_{3c} M^\alpha t^k, \tag{29}$$

for flows governed by the full equations, by

$$x_c = \xi_{1c} M^\alpha t^k, y_c = \xi_{2c} \sqrt{\nu t}, z_c = \xi_{3c} \sqrt{\nu t}, \tag{30}$$

for flows governed by the boundary-layer equations, and by

$$x_c = u_\infty t + \xi_{1c} M^\alpha t^k, y_c = \xi_{2c} \sqrt{\nu t}, z_c = \xi_{3c} \sqrt{\nu t}, \tag{31}$$

for flows governed by the Oseen equation.

If we take $\delta = \sqrt{y_c^2 + z_c^2}$ as a cross-stream length scale and $u_0 = u_\infty - \dot{x}_c$ as a characteristic streamwise velocity scale, then the four cases listed in Table 1, item 5, may be distinguished. Using the relations in Table 1, plus the parameters that are used to characterize various shear flows, one can construct Table 2. The important point in all of the above is the connection between turbulence structure and specific structural details of the velocity field. The concept of organized structure is advanced to a description in terms of critical points in the entrainment diagram and their relationship to the propagation and decay of the flow.

For turbulent flows the time evolution of scales is given by

$$\delta \sim M^\alpha t^k, u_0 \sim k M^\alpha t^{k-1}. \tag{32}$$

Essential to the self-similar development of turbulent flows is the

**Table 1** Equations and self-similar forms

| Full equations of motion | Boundary-layer equations | Oseen equations |
|---|---|---|
| | $\frac{\partial p}{\partial y}=0$ | $\frac{\partial p}{\partial y}=0$ |
| $\frac{\partial u_i}{\partial x_i}=0$ | $\frac{\partial u}{\partial x}+\frac{\partial v}{\partial y}=0$ | $\frac{\partial u}{\partial x}+\frac{\partial v}{\partial y}=0$ |
| $\frac{\partial u_i}{\partial t}+u_j\frac{\partial u_i}{\partial x_j}=-\frac{1}{\rho}\frac{\partial p}{\partial x_i}+\frac{1}{\rho}\frac{\partial \tau_{ij}}{\partial x_j}$ | $\frac{\partial u}{\partial t}+u\frac{\partial u}{\partial x}+v\frac{\partial u}{\partial y}=-\frac{1}{\rho}\frac{\partial p}{\partial x}+\nu\frac{\partial^2 u}{\partial y^2}$ | $\frac{\partial u}{\partial t}+u_\infty\frac{\partial u}{\partial x}=\nu\frac{\partial^2 u}{\partial y^2}-\frac{1}{\rho}\frac{\partial p}{\partial x}$ |
| (1) Similarity variables | | |
| $\xi_i=\frac{x_i-V_i t}{M^{\alpha}t^{k}}$ | $\xi=\frac{x}{M^{\alpha}t^{k}}$ | $\xi=\frac{x-u_\infty t}{M^{\alpha}t^{k}}$ |
| $U_i(\underline{\xi})=\frac{u_i-V_i}{M^{\alpha}t^{k-1}}$ | $\eta=\frac{y}{\sqrt{\nu t}}$ | $\eta=\frac{y}{\sqrt{\nu t}}$ |
| $P(\underline{\xi})=\frac{p}{\rho M^{2\alpha}t^{2k-2}}$ | $U(\xi,\eta)=\frac{u}{M^{\alpha}t^{k-1}}$ | $U(\xi,\eta)=\frac{u-u_\infty}{M^{\alpha}t^{k-1}}$ |
| $T_{ij}(\underline{\xi})=\frac{\tau_{ij}}{\rho M^{2\alpha}t^{2k-2}}$ | $V(\xi,\eta)=\frac{vt^{1/2}}{\nu^{1/2}}$ | $V=\frac{vt^{1/2}}{\nu^{1/2}}$ |
| | $P(\xi,\eta)=\frac{p}{\rho M^{2\alpha}t^{2k-2}}$ | $P(\zeta,\eta)=\frac{p}{\rho M^{2\alpha}t^{2k-2}}$ |
| (2) Reduced equations | | |
| | $\frac{\partial P}{\partial \eta}=0$ | $\frac{\partial P}{\partial \eta}=0$ |
| $\frac{\partial U_i}{\partial \xi_i}=0$ | $\frac{\partial U}{\partial \xi}+\frac{\partial V}{\partial \eta}=0$ | $\frac{\partial U}{\partial \xi}+\frac{\partial V}{\partial \eta}=0$ |
| $(k-1)U_i+(U_j-k\xi_j)\frac{\partial U_j}{\partial \xi_j}=-\frac{\partial P}{\partial \xi_i}+\frac{\partial T_{ij}}{\partial \xi_j}$ | $(k-1)U+(U-k\xi)\frac{\partial U}{\partial \xi}+\left(V-\frac{\eta}{2}\right)\frac{\partial U}{\partial \eta}=-\frac{\partial P}{\partial \xi}+\frac{\partial^2 U}{\partial \eta^2}$ | $(k-1)U-k\xi\frac{\partial U}{\partial \xi}-\frac{\eta}{2}\frac{\partial U}{\partial \eta}=\frac{\partial^2 U}{\partial \eta^2}-\frac{\partial P}{\partial \xi}$ |

| | | |
|---|---|---|
| (3) Pathline equations | | |
| $\frac{d\xi_i}{d\tau}=U_i(\underline{\xi})-k\xi_i$ | $\frac{d\xi}{d\tau}=U(\xi,\eta)-k\xi$ | $\frac{d\xi}{d\tau}=U(\xi,\eta)-k\xi$ |
| | $\frac{d\eta}{d\tau}=V(\xi,\eta)-\frac{\eta}{2}$ | $\frac{d\eta}{d\tau}=V(\xi,\eta)-\frac{\eta}{2}$ |
| $\tau=\ln t$ | $\tau=\ln t$ | $\tau=\ln t$ |
| (4) Temporal scaling laws for $\delta$ and $u_0$ | | |
| $\delta\sim t^k$ | $\delta\sim t^{1/2}$ | $\delta\sim t^{1/2}$ |
| $u_0\sim t^{k-1}$ | $u_0\sim t^{k-1}$ | $u_0\sim t^{k-1}$ |
| (5) Spatial scaling laws for $\delta$ and $u_0$ | | |
| Case 1. Jet-like $V=(0,0,0)$ | Case 3. Jet-like boundary layers | Case 4. Wake-like boundary layers |
| $\delta\sim x$ | $\delta\sim x^{1/2k}$ | $\delta\sim x^{1/2}$ |
| $u_0\sim x^{1-1/k}$ | $u_0\sim x^{1-1/k}$ | $u_0\sim x^{k-1}$ |
| Case 2. Wake-like $V=(u_\infty,0,0)$ | | |
| $\delta\sim x^k$ | | |
| $u_0\sim x^{k-1}$ | | |

Note: $M$ is a parameter of the motion with units $L^mT^{-n}$ and $\alpha=1/m$, $k=n/m$.

Note: $\delta$ is a characteristic cross-stream length scale. $u_0$ is a characteristic stream-wise velocity scale.

**Table 2** Growth and decay laws for shear flows

| Class/Flow | Invariant | Parameter | Units | Exponent in time $k$ | Exponent in space $\delta$ | Exponent in space $u_0$ |
|---|---|---|---|---|---|---|
| **Jet-like** | | | | | | |
| Plane mixing layer | $u_0$—velocity difference | $u_1 - u_2$ | $LT^{-1}$ | 1 | 1 | 0 |
| Turbulent spot | $u_0$—free stream velocity (neglect variation in $C_f$) | $u_\infty$ | $LT^{-1}$ | 1 | 1 | 0 |
| Turbulent boundary layer | $u_0$—free stream velocity (neglect variation in $C_f$) | $u_\infty$ | $LT^{-1}$ | 1 | 1 | 0 |
| Plane plume | $u_0^3$—2-D buoyancy flux | $B$ | $L^3T^{-3}$ | 1 | 1 | 0 |
| Round plume | $u_0^3\delta$—3-D buoyancy flux | $B$ | $L^4T^{-3}$ | 3/4 | 1 | −1/3 |
| Vortex sheet rollup | $u_0\delta^{1/2}$—potential flow parameter | $\gamma$ | $L^{3/2}T^{-1}$ | 2/3 | 1 | −1/2 |
| Plane jet | $u_0^2\delta$—2-D momentum flux | $J$ | $L^3T^{-2}$ | 2/3 | 1 | −1/2 |
| Round jet | $u_0^2\delta^2$—3-D momentum flux | $J$ | $L^4T^{-2}$ | 1/2 | 1 | −1 |
| Laminar round jet | $u_0^2\delta^2, \nu$—3-D momentum flux, viscosity | $\nu$ | $L^2T^{-1}$ | 1/2 | 1 | −1 |
| Radial jet | $u_0^2\delta^2$—3-D momentum flux | $J$ | $L^4T^{-2}$ | 1/2 | 1 | −1 |
| Laminar radial jet | $u_0^2\delta^2, \nu$—3-D momentum flux, viscosity | $\nu$ | $L^2T^{-1}$ | 1/2 | 1 | −1 |
| Line vortex | $u_0\delta$—circulation | $\Gamma$ | $L^2T^{-1}$ | 1/2 | 1 | −1 |
| Laminar line vortex | $u_0\delta, \nu$—circulation, viscosity | $\nu$ | $L^2T^{-1}$ | 1/2 | 1 | −1 |
| Vortex pair | $u_0\delta^2$—2-D impulse | $I$ | $L^3T^{-1}$ | 1/3 | 1 | −2 |
| Vortex ring | $u_0\delta^3$—3-D impulse | $I$ | $L^4T^{-1}$ | 1/4 | 1 | −3 |
| **Wake-like** | | | | | | |
| Plane wake | $u_0\delta$—2-D far wake drag | $D$ | $L^2T^{-1}$ | 1/2 | 1/2 | −1/2 |
| Round wake | $u_0\delta^2$—3-D far wake drag | $D$ | $L^3T^{-1}$ | 1/3 | 1/3 | −2/3 |
| Plane momentumless wake | $u_0\delta^3$—requires a model | — | $L^4T^{-1}$ | 1/4 | 1/4 | −3/4 |
| Round momentumless wake | $u_0\delta^4$—requires a model | — | $L^5T^{-1}$ | 1/5 | 1/5 | −4/5 |
| Grid turbulence initial decay | $u_0^2\delta^3$—Saffman invariant | $S$ | $L^5T^{-2}$ | 2/5 | 2/5 | −3/5 |
| Grid turbulence initial decay | $u_0^2\delta^5$—Loitsianskii invariant | $L$ | $L^7T^{-2}$ | 2/7 | 2/7 | −5/7 |

| | | | | | | |
|---|---|---|---|---|---|---|
| Jet-like boundary layer | | | | | | |
| Laminar plane mixing layer | $u_0, \nu$—velocity difference, viscosity | $u_1 - u_2$ | $LT^{-1}$ | 1 | 1/2 | 0 |
| Laminar plane plume | $u_0^3, \nu$—2-D buoyancy flux, viscosity | $B$ | $L^3T^{-3}$ | 1 | 1/2 | 0 |
| Laminar round plume | $u_0^3\delta, \nu$—3-D buoyancy flux, viscosity | $B/\sqrt{\nu}$ | $L^3T^{-5/2}$ | 5/6 | 3/5 | −1/6 |
| Laminar plane jet | $u_0^2\delta, \nu$—2-D momentum flux, viscosity | $J/\sqrt{\nu}$ | $L^2T^{-3/2}$ | 3/4 | 2/3 | −1/3 |
| Falkner Skan boundary layers | $u_0 x^{-\beta}, \nu$—free stream parameter, viscosity | $M$ | $L^{1-\beta}T^{-1}$ | $1/(1-\beta)$ | $(1-\beta)/2$ | $\beta$ |
| $\beta=0$ Blasius layer | $u_0, \nu$—free stream velocity, viscosity | $u_\infty$ | $LT^{-1}$ | 1 | 1/2 | 0 |
| $\beta=-1$ Jeffrey-Hamel flow | $u_0\delta, \nu$—area flux, viscosity | $Q$ | $L^2T^{-1}$ | 1/2 | 1 | −1 |
| Wake-like boundary layer | | | | | | |
| Laminar round wake | $u_0\delta^2, \nu$—3-D far wake drag, —viscosity | $D/\nu$ | L | 0 | 1/2 | −1 |

observation discussed earlier, drawn from the study of homogeneous and isotropic turbulence and confirmed empirically for shear flows, that the energy-containing flow structure does not depend directly on the value of the fluid viscosity; one can assume Reynolds number invariance. This assumption is intimately connected to the dependence of the flow Reynolds number on time

$$\mathrm{Re}_\delta = \frac{u_0 \delta}{\nu} = \frac{kM^{2\alpha}}{\nu} t^{2k-1}. \tag{33}$$

For flows with $k = 1/2$, inertial and viscous times scale together and the assumption is unnecessary. For flows with $k > 1/2$ the inertial time will dominate at all but the smallest scales. However, flows with $k < 1/2$ will tend to follow a viscous scale as time increases.

Now let us examine the behavior of small-scale motions. If we assume that production and dissipation scale together then dimensional analysis leads to

$$\frac{\lambda}{\delta} \sim \mathrm{Re}_\delta^{-1/2} \text{ and } \frac{\eta}{\delta} \sim \mathrm{Re}_\delta^{-3/4} \tag{34}$$

upon substitution of (32)

$$\lambda \sim \left(\frac{\nu t}{k}\right)^{1/2} \text{ and } \eta \sim \nu^{3/4} M^{-\alpha/2} k^{-3/4} t^{3/4 - k/2}. \tag{35}$$

The main result here is that the time evolution of the Taylor microscale is always independent of the global parameter $M$.

The entrainment diagram is always invariant for moving observers. Under the assumption of nonsteady similarity, the global parameter $M$ determines the appropriate value of $k$. Once $k$ has been determined, the rate of convection and growth of structural features in the flow (critical points, turbulent interfaces, etc) is determined. If we choose to move (nonuniformly) with a coordinate system that remains attached to some preferred feature, then, in the moving coordinate system,

$$x_i' = x_i + a_i M^\alpha t^k,\ t' = t,\ u_i' = u_i + a_i k M^\alpha t^{k-1}, \tag{36}$$

and the similarity variables in moving coordinates are

$$\xi_i' = \xi_i + a_i,\ U_i(\xi') = U_i(\xi) + k a_i, \tag{37}$$

where $a_i$ is a dimensionless rate of motion in the $x_i$ direction.

It is clear from the above that the pattern formed by the velocity vector field will depend on the $a_i$. This is true whether one plots the $u_i$ field in physical coordinates or the $U_i$ field in similarity coordinates.

Similarly, the pattern of particle displacements, $dx_i$, in physical coordinates will depend on the $a_i$. However, the pattern of particle displacements in similarity coordinates, $d\xi_i$, is independent of the $a_i$. This follows from the form of the pathline equations in Table 1, item 3.

$$U_i(\xi)-k\xi_i=U_i'(\xi')-ka_i-k\xi_i'+ka_i=U_i'(\xi')-k\xi_i'. \tag{38}$$

Equation (38) is an important result for it states that the location and character of a critical point in similarity coordinates is fixed by the dynamics governing the flow and by the choice of a value for $k$ (which is a consequence of the units of $M$) and not by the incidental choice of speed for a moving observer.

The above results form the framework for a powerful method for analyzing the *dynamics* of fluid motion. To illustrate this, let us examine the Reynolds-number dependence of an impulsively started, axisymmetric, laminar jet produced by a point momentum source of strength $J/\rho$. The Reynolds number of the jet is $\mathrm{Re}=(J/\rho)^{1/2}/\nu$. Dimensional considerations lead to a formulation of the problem in terms of similarity variables

$$\psi=\nu^{3/2}t^{1/2}g(\xi,\theta;\mathrm{Re}),\ \xi=r/\sqrt{\nu t} \tag{39}$$

where $\psi$ is the Stokes stream function and $r$ and $\theta$ are the radial distance and azimuthal angle in spherical polar coordinates. A solution for the creeping-flow limit $\mathrm{Re}\to 0$ due to Sozou (1979) is

$$g(\xi,\theta)=\frac{\mathrm{Re}^2}{16\pi}\sin^2\theta\left\{2\xi-\frac{4}{\sqrt{\pi}}e^{-\xi^2/4}-\left(2\xi-\frac{4}{\xi}\right)\mathrm{erf}\left(\frac{\xi}{2}\right)\right\}. \tag{40}$$

By all conventional measures (40) would appear to exhibit only a trivial dependence on Reynolds number. However, an examination of the entrainment diagram of (40) reveals a remarkably complex structure (Cantwell 1980). The equations for particle paths are

$$\frac{dr}{dt}=u(r,\theta,t;\mathrm{Re});\ \frac{d\theta}{dt}=\frac{v(r,\theta,t;\mathrm{Re})}{r} \tag{41}$$

or, in terms of similarity variables,

$$\frac{d\xi}{d\tau}=U(\xi,\theta;\mathrm{Re})-\frac{\xi}{2};\ \frac{d\theta}{d\tau}=\frac{V(\xi,\theta;\mathrm{Re})}{\xi} \tag{42}$$

where $\tau=\ln t$ and

$$U=\frac{1}{\xi^2\sin\theta}\frac{\partial g}{\partial\theta};\ V=-\frac{1}{\xi\sin\theta}\frac{\partial g}{\partial\xi}. \tag{43}$$

Substitution of (40) into (42) leads to

$$\frac{d\xi}{d\tau}=\frac{\mathrm{Re}^2}{2\pi}\frac{\cos\theta}{\xi^2}\left\{\frac{\xi}{2}-\frac{1}{\sqrt{\pi}}e^{-\xi^2/4}-\left(\frac{\xi}{2}-\frac{1}{\xi}\right)\mathrm{erf}\left(\frac{\xi}{2}\right)\right\}-\frac{\xi}{2}, \tag{44}$$

$$\frac{d\theta}{d\tau}=-\frac{\mathrm{Re}^2}{4\pi}\frac{\sin\theta}{\xi}\left\{\frac{1}{2}+\frac{1}{\xi\sqrt{\pi}}e^{-\xi^2/4}-\left(\frac{1}{2}+\frac{1}{\xi^2}\right)\mathrm{erf}\left(\frac{\xi}{2}\right)\right\}. \tag{45}$$

The structure of the flow is examined by finding and classifying critical points of (44) and (45); points $(\xi_c, \theta_c)$ at which both right-hand sides are equal to zero. The zeros of (45) are at $(\theta=0, \pi, \text{all } \xi)$ and $(\xi=1.7633, \text{all } \theta)$ and are clearly the same for all Reynolds numbers. Setting the right-hand side of (44) equal to zero gives

$$\mathrm{Re}^2=\frac{\pi\xi_c^3}{\left(\frac{\xi_c}{2}-\frac{1}{\sqrt{\pi}}e^{-\xi_c^2/4}-\left(\frac{\xi_c}{2}-\frac{1}{\xi_c}\right)\mathrm{erf}\left(\frac{\xi_c}{2}\right)\right)\cos\theta_c}. \tag{46}$$

Equation (46) defines a family of curves in the $(\xi, \theta)$ plane for various values of the Reynolds number. Intersections between (46) and the zeros of (45) locate critical points in the entrainment diagram of the solution (40). From the above discussion, it is clear that, in spite of the apparently trivial Reynolds-number dependence of the streamline pattern of (40), the entrainment diagram may exhibit a Reynolds-number dependence that is quite complex. Figure 19 shows the entrainment diagram of (40) at three values of the Reynolds number. For sufficiently small Reynolds number, pathlines converge to a single stable node which lies on the axis of the jet. At a Reynolds number of 6.7806 the pattern bifurcates to a saddle lying on the axis of the jet, plus two stable nodes lying symmetrically to either side of the axis. At a Reynolds number of 10.09089 the pattern bifurcates a second time to form a saddle and two stable foci. Two points should be made here. The first is that the diagrams in Figure 19 depict the behavior of (40) at Reynolds numbers that lie outside of its region of validity, although one may expect the nonlinear solution to behave in a similar fashion. The second point is that the rollup of particle trajectories depicted in Figure 19*c* occurs *entirely without any local concentration of vorticity*.

It is of interest to speculate on the high-Reynolds-number (non-axisymmetric) flow from a point momentum source. In Section II we saw a model of the plane mixing layer by Corcos (1979) in which the flow was treated as deterministic at all scales. For the mixing layer, dimensional analysis leads to $\delta \sim t$, $\lambda \sim t^{1/2}$, $\eta \sim t^{1/4}$. The disparity between large and small scales increases with time. It is not obvious that in the face of such a dynamic the small scales can retain their identity for any significant length of time. However, in the case of the three-dimensional jet all three scales vary like $t^{1/2}$. The flow is self-similar at all scales. This suggests that there ought to exist a high-Reynolds-number solution which is enormously complex (with critical points in a three-dimensional phase space) but wholly deterministic and which would mimic all of the important features (spreading rate, entrainment rate, dissipation at small scales, etc) of a fully turbulent jet.

Entrainment diagrams can be worked out for a wide variety of flows including those listed in Table 2. Figure 20 illustrates three examples. Figure 20*c* shows the entrainment diagram for the Oseen vortex, an explicitly unsteady flow in the variable $r/\sqrt{\nu t}$ . Figures 20*a* and *b* show entrainment diagrams derived from the stationary mean-velocity profiles of the plane turbulent mixing layer and plane turbulent jet (Cantwell 1979). Both of these flows can be formulated as self-similar in time. Normally they are measured by a laboratory observer who takes a long time average at a fixed position in physical coordinates $(x, y)$ with the mean profiles illustrated in Figure 20, the empirically measured result. However, it is clear that another kind of average is possible. Operationally this would be a long time average referred to a receding observer who looks at the flow quite literally through the zoom lens of a camera. The rate of zoom is adjusted to match the global parameter that governs the motion and the averaging time of the experiment is limited by the physical size of the apparatus that contains the flow. Fluctuations in the evolving flow are assumed to follow the same time scale as the coherent motion and are averaged out by the receding observer. Given the complexity of the entrainment diagram for the round jet, one may conjecture that, in general, entrainment diagrams based on through-the-zoom-lens averaging will exhibit a rich structure not found in the average referred to a fixed laboratory observer.

The entrainment diagram has several useful features. It gives a compact, invariant, visual impression of the flow pattern with easily accessible information about the motion of fluid particles. The dominant physical picture of eddies as spirals comes through in the well-defined form of stable foci. The entrainment diagram can be used to analyze the dynamics of fluid motion, revealing, in some cases, a dependence on Reynolds number that may remain hidden in a pattern of streamlines. The correspondence between the near-wall structure of the entrainment diagram for the spot (Figure 11*c*), and the space-time correlation results of Wygnanski et al (Figure 11*b*) suggests that both methods somehow focus on the same energetic motions in the flow. The prospect that space-time correlation measurements could be connected to flow variables through critical points in the entrainment diagram is an intriguing possibility that needs further study. Perhaps some hint for approaching this issue can be found in the vector field of correlations introduced by Willmarth & Wooldridge (1963).

# VI CONCLUDING REMARKS

Our understanding of the physics of turbulent motion has increased tremendously in recent years. The major new fact is that turbulence is characterized by a remarkable degree of order. But along with order has come new complication for it appears that many turbulent flows retain a permanent imprint of their infinitely variable initial state. Approximately twenty years have passed since the earliest observations of organized structure. Yet progress in incorporating this structure into practical engineering methods has been slow and the connection to a truly predictive theory has not yet been made. Turbulence remains a major unsolved problem of classical physics.

ACKNOWLEDGMENT

I would like to express thanks to Garry Brown, Luis Bernal, and John Konrad who were involved in various aspects of making the film used for the animation of vortex pairing. I would also like to acknowledge the support of NASA Ames Research Center under NASA Grant NSG2392.

*Literature Cited*

Abernathy, F. H., Kronauer, R. E. 1962. The formation of vortex streets. *J. Fluid Mech.* 13:1–20

Ashurst, W. T. 1977. Numerical simulation of turbulent mixing layers via vortex dynamics. *Sandia Lab. Rep. SAND 77-8612*

Bakewell, H. P., Lumley, J. L. 1967. Viscous sublayer and adjacent region in turbulent pipe flow. *Phys. Fluids* 10:1880–89

Batt, R. G. 1975. Some measurements on the effect of tripping the two-dimensional shear layer. *AIAA J.* 13:245–47

Bevilaqua, P. M., Lykoudis, P. S. 1971. Mechanism of entrainment in turbulent waves. *AIAA J.* 9:1657–59

Birch, S. 1980. *Evaluation of Shear-Layer Data—Rep. to the Organizing Committee, 1980–81 Conf. on Computation of Complex Turbulent Flows*. Stanford Univ.

Blackwelder, R. F. 1978. The bursting process in turbulent boundary layers. *Lehigh Workshop on Coherent Structure in Turbulent Boundary Layers*. ed. C. R. Smith, D. E. Abbott, pp. 211–27

Blackwelder, R. F., Eckelmann, H. 1979. Streamwise vortices associated with the bursting phenomenon. *J. Fluid Mech.* 94:577–94

Blackwelder, R. F., Kaplan, R. E. 1971. Intermittent structures in turbulent boundary layers. *AGARD Conf. Proc. No. 43*, London, 5-1/5-7

Blackwelder, R. F., Kovasznay, L. S. G. 1972. Time scales and correlations in a turbulent boundary layer. *Phys. Fluids* 15:1545–54

Bouchard, G. E., Reynolds, W. C. 1978. Control of vortex pairing in a round jet. *Bull. Am. Phys. Soc.* 23(8):1013

Boussinesq, J. 1877. Théorie de l'écoulement tourbillant. *Mém. Prés. Acad. Sci.* 23:46

Briedenthal, R. E. 1978. *A chemically reacting turbulent shear layer*. PhD thesis. Calif. Inst. Tech., Pasadena, Calif.

Brown, G. L., Roshko, A. 1974. On density effects and large structure in turbulent mixing layers. *J. Fluid Mech.* 64:775–816

Brown, G. L., Thomas, A. S. W. 1977. Large structure in a turbulent boundary layer. *Phys. Fluids* 20(10):S243–52

Cantwell, B. J. 1978. Similarity transformations for the two-dimensional unsteady stream function equation. *J. Fluid Mech.* 85:257–71

Cantwell, B. J. 1979. Coherent turbulent structures as critical points in unsteady flow. *Arch. Mech. Stosow.* 31:707–21

Cantwell, B. J. 1980. Transition in the axisymmetric jet. (*SUDAAR*) *Rep. No. 521*, Stanford Univ., Dept. Aeronaut. and Astronaut. *J. Fluid Mech.* In press

Cantwell, B. J., Coles, D. E., Dimotakis, P. E. 1978. Structure and entrainment in the plane of symmetry of a turbulent spot. *J. Fluid. Mech.* 87:641–72

Chandrasekhar, S. 1949. On Heisenberg's elementary theory of turbulence. *Proc. R. Soc. London Ser. A.* 200:20–33

Chandrasuda, C., Mehta, R. D., Weir, A. D., Bradshaw, P. 1978. Effect of free shear stream turbulence on large structure in turbulent mixing layers. *J. Fluid Mech.* 85:693–704

Chapman, D. R. 1979. Computational aerodynamics development and outlook. *AIAA Dryden Lect. 79-0129*

Chorin, A. J., Bernard, P. S. 1973. Discretization of a vortex sheet with an example of roll-up. *J. Comput. Phys.* 13:423–29

Clark, R., Ferziger, J. H., Reynolds, W. C. 1979. Evaluation of subgrid-scale models using an accurately simulated turbulent flow. *J. Fluid Mech.* 91:1–16

Clements, R. R. 1973. An inviscid model of two-dimensional vortex shedding. *J. Fluid Mech.* 57:321–36

Coles, D. E. 1956. The law of the wake in the turbulent boundary layer. *J. Fluid Mech.* 1:191–226

Coles, D. E. 1978. A model for the flow in the viscous sublayer. *Lehigh Workshop on Coherent Structure in Turbulent Boundary Layers*, ed. C. R. Smith, D. E. Abbott, pp. 462–75

Coles, D. E., Barker, S. J. 1975. Some remarks on a synthetic turbulent boundary layer. In *Turbulent Mixing in Nonreactive and Reactive Flows*, ed. S. N. B. Murthy, pp. 285–92

Coles, D. E., Van Atta, C. W. 1967. Digital experiment in spiral turbulence. *Phys. Fluids Suppl. Part II* 10:S120–21

Corcos, G. 1979. The mixing layer: deterministic models of a turbulent flow. *Dept. Mech. Engrg. Rep. No. FM-79-2*. Univ. California, Berkeley

Corino, E. R., Brodkey, R. S. 1969. A visual investigation of the wall region in turbulent flow. *J. Fluid Mech.* 37:1–30

Corrsin, S. 1943. Investigations of flow in an axially symmetric heated jet of air. *NACA Adv. Conf. Rep. 3123*

Corrsin, S., Kistler, A. 1954. The free stream boundaries of turbulent flows. *NACA Tech. Note No. 3133* (also *NACA Tech. Rep. No. 1244*, 1955)

Crow, S. C., Champagne, F. H. 1971. Orderly structure in jet turbulence. *J. Fluid Mech.* 48:547–91.

Davies, M. E. 1976. A comparison of the wake structure of a stationary and oscillating bluff body, using a conditional averaging technique. *J. Fluid Mech.* 75:209–23

Dimotakis, P. E., Brown, G. L. 1976. The mixing layer at high Reynolds number: large structure dynamics and entrainment. *J. Fluid Mech.* 78:535–60

Dinkelacker, A., Hessel, M., Meier, G., Schewe, G. 1977. Investigation of pressure fluctuations beneath a turbulent boundary layer by means of an optical method. *Phys. Fluids* 20(10):S216–24

Elder, J. W. 1960. An experimental investigation of turbulent spots and breakdown to turbulence. *J. Fluid Mech.* 9:235–46

Emmons, H. W. 1951. The laminar-turbulent transition in a boundary layer. Part I. *J. Aeronaut. Sci.* 18:490–98

Falco, R. E. 1977. Coherent motions in the outer region of turbulent boundary layers. *Phys. Fluids* 20(10):5124–32

Favre, A. J., Gaviglio, J. J., Dumas, R. 1957. Space-time double correlation and spectra in a turbulent boundary layer. *J.*

*Fluid Mech*. 2:313–41

Freymuth, P. 1966. On transition in a separated boundary layer. *J. Fluid Mech*. 25:683–704

Furuya, Y., Osaka, H. 1976. *Three-dimensional structure of nominally two-dimensional turbulent boundary layer*. Presented at the 14th IUTAM Congress, Delft. *Abstr. No. 160*

Gad-el-Hak, M., Blackwelder, R. F. 1979. A visual study of the growth and entrainment of turbulent spots. *Proc. IUTAM Symp. on Transition*, Stuttgart, Germany

Grant, H. L. 1958. The large eddies of turbulent motion. *J. Fluid Mech*. 4:149–90

Grass, A. J. 1971. Structural features of turbulent flow over smooth and rough boundaries. *J. Fluid Mech*. 50:223–56

Griffin, O. M., Ramberg, S. E. 1974. The vortex street wakes of vibrating cylinders. *J. Fluid Mech*. 66:553–78

Gupta, A. K., Laufer, J., Kaplan, R. E. 1971. Spatial structure in the viscous sublayer. *J. Fluid Mech*. 50:493–512

Hama, F. R., Nutant, J. 1963. Detailed flow field observations in the transition process in a thick boundary layer. *Proc. 1963 Heat Transfer and Fluid Mech. Inst*., pp. 77–94

Hanratty, T. J., Chorn, L. G., Hatziavramidis, D. T. 1977. Turbulent fluctuations in the viscous wall region for Newtonian and drag reducing fluids. *Phys. Fluids* 20(10):S112–19

Head, M. R., Bandyopadhyay, P. 1978. Combined flow visualization and hot wire measurements in turbulent boundary layers. *Lehigh Workshop on Coherent Structure in Turbulent Boundary Layers*, ed. C. R. Smith, D. E. Abbott, pp. 98–129

Heisenberg, W. 1948a. Zur statistischen theorie der turbulenz. *Z. Phys*. 124:628–57

Heisenberg, W. 1948b. On the theory of statistical and isotropic turbulence, *Proc. R. Soc. London Ser. A* 195:402–6

Hinze, J. O. 1959. *Turbulence*. New York: McGraw-Hill. 1st ed.

Hunt, J. C. R., Abell, C. J., Peterka, J. A., Woo, H. 1978. Kinematic studies of the flows around free or surface mounted obstacles; applying topology to flow visualization. *J. Fluid Mech*. 86:179–200

Hussain, A. K. M. F., Reynolds, W. C. 1975. Measurements in fully developed turbulent channel flow. *J. Fluid Engrg*. 97:568–80

Hussain, A. K. M. F., Zaman, K. B. M. Q. 1978. The free shear layer tone phenomenon and probe interference. *J. Fluid Mech*. 87:349–84

Kim, H. T., Kline S. J., Reynolds, W. 1971. The production of the wall region in turbulent flow. *J. Fluid Mech*. 50:133–60

Klebanoff, P. S. 1954. Characteristics of turbulence in a boundary layer with zero pressure gradient. *NACA Tech. Note No. 3178*

Kline, S. J. 1978. The role of visualization in the study of the structure of the turbulent boundary layer. *Lehigh Workshop on Coherent Structure of Turbulent Boundary Layers*, ed. C. R. Smith, D. E. Abbott, pp. 1–26

Kline, S. J., Falco, R. E. 1980. Summary of the AFOSR/MSU research specialists workshop on coherent structure in turbulent boundary layers. *AFOSF-TR-80-0290 ADA 083717*

Kline, S. J., Runstadler, P. W. 1959. Some preliminary results of visual studies of the flow model of the wall layers of the turbulent boundary layer. *Trans. ASME (Ser. E)* 2:166–70

Kline, S. J., Reynolds, W. C., Schraub, F. A., Runstadler, P. W. 1967. The structure of turbulent boundary layers. *J. Fluid Mech*. 30:741–73

Kolmogorov, A. N. 1941. The local structure of turbulence in incompressible flow for very large Reynolds number. *C. R. Acad. Sci. U.R.S.S*. 30:301

Kovasznay, L. S. G. 1948. Spectrum of locally isopropic turbulence. *J. Aeronaut. Sci*. 15:745–53

Kovasznay, L. S. G., Kibens, V., Blackwelder, R. 1970. Large scale motion in the intermittent region of a turbulent boundary layer. *J. Fluid Mech*. 41:283–325

Laufer, J. 1954. The structure of turbulence in fully developed pipe flow. *NACA Tech. Note No. 2954*

Laufer, J., Narayanan, M. A. B. 1971. Mean period of the turbulent production mechanism in a boundary layer. *Phys. Fluids* 14:182–83

Lee, M. K., Eckelmann, L. D., Hanratty, T. J. 1974. Identification of turbulent wall eddies through the phase relation of the components of the fluctuating velocity gradient. *J. Fluid Mech*. 66:17–34

Leonard, A. 1979. Vortex simulation of three-dimensional, spotlike disturbances in a laminar boundary layer. *NASA Tech. Memo. 78579* (also in *Turbulent shear flows II*, ed. L. J. S. Bradbury, pp. Berlin:

Springer 67:77

Liepmann, H. W. 1952. Aspects of the turbulence problem. *J. Appl. Math. and Phys. (ZAMP)* 3:321–426

Liepmann, H. W. 1962. Free turbulent flows. In *Colloques Internationaux du Centre National de la Recherche Scientifique (C.N.R.S.), Mechnique de la Turbulence, Marseilles*, 108:211–27

Liepmann, H. W. Laufer, J. 1947. Investigation of free turbulent mixing. *NACA Tech. Note No. 1257*

Lighthill, M. J. 1963. In *Laminar Boundary Layers*, ed. L. Rosenhead, pp. 48–88. Oxford Univ. Press

Lindgren, E. R. 1969. Propagation velocity of turbulent slugs and streaks in transition pipe flow. *Phys. Fluids* 12:418–25

Lu, S. S., Willmarth, W. W. 1973. Measurements of the Reynolds stress in a turbulent boundary layer. *J. Fluid Mech.* 60:481–511

Lugt, H. J. 1979. The dilemma of defining a vortex. In *Recent Developments in Theoretical and Experimental Fluid Mechanics*, ed. U. Muller, K. G. Roesner, B. Schmidt, pp. 309–21. Berlin/Heidelberg/New York: Springer

Meijer, M. C. 1965. Pressure measurements on flapped hydrofoils in cavity flows and wake flows. *Hydrodyn. Lab. Rep. No. E-133.2.* Calif. Inst. Tech.

Mitchner, M. 1954. Propagation of turbulence from an instantaneous point disturbance. *J. Aeronaut. Sci.* 2(5):350–51

Moin, P., Reynolds, W. C., Ferziger, J. H. 1978. Large eddy simulation of incompressible turbulent channel flow. *Dept. Mech. Engrg. Rep.* 91:1–16. Stanford Univ.

Narayanan, B., Marvin, J. 1978. On the period of the coherent structure in boundary layers at large Reynolds numbers. *Lehigh Workshop on Coherent Structure in Turbulent Boundary Layers*, ed. C. R. Smith, D. E. Abbot, pp. 380–86

Nychas, S. G., Hershey, H. C., Brodkey, R. S. 1973. A visual study of turbulent shear flow. *J. Fluid Mech.* 61:513–40

Offen, G. R., Kline, S. J. 1974. Combined dye-streak and hydrogen-bubble visual observations of a turbulent boundary layer. *J. Fluid Mech.* 62:223–39

Offen, G. R., Kline, S. J. 1975. A proposed model of the bursting process in turbulent boundary layers. *J. Fluid Mech.* 70:209–28

Oldaker, D. K., Tiederman, W. G. 1977. Spatial structure of the viscous sublayer in drag reducing channel flows. *Phys. Fluids* 20(10): *S*133–44

Onsager, L. 1945. The distribution of energy in turbulence. *Phys. Rev. Abstr.* 68:286

Onsager, L. 1949. Statistical hydrodynamics. *Nuovo Cimento Suppl.* 6:279–87

Oster, D., Wygnanski, I., Dziomba, B., Fiedler, H. 1977. On the effect of initial conditions on the two dimensional turbulent mixing layer. In *Lecture Notes in Physics, Vol. 75*, p. 48. Berlin: Springer (*Structure and Mechanisms of Turbulence I*, ed. H. Fiedler)

Oswatitsch, K. 1958. *Die Ablösungsbedingung von Grenzschichten*, Grenzschicht Forschung, ed. H. Goertler, pp. 357–67. Berlin/New York: Springer

Owen, F. K., Johnson, D. A. 1978. Measurements of unsteady vortex flow fields. *AIAA Paper 78–18*

Payne, F. R., Lumley, J. L. 1967. Large eddy structure of the turbulent wake behind a circular cylinder. *Phys. Fluids Suppl.* 10:S194–96

Peake, D., Tobak, M. 1980. Three-dimensional interactions and vortical flows with emphasis on high speeds. *NASA Tech. Memo. 81169*

Perry, A. E., Lim, T. T. 1978. Coherent structures in coflowing jets and wakes. *J. Fluid Mech.* 88:451–64

Perry, A. E., Lim T. T. 1978. Coherent structures in coflowing jets and wakes. *J. Fluid Mech.* 88:451–64

Perry, A. E., Watmuff, J. H. 1979. The phase-averaged large-scale structures in three-dimensional turbulent wakes. *Dept. Mech. Rep. FM-12.* Univ. Melbourne

Perry, A. E., Lim, T. T., Chong, M. S., Teh, E. W. 1980. The fabric of turbulence. *AIAA Pap. 80-1358*

Perry, A. E., Lim, T. T., Chong, M. S. 1980. Instantaneous vector fields in coflowing jets and wakes. *J. Fluid Mech.* In press

Prandtl, L. 1925. Bericht uber untersuchungen zur ausgebildeten turbulenz. *Z. Angew. Math. Mech.* 5:136–39

Praturi, A. K., Brodkey, R. S. 1978. A stereoscopic visual study of coherent structures in turbulent shear flow. *J. Fluid Mech.* 89:251–72

Pullin, D. I. 1978. The large-scale structure of unsteady self-similar rolled-up vortex sheets. *J. Fluid Mech.* 80:401–30

Rao, K. N., Narasimha, R., Narayanan, M. A. B. 1971. Bursting in a turbulent boundary layer. *J. Fluid Mech.* 48:339–52

Reynolds, O. 1895. On the dynamical theory of incompressible viscous fluids and the determination of the criterion. *Philos. Trans. R. Soc. London Ser. A* 186:123

Reynolds, W. C. 1976. Computation of turbulent flows. *Ann. Rev. Fluid Mech.* 8:183–208

Roshko, A. 1976. Structure of turbulent shear flows: a new look. Dryden Research Lecture. *AIAA J.* 14:1349–57

Sabot, J., Saleh, I., Comte-Bellot, G. 1977. Effects of roughness on the intermittent maintenance of Reynolds shear stress in pipe flow. *Phys. Fluids* 20(10):S150–55

Saffman, P. G., Baker, G. R. 1979. Vortex interactions. *Ann. Rev. Fluid Mech.* 11:95–122

Sarpkaya, T. 1975. An inviscid model of two-dimensional vortex shedding for transient and asymptotically steady separated flow over an inclined flat plate. *J. Fluid Mech.* 68:109–28

Savas, O. 1979. *Some measurements in synthetic turbulent boundary layers.* PhD thesis. Calif. Inst. Tech., Pasadena, Calif.

Schubauer, G. B., Klebanoff, P. S. 1955. Contributions on the mechanics of boundary layer transition. *NACA Tech. Note No. 3489* (See also *NACA Tech. Rep. No. 1289*, 1956)

Smith, C. R. 1978. Visualization of turbulent boundary layer structure using a moving hydrogen bubble wire probe. *Lehigh Workshop on Coherent Structure in Turbulent Boundary Layers*, ed. C. R. Smith, D. E. Abbott, pp. 48–97

Sozou, C. 1979. Development of the flow field of a point force in an infinite fluid. *J. Fluid Mech.* 91:544–54

Spalding, D. B. 1961. A single formula for the law of the wall. *J. Appl. Mech.* 28:455–57

Spencer, B. W., Jones, B. G. 1971. Statistical investigation of pressure and velocity fields in the turbulent two-stream mixing layer. *AIAA Pap. No. 71-613*

Taylor, G. I. 1915. Eddy motions in the atmosphere. *Philos. Trans. R. Soc. London Ser. A* 215:1–26

Taylor, G. I. 1932. Note on the distribution of turbulent velocities in a fluid near a solid wall. *Proc. R. Soc. London Ser. A* 135:678–84

Taylor, G. I. 1935. The statistical theory of turbulence, Parts I–IV, *Proc. R. Soc. London Ser. A* 151:421–511

Theodorsen, T. 1955. The structure of turbulence. In *50 Jahre Grenzschichtforsung*, ed. H. Görtier, W. Tollmein, p. 55. Braunschweig, Veiweg & Sohn

Townsend, A. A. 1947. Measurements in the turbulent wake of a cylinder. *Proc. R. Soc. London Ser. A* 190:551–61

Townsend, A. A. 1956. *The Structure of Turbulent Shear Flow.* Cambridge Univ. Press. 1st ed.

Townsend, A. A. 1970. Entrainment and the structure of turbulent flow. *J. Fluid Mech.* 41:13–46

Townsend, A. A. 1976. *The Structure of Turbulent Shear Flow.* Cambridge Univ. Press. 2nd ed.

Tu, B. J., Willmarth, W. W. 1966. An experimental study of turbulence near the wall through correlation measurements in a thick turbulent boundary layer. *Tech. Rep. No. 02920-3-T.* Dept. Aerosp. Engrg., Univ. Michigan, Ann Arbor

Turner, J. S. 1964. The flow into an expanding spherical vortex. *J. Fluid Mech.* 18:195–208

Von Weizsäcker, C. F. 1948. Das spectrum der turbulenz bei grossen Reynolds-Schen zahlen. *Z. Phys.* 124:614–27

Willmarth, W. W. 1975. Pressure fluctuations beneath turbulent boundary layers. *Ann. Rev. Fluid Mech.* 7:13–37

Willmarth, W. W., Bogar, T. J. 1977. Survey and new measurements of turbulent structure near the wall. *Phys. Fluids* 20(10):S9–21

Willmarth, W. W., Lu, S. S. 1972. Structure of the Reynolds stress near the wall. *J. Fluid Mech.* 55:65–69

Willmarth, W. W., Wooldridge, C. E. 1962. Measurements of the fluctuating pressure at the wall beneath a thick turbulent boundary layer, *J. Fluid Mech.* 14:187–210

Willmarth, W. W., Wooldridge, C. E. 1963. Measurements of the correlation between the fluctuating velocities and the fluctuating wall pressure in a thick turbulent boundary layer. *AGARD Rep. 456*

Willmarth, W. W., Yang, C. S. 1970. Wall pressure fluctuations beneath turbulent boundary layers on a flat plate and a cylinder. *J. Fluid Mech.* 41:47–80

Winant, C. D., Browand, F. K. 1974. Vortex pairing, the mechanism of turbulent mixing layer growth at moderate Reynolds number. *J. Fluid Mech.* 63:237–55

Wygnanski, I., Fiedler, H. E. 1970. The two dimensional mixing region, *J. Fluid Mech.* 41:327–61

Wygnanski, I. J., Champagne, F. H. 1973.

On transition in a pipe. Part I. The origin of puffs and slugs and the flow in a turbulent slug. *J. Fluid Mech*. 59:281–335

Wygnanski, I. J., Sokolov, N., Friedman, D. 1976. On the turbulent "spot" in a boundary layer undergoing transition. *J. Fluid Mech*. 78:785–819

Wygnanski, I., Haritonidis, J. H., Kaplan, R. E. 1979. On a Tollmien-Schlichting wave packet produced by a turbulent spot. *J. Fluid Mech*. 92:505–28

Zakkay, V., Barra, V., Wang, C. R. 1978. Coherent structure of turbulence at high subsonic speeds. *Lehigh Workshop on Coherent Structure in Turbulent Boundary Layers*, ed. C. R. Smith, P. E. Abbott, pp. 387–95

Zilberman, M., Wygnanski, I., Kaplan, R. E. 1977. Transitional boundary layer spot in a fully turbulent environment. *Phys. Fluids* 20(10):S258–71

# AUTHOR INDEX

Z

# CUMULATIVE INDEXES

## CONTRIBUTING AUTHORS, VOLUMES 9–13

# CHAPTER TITLES, VOLUMES 9–13